practical transistor circuit design and analysis

practical transistor circuit design and analysis

Gerald E. Williams
Riverside City College

McGraw-Hill Book Company

New York
St. Louis
San Francisco
Düsseldorf
Johannesburg
Kuala Lumpur
London
Mexico
Montreal
New Delhi
Panama
Rio de Janeiro
Singapore
Sydney
Toronto

Library of Congress Cataloging in Publication Data

Williams, Gerald E
Practical transistor circuit design and analysis.

1. Transistor circuits. 2. Electronic circuit design. I. Title.
TK7871.9.W53 621.3815'3'0422 73-7973
ISBN 0-07-070398-1

PRACTICAL TRANSISTOR CIRCUIT DESIGN AND ANALYSIS

34567890 VHVH 798

The editors for this book were Alan W. Lowe and Alice V. Manning, the designer was Marsha Cohen, and its production was supervised by James E. Lee. It was set in Melior by Progressive Typographers.
It was printed and bound by Kingsport Press, Inc.

To my wife Patty for turning a long
and difficult project into a loving
and sharing experience

contents

preface

Let me introduce this book by explaining that this is a totally different approach to transistors. Designers who are familiar with the intricacies of hybrid and other device parameters, and device-oriented procedures, may find it a little strange at first, but a little study will convince them that the book provides a new breed of methods and procedures that are not only far simpler than the older established procedures, but are also more powerful, in light of modern technological developments.

Those of us who have gone through the agony of learning to design around hybrid parameters are bound to resent learning a new system and junking all that effort, but I personally have gone entirely to the system presented in this book, and the results have been so spectacular that I no longer have any regrets about it. I have also found that the effort of learning those parameters was not entirely wasted, because that knowledge helped me to develop the methods I present here, and has provided valuable insights into why these methods work so well. In this book I offer you a simple, practical approach to the design and analysis of transistor circuits. What is equally important, I hope to give you an intuitive "feel" for transistors.

This book is designed for anyone who must make transistor circuits *work,* either as a primary part of his profession or simply as something necessary to get on with the job at hand. The methods I offer you here have been laboratory-tested by hundreds of students over an eight-year period, and they work consistently and well. In addition to time-proven results, these methods of design and analysis have the added advantage of much greater simplicity than any other methods I know of. The approach is unique in that it is circuit-oriented, not device-oriented, and there are several advantages to a circuit-oriented scheme.

To begin with, the parameters of a transistor vary so greatly both in manufacturing tolerances and with changes in temperature that knowledge of the parameters of a typical transistor of a given type number is of dubious value. Secondly, any design process based on device parameters must ultimately *design out* the transistor's parameters, if the circuit is to function and be stable with changes in temperature. The desired end result is always a *circuit* which performs a particular job as the designer intended, and whatever the design process (and there are several), it eventually, often after much effort, becomes circuit-oriented.

A circuit-oriented approach does not require precise knowledge of transistor parameters; this simplifies things considerably. The methods of analysis in this book are intimately related to circuit-oriented methods of design (a logical extension of them) and as such *do not* require a knowledge of the parameters of the transistors involved to analyze a transistor circuit *quantitatively.* This is of great practical advantage, because circuits encountered in the field which must be analyzed often contain transistors about which nothing at all is known. Integrated circuitry is part of the current technical scene, and *device*-oriented analysis techniques are almost certain to fall short as tools for the analysis of integrated circuits. It is almost a certainty that parameters of individual transistors on a chip will *not* be available to the technician or designer.

And, if history is any guide, even overall integrated circuit performance data will often be lacking when the IC is a part of a commercial system. If a schematic diagram of the IC is available, the important performance data can be analyzed out of the schematic. Device-oriented methods could be used by making educated guesses about parameters and plugging those into the appropriate equations. At best this is a makeshift procedure. What I offer you here is a tool which is appropriate to the new technology, and one which is far faster and simpler to use than any of the device-oriented approaches.

The methods of analysis presented in the book are tools for today's technology, equally useful for integrated circuits and discrete systems.

With this circuit-oriented design approach it is sufficient, for most design purposes, to know all maximum ratings and to have a rough idea of the transistor's β.

Now, before you get the impression that the approach in this book is some newfangled, limited, and possibly totally unscientific approach, let me assure you that is not the case at all. Virtually every technique in this book has been used for years by those people who work with transistors for a living, but they have, until now, been used sporadically, pragmatically, inconsistently, and perhaps somewhat unscientifically. What I have done in this book is to take those practical techniques and integrate them into a coherent, logical, and consistent scheme for designing and analyzing transistor circuits. But more than that, I have also tried to explain the theory behind each equation and the system in general. When I have taken the liberty of dropping some variable, representing some small-order effect, from an equation, I have explained the reasoning behind that liberty. Those things that I cannot adequately explain I have tried to be honest about.

The book has grown out of a one-semester *second* course in semiconductors with heavy emphasis on practical applications in the laboratory. A companion book, *Student Exercises*, is available. Although the book is complete in the sense that all the background required for design and analysis is contained in the earlier chapters, this background is fairly brief and primarily intended for review and orientation. It is not practical to try to cover all the material in the book in a single semester, unless the student has had some previous transistor theory or unless unusually large time blocks are used.

There will, of course, be some mathematics involved. Without mathematics you could not design a bread box or a birdhouse, for design is by definition a process by which the details of something to be constructed are developed "on paper." If the design is good, all elements will fit together and the finished product will perform the task intended for it. That is what this book is all about. The methods presented here work! They have been tested and modified in the laboratory until they were right, and results have been reproduced by hundreds of students. Some basic electronics math and a little algebra are necessary. An understanding of network theory will help in the sense that less will have to be taken on faith, but it is not absolutely essential.

THE ORGANIZATION

The book consists of three major divisions: theory, design, and analysis. Chapters 1 to 4 ("Foundations") serve the dual purpose of reviewing basic theory and reorienting the student to the kind of thinking necessary for understanding the particular design and analysis procedures which represent the main thrust of the book.

Because of the intimate relationship between the techniques of analysis and those of design, design and analysis are interdigitated throughout remaining chapters of the book ("Techniques").

One of the outstanding features of the book is the branching linear design program and its associated linear branching analysis program, which evolved after years of trying to unify the system. *The book is not a programmed text*, but the fact the design method evolved into such an iterative program made possible a kind of highly effective cross-referencing among explanations, worked examples, and design problems. I have devised a set of design sheets where a few basic steps with minor branches permit the design of any common transistor circuit (I have done the same for analysis problems). For those design (and analysis) sheets I have adopted the following terminology:

Field: A linear segment in the design procedure (a short of graphic step-by-step procedure).
Frame: A single step in the design procedure.
Branch: A short sequence of design (or analysis) steps which departs from the main program. It is these branches which accommodate individual differences among various transistor circuits.

I have used the same headings, *field, frame*, and *branch*, for worked examples, explanations, and problems, so that cross-referencing is automatic and easy. The iterative design and analysis programs also allow a significant *spiral* reinforcement situation that is pedagogically desirable, but not normally easy to obtain. This basic structure makes individualized instruction a practical approach, because of its inherent flexibility.

TABLES OF POSSIBILITIES

One of the most difficult areas in design is that of designing within practical limits. When there are as many different parameters involved as there are in a transistor circuit, it is often possible that the finished circuit will require some voltage, current, or impedance which is either impossible or impractical. One significant contribution of this book is a set of tables of possibilities. These tables substitute for years of experience in that even the most inexperienced designer can, with their aid, always come up with not only a workable circuit, but also a convenient circuit design, one which meets all specifications without demanding inconvenient voltages, currents, or impedances.

In a sense, the tables of possibilities allow the designer to look into the future and to be certain of a *practical* finished product. The absence of well-defined working limits has always been (in my view) one of the most serious omissions of design-oriented textbooks. I have tried to remedy that fault in this book. Perhaps the most important aspect of the approach in this book is that it works so well with students. My colleagues and I have been delighted with the results. In the end, the crucial question about any textbook is, Does it teach, and do the students learn from it?'' The answer here is a most emphatic *yes*.

Tables of possibilities and design (and analysis) sheets appear where appropriate in the text, but they are used many times throughout the text. To make reference to these essential items easy, I have put them all together as an appendix at the back of the book. You will also find other often-used tables there.

Gerald E. Williams

to the instructor

Behavioral objectives are included at the beginning of each chapter along with one or more self-evaluation methods. In Chaps. 5 through 12 the emphasis is on the student's ability to design or analyze entire circuits, and so there are fewer stated objectives in each of these chapters and the final real-world self-evaluation comes in the laboratory. Problems in design (or analysis) are provided at the end of these chapters, but are intended to be used as a preliminary student self-evaluation before he tries his hand in the laboratory. The laboratory is the proving ground: Can the student design circuits that really work according to specifications? I have tried in both the text and Student Exercises to provide some guidelines for acceptable performance. These guidelines cannot be absolute, because the method espoused in this text is less than perfect, like everything in this world. I must depend upon your judgment, but I have tried to supply statistical information to help you form realistic evaluation standards.

There is much emphasis these days, and it is long overdue, on individualized instruction. This text in conjunction with the laboratory manual (which I consider to be an essential part of the package) can provide an unusual degree of flexibility in instruction. I can't honestly claim credit for any unusual foresight in this direction, but the design sheets and tables of possibilities make this text valuable as a primary resource for individualized instruction.

As soon as I recognized the potential of the combined text and Student Exercises in the direction of individualized instruction, I did what I could to augment that potential. I have made the statement that this book is devoted to the student. For that reason, although I have designed the text with the latest educational techniques in mind, I have omitted nearly all the jargon and special educational terminology involved. Answers to the questions are not provided, and they (the questions) are intended to function as a modified form of adjunct programming where the student must search out answers to those questions he cannot answer. Design and analysis problems are primarily intended to help the student evaluate how well he has met the stated behavioral objectives, and selected answers are provided in the back of the book.

acknowledgments

I would like to express my gratitude to those who have contributed so much of themselves in helping me with this book:

To my wife Patty for her contribution in making all the many line drawings and schematic diagrams, which are so essential to a work of this kind.

To Professor Denton Titus for his faith and encouragement.

To Edward Young for his careful and excellent laboratory work.

To Professor W. Paul Matthews for his courage in teaching the ideas in this book when they were only half-baked, for his faith that those ideas would someday become fully baked, for his ability always to teach a good course in spite of constantly changing and evolving ideas, and for his many ideas which helped so much to make this text possible.

To Toni Ducas for all her help.

To Lenore Gibson for outstanding typing of a difficult manuscript.

To my children Geoffrey and Kelly for being so patient, understanding, and helpful when their dad wasn't giving them the attention they so deserve.

To Professor R. T. Burchell for his computer work in correcting the problems in the split emitter table of possibilities.

To Mr. Roy Carr for translating the design and analysis sheets into computer programs.

book one

foundations

1

a bit of physics

SELF-EVALUATION

In these first few chapters I shall often ask you to make some sketches, define some terms, and perhaps do a few other things, and then I will ask you to explain something to another person, using your drawings, definitions, etc. You probably will be reluctant, at first, to involve someone else in this way, and you will find any number of excuses to avoid doing it, or will try to convince yourself that writing the explanation or explaining it to yourself will do just as well. This kind of reluctance is almost universal, and it is not because we don't want to bother another person, but rather that we are *afraid* of making errors, faltering, getting confused, showing our ignorance, or seeming foolish. The reason for trying to explain things to someone is to help spot those things that you do not really understand. Any teacher can tell you that there is no better way to learn something than trying to explain it to someone.

Every time I lecture on a subject for the first time, I spot holes and inconsistencies in my lecture and find myself doing some additional studying on the subject. Fortunately for my ego the students seldom catch me in these blunders, and I correct the problems before I give that lecture again. (They get only one shot at me.) If I didn't do some additional study and give the subject more thought, sooner or later, some bright student would catch me, and my ego would get bruised. It is not necessary that the person you use should know anything about the subject; he must simply be a good listener. You will catch your own deficiencies, and if you don't follow up in correcting them, you too will someday get caught.

An alternative procedure (although not as good) is to lecture to a tape recorder and then play it back as though you were a student and this was a particularly disliked instructor, whom you would like to catch in a mistake—any mistake. Ask yourself, "If I were paying $40 per unit to hear this guy lecture, would I be satisfied with this performance?" Try these methods and see if they don't provide you with a good self-evaluation of how well you understand what you have learned. I have also prepared some questions for you to answer as a self-test. You may combine these methods in any way that suits your particular learning style.

OBJECTIVES:

(Things you should be able to do upon completion of this chapter.)

1. List the three classifications of matter according to conductivity.
2. Define *semiconductor.*
3. Name the primary factor which governs the amount of current flow through an intrinsic semiconductor material.
4. List the two ways in which semiconductor circuits can be made less temperature-dependent.
5. Define *doping.*
6. Define *n-type semiconductor.*
7. Define *p-type semiconductor.*
8. Define *valence electrons.*
9. Define *covalent bonds.*
10. Define *hole.*
11. Define *donor.*
12. Define *acceptor.*
13. Define *electron-hole pair.*
14. Define *dangling bond.*
15. Define *majority carrier.*
16. Define *minority carrier.*
17. Define *intrinsic semiconductor.*
18. List the common dopant elements.

SELF-TEST

Describe conduction in an intrinsic semiconductor block, in a *p*-type semiconductor, and in an *n*-type semiconductor to someone, using the lists, definitions, etc., in objectives 1 to 18.

INTRODUCTION

The physics of semiconductor materials is a frighteningly complex subject, but no matter how complex you make the explanation, it is bound to be only partly the truth. If we must be liars by omission, let us try to adopt those half-truths that will be the most useful. I shall hit the high points and leave you to spend as little or as much time as you like in filling in the valleys I leave between. What I shall try to do is provide you with the shape and form of those concepts which will be essential later on, along with a few necessary details.

There are two aspects of semiconductor physics which are of vital importance: the nature of conduction in semiconductors and the effect of heat upon that conduction. Our preoccupation with temperature may seem at times to border on paranoia; this is only natural because of the tremendous influence heat has upon the operating characteristics of semiconductor devices.

1-1 THE STRUCTURE OF SEMICONDUCTOR CRYSTALS

There are three more or less distinct classes of materials according to their conductivity:

1. Conductors, which are mostly metals
2. Insulators, which include many kinds of materials from wood, a complex organic material, to glass, which is largely silicon and oxygen
3. Semiconductors, which are really insulators driven into conduction by heat

At absolute zero (−273°C) semiconductors would be excellent insulators. From there on up, the higher the temperature, the better conductors semiconductor materials become.

Within the range of temperature in which human life can exist, semiconductors range from poor to fair conductors. And yet the picture is not quite complete because this change in conductivity with temperature is not a simple linear relationship. By this I mean that doubling the temperature does not double the conductivity of a piece of semiconductor material. In fact, in germanium an increase in temperature of only 10°C will double its conductivity.[1]

If we take a block of silicon or germanium, the two currently most important semiconductors, and connect it in the circuit in Fig. 1-1, we will find that the current in the circuit will gradually increase as the temperature of the block increases. We would also expect the current to increase if the voltage is increased, and it does. But increasing the voltage increases the current only proportionally, obeying Ohm's law, while increasing the temperature increases the current (without changing the voltage) at an exponential rate.

So you see, the temperature, over which we have no control, exerts more influence upon the current than the voltage, which we customarily do control.

There are two ways in which we try to cope with this temperature problem and make the current more voltage-dependent than temperature-dependent. First, free electrons are made available in large quantities by

[1] In silicon each 10°C increase in temperature results in three times the conductivity of the 10° lower temperature.

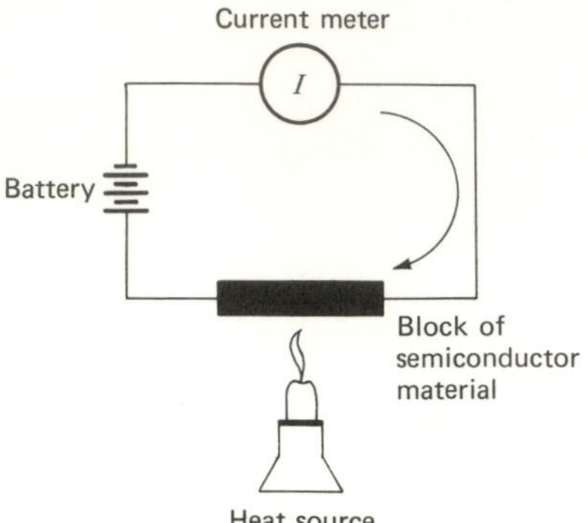

Fig. 1-1 Heat and conductivity in a semiconductor

adding certain impurities in controlled amounts to the semiconductor material. These impurity-produced electrons are not temperature-dependent, but voltage-dependent, and they are under our control. The device manufacturer takes care of this part; we, as users of the device, have no responsibility for it. The second method of restricting temperature-controlled current will be the main topic of the rest of this book; it simply involves proper circuit design.

1-2 THE STRUCTURE OF SEMICONDUCTOR MATERIALS

Two semiconductor materials have been the mainstay of the transistor industry for as long as there has been a transistor industry. These are silicon and germanium, and they are both elements. Compounds such as gallium arsenide and silicon carbide have been used, but they have not yet presented any real threat to germanium and silicon. Germanium and silicon are both tetra-valent elements. This simply means that there are four electrons in the atom's outermost orbit. These outlander electrons are the only ones that can leave the atom.

In whatever "model" we choose to adopt, we shall find that electrons behave differently when there are many atoms in close proximity. In the case of metals, the valence electrons retain nearly complete freedom of movement all through the structure. This high degree of freedom of valence electrons makes metals good conductors. Insulators, on the other hand, tend to form relatively strong electrostatic forces (bonds) which prevent valence electrons from leaving specific grain boundaries within the solid. Aggregates of semiconductor atoms form a crystalline structure in which the electrostatic forces that tend to restrict electron movement within specific grain boundaries are only moderately strong. These forces in semiconductors are great enough to require quite large external electric fields to force valence electrons beyond their normal prescribed boundaries, but small enough to enable thermal energy to move electrons beyond the normal boundaries. These so-called valence electrons are only lightly bound to the atom. These inertial forces are small (about 0.05 electron volt), but it is not natural for the electrons to be bound within a solid by these minimal forces alone. Large numbers of atoms, especially in certain crystalline solids, tend to form additional electrostatic bonds which keep the electrons belonging to a specific "grain" within the boundaries of that grain. The behavior of electrons in this situation is too complex for easy comprehension. At this point we reach a fork in the road. One fork leads in the direction of nearly unending complexity, where human imagination becomes strained and understanding is traded for "rigor." The other path leads to a model which, though far from complete, has the distinct advantage of being easy to visualize. We shall take this more understandable route, for it is quite adequate for our purposes—it will serve.

Germanium and silicon are both tetra-valent. That is, they each have four free electrons in their outer orbits. When there are many atoms of silicon or germanium in their natural crystalline form, an electron-sharing phenomenon occurs. The result of all this complex sharing behavior is as though the formerly free valence electrons had become bound to their particular places in the crystal structure. Figure 1-2 shows a stylized diagram of this so-called covalent bond. An electron around an individual silicon atom would require only 0.05 electron volt to free it from the atom. A large number of atoms form covalent bonds which bind the electron to a pair of atoms so tightly that it takes about 0.7 eV to break the electron free.

These covalent bonds, in semiconductors, are still so weak that many of them are broken at any temperature above absolute zero. The higher the tem-

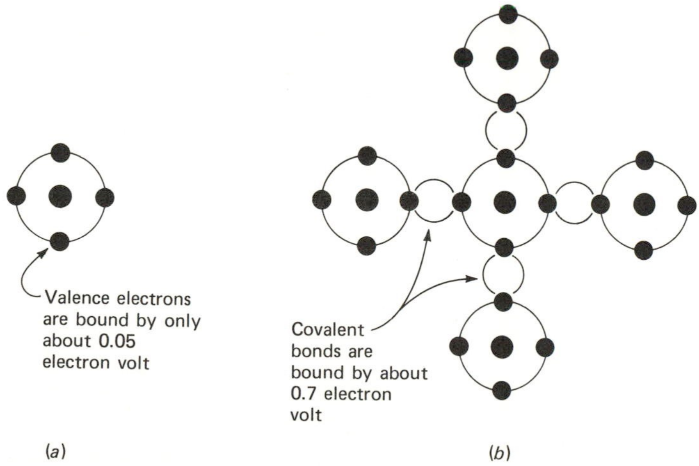
Valence electrons are bound by only about 0.05 electron volt
Covalent bonds are bound by about 0.7 electron volt
(a)
(b)

Fig. 1-2 Covalent bonds. (a) *Isolated germanium atom showing the nucleus and valence electrons;* (b) *a group of germanium atoms with additional covalent bonds*

perature, the more covalent bonds are broken. Whenever a covalent bond is ruptured by thermal energy, an electron becomes free to wander and to be moved by an electric field, if one is present. When the electron leaves its home in the lattice,[2] it leaves behind a dangling bond. This dangling bond has a definite attraction for any electron passing close by, and may capture it if it strays too near. The dangling bond is called a hole, in semiconductor parlance, and because of its attraction for electrons it is conveniently treated as though it were a positively charged body. Once we agree to treat the hole as a bona fide positively charged particle, we must accord it all the rights and privileges of a duly authorized, charged body. Strange as it may seem, one of these privileges is mobility; the hole must have the freedom to move through the structure, just as the electron does.

This concept of hole mobility is, of course, a fiction, but a very convenient bit of fiction. Although we cannot really visualize holes moving under the influence of an electric field, we *can* easily visualize changing concentrations of holes.

Suppose we make a simple sketch to demonstrate. See Fig. 1-3, in which *B* is an electron whose covalent bond has been broken and has gained its freedom. The dangling bond *A* is the hole.

[2] The geometric structure of a crystal.

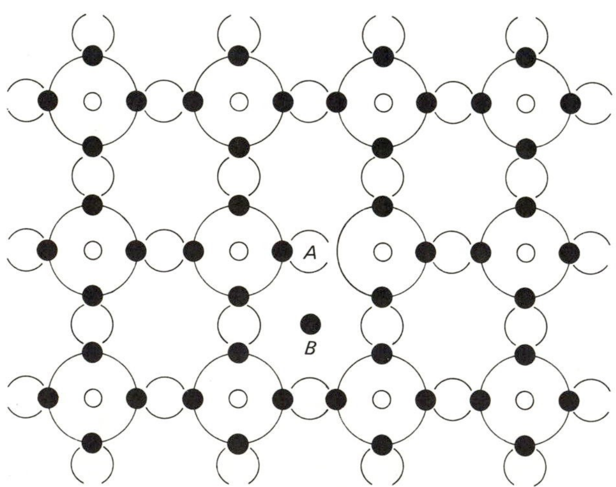

Fig. 1-3 The forming of electron-hole pairs

1-3 CONDUCTION IN SEMICONDUCTORS

When there is no electric field involved, free electrons are pretty evenly distributed throughout the block of semiconductor. Electrons are also being captured by dangling bonds (holes), and each capture results in the annihilation of an electron-hole pair. You see, when an electron is captured by a hole, the hole no longer exists and the electron is no longer a *free* electron. Electron-hole pairs are constantly created and destroyed in about the same numbers. A kind of equilibrium is established.

This uniform distribution is upset when an electric potential is applied across the block. Electrons zip to the positive end of the block as soon as they become free. Now, electron-hole pairs are being formed at both ends of the block (in the middle, too) at about the same rate, but owing to the electric field, there is a heavy concentration of electrons at the positive end and a deficiency of electrons at the negative end. Meanwhile, those holes which are being created at the negative end of the block are not capturing electrons, for there are almost none available to capture. They have defected to the other side. So, at the negative end of the block we have an excess of holes. At the positive end of the block there is such an abundance of electrons that holes are filled as soon as thermal energy generates them.

We now have a block of semiconductor material with a high concentration of electrons at one end and a high concentration of holes at the other. Is it so farfetched to say that electrons have moved to the positive end of the block while holes have moved to the negative end? Well, perhaps it is, but it is a very convenient way to view the situation. It is so convenient that everybody does it. As a matter of fact, we shall find it a hard view to do without.

As I said before, in the pure semiconductor all available carriers, holes and electrons, are the result of heat-ruptured bonds. Conduction characteristics can be controlled by providing controlled amounts of electrons (or holes) which do not depend upon heat for their freedom.

1-4 *n*-TYPE SEMICONDUCTORS

If a small quantity of some impurity with five valence electrons is introduced into the crystal, we gain one free electron. Figure 1-4 is a sketch of how this comes about.

The atom in the center is called a donor, because it donates a free electron to the cause. In addition to having five valence electrons, the donor atoms must fit into the crystal structure comfortably. This restricts available donors to a few members of the ninety-odd natural elements. Two of the most common donor atoms are arsenic and phosphorus. There may be somewhere in the neighborhood of one impurity atom to each one hundred thousand to one million atoms of silicon or germanium. If this relatively small

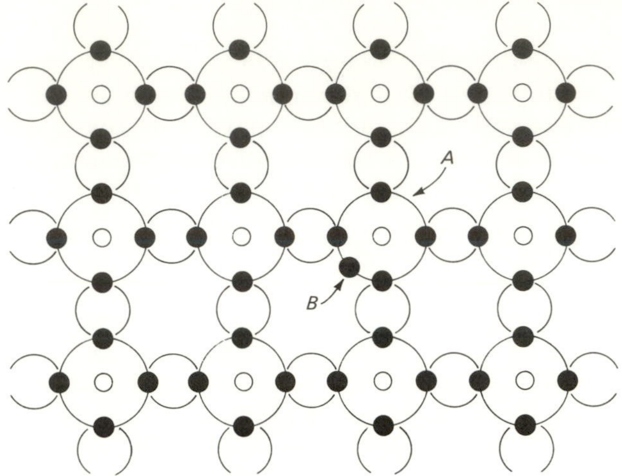

A–A five-valent impurity atom

B–A free electron (not bound by a covalent bond)

Fig. 1-4 The n-*type semiconductor crystal*

quantity of donor atoms seems hardly worthwhile, it is because your perspective is shaped by the macroscopic world in which we live—a world where a million dollars is a lot of money, and where a dollar profit for a million invested would be sloppy business indeed. In the realm of the atom, the perspective is different. For example, there are roughly 10^{22} atoms in a cube of silicon 1 by 1 by 1 cm. That is 10,000,000,000,000,000,000,000 atoms in 1 cm^3.

At the rate of one donor atom per million silicon atoms, our 1-cm^3 block would have received 10^{16} free electrons from the donors. In addition to the donated electrons, there are perhaps some 10^{10} electrons which have been freed owing to heat-ruptured bonds between silicon atoms. The majority of the electrons available to participate in conduction are derived from the donors and are free at all temperatures above absolute zero. In fact, at room temperature we have a system with 10^{16} donor electrons and about one-millionth of that number which are the result of broken bonds. Donor electrons are in the majority, and their number is essentially constant at all usual temperatures; this is because they are not part of a covalent bond and only 0.05 eV is required to free them. So far all we have accomplished by adding donor atoms, a process called doping, is to make a less temperature-sensitive semiconductor with an abundance of electrons. Because electrons are the majority carriers—there being more electrons than holes—and because the electrons carry a negative charge, a semiconductor doped with donor atoms is called *n* type (*n* for negative). The *n*-type semiconductor, if we ignore the thermally generated electron-hole pairs, has only free electrons. At absolute zero this would be a true picture, but at normal temperatures heat-ruptured bonds yield electron-hole pairs. The thermally produced electrons simply join the donor electrons and become members of the group. The holes are true nonconformists; they form a small minority that always takes a direction in exact opposition to the direction taken by the majority.

1-5 *p*-TYPE SEMICONDUCTORS

In order to make practical semiconductor devices, we need another type of semiconductor which is the complement of the *n* type. This complementary type is called the *p* type; it has only free holes (at absolute zero) and no free electrons.

I realize that this all sounds more like alchemy than science, but let me explain the details of our "model" for *p*-type semiconductors and you will see that it is at least a convenient model of reality. *p*-type semiconductors are also made by doping the pure semiconductor material. In this case, however, the doping atoms have only *three* valence electrons. The trivalent doping atoms are called acceptors, because each one contributes a dangling bond (or hole) to the structure. These holes can *accept* electrons supplied from outside the block. Again, we shall resort to a simple sketch. See Fig. 1-5.

In doping the semiconductor crystal with trivalent atoms, such as aluminum, we have created a crystal which has, at absolute zero, no free electrons but only an abundance of holes (dangling bonds). Those dangling

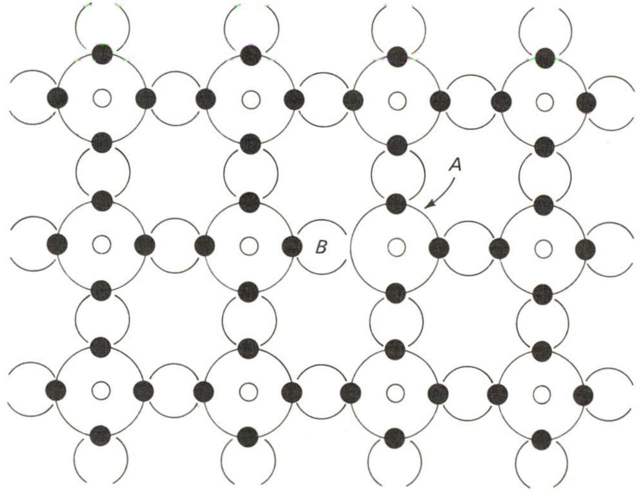

A–A tri-valent impurity atom
B–A dangling bond (hole)

Fig. 1-5 The p-type semiconductor crystal

bonds can accept electrons, which must be supplied by an external battery or other source of electrons if current is to flow through a *p*-type crystal.

If we connect a battery across a *p*-type crystal, electrons jump off the negative electrode of the battery into nearby holes in the crystal because they are attracted by the positive field. However, the positive end of the block not only has holes to attract the electrons, but also the positive field of the battery. The net result is that electrons rush to the positive end of the block, filling nearly all the holes at the positive end of the block but leaving many unfilled holes at the negative end of the block.

Not only are there enough electrons at the positive end of the block to fill all available holes, there are also many electrons that find no hole to fall into. The result is a high concentration of free electrons at the positive end of the block. At the negative end of the block the electrons that are injected race to the positive end of the block, bouncing lightly in and out of the holes at the negative end of the block.

The result is a high concentration of holes at the negative end of the block and a high concentration of electrons at the positive end. It is as though the electrons had migrated to the positive end of the block, while the holes had migrated to the negative end of the block. The concept of hole movement is crucial to the forthcoming discussion of junction diodes and transistors. Again, in the *p*-type semiconductor, electron-hole pairs are being formed by heat-ruptured bonds. The holes that are generated by thermal activity join the *p* (positive) carriers (holes), which are the majority carriers. The electrons so generated join the injected electrons. In the *p*-type semiconductor, electrons are called minority carriers, because all electrons either originate from outside or are the result of thermally broken bonds. There are no free electrons at absolute zero in an electrically isolated crystal of *p*-type semiconductor.

Of course, what I have just said is a small fabrication, for in practice it is rare to have such ideal control as to have no *p*-type impurities in an *n*-type crystal, and vice versa. Such mixed impurities may bend our imaginations a bit, but in point of fact, when one kind of impurity is sufficiently dominant, the *minor* impurity may be ignored.

Table 1-1 shows the most common dopant elements. Figure 1-6 shows a pictorial summary of conduction in semiconductors.

TABLE 1-1. COMMON DOPANT ELEMENTS FOR SILICON AND GERMANIUM

p **Type**	*n* **Type**
Aluminum (Al)	Phosphorus (P)
Boron (B)	Arsenic (As)
Gallium (Ga)	Antimony (Sb)
Indium (In)	

Fig. 1-6 A summary of conduction in semiconductors

a. Intrinsic semiconductor with no field applied.

1. All carriers, electrons and holes, are thermally generated.
2. Electrons and holes are generated in pairs, and are distributed evenly throughout the block.

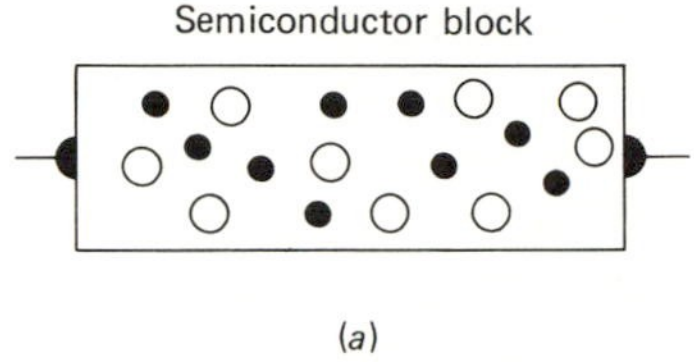

(*a*)

NOTE: The term *intrinsic* refers to a pure crystal which has not been altered in any way except to remove impurities that might have been there.

b. Intrinsic semiconductor with field applied:

1. All carriers, electrons and holes, are thermally generated.
2. Electrons and holes are equal in number because they are always formed as electron-hole pairs.
3. Free electrons from within the block are attracted to the positive end of the block, filling nearly all the holes in the positive end of the block. Other electrons in the positive end of the block, which have not been captured by holes, are returned to the positive plate of the battery.
4. Electrons enter the negative end of the block from the battery and rush from hole to hole toward the positive end of the block.
5. The amount of current, for a given supply voltage, is a function of temperature and increases exponentially as the temperature increases.

The arrow shows the direction of electron travel

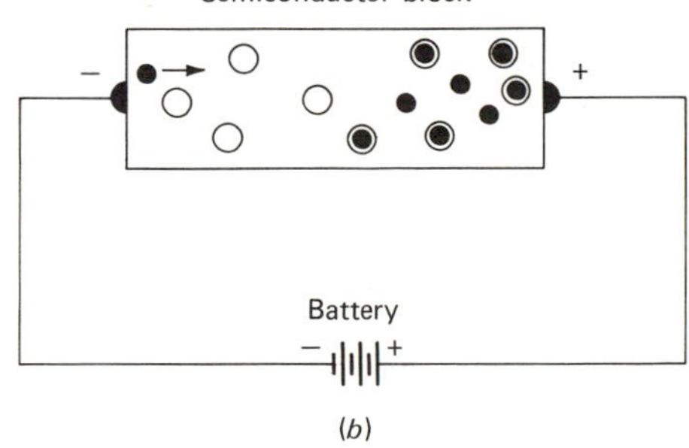

(*b*)

● – Electron

○ – Hole (dangling bond)

◉ – A hole which has captured an electron

c. *n*-type semiconductor with no field applied;

1. The *n*-type semiconductor has only free electrons, no holes at absolute zero. Even at room temperature there would be only one thermally generated hole for each million or so dopant-supplied electrons.
2. These free electrons are free at most ordinary temperatures, and conduction is not dependent on temperature, when a potential is applied.

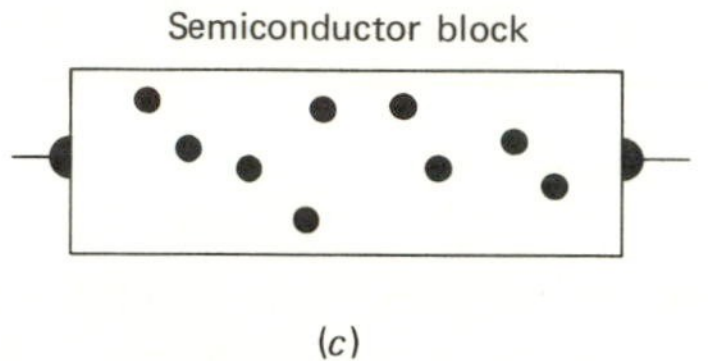

(*c*)

d. *n*-type semiconductor with field applied:

1. Electrons are attracted to the positive end of the block. Electrons in the block are supplied by five-valent impurity atoms and are so lightly bound that they are free at most temperatures much above absolute zero.
2. Conduction is not much influenced by temperature.
3. A few holes, perhaps one for each million electrons, exist as the result of thermally ruptured covalent bonds. The number of holes is an exponential function of temperature.

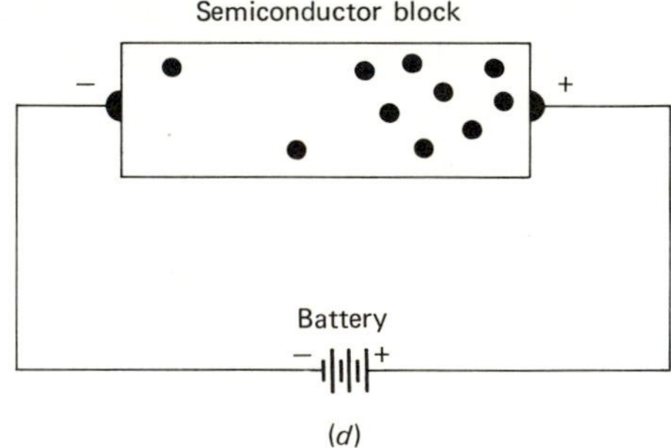

(*d*)

e. *p*-type semiconductor with no field applied:

1. The *p*-type semiconductor has only holes, no electrons, at absolute zero. Holes are provided by trivalent impurity atoms, and their number is not a function of temperature.
2. Some electrons exist in the free state owing to thermally ruptured covalent bonds, but the number is very small compared to the quantity of holes supplied by the dopant atoms.

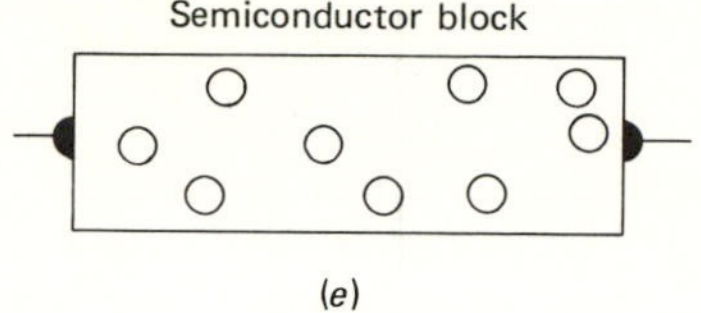

(e)

f. *p*-type semiconductor with field applied:

1. All the electrons involved in conduction through the block must be supplied from outside the block, in this case by the battery.
2. Electrons injected into the block move freely in and out of available holes. Because the holes are supplied by trivalent dopant atoms, there is no covalent bond involved and electrons captured by holes are only very lightly bound.
3. Conductivity is not a function of temperature, except that owing to additional carriers provided by thermally ruptured covalent bonds. This is a relatively small-order effect.

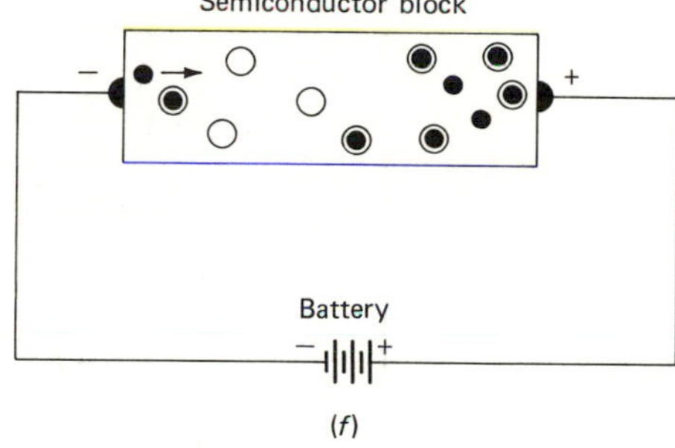

(f)

|2|

the *p-n* junction

OBJECTIVES:

(Things you should be able to do upon completion of this chapter.)

1. Describe the formation of a *p-n* junction.
   a. Make a drawing showing carrier movement when the junction is newly formed.
   b. Define *transition zone*.
   c. List the transition zone potentials for silicon and germanium.
2. Describe the forward-bias condition of a *p-n* junction according to the following directions:
   a. Make a sketch showing the distribution and direction of travel in the carriers in a forward-biased *p-n* junction. Show bias potential as applied to the *n* and *p* ends of the block.
   b. Define *recombination*.
   c. Draw a graph showing the relationship between current (vertical axis) and voltage (horizontal axis). Draw the graph to show conduction in both reverse- and forward-bias directions. Label each major area of the curve and each point where the curve makes an abrupt change in direction.
   d. Define *recovery time*.
   e. List the forward-bias failure mechanisms in a junction diode.
3. Describe the reverse-bias condition of a *p-n* junction.
   a. Make a sketch showing the junction, *p* and *n* zones, carrier distribution, external bias potentials, direction of carrier movement (both holes and electrons), and carrier distribution in the depletion zone.
   b. Define *depletion zone*.
   c. Identify the capacitor plates and the dielectric which form the reverse-biased junction capacitance.
   d. Write the approximate numerical relationship between temperature increase and leakage current.
   e. Make a sketch showing the formation of an electron-hole pair on each side of the junction. Show the direction of travel for all four carriers.
4. List and describe the limiting factors in *switching* time in a diode junction.
   a. Define *storage time*.

b. Define *recovery time.*
c. Define *junction capacitance.*

5. Write the equation for the resistance of a forward-biased *p-n* junction and relate the equation to the *E* vs. *I* conduction curve in objective 2*c*.
6. Define *dynamic impedance* and *static impedance.*
7. Complete the following self-test with a score of at least 90 percent.

SELF-EVALUATION

1. Explain junction formation to someone, using your drawing, definitions, etc.
2. Explain to someone how conduction takes place in a forward-biased diode. Start your explanation by using the *E* vs. *I* conduction curve (as a visual aid) at $E = 0$ V, and describe conduction events as the forward-bias voltage is increased. Explain the failure mechanism(s), and how failure can be avoided. Use all the appropriate definitions and sketches you have prepared for objective 2, *a* through *e*.
3. Explain to someone how reverse bias keeps large currents from flowing in the external circuit. Also explain leakage current, relating it to temperature. Describe what goes on in the depletion zone when electron-hole pairs are formed by thermal energy. Explain the relationship between temperature and leakage current. Use the sketches and definitions from objective 3*a*, *b*, and *c*. Explain junction capacitance.
4. Using the definitions in objective 4*a*, *b*, and *c*, describe and explain the factors which require a finite time (greater than simple carrier transit time) to make a diode *switch* from a forward- to a reverse-bias condition, and from a reverse-bias to a forward-bias condition.

SELF-TEST:

(Watch for the answers to these questions.)

1. Describe the formation of a *p-n* junction.
2. What is the hook voltage or potential hill at a *p-n* junction?
3. How does a semiconductor junction prevent current flow in the reverse-bias state?
4. What is the depletion zone?
5. What happens to thermally generated electron-hole pairs when the junction diode is reverse-biased?
6. How does conduction take place in the forward-bias condition?
7. What happens to thermally generated electron-hole pairs when the junction is forward-biased?

8. How does temperature affect the amount of reverse current in an operating junction-diode circuit?
9. How does temperature affect the amount of forward current in an operating junction-diode circuit?
10. How does voltage affect the amount of current through a reverse-biased junction-diode circuit?
11. How does voltage affect the amount of current through a forward-biased junction-diode circuit?
12. What is junction capacitance?
13. What is carrier storage, and what is *storage time?*
14. Define *recovery time.*
15. How does reverse bias influence the junction capacitance?
16. Write the Shockley relationship and explain its meaning.
17. What is the zener region and what is the zener knee in a junction diode?
18. What is an avalanche condition?
19. What is controlled avalanche?
20. What are the most important electrical ratings of a junction diode?
21. What are the chief failure mechanisms in a forward-biased diode?
22. What are the chief failure mechanisms in a reverse-biased diode?
23. How are forward- and reverse-bias resistances computed?
24. What is the difference between static and dynamic resistance?

INTRODUCTION

You have, perhaps, read the long list of preview objectives and wondered why this chapter gives so long and involved an account of diodes when the avowed purpose of this book is transistor circuit design. The reason is simply that the common transistor is composed of two diode junctions back to back in a continuous crystal structure. In normal operation one junction is in a forward-biased or conducting state, while the other junction is in the reverse-biased or nonconducting state. It is important to understand independent junction behavior as thoroughly as possible before tackling the transistor. For, you see, the transistor is more than the simple sum of its parts, more than two junctions back to back. It has a property called gain or amplification, which greatly magnifies some of the effects of normal junction behavior. In addition, the two junctions in a transistor are not quite independent of each other.

If we are not *very* familiar with normal junction properties, we may easily mistake magnified normal junction behavior for some new entity peculiar to the transistor, or be confused by the relative magnitudes of amplified and unamplified quantities. We shall first examine the basic theme and later study the variations upon the theme, rather than attempt to study all the variations and hope that the theme will somehow emerge. So what we are really doing in this chapter is studying junction behavior rather than the device called a junction diode. We shall examine the simplest two-element semiconductor device, the junction diode, but our interest will be more in junction properties than in the device proper.

2-1 THE JUNCTION DIODE

The junction diode is made by joining a block of *p*-type semiconductor with a block of *n*-type semiconductor in a *continuous* crystalline structure.[1]

There are literally dozens of manufacturing methods for accomplishing this, but most of them start by diffusing donor and acceptor impurities into a block of "pure" silicon or germanium. The diffusion is accomplished at temperatures near 1000°C. More *p*-type (acceptor) impurities are introduced into one end while *n*-type (donor) impurities dominate the other end of the block. Within the block there is a more or less abrupt transition from *p* type to *n* material. This transition is called a junction. Figure 2-1 diagrams the junction diode. The junction diode is essentially a one-way electrical valve, a rectifier in electronics jargon.

Of course, our primary interest is in the junction properties rather than the device properties; but device properties (with the exception of some power considerations) are the junction properties. Therefore any distinction between device characteristics and junction characteristics is largely a semantic one.

2-2 THE TRANSITION ZONE

If you will now glance at Fig. 2-1, you will notice that I have drawn a vertical line between the *n*-type area and the *p*-type area. The line seems to imply

[1] This requirement precludes mechanical joining methods.

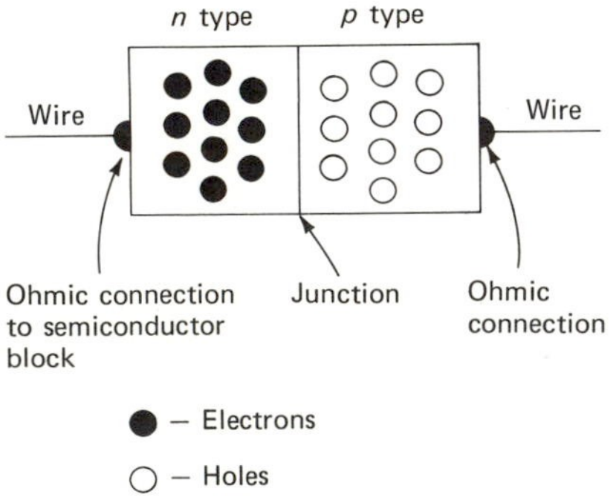

Fig. 2-1 The semiconductor junction diode

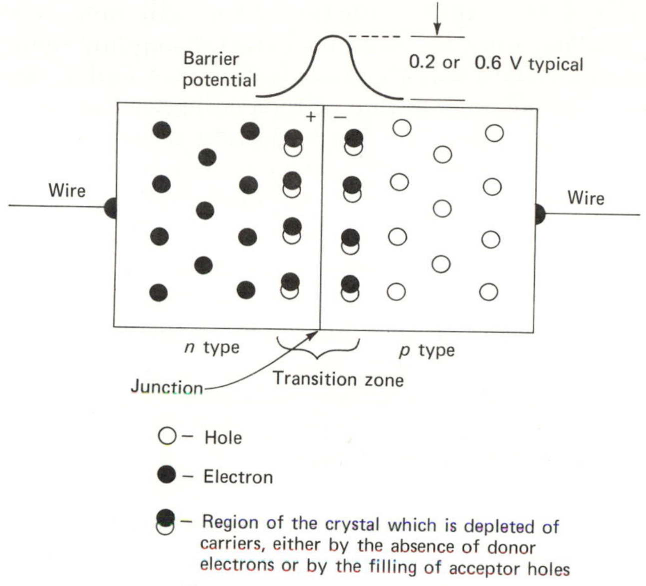

Fig. 2-2 The transition zone

some sort of barrier between the two regions; and indeed there must be such a barrier, for without one what is to prevent all the free electrons in the *n*-type region from gradually diffusing into the holes in the *p* region, leaving neither free electrons nor holes? The question now concerns the nature of this barrier and its genesis.

In the beginning, when the crystal was formed, it was without a barrier and electrons began to diffuse into holes, forming a transition zone which had neither free electrons nor holes. But the process was self-arresting and ceased upon completion of a very narrow transition zone.

If you look at the sketch in Fig. 2-2, you can see that, as the transition zone grows wider, electrons must cross a widening area where the carriers (electrons and holes) are few and far between. The amount of recombination decreases exponentially as the distance from the junction increases. The exponential gradient is shown by the curve in Fig. 2-2.

Both the *n* and *p* regions are electrically neutral in the beginning, but as the crystal is formed, electrons diffuse across the junction and fall into holes on the other side. The *n* side loses some of its free electrons to holes on the *p* side, leaving the *n* side slightly positive. This makes the area next to the junction on the *p* side slightly negative because of the electrons that have come over from the *n* side. In silicon a potential barrier equal to approximately 0.6 V is formed, and further movement across the junction is prohibited until an external potential greater than the 0.6 V barrier potential is applied to the device. The barrier potential is largely a function of the semiconductor material used, and *all* silicon devices will have a barrier potential of 0.6 V, give or take a few hundred millivolts depending upon the temperature. Germanium devices will all have a barrier potential of about 0.2 V. We cannot measure this potential with a voltmeter, but we can verify its existence by applying a voltage across the device starting at zero volts, increasing the voltage to a little more than 0.6 V, and plotting current against voltage on a graph. A little later I will show you what that graph looks like.

The important thing to remember about the transition zone, from a design standpoint, is that it establishes a minimum threshold potential. Current cannot flow through the device until a potential of approximately 0.6 V is applied across the silicon junction. This critical voltage is sometimes called the *hook voltage* or *potential hill.*

2-3 THE REVERSE-BIASED JUNCTION

The reverse-bias condition is the "off" or nonconducting condition. Figure 2-3 shows the reverse-bias battery polarity and the hole and electron distributions in a reverse-biased junction diode.

As the diagram (Fig. 2-3) shows, electrons are attracted to the positive field (provided by the battery) at the right end of the drawing. Meanwhile, the holes in the *p* end of the block are attracted to the negative pole of the battery, which is connected to the left end of the drawing. The carriers, both holes and electrons, withdraw from the junction, leaving a kind of no-man's-

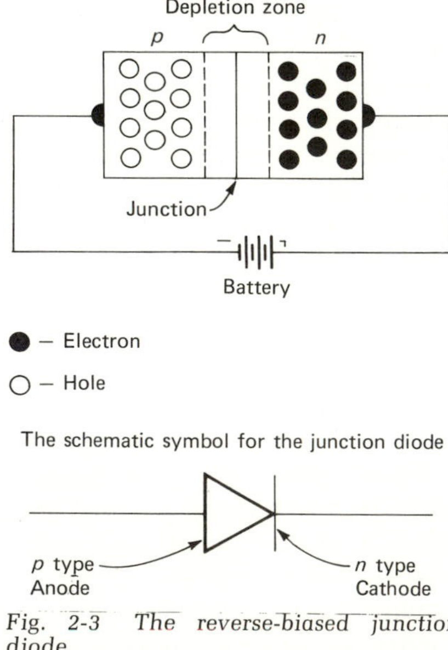

Fig. 2-3 The reverse-biased junction diode

land which is almost completely depleted of carriers. This area is known as the depletion zone. The higher the reverse-bias voltage, the wider the depletion zone becomes. The depletion zone is essentially an insulator; it has no available carriers. No current flows through the junction area in the reverse-bias condition.

2-4 JUNCTION CAPACITANCE

A capacitor consists of two (or more) conductive plates separated by an insulating material called the dielectric. Although the physical dimensions are small, a reverse-biased diode meets all the requirements to become a full-fledged capacitor. It has two areas which are rich in free carriers (conductors by definition), separated by the depletion zone, which has no available carriers and is therefore, by definition, an insulator. The dielectric constant of a silicon depletion zone is approximately 12. The amount of capacitance in a junction diode is small enough so that it becomes a factor only at quite high frequencies. Special junction diodes are manufactured in such a way as to exhibit fairly high capacitances (in the order of 10 pF, more or less).

These diodes are used as voltage variable capacitors in oscillator tank circuits and other applications. The capacitance of a reverse-biased junction is varied by varying the bias voltage. Increasing the voltage increases the width of the depletion zone (dielectric), decreasing the capacitance.

This relatively small capacitance in a diode junction can become the equivalent of a fairly large capacitor in a transistor, because capacitive effects are amplified along with the signal and some other intrinsic parameters.

2-5 THE FORWARD-BIASED DIODE

The forward-bias condition is the conducting condition. Figure 2-4 shows the battery polarity and the distribution of electrons and holes in the forward-bias connection. Perhaps the best way to explain the action taking place is to list the events involved.

1. The negative field at the *n* end of the device repels the free electrons toward the junction (the voltage must be higher than the hook voltage or the process ends at the transition zone).
2. The positive field at the *p* end of the block repels the holes toward the junction. The arrows within the block show the direction of carrier movements within the block. The arrows outside the block show the direction of electron movement.
3. If the battery voltage is higher than the hook voltage (0.6 V for silicon), electrons are forced through the transition zone across the junction and

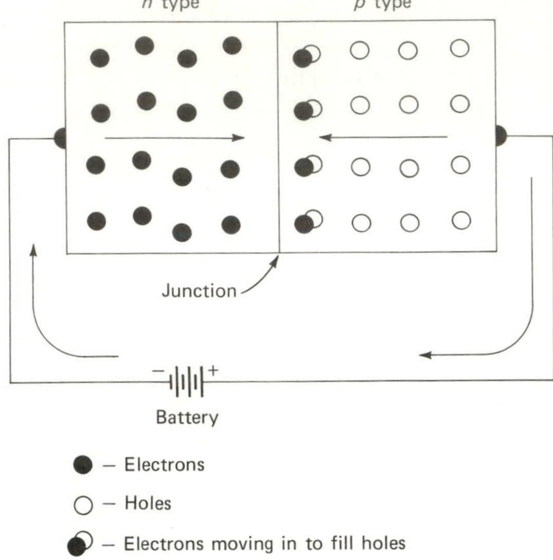

Fig. 2-4 The forward-biased junction diode

into the *p* region. An electron crossing the junction into the *p* region is promptly captured by a hole. The electron has left the *n* region, leaving it one electron short. The *n* crystal, which was electrically neutral, now becomes one unit positive by virtue of the fact that the crystal now has one more proton than it has electrons. This positive charge draws an electron from the negative terminal of the battery. One electron has left the battery.

4. Meanwhile, back at the *p* side of the block, our wandering electron has dropped into a hole. Once the hole is filled, it is, of course, no longer a hole. For all practical purposes, one hole has been lost while an electron has been gained by the *p* crystal. The *p* part of the crystal now has an excess electron and consequently a unit negative charge. The positive terminal of the battery draws an electron out of the righthand end of the block to restore the charge balance in the *p* end of the crystal (see Fig. 2-4). One electron has now left the negative battery terminal, and one electron has returned to the positive battery terminal. If we can visualize some 10^{14} to 10^{19} electrons crossing the junction and jumping into the holes in the *p* region near the junction, while a like number of electrons leave the negative terminal of the battery and a like number are returned to the positive pole of the battery, we have a rough picture of a forward-biased diode in action. Forward bias is the conducting condition for a junction diode.

2-6 REVERSE-BIAS LEAKAGE CURRENT

So far, the description of both forward- and reverse-bias conditions is valid only for the donor- and acceptor-supplied carriers. It is time now to examine the behavior of the thermally generated electron-hole pairs. Remember that an electron which has been freed by thermal energy leaves a hole behind. Thermal energy *always* generates an electron-hole pair. This is true in both *p* and *n* ends of the diode crystal. Figure 2-5 shows a reverse-biased diode with electron-hole pairs being thermally generated within the depletion zone.

The arrows in the drawing in Fig. 2-5 show the direction of carrier movement under the influence of the field produced by the battery. If you will examine electron-hole pair *A*, you will see that the electron moves away from the junction to join the rest of the electrons in the *n* group. The hole is repelled toward the junction. In electron-hole pair *B* the hole is attracted away from the junction to join the rest of the holes in the *p* group. The electron, on the other hand, is repelled toward the junction, where it combines with the hole from pair *A*. This upsets the balance in each side of the crystal, and the negative pole of the battery delivers an electron to the *p* end of the crystal and the positive pole draws one from the *n* side of the crystal. One half of the electron-hole pair *A* is forward-biased, and these carriers (the electron on the right and the hole on the left in Fig. 2-5) constitute a current flow called *leakage current*. The other half of each *pair* is reverse-biased and does not cause any current flow.

This leakage current is the sole current in the reverse-bias direction; it

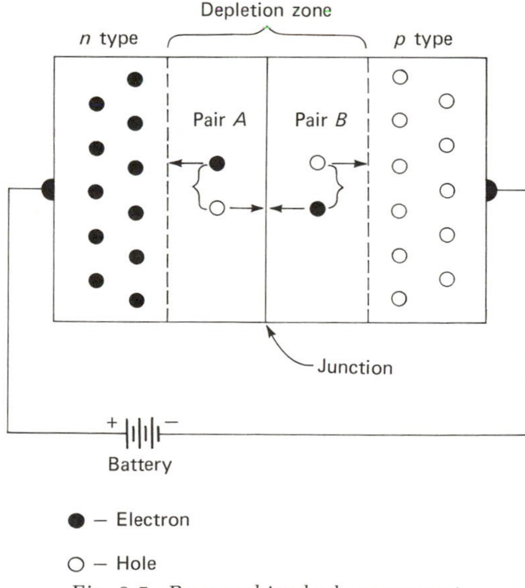

Fig. 2-5 Reverse-bias leakage current

is normally from 1/10,000 to 1/10,000,000 as much as the forward-bias current. The amount of leakage current increases exponentially with temperature. In germanium the number of electron-hole pairs (and consequently the leakage current) doubles with each 10°C rise in temperature. In silicon it triples with each 10°C rise in temperature.

However, a germanium device at 20°C has about 1,000 times the leakage current of a similar silicon device; therefore, a silicon device could operate at over 80°C before it would have the same leakage current as a germanium device at a little above room temperature (water boils at 100°C).

Since leakage currents are generally an undesirable fact of life, silicon is gradually displacing germanium in most applications. Production purification and doping are more difficult to control with silicon than with germanium. As a result, silicon devices have only fairly recently become economically competitive with germanium.

2-7 FORWARD-BIAS LEAKAGE CURRENT

I have attempted to make several sketches illustrating leakage currents in the forward-bias condition. The result has been a compendium of confusion, and perhaps I have discovered the outside exception where one word is worth a thousand pictures.

Curiously enough, it is almost as difficult to measure leakage currents in the forward direction as it is to illustrate them. As a result, leakage currents are always measured in the reverse-bias condition, and the magnitude of the forward-bias leakage is assumed, with good theoretical reasons, to be equal to the reverse-bias leakage (at any given temperature).

The number of thermally generated electron-hole pairs is a function of temperature, not bias condition. In the forward-bias condition all electrons in the *n* region and all the holes in the *p* region participate in forward conduction, regardless of whether they are supplied by dopant atoms or are thermally generated. All minority carriers, electrons in the *p* area and holes in the *n* area, are reverse-biased and do not participate in conduction in the forward-biased condition. As a result, each time there is one electron-hole pair generated in the *p* area and one pair generated in the *n* area, there is one electron leaving the negative terminal of the battery and one returning to its positive terminal. And so the leakage current is simply added to the forward-conduction current. When there is perhaps 1 A of current flowing in the forward-bias direction, the addition of 1 μA or even 1 mA of leakage current will hardly be significant.

In a junction diode leakage currents are generally of little importance, but the transistor has the ability to amplify, and leakage currents amplified by a factor of 100 or more can be a significant problem.

It is time to take a quantitative look at the nature of forward-conduction current and leakage current. Nature being what it is, it is too much to expect the current-voltage relationship to be a simple one.

The best way to describe these relationships is to draw a graph, Fig. 2-6. If you will examine the upper righthand quadrant, you will notice that almost no current flows until the hook voltage is reached. The small current that does flow is mostly due to thermally generated carriers within the transition zone. As soon as the hook voltage is reached, a small increase in forward-bias voltage results in a large increase in current. But you will notice that the curve is not quite straight, and that means that the resistance of the junction changes in a nonlinear fashion as the voltage across the junction changes.

If our graph had numerical values and we were to calculate junction resistance ($R = E/I$) at several points along the graph, we would find that the junction resistance decreases as the voltage increases. Again this change in junction resistance is of little concern in a junction diode; but when it is the forward-biased junction in a transistor that we are concerned with, we find that the small *effect* is magnified many times by the amplification of the transistor.

You will notice that the forward-conduction curve rises steeply and shows no sign of leveling off, implying that there is no limit to the current. If

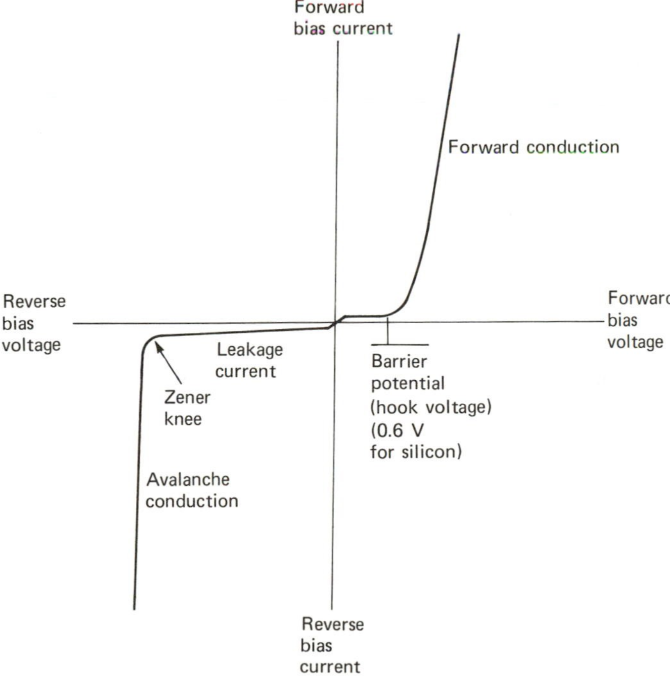

Fig. 2-6 The curve showing the diode conduction behavior

we could keep the junction cool, we could run the current up quite high; and to some extent we do provide for junction cooling in the form of heat-radiating fins and such. But in spite of all practical cooling efforts, junction temperature determines the upper limits of current flow. The principal failure mechanism in a forward-biased junction is excessive junction temperature. Excessive junction temperature may occur before it is noticeable on the outside of the device.

To avoid excessive junction temperature, we must pay attention to the manufacturer's instructions concerning maximum voltage, current, and mounting arrangements. The manufacturer has probably burned out a number of the devices in determining these ratings, and it is foolish for us to repeat his efforts. However, it is wise to assume that his tests were conducted in outer space or under some other conditions which we are not likely to encounter and to operate only up to 80 percent or so of the manufacturer's maximum ratings. How conservative we can be is often determined by cost factors; and although a 100 percent safety factor might be desirable, it might also cost too much. The principal points of interest on the forward-conduction curve are:

1. The hook voltage
2. The variation in junction resistance (the nonlinearity and slope of the curve)
3. The failure mechanism: excessive junction temperature

Now, let us take a look at the reverse-bias part of the curve in the lower left quadrant of Fig. 2-6. You will notice that increasing the reverse-bias voltage increases the reverse current only slightly until a critical voltage called the zener point is reached. At the zener knee of the curve the current suddenly rises from microamperes to milliamperes or amperes. There are actually two principal mechanisms involved in production of this current: field emission and ionization by collision.

As we increase the voltage in the reverse-bias direction, the depletion zone becomes wider and wider. At some critical point the field is large enough and the path a carrier can travel without colliding with an atom is long enough so that an electron can gain enough momentum to strip electrons from atoms with which it collides. The dislodged electrons pick up speed until they collide with other atoms, liberating still more electrons. Figure 2-7 is a sketch of this process, which is often called an avalanche because unless it is controlled, it will continue building up like an avalanche until the device is destroyed.

The carriers that initiate the avalanche process are thermally generated minority carriers in the depletion zone. The increase in current at the beginning of the avalanche process causes an increase in temperature, because any increase in current through a given resistance causes a larger voltage drop and

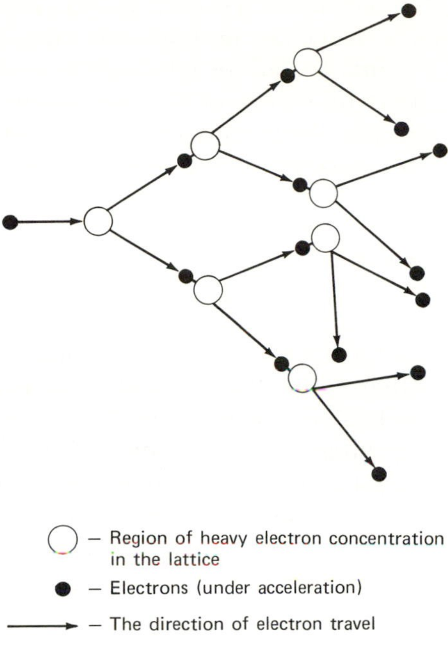

Fig. 2-7 The avalanche process

an increase in heat. As the temperature increases, more thermally produced carriers appear, the current increases and produces more heat, and on and on it goes. This is called thermal regeneration or thermal runaway.

The two kinds of avalanche processes, thermal and collision, do not simply augment one another; the increased number of carriers lowers the resistance and makes the whole process a dynamic and complex one. However, the picture conjured up by the thermal and collision avalanche mechanisms is adequate for our purposes. We need not complicate it with further details.

Field emission was hypothesized by Zener as *the* mechanism. It has since been given a secondary place by most authorities. Field emission is simply the concept that atoms can have their electrons literally torn out of orbit if the field is large enough. Field emission is known to occur, but how important it is to the avalanche condition is still the center of some debate.

Some silicon devices are designed to operate in a *limited* avalanche mode. Limited avalanche operation is permissible as long as the junction temperature is not permitted to go too high. This is normally accomplished by restricting the current through the device by placing a limiting resistance in the circuit and by removing the heat that is generated.

This same limited avalanche operation can be one of the chief failure mechanisms in a transistor's reverse-biased junction. This is the case because of a tendency for the avalanche breakdown to occur first in a limited area forming a sort of tunnel of avalanche current and, more often than not, a fatal hot spot. This was a principal failure mechanism in early controlled avalanche devices, too. This tunneling avalanche failure mechanism is often called secondary[2] breakdown. Secondary breakdown is no longer the threat it once was, and it may soon become an obsolete problem altogether.

2-8 JUNCTION SWITCHING TIME

One of the specifications sometimes given for certain junction devices is recovery time. It takes a finite time to switch a junction from a forward- to a reverse-bias condition, that is, from a conducting to a nonconducting state. (The reverse is true also.) This specification is important to computer-destined devices and others whose functions involve rapid turn-on or turn-off.

When a junction is forward-biased, the depletion zone is saturated with carriers because of the large number of recombinations (holes and electrons) taking place at the junction. If the bias is suddenly removed, conduction does not cease instantly, because the nonconducting (reverse-bias) state cannot be obtained until carriers have been cleared out of the depletion zone.

This short period of current flow after the bias has dropped to zero is called the storage phenomenon. The time required to clear the carriers out of the transition zone and restore the potential hill (hook) is called the storage time. You will hear more of this later.

There is also a delay involved in switching from a reverse-bias to a forward-bias state. This delay is due to the fact that the junction capacitance soaks up the initial forward-bias current in the process of taking on a charge. After the charging is "complete," current becomes available for normal conduction.

2-9 THE RESISTANCE OF A JUNCTION DIODE

The resistance of a forward-biased junction diode is composed of two principal components, the bulk resistance and the junction resistance. The bulk resistance is simply the ohmic resistance of the *p* and *n* blocks of silicon, and with the exception of some variation with temperature, this resistance is fairly constant. The junction resistance is the effective resistance of the transition zone, and it is dependent not only upon temperature but also upon the operating current. An empirical equation derived by Shockley gives the junction resistance as

[2] Although it is grammatically incorrect, the term *second* breakdown is also common in the literature.

$$r_j \approx \frac{25 \text{ mV}}{I \text{ (mA)}}$$

In junction diodes used at higher currents, the junction resistance drops to a very low value and the bulk resistance becomes dominant. In low-current diodes and transistors the junction resistance is fairly high and the bulk resistance is normally small enough to ignore. The complete equation for the resistance of a forward-biased junction diode is $R \approx r_j + r_B$, where r_j is the junction resistance and r_B is the bulk resistance.[3]

The Shockley relationship is really only a useful approximation, but it will prove to be a very important approximation in many design and analysis problems. The Shockley relationship is not overwhelmingly precise, but it is the best quantitative description of the resistance of a forward-biased junction that anyone has come up with.

Dynamic and Static Resistances

A semiconductor junction is a complex system; a part of its complexity is that it has two distinct resistances in both the forward- and reverse-bias directions. In one sense these two resistances are dependent upon how the resistance is measured, but in a broader sense the two resistances reflect different reactions to different operating conditions. Static resistance is easy to understand because it involves the simplest form of Ohm's law, $R = E/I$.

Dynamic resistance is a little more difficult to understand because it must be stated in terms of the *changes* in current caused by *changes* in voltage. In mathematical form,

$$R_D = \frac{\Delta E}{\Delta I}$$

where R_D is the dynamic resistance (in ohms), ΔE is the *change* in voltage (in volts), and ΔI is the *change* in current (in amperes).

In the case of the forward-biased diode, for operation anywhere beyond the hook in the diode conduction curve (Fig. 2-6), we can consider the static and dynamic resistances to be the same. This is a satisfactory assumption under the conditions encountered in transistors (where Δ is assumed to be small).

In a reverse-biased junction (such as the collector-base junction in a transistor) it is quite a different story. The static resistance of a reverse-biased junction is simply $R = E/I$, where E is the voltage impressed across the junction. I is the current through the junction, which in a simple diode is only

[3] An excellent treatment of this topic can be found in Albert P. Malvino, *Transistor Circuit Approximations*, McGraw-Hill Book Company, New York, 1968.

the leakage current. In a transistor I is made up of two components, the leakage current and a second current which is *injected* across the junction by the transistor's emitter. The static impedance (resistance) controls the quiescent current and voltage drops, the ones we normally measure with a meter.

The dynamic resistance is a measure of how *changes* in voltage affect *changes* in current. If you will look again at the reverse-bias part of the diode conduction curve (Fig. 2-6), you will notice that the reverse leakage current increases only very slightly as the reverse-bias voltage increases. In a more or less typical case we would expect no more than a 10 percent increase in current for a 100 percent increase in the applied voltage.

Suppose we try a numerical example. Let us assume that we have a diode connected in the reverse-bias direction and that, with 10 V applied, we measure a current of 10 μA.

Compute the *static* resistance of the junction.

$$R_s = \frac{E}{I}$$

$$R_s = \frac{10\ \text{V}}{10\ \mu\text{A}}$$

$$R_s = 1\ \text{M}\Omega$$

Now, assume that a doubling of the voltage results in a current increase of 10 percent, and compute the *dynamic* resistance of the reverse-biased diode when the voltage is increased to 20 V.

1. 10 percent of $10\ \mu\text{A} = 1\ \mu\text{A}$
2. $\Delta E = 10\ \text{V}$
3. $\Delta I = 1\ \mu\text{A}$

$$R_D = \frac{\Delta E}{\Delta I}$$

$$R_D = \frac{10\ \text{V}}{1\ \mu\text{A}}$$

$$R_D = 10\ \text{M}\Omega$$

Of course this is just a made-up example, but the results are reasonable enough. The absolute values will vary over a wide range, but the implication that the dynamic resistance is always greater than the static resistance in a junction diode in the reverse-bias state is a valid one. The *static* resistance of a reverse-biased *diode* is controlled by temperature. (In a transistor temperature plays a lesser part.) The *dynamic* resistance is mostly independent of temperature.

So far I have only described the dynamic resistance of a reverse-biased junction; I have not explained it. The explanation will have to be a bit oversimplified, but here it is. We would expect an increase in potential to scavenge available electrons more efficiently from the system, resulting in an increase in current. However, working against the increased potential is the fact that the depletion zone is widened by the increased potential. The longer path through the high-resistance depletion zone that the carriers must travel very nearly balances out the effect of the increased potential. The result is that the current does not increase as much as Ohm's law would lead us to expect for a given increase in voltage. The static and dynamic resistances of an ordinary resistor are the same, but a reverse-biased diode is not at all like an ordinary resistor.

SUMMARY

The junction diode is composed of a continuous crystal, part of which is doped to be a *p* type, and part of which is doped to be an *n* type. The point of transition from *n* to *p* is called the junction.

During the crystal formation electrons cross the junction from the *n* side to the *p* side, developing a small charge zone on either side of the junction. This region is free of active carriers and is called the transition zone. The amount of charge accumulated in the transition zone is characteristic of the particular kind of semiconductor. In silicon it is approximately 0.6 V, and in germanium about 0.2 V. The junction must be forward-biased by a voltage which is higher than this barrier potential before forward conduction can take place.

In the forward-conduction condition an increase in current causes the junction resistance to drop in such a way that the voltage drop across the junction remains at the same (approximate) value as the barrier potential for the entire range of normal currents. The Shockley relationship describes this resistance drop in a quantitative fashion as

$$r_j \approx \frac{25 \text{ mV}}{I \text{ (mA)}}$$

The static and dynamic resistances in a forward-biased junction may be considered equal if the increment (Δ) is small.

When the diode is reverse-biased, holes and electrons withdraw from the junction, leaving an area on either side which is cleared of carriers. This area is called the depletion zone, and it forms the dielectric of a small capacitor. Increasing the reverse-bias voltage widens the depletion zone, decreasing the capacitance. In a reverse-biased junction, the dynamic resistance is higher than the static resistance, even for small increments.

Leakage current is caused by thermal rupture of covalent bonds, which always results in the generation of electron-hole pairs. Leakage current is a function of temperature. It increases by a factor of 3 in silicon and by a factor of 2 in germanium for each 10°C increase in temperature. Silicon devices have only about 0.001 of the leakage current (at 20°C) that we would expect from a germanium device at the same temperature. Leakage current is measured in the reverse-bias condition, where it constitutes the only current in the system. Leakage current is only slightly influenced by changes in the reverse-bias voltage.

The chief failure mechanism in a junction diode is excessive junction temperature. In the forward-bias condition overheating is caused by excessive current, inadequate heat dissipation, or both. In the reverse-bias condition excessive junction temperature results from unrestricted avalanche current or from tunneling avalanche and its resulting hot spots.

Now go back to the questions at the beginning of the chapter and see how well you can answer them.

3

the transistor as a practical amplifier

OBJECTIVES:

(Things you should be able to do upon completion of this chapter.)

1. Explain the amplifying mechanism in a transistor.
   a. Draw a block diagram of an *npn* transistor showing the two junctions, the terminals, and the external bias potentials. Label collector, base, and emitter. Draw arrows in the external circuit showing the direction of current (electron) flow in each leg of the device. Draw arrows inside the device showing directions and relative magnitudes of electron flow.
2. Define the following terms: I_c, I_b, I_e, I_{co}, I_c max, E_c max, Z_{in}, β, I_g, R_o, R_{co}, V_g, R'_e, R_{in}, R_{Lw}, R_{Lac}, R_L.
3. Define the *Shockley relationship*.
4. Draw the schematic diagram of each of the three configurations.
5. Write the equations for the following parameters for each circuit in objective 1.
   a. The equation for the voltage gain of a complete circuit, including the working load R_{Lw}.
   b. The equation for the input impedance, including an external emitter resistor.
6. Compute important parameters.
   Given the schematic diagram in Fig. 3-16, compute the parameters requested in Problems 3-1 to 3-5.

SELF-EVALUATION

Explain to someone how a small base-emitter current controls a larger collector-emitter current. Use the drawing in objective 1*a* to aid your explanation.

SELF-TEST

1. Why are device parameters, such as the hybrid parameters, more important to the manufacturer than they are to the designer or technician?
2. Define I_b.
3. Define I_{co}.
4. Define I_c.
5. Define I_e.
6. Define R_o.
7. Define R_{co}.
8. Define I_g.
9. Define V_g.
10. Define Z_{in}.
11. Define R_e' in the form of an equation.
12. What is the range of voltage gains for the common-emitter circuit?
13. What is the range of voltage gains for the common-collector circuit?
14. What is the range of voltage gains for the common-base circuit?
15. If a manufacturer lists the maximum collector current I_c (max) at 100 mA, what is the highest steady-state current that the transistor should be operated at?
16. What is a heat sink? ANS: It is a large-area high-thermal-conductivity metal (usually copper or aluminum) structure, which is clamped to a transistor to carry away internally generated heat. The heat sink often has heat-radiating fins to compress a large area into a small volume.
17. Define *avalanche* and *zener breakdown*.
18. What is punch-through? Is it always a failure mechanism?
19. In normal transistor operation, which junction is forward-biased and which is reverse-biased?
20. Define *power gain*.
21. Define *barrier potential*. What causes it?
22. Define *depletion zone*.
23. Write the nearest hybrid-parameter symbol for each of the following: α, β, Z_{in}, R_o.

INTRODUCTION

Transistor technology has provided a bewildering jungle of electrical and mathematical models to describe how transistors behave. There are three basic types of models, each of concern to different people. The physicist is interested in models and descriptions of the complex happenings within the device itself. The manufacturer needs models which lend themselves to practical measurements (of device parameters) which are reasonably independent of the circuits in which they are measured. The engineer and technician require models which describe how the transistor will behave in various kinds of circuits.

Although the models used by the manufacturers are largely of academic interest to the designer, it is important that he know how the important parameters, derived from these models and listed in the manufacturer's specification sheets and manuals, apply to his problems.

In this chapter we will examine the mechanism by which amplification takes place in the transistor, examine its structure, and discuss important electrical parameters of the device. It is customary to present the transistor's electrical parameters in great detail without comment about their importance (or lack of it) in actual circuits.

Here we shall not only present the device's parameters, but also discuss the relationships of those parameters to circuit parameters. This is a realistic approach, because most of the transistor's important parameters are far from independent of circuit values. In addition, there are three basic circuits—common-emitter, common-collector, and common-base, listed in the order of popularity—and each configuration provides different values for input impedance, gain, output impedance, and so on. These important parameters are β, α, maximum collector current, maximum collector voltage, and leakage current. The rest of the parameters cannot really be discussed at a practical level without including basic circuit structure. To that end we will discuss device parameters in conjunction with each of the three basic circuit configurations. We will consider the common-emitter circuit to be the basic circuit throughout the book, and so we will treat it first. There is considerable overlap of equations among the three configurations, so I will *explain* each one only once and when it comes up again, I will simply write the equation, in case you have forgotten it. The segment dealing with the common-emitter circuit will be the longest because it will contain all the basic explanations while the other two sections will list all applicable equations and explain only those which differ from common-emitter equations.

Current manufacturing technology does not permit very close tolerances for intrinsic parameters, and if it did, there would be but little gained by it. Large variations in these important parameters due to only moderate temperature variations are a problem every bit as troublesome as broad manufacturing tolerances. Even if manufacturers could hold closer tolerances, we would still have to cope with temperature sensitivity; it is an inherent property of semiconductors, but as it happens, the techniques used to deal with temperature problems also serve to accommodate large manufacturing tolerances. By the time the circuit designer has compensated for those inherent temperature problems, he has also compensated for very large manufacturing tolerances. As a result the designer does not press the manufacturer for closer tolerances, and so the manufacturer has no incentive to hold closer tolerances. In addition to tolerance and temperature variations, some important parameters vary with the operating current, which is selected according to the needs of the designer, but these variations also yield to the same basic techniques which allow for large tolerance and temperature variations.

The military electronics industry often demands close tolerances, but then it is a case of *selecting devices* that fall within military standards, a process that is far too expensive for commercial purposes, and one which actually allows designers to produce designs which are possibly less reliable, and less serviceable in the field, than they might be. Military procedures have always been strange to all but those people who are oriented to them, and transistor electronics is no exception.

I am trying to prepare you for the *idea* that specifications in the transistor manual are only "helpful" guides (they are not to be taken too seriously), and at the same time to prepare you for the even more important idea that "exact" specifications of transistor parameters are *not* necessary, nor would they even be very helpful. Exact specifications are never likely to become an economic possibility anyway, but we don't really care.

Important Symbols and Terminology

$\approx$: approximately equal to
Example: $A \approx B$, A is approximately equal to B
Δ: A change or variation in some quantity
Example: ΔI, a change (or variation) in the current I
$\|$ or //: in parallel with
Example: $R_1 \| R_2$, resistor R_1 in parallel with resistor R_2
Typ.: Typically, or typical value

3-1 THE STRUCTURE OF THE TRANSISTOR

The transistor consists of two junctions in a continuous crystal. The configuration may be *pnp* or *npn*. See Fig. 3-1. An ohmic contact is made to each crystal area, making a three-terminal device. One "end" block is called the emitter, the middle block is called the base, and the other "end" block is called the collector. The middle section is always called the base. Theoretically either "end" could be called the emitter or collector. In practice, one "end" is designated by the manufacturer as the collector and one "end" as the emitter. If you will take a real transistor and exchange the emitter and collector leads, you will still have a functional transistor.

It will not conform to the specifications published by the manufacturer, but it will still be a transistor. The differences are due to the fact that the manufacturer has, perhaps, doped one "end" block more heavily than the other, and he has definitely made certain that the block he calls the collector has been arranged in such a way that it is easily cooled. By this I mean that there is a good *thermal* connection between the area designated "collector"

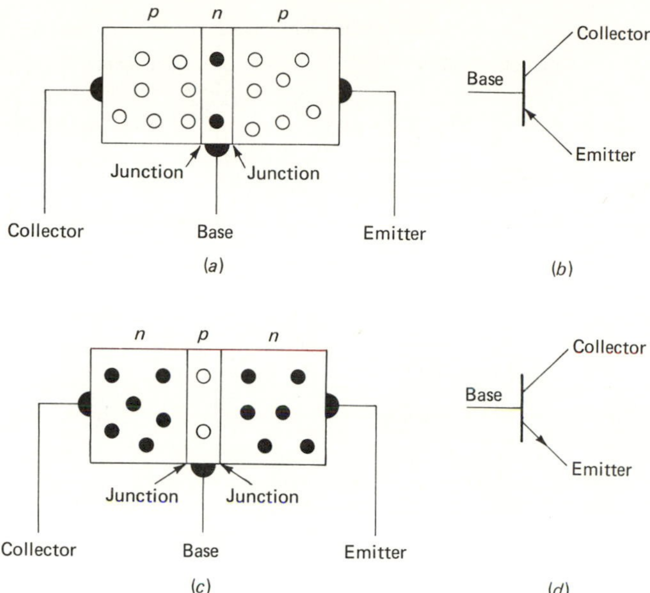

Fig. 3-1 pnp *and* npn *transistor structures.* (a) *The* pnp *structural diagram;* (b) *the* pnp *schematic diagram;* (c) *the* npn *structural diagram;* (d) *the* npn *schematic diagram*

and the case, which is exposed to the open air. Convection currents moving around the case carry the heat away into the atmosphere. The area designated "emitter" is less favored in its ability to get rid of internally generated heat. This means that the power rating of an inverse-connected transistor is most likely lower than a "normally" connected transistor. Other parameters will be different also but less predictably so.

The base region of the crystal is very thin, 0.8 microns or less. This thin base region is essential to normal transistor operation, as we shall see shortly. Twenty-five microns is approximately equal to 0.001 in. Manufacturing techniques and physical geometry could well provide material for at least one book. I shall not go into that aspect here, but I shall provide you with a reference source for this information.[1]

3-2 THE NATURE OF TRANSISTOR AMPLIFICATION

During the following discussion please keep these two things in mind:

First, the base region is very thin and lightly doped; as a result, it has

[1] *The General Electric Transistor Manual*, Semiconductor Products Department, Electronics Park, Syracuse, N.Y.

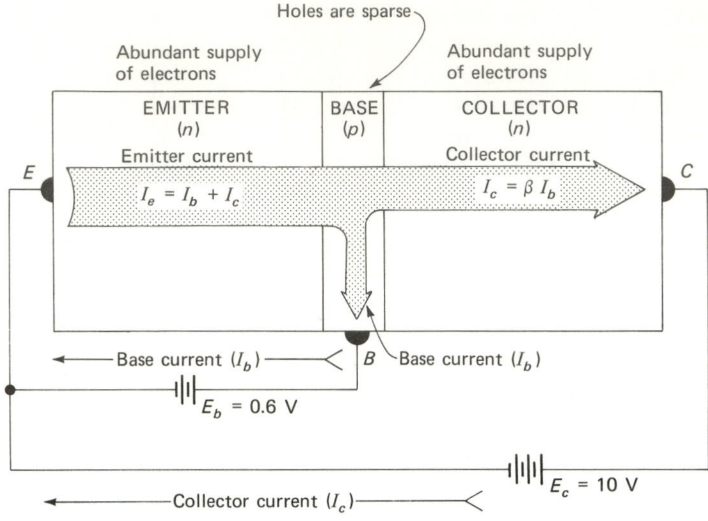

Fig. 3-2 Transistor amplification

very few dangling bonds (holes) as compared to the vast number of free electrons available in both emitter and collector regions.

Second, the collector-base junction is reverse-biased.

Now, refer to the sketch (Fig. 3-2) often as we follow through the operation.

1. Assume that the base-emitter bias is, at first, less than the hook voltage (the transition zone potential) of the base-emitter junction. The base-emitter junction is therefore reverse-biased, and there is no base-emitter current.
2. The collector-base junction is reverse-biased by, let's say, 10 V. Thus, there is no collector-base current flowing. There is also no emitter-to-collector current flowing because both junctions are reverse-biased.
3. So far the transistor has been in the off, or nonconducting, state. Now, let us bring the emitter-base bias voltage up above the hook voltage, forward-biasing the junction and starting base-emitter circuit conduction. A superabundance of electrons rushes through the emitter block toward the junction. These electrons rush to the junction only to find, once they get into the base region, that holes are sparse. Those electrons which do find holes in the base region combine with them and result in base current in the fashion of any forward-biased junction. For each 100 electrons drawn into the base region (at any given instant), the holes in the base can capture only a fraction (1 percent or less) of them. The rest of them cannot go back to the emitter, because the field is pushing them away from the emitter. These surplus electrons congregate near the collector-base

junction, where they are attracted by the positive field provided by the collector battery. They are drawn into the collector, forming an emitter-to-collector current loop. Thus, we find that electrons leaving the emitter block take two paths. A small percentage combine with holes in the base to form base-emitter current, while the majority of them move through the collector block to form an emitter-collector current. This relatively large emitter-collector current does not begin to flow until emitter-base current begins to flow. To state it a little differently, the large emitter-collector current is dependent upon the flow of the much smaller emitter-base current. If the base-emitter voltage is increased slightly, more electrons are drawn into the base region; far more than the base can use. The result is a small increase in base current and a much larger increase in collector current. This current gain is called beta (β).

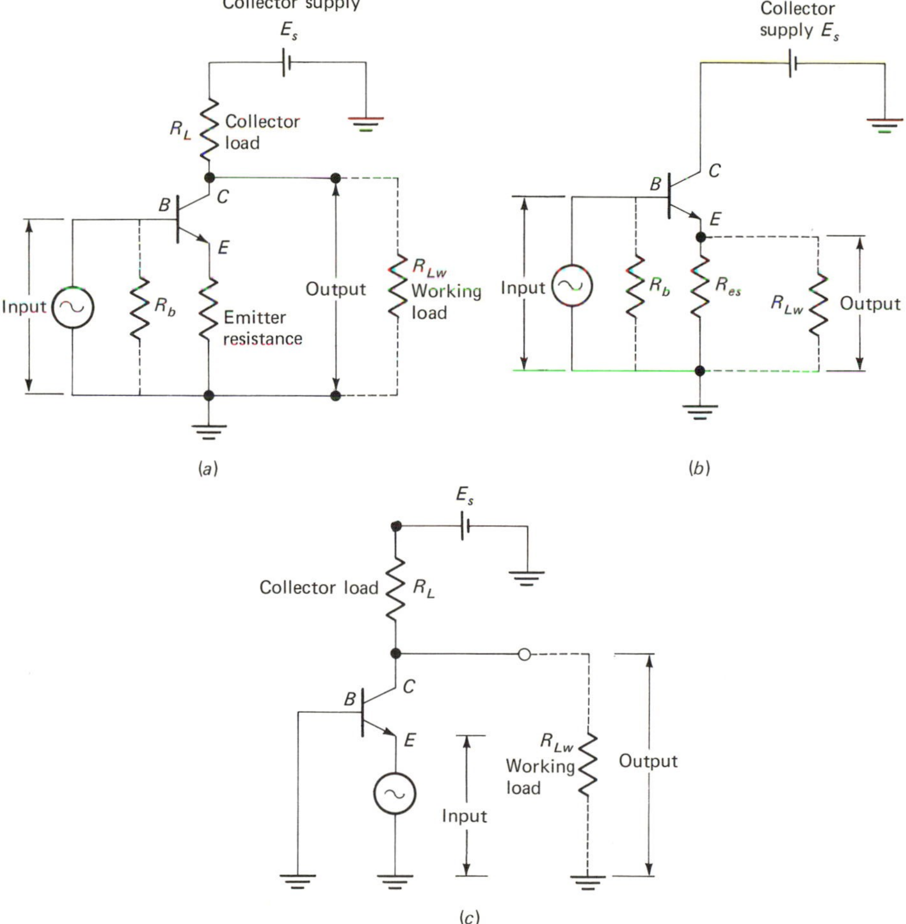

Fig. 3-3 Common-emitter, common-collector, and common-base configurations. (a) *The common-emitter circuit;* (b) *the common-collector circuit;* (c) *the common-base circuit*

This is of course a much simplified explanation, but it is adequate for our purposes here.[2]

Before we go on to discuss β and other parameters in greater detail, I will list and briefly define the parameters we will be discussing, and show you the schematic diagrams for the three basic configurations we will be discussing. See Fig. 3-3.

A Preview of the Parameters to Be Discussed in This Chapter

I_c: The collector current
NOTE: The collector and emitter currents are approximately equal.
I_b: The base current
I_e: The emitter current

$$(I_e \approx I_c \approx \beta I_b)$$

I_c (max): The manufacturer's absolute maximum steady-state collector current. Normal operation is at *less than* 40 percent of this figure.

E_c (max): The manufacturer's absolute maximum rating for the reverse-bias voltage across the collector-base junction. Even an instantaneous excess can sometimes be fatal to the device.

I_{co}: The reverse-bias leakage current of the collector-base junction. Because the leakage current results from thermally produced carriers, the temperature is normally stated as part of the parameter. The emitter is not connected. The symbol I_{cbo} is also used for I_{co}.

α (alpha): The intrinsic current gain in the *common-base* configuration. Range: from 0.9 to a limit of 1.

$$\alpha = \frac{I_c}{I_e} \quad \text{a simple ratio, no dimension}$$

β (beta): The intrinsic current gain of the transistor in the common-emitter and common-collector configurations. Range: Typ. 10 to 200.

$$\beta = \frac{I_c}{I_b} \quad \text{a simple ratio, no dimension}$$

I_g: The current gain of the complete circuit. I_g is nearly always less than β.

[2] See M. V. Joyce and K. K. Clarke, *Transistor Circuit Analysis*, Addison-Wesley, Reading, Mass., 1961.

(Dimensionless)
Range:
a. For the *common-emitter circuit*, 1 to β
b. For the *common-collector circuit*, 1 to β
c. For the *common-base circuit*, less than 1

R_o: The intrinsic output resistance of the transistor. The dynamic resistance of the reverse-biased collector-base junction. Range $\approx$ 200 kΩ to 2 MΩ – not normally predictable.

$$R_o = \frac{\Delta E_c}{\Delta I_c}$$

R_o is normally shunted by a comparatively low resistance load which governs the output resistance of the complete circuit.

R_{co}: The output resistance of the complete circuit.
a. *The common-emitter circuit*
$R_{co} \approx R'_L$, where R'_L is the effective load resistance (later to be called R_{Lac}). Range: Typ. 100 Ω to 50 kΩ.
b. *The common-collector circuit*
$R_{co} \approx R_L \| (R_b/\beta)$ R_L and R_{es} are the same.
(This is a bit complex and will be explained later.)
c. *The common-base circuit*

$$R_{co} \approx R'_L$$

R'_e: The forward-biased base-emitter junction resistance.

$$R'_e \approx \frac{25\ (\text{mV})}{I_c\ (\text{in mA})} \qquad \text{the Shockley relationship}$$

R'_e is in ohms.

R_{in}: The intrinsic input resistance of the *operating* transistor, ignoring any resistances from base to ground (common). Range: from a few ohms to a *practical* limit of approximately 0.5 MΩ.
a. *Common-emitter circuit*

$$R_{in} \approx \beta \times (R'_e + R_e)$$

where R'_e is the base-emitter junction resistance, and R_e is any external resistance in series with the emitter lead.
b. *Common-collector circuit*

$$R_{in} \approx \beta \times (R'_e + R_e)$$

c. Common-base circuit

$$R_{in} \approx R_e \| R'_e$$

in most practical situations;

$$R_{in} \approx R'_e$$

(the resistance from base to ground in this configuration is considered to approach 0 Ω).

V_g: The circuit voltage gain. Range (dimensionless):

a. For the *common-emitter circuit,* 1 to 500
b. For the *common-collector circuit,* $V_g \approx 1$ (but less than 1)
c. For the *common-base circuit,* 1 to 500

A WORD ABOUT IMPEDANCE AND RESISTANCE. The term *resistance* generally applies to purely ohmic values; the term *impedance* carries the implication of reactances and phase angles. We do use capacitors frequently for coupling and bypass purposes, but when we do, we try to make the reactance at all frequencies of interest approach 0 Ω. This means that phase angles also approach 0°. Also, in a less familiar form, the term *reactance* can be used to apply to anything that opposes the flow of current in a circuit, including opposing dc voltages and varying negative feedback voltages. If this use of the term *reactance* is adopted, we should always designate the input resistance of a transistor (even at very low frequencies) as an input impedance, because the junction potential is involved along with ordinary resistances. The distinction is a small one and not universally agreed upon. For our purposes here I will use the term *impedance* to indicate the possibility or probability of a capacitor's being used as a coupling element, but I will assume that the reactance of the coupling capacitor and the resulting phase angle are approximately zero. The distinction between resistance and impedance is simply this: resistance applies to dc parameters and impedance applies to signal parameters. In some cases there will be no numerical difference between the impedance and the resistance, and in other cases there will be a numerical difference. Later, when we look at high-frequency phenomena, I will be careful to indicate the vector character of the impedances involved.

3-3 MAXIMUM COLLECTOR CURRENT [I_c (max)]

When the manufacturer states that the maximum collector current for a particular transistor is 100 mA, believe him. If you run a transistor at a steady-state collector current larger than what the manufacturer specifies, you may rest assured that it will come to an inglorious, catastrophic *end.* Exceeding the maximum collector current rating causes excessive junction temperatures which will, at best, tend to shift the operating point upward and shorten the life of the device, and probably result in catastrophic failure.

Most manufacturers base this rating on free air cooling for general-purpose devices and on an infinite heat sink for power devices. Derating is required wherever high temperatures prevail, airflow is restricted, or the heat sink is inadequate. Small press-fit or snap-on heat sinks are available for low-power transistors to help ensure adequate heat removal for restricted airflow environments. Larger heat sinks are available for power transistors.

Manufacturers of early transistorized car radios learned this lesson the hard way. Heat sinks which were adequate in the open failed to do the job when the radio was mounted under the dashboard, where airflow was severely restricted. The result was very short-lived power transistors, many of which failed within the warranty period.

In practice it is never necessary to operate with a quiescent collector current greater than 50 percent of the manufacturer's specified maximum. For anything but power transistors there is no need to operate above 40 percent of I_c (max).

In most cases it is advantageous to operate at as low a quiescent collector current as is practical. This often means 0.5 to 10 percent of I_c (max). The load requirements will often set the lower limit, but about equally often the lowest practical quiescent current will be dictated by such things as the tendency of β to drop at lower currents, noise considerations, and stability considerations.

3-4 MAXIMUM COLLECTOR VOLTAGE [E_c (max)]

The maximum collector voltage is limited primarily by base thickness, structure, and doping. There are three basic breakdown mechanisms:

1. *Surface breakdown.* Surface contaminants and surface dangling bonds provide a breakdown path, which at one time was a principal limiting factor for collector voltage, but better geometry, cleaner handling techniques, and surface passivation methods have considerably reduced surface breakdown problems.
2. *Zener and avalanche breakdown.* The base-collector junction is reverse-biased, and hence subject to being pushed over the zener knee. In a germanium device this is nearly always catastrophic. Silicon devices may or may not recover, depending on how much current the external circuit will allow, adequacy of heat transfer away from the junction, and specific junction geometry. There is also the danger of a localized avalanche developing into a kind of avalanche tunnel and generating a hot spot, which can destroy the junction. Because the hot spot can funnel a lot of carriers through a small cross section, the current density in the tunnel

can be extremely high even though the gross external current is relatively small. This problem, called second breakdown, is very destructive. Some degree of avalanche behavior often occurs at high collector current, voltage, or a combination of the two, and represents a minor instability which can easily develop into a full-fledged avalanche breakdown. The problem is especially pronounced in power transistors where collector voltage and current are both necessarily high. In small-signal transistors, operating currents are typically only a small fraction of the maximum allowable; and the problem can be greatly minimized by keeping the collector-emitter voltage well below the manufacturer's specified maximum. The manufacturer generally specifies the maximum collector-to-base voltage. If we assume that to be equal to the collector-to-emitter voltage, we will have a more convenient parameter and one which is a little conservative at that.

3. *Punch-through* (see Fig. 3-1). This mechanism is peculiar to transistors. The other mechanisms I have discussed are common to diodes as well. If you remember that an increase in voltage across a reverse-bias junction widens the depletion area, it is not hard to visualize the collector-base junction depletion zone becoming wider and wider until it goes all the way through the narrow base. As soon as the base depletion zone spreads into the emitter, we have the equivalent of a collector-emitter short. The transistor often recovers as soon as the depletion zone narrows down to normal, but it is poor practice to count on its recovery.

3-5 LEAKAGE CURRENT (I_{co})

There are three basic leakage current mechanisms in a transistor. In any junction device electron-hole pairs are generated by thermal energy. In the case of a reverse-biased junction, the only current flowing is that due to thermally generated carriers. This reverse-bias current is called leakage current, and in a transistor it is measured in the reverse-biased collector-base junction with the emitter lead left open.

In the case of a forward-biased junction the thermally generated carriers are few compared to the dopant-supplied carriers, and the small leakage current is a negligible part of the total current. It is not practical to measure the leakage current in a forward-biased junction, but it is safe to assume it to be of the same magnitude as the reverse-bias leakage. The problem with I_{co} in a transistor is that it appears in the collector circuit as a much larger quantity because of the current gain of the transistor.

I_{co} is temperature-dependent but not very voltage-dependent. In a silicon device we can expect a threefold increase in I_{co} for each 10°C increase in temperature. In a germanium device, I_{co} doubles for each 10° increase. Even though silicon is more temperature-sensitive than germanium, silicon devices cause fewer I_{co} problems because their room-temperature ($\approx 20°C$) leakage is so much smaller than in a germanium device. We can operate a silicon device at about 80°C with about the same I_{co} we would expect from a germanium device at 20°C.

The second kind of leakage is due to surface contamination; it is largely resistive in nature. Improved manufacturing methods have all but done away with this problem.

The third leakage mechanism involves surface electrical effects. The crystal lattice ends at the surface, leaving some dangling bonds to accumulate charges. These surface charges contribute to the total leakage, but improved geometry and surface-passivation techniques have nearly eliminated this form of leakage.

Only I_{co} continues to be a circuit problem, and we shall see how to deal with it in Chap. 4.

3-6 CURRENT AMPLIFICATION IN THE COMMON-EMITTER CIRCUIT

If this chapter seems to contain a discouraging undercurrent, it is only because I have taken the literary liberty of pointing out the fire before demonstrating how to put it out.

The uncertainties that will seem to be the norm do indeed present some problems (disturbing ones for anyone with a little scientific background), partly because of the seemingly large gap between theory and practice, and partly because most texts make little or no effort to bridge this gap; and herein lies one of my principal excuses for writing this book. Be assured that there are methods for overcoming these difficulties, and most of this book is devoted to showing you how to do just that.

It is poor economy to hold manufacturing tolerances for a given transistor any closer than about +100, −50 percent of the specified value. Even if closer tolerances could be held in manufacturing, economically, we would still be faced with the fact that β (a key parameter) varies by a similar amount (100 percent) over the range of normal operating collector currents. In addition, β is a very temperature-dependent parameter and varies widely over the normal temperature range. Fortunately, we can compensate for these unusual variations by proper circuit design. The only restriction is that we must design the circuit around the lowest expected β value. As long as the transistor's *real* β never drops below the design β value, a properly designed circuit will perform well in spite of these large, and not very predictable, variations. There are other temperature-sensitive parameters with which we will have to deal, but proper circuit design can control them as well.

Proper circuit design generally hinges upon making the transistor's inherent parameters as insignificant as possible. Proper perspective can be maintained only if both intrinsic device and extrinsic circuit parameters are taken into account. It is the intent of this chapter to present a concise summary and discussion of these various intrinsic and extrinsic parameters, and the relative importance of each. Both kinds of parameters will be encountered later in conjunction with design and analysis procedures.

Normal transistor operation depends upon the emitter-base junction being forward-biased, while the collector-base junction is reverse-biased. Once these conditions are established, the device is a lot more than simply two junctions back to back; it is an amplifier. An amplifier is defined as any device in which a larger power can be controlled by a smaller power. The power gain is the ratio of the amount of controlled power to the amount of power required to control it (controlled power/control power).

All amplifiers are power amplifiers, but often either voltage amplification or current amplification predominates. Even when voltage and current gains are of the same order of magnitude, we may be interested only in the voltage gain and have no use for the current gain. What we call a particular circuit is often dictated by the kind of job we want it to do more than by the kind of circuit it is. The transistor is "called" a current amplifier because its amplifying mechanism essentially involves the control of a larger current by a smaller current. It can, however, be used for voltage gain, current gain, or power gain, depending upon the external circuitry. We will briefly examine each of these kinds of amplification in this chapter. (Power gain is the product of voltage gain times current gain.)

β, Common-emitter Forward Current Transfer Ratio (Current Gain)

$$\beta = \frac{I_c}{I_b}$$

This parameter is temperature-dependent, and β generally increases with a rise in temperature at a rate of from 0.5 percent/C° to 2 percent/C°. An average value of 1 percent/C° is often used for practical estimates.

β range: 10 to 500 (with some composite devices having β's as high as 40,000). Typical value is *approximately* 100 for silicon general-purpose transistors.

Beta is the current gain in the common-emitter (and the common-collector) circuit. Beta is also known as the "forward current transfer ratio," and the symbols β and h_{fe} are used to denote this parameter. The equation for β is

$$\beta = \frac{I_c}{I_b}$$

where I_c is the collector current and I_b is the base current. Values for β may be obtained from the transistor manual, but listed values must be taken as approximate.

The curve formed by plotting I_c (the collector current) vs. I_b (the base current) is the β-transfer curve. It is this transfer curve that will guide our design procedures and our choice of operating points for the various classes of operation. Collector voltage vs. collector current curves and static load lines, though often published, are of little use. Figure 3-4*a* is a typical β-transfer curve.

This curve is useful in a general way, because it helps to visualize certain concepts about operating conditions; but it is not normally published, and it isn't important to have such curves available for individual transistors.

There is also an ac β, but because there is very little difference (except at high frequencies) between the two, and because of inherent uncertainties in β values, a distinction between the two is seldom made. Figure 3-4*b* is a graph of β vs. collector current. As you can see, β falls off badly at currents below 1mA. There is nothing sacred about the 1-mA figure, except that it is typical of a large variety of small-signal transistors. Transistors designed for very low collector-current operation will likely have the β peak at much less than 1 mA, whereas power transistors may have the peak about halfway between zero and maximum collector current. So, since the manufacturer seldom provides curves like that in Fig. 3-4*b*, how do we know at what current maximum β occurs? In practice we have only one reliable guide: experience. The curve in Fig. 3-4*b* is hardly the sort of thing a manufacturer would

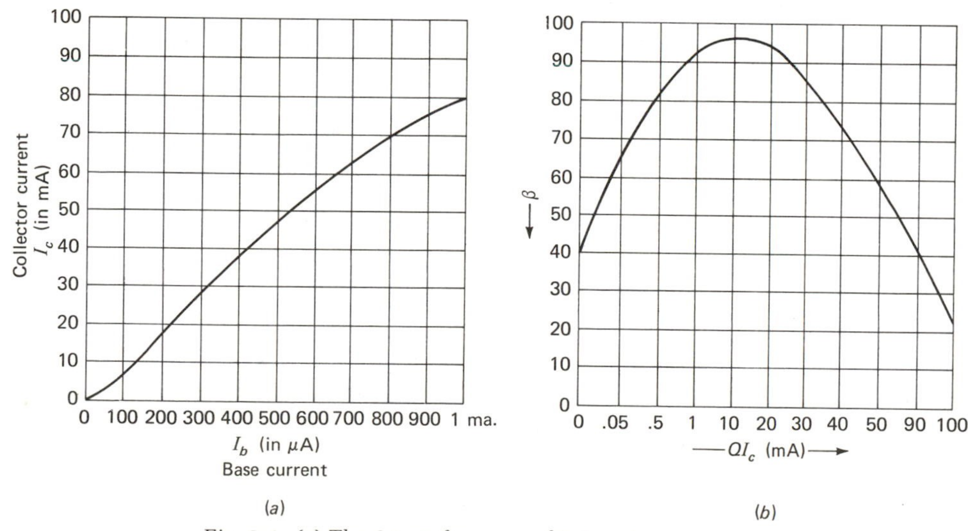

Fig. 3-4 (a) The β transfer curve; (b) β vs. I_c curve

be inclined to crow about; something better left unsaid. On the other hand, you would not expect him to measure β at some place which is so low on the curve as to seem unsatisfactory.

Now if it were me, I would find out what current yielded the maximum β, and I would take pains to measure β at that current. But as a scientist, I would be violating my tradition if I did not specify the current at which the β figure was measured.

Most manufacturers specify the current at which β is measured, and it generally turns out that the current at which they make the measurement is that current at which β is at its maximum. My advice to you is: Assume that the current at which β is measured represents that current at which β is at its maximum.

To give the manufacturers their due, they frequently do give β figures for more than one operating current, when such information cannot well be done without. Power transistors are a particular example. When such information *can* be done without, be prepared to do without it. I shall try to provide you with enough guidelines so that you will find such shortages of data to be no real problem.

The current amplification, or current gain, is the ratio of the input current to the output current (I_{out}/I_{in}). For the transistor alone this is *called* the beta (β) value. $\beta = I_c/I_b$ (collector current/base current), but in a *circuit* the current gain will nearly always be something less than β. The base-emitter junction must always be forward-biased for normal transistor action, and this means resistances (R_b) from base to ground which constitute a parallel circuit. This parallel circuit is composed of the transistor input resistance in parallel with the bias resistors. As a result there are two current paths: one through the base-emitter circuit, which gets amplified in the collector circuit, and one through the bias resistors to ground, which does not get amplified. Figure 3-5 illustrates this situation.

In Fig. 3-5, I_1 is the total actual input current into the circuit, I_2 is the current through the base bias resistances (R_b), I_3 is the base-emitter current, and I_4 is the output current. For the sake of illustration, let us assume some (not necessarily typical) values for the circuit in Fig. 3-5. Let us assume that $\beta = 10$, $I_1 = 1.0$ mA, and R_b and the transistor input resistance are equal. Under these conditions one-half of the 1-mA input current (I_1) will be diverted through R_b, leaving 0.5 mA flowing through the base-emitter junction. With a β of 10, we will get 5 mA of output current (I_4). Therefore, we are putting in 1 mA (I_1) and getting out 5 mA, a circuit current gain of 5, which is just half of the transistor's current gain (β) of 10. You must remember that the circuit current gain is generally less than β. We will discuss this in detail in Chap. 4.

It is a popular misconception that the transistor is a current amplifier and must always be treated as such. The gain mechanism is indeed largely a

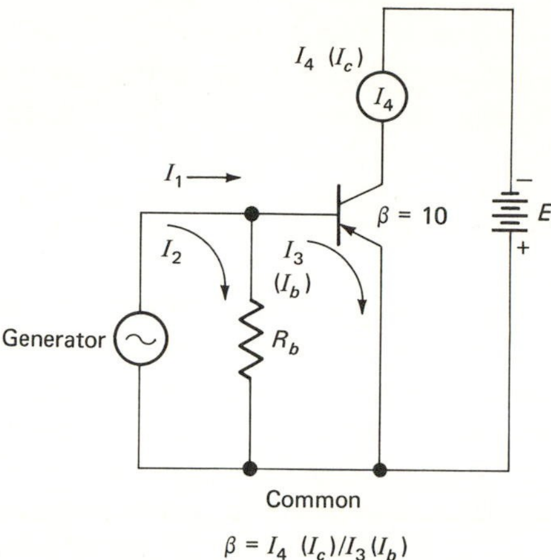

Fig. 3-5 Circuit current gain

current phenomenon, but it is the circuit design and the application to which we put the transistor that determine whether we have a current amplifier, a voltage amplifier, or a power amplifier.

Transistors are capable of respectable gains in all three cases.

3-7 OUTPUT RESISTANCE IN THE COMMON-EMITTER CIRCUIT

Output resistance $R_o = \dfrac{\Delta E_c}{\Delta I_c}$

For common-emitter circuits, the output resistance of a reverse-biased junction is quite high; and since it will always be very high compared to the load resistance, it can be assumed to approach an infinite value; 500 kΩ to 2 MΩ is typical. R_o is the dynamic junction resistance.

$$R_o = \frac{\Delta E_c}{\Delta I_c}$$

where E_c = collector voltage and I_c = collector current. *Collector current is only slightly dependent upon collector voltage.* A 100 percent increase in the collector voltage may cause a 1.0 percent increase in collector current.

Example:

Let us try some numbers. Let E_c change from 10 V to 20 V (a 100 percent change). $\Delta E_c = 10$ V. And let I_c change by 1 percent, from 1 mA to 1.01 mA. $\Delta I_c = 0.01$ mA.

$$R_o = \frac{\Delta E_c}{\Delta I_c} = \frac{10\text{ V}}{0.01\text{ mA}} = \frac{10\text{ V}}{1 \times 10^{-5}\text{ A}} = 1 \times 10^6\ \Omega\text{, or } 1\text{ M}\Omega$$

It is this relative independence of collector current on collector voltage that makes some of our design procedures as simple as they are. The collector simply scavenges carriers which are drawn from the emitter into the base region and are not captured there. The collector field is normally high enough to scavenge nearly all the carriers which are available. Increasing the collector field by increasing collector voltage doesn't draw additional carriers to the collector, because it is already collecting all that are available. To increase the collector current, more carriers must be provided by an increase in base current. Increasing the collector voltage does slightly increase the collector's scavenging efficiency, so that there is a small increase in collector current. There are also some thermally generated carriers in the collector-base depletion zone. Therefore, collector current is not completely independent of collector voltage (R_o is not infinite), but R_o is close enough to infinite for most practical purposes.

In the common-emitter circuit the output resistance of a *transistor* is essentially the dynamic resistance of the reverse-biased collector-base junction; its principal importance, from a designer's standpoint, is certainly not its influence on circuit output impedance, but rather the fact that such a high dynamic resistance makes the collector current relatively independent of collector voltage (constant current source). It is this high dynamic collector-base junction resistance which allows us to make the practical assumption that *collector current is independent of collector voltage.* And it is this assumption which simplifies several phases of the designer's problem. *Remember that the collector-base junction is reverse-biased in normal operation.*

The output resistance of the *complete circuit* is nearly always dependent upon the *load resistance* rather than the transistor's output resistance. In the common-emitter circuit, the circuit output resistance is almost exactly equal to the load resistance. The circuit output impedance consists of the transistor's output impedance (resistance) in *parallel* with the load impedance, where the transistor's output impedance ($\Delta E_c/\Delta I_c$) is normally 0.5 to 1 MΩ, and the load impedance seldom exceeds 20,000 Ω and is generally somewhat lower. 500 kΩ in parallel with 20 kΩ yields a little over 19 kΩ, and so, as you can see, the 20-kΩ load is the determining factor.

The statement that the circuit collector load (R_L) is considered to be in parallel with the transistor's output impedance perhaps needs some explanation. R_o is the collector-base junction impedance (resistance), which is very high, because it is a reverse-biased junction. The collector-to-emitter circuit

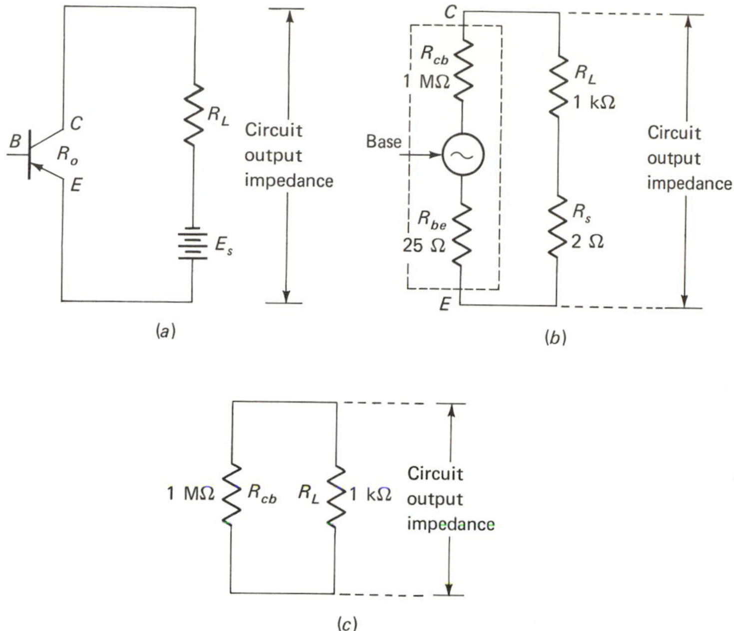

Fig. 3-6 The output resistance of a common-emitter amplifier. (a) *The transistor circuit;* (b) *the equivalent circuit in terms of impedances;* (c) *the simplified equivalent circuit in terms of impedances*

of the transistor is considered to be the generator in Fig. 3-6*b*, because it is the transistor which causes the signal variations. The battery supplies the power to make the transistor work, but it does *not* cause the signal variations; the transistor does that.

In the circuit in Fig. 3-6*b*, there are resistances: R_{be}, which is the base-emitter junction resistance with a typical value of 25 Ω; R_{cb}, which is a reverse-biased junction with a resistance of *about* 1 MΩ; R_s, which is the internal power-supply resistance of perhaps 2 Ω; and R_L, which we will give a value of 1 kΩ for this example. The box (in Fig. 3-6*b*) encloses the collector-base and emitter-base junction impedances (resistances) which are inside the transistor. Point *C* is the transistor collector terminal and point *E* is the transistor emitter terminal. The output impedance is measured between point *C* and point *E*, that is, between collector and emitter, as shown in Fig. 3-6*b*. When we measure the output impedance now, with R_L and the power supply in the circuit, we are looking into two parallel branches. The left branch consists of the collector-base junction (R_{cb}) in series with the forward-biased base-emitter junction whose 25 Ω adds almost nothing to the 1 MΩ of R_{cb}. And so, for all practical purposes, the left branch is simply the 1-MΩ collector-base impedance. The right branch consists of the 1-kΩ load resistor in series with the 2 Ω of internal power-supply impedance (R_s). If we leave out the 25

Ω of R_{be} and the 2 Ω of R_s because of their relative insignificance, we get something that looks like Fig. 3-6*c*.

Now, when we consider that the practical values of R_L range from 0 to about 25 kΩ, we can see that the value of R_L dominates the circuit. Even if R_L approached 100 kΩ (which is most unusual), ignoring the transistor's collector-base junction impedance would cause only a 10 percent (or so) error. For all practical purposes the *circuit* output impedance is equal to the value of R_L. In the previous example I assigned the power supply an internal impedance of 2 Ω. I did this because the power supply *must always* have an internal impedance that is very low at all signal frequencies, preferably down to 0 Hz. I will say more about this requirement later.

In many practical circuits we must know the total load impedance. In a few instances R_L (in Fig. 3-6*a*) is a relay, lamp, or some other *working* load. But in a larger number of cases the actual *working* load (R_{Lw}, resistance–load-working) is coupled to the collector by some coupling device such as a capacitor or diode, and R_L serves simply as a resistance across which a varying voltage is produced. This varying voltage serves as part of the *generator* to drive the actual load, which is often the input resistance of another stage. Figure 3-7 shows this situation.

The coupling device is usually selected to have *near* 0 Ω impedance at signal frequencies. The total output impedance (load impedance, called R_{Lac}) consists of R_c in parallel with R_L in parallel with R_{Lw}. But because R_c is so high compared to normal values of R_L and R_{Lw}, R_{Lac} (the total signal load impedance) is considered to be R_L in parallel with R_{Lw}.

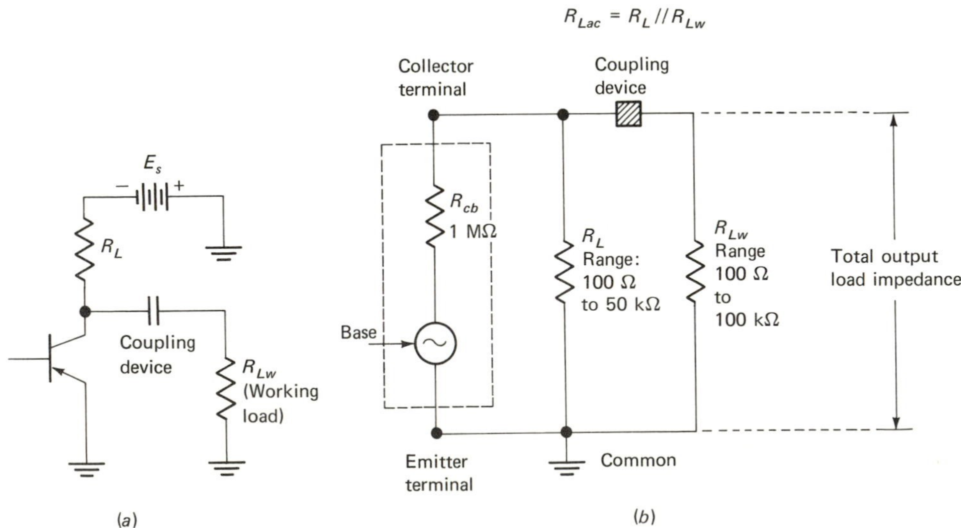

Fig. 3-7 The addition of a coupled working load. (a) *The actual circuit;* (b) *the equivalent circuit model*

$R_{Lac} = R_L \| R_{Lw}$

where R_{Lac} is the total load output impedance at signal frequencies.

R_L is the collector load impedance and R_{Lw} is the working load; R_c is generally ignored. R_c, the collector-base dynamic impedance, is not a predictable parameter and in practice its actual numerical value is rarely known. Since we almost never have a reliable figure for R_c, it is fortunate that it is high enough to be left out of most practical design and analysis equations.

3-8 INPUT IMPEDANCE R_{in} (COMMON EMITTER AND COMMON COLLECTOR)

Input resistance is a most illusive parameter, because it is a function of so many parameters that are difficult to predict. The base-emitter diode behaves basically like any other forward-biased diode. Its resistance consists of the two bulk resistances, base and emitter, and a junction resistance; but unlike the ordinary diode, this combination of resistances in a transistor appears to be β times as large, looking into the base, as it would in a diode. Before I analyze the increase of input impedance by the factor β, let me take a moment to make some comments about the base-emitter diode.

If you will remember, the base is only 1 micron or so thick; it would seem safe to assume that the bulk resistance of the base would be negligible, and for most practical purposes it is. But it is higher than might be expected because of a phenomenon called spreading. The bulk resistance of the base is called the base spreading resistance. If you can visualize a p-type base region as a thin sheet with relatively few holes for electrons to fall into, you can see that where there is an abundance of electrons near the surface of the sheet, and most of the holes will be occupied at any given time, any given electron may wander quite a distance over the "surface" before being captured. Between long paths and the relative scarcity of available holes, the base spreading resistance is much higher than we might suspect.

Voltage Drop across the Base-emitter Junction

This junction voltage is the barrier-potential transition voltage, and to see how it can appear as a resistance, let us take a look at Fig. 3-8. Having examined the illustration, can you figure out what is in the box?

If you used a little Ohm's law and came up with a 10-Ω resistor in the box, you could be right, and then again you might be wrong. For example, the circuit in Fig. 3-9 would yield exactly the same current and voltage drops if the box contained a 5-V source (a battery in this case). The battery is connected in series opposing so that the *effective* supply voltage is only 5 V (10 V − 5 V). Five volts in series with a 10-Ω resistor still produces 0.5 A of current. However, if we look at the circuit in Fig. 3-8 in terms of total imped-

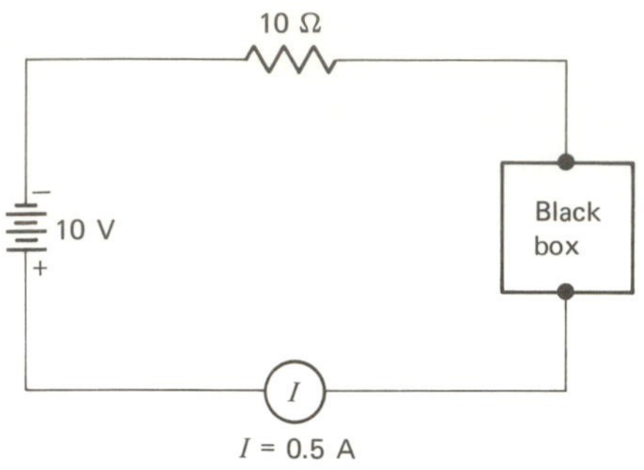

Fig. 3-8 The black box

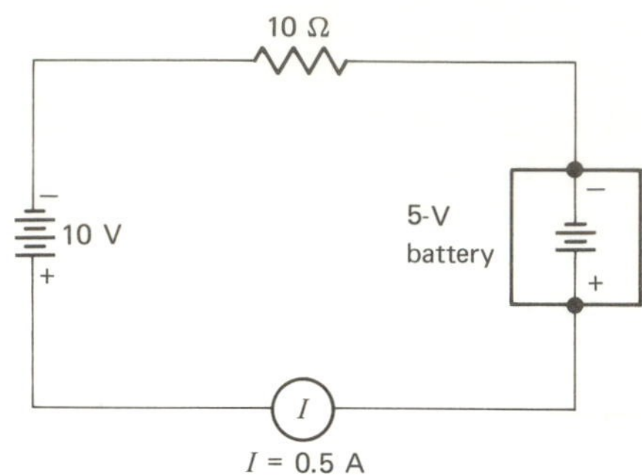

Fig. 3-9 The black box revealed

ance (resistance), we can only conclude that there is a total of 20 Ω circuit resistance. Whatever is in the box, it *looks* like 10 Ω of resistance, and in this case it has increased the circuit impedance by a factor of 2. The barrier potential in a transistor opposes base-emitter current in the same way our battery did in Fig. 3-9. It behaves as a 0.25-V source in a germanium device and as a 0.6-V source in a silicon device. The junction voltage decreases by about 2.5 mV/C° increase in temperature.

The relationship of the barrier potential's behavior as junction resistance can be taken directly from the volt-ampere diode conduction curve. However, such curves are not available for individual transistors; they shift with temperature and vary from unit to unit. Because we cannot know this junction resistance accurately in practice, we must depend upon an approximate equation called the Shockley relationship.

$$R_j \approx \frac{25 \text{ mV}}{I \text{ (mA)}} \qquad R = \frac{E}{I}$$

where 25 mV is a constant and I is the junction current *in milliamperes*.

In the transistor, this junction resistance is the most important part of the base-emitter diode resistance. Base spreading resistance and emitter block bulk resistance are both rather insignificant. We will meet the Shockley relationship many times in chapters to follow. In transistors (at least in this book) the junction resistance is called R_e'.

Analyzing the Transistor Circuit Input Impedance

A simple loop analysis demonstrates that collector current, *which is always β times as large as the base current*, causes the input impedance to appear as β times the junction resistance. Figure 3-10*a* shows the transistor circuit, and Fig. 3-10*b* shows the equivalent circuit for analysis. Let us follow the loop analysis through.

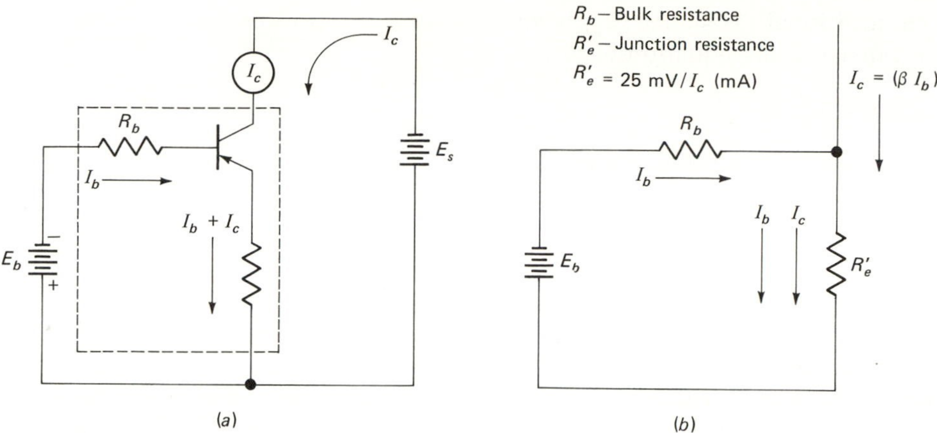

Fig. 3-10 Analyzing the input impedance. (a) *The transistor circuit;* (b) *the equivalent circuit for analysis*

1. Writing the loop equation (see Fig. 3-10b):
   a. $E_b = I_b R_b + (I_c + I_b) R'_e$
   b. Because $I_c = \beta \times I_b$, we can write
      $E_b = I_b R_b + (I_b + \beta I_b) R'_e$
   c. Clearing the parentheses, we have
      $E_b = I_b R_b + I_b R'_e + \beta I_b R'_e$
   d. Dividing through by I_b, we get

      $$\frac{E_b}{I_b} = R_b + R'_e + \beta R'_e$$

   e. $\dfrac{E_b}{I_b} = R_{in}$
   f. $R_{in} = R_b + R'_e + \beta R'_e$

2. Let us try some numbers.
   Let $\beta = 100$.
   Let $R'_e = 25\ \Omega$.
   Let $R_b = 5\ \Omega$.
   These are more or less typical values.
   $R_{in} = 5 + 25 + (100 \times 25) = 2{,}530\ \Omega$

Because the bulk resistance of the base is seldom known and because it has little effect anyway, R_b is not generally included in the equation. And because R'_e is nearly always *very* small compared to $\beta R'_e$, the equation is usually written as

$$R_{in} \approx \beta \times R'_e$$

If an additional resistance (R_e) is added in series with the emitter, external to the transistor, the equation becomes

$$R_{in} \approx \beta \times (R_e' + R_e)$$

And in cases where the external resistance (R_e) is many times larger than R_e', we can write

$$R_{in} \approx \beta \times R_e$$

Because variations in R_e' are magnified by the factor β, small changes in R_e' due to temperature variations cause *large* changes in the input resistance of the transistor. If you consider this along with the fact that β itself varies with temperature and operating current, you can understand why the input impedance of a transistor is such an unpredictable parameter. The Shockley relationship is really only an approximation, but under the circumstances it proves to be a very good one.

If we use the Shockley relationship along with a reasonably good approximation of β, we can come up with fairly acceptable estimates of the transistor's input resistance, and short of exact measurements in the circuit, we can do no better. Of course we cannot build the circuit first, measure it, and then, afterwards, design it; so we shall have to lean rather heavily upon Dr. Shockley's little gem.

3-9 VOLTAGE AMPLIFICATION IN THE COMMON-EMITTER CIRCUIT

The transistor can be used as a voltage amplifier because of the large permissible, and in most cases essential, large collector supply voltage. The base-emitter circuit requires, because of its relatively low impedance, a range of only a few tenths of a volt to vary the base-emitter junction current from zero to maximum. The collector-base junction may be capable of handling a voltage drop of from 10 to 50 V (sometimes more), depending upon the particular transistor. In Fig. 3-11*a* and *b*, we see a transistor which is properly biased and has a collector supply voltage of 10 V (a typical value) and a collector load resistor of 1 kΩ.

In Fig. 3-11*a*, we have 0 V of base-emitter bias voltage. This means that there is no base-emitter current, and consequently no collector current. Under these conditions we will have no voltage drop across the collector load resistor (R_L). (Assume a second *fixed* bias voltage which is within a millivolt of overcoming the barrier potential.)

$$E = IR \qquad E = 0 \times 1\ \text{k}\Omega = 0\ \text{V}$$

Therefore, the output voltage will be equal to the supply voltage, in this case 10 V.

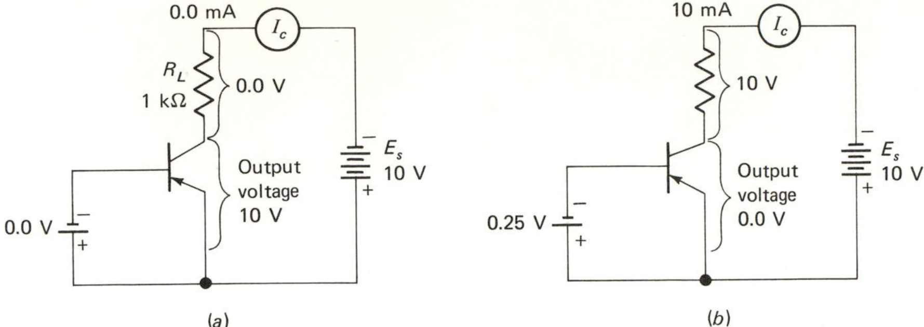

Fig. 3-11 Voltage amplification. (a) *With 0 V input;* (b) *with 0.25 V input*

For Fig. 3-11*b*, let us assume that increasing the base bias battery voltage from 0 V to 0.25 V causes enough base-emitter current to flow to result in 10 mA of collector current. These figures are fairly typical for a germanium device.

With 10 mA of current flowing through the 1-kΩ collector load resistor (R_L), a little Ohm's law tells us that the voltage drop across that resistor is 10 V.

$$E = IR \qquad E = 1.0 \text{ k}\Omega \times 10 \text{ mA} = 10 \text{ V}$$

Now, if the entire supply voltage is dropped across the collector load resistor, the voltage drop across the transistor, and consequently the output voltage, must be 0 V. The voltage gain is equal to the change in output voltage divided by the change in input voltage.

$$\text{Voltage gain} = \frac{\Delta \text{ output voltage}}{\Delta \text{ input voltage}} = \frac{\Delta E_o}{\Delta E_{in}}$$

In this example the change in input voltage is 0.25 V, $\Delta E_{in} = 0.25$ V. The change in output voltage is 10 V, $\Delta E_o = 10$ V. The voltage gain of the circuit is

$$V_g = \frac{E_o}{E_{in}} = \frac{10 \text{ V}}{0.25 \text{ V}} = 40$$

Possible voltage gains range from unity to several hundred, depending upon the circuit design.

This discussion pertains to the voltage gain of the common-emitter circuit and all its variations. Later on you will be able to see, by inspection, that it is also valid for the common-base circuit (the common-collector circuit always has a voltage gain of about 1).

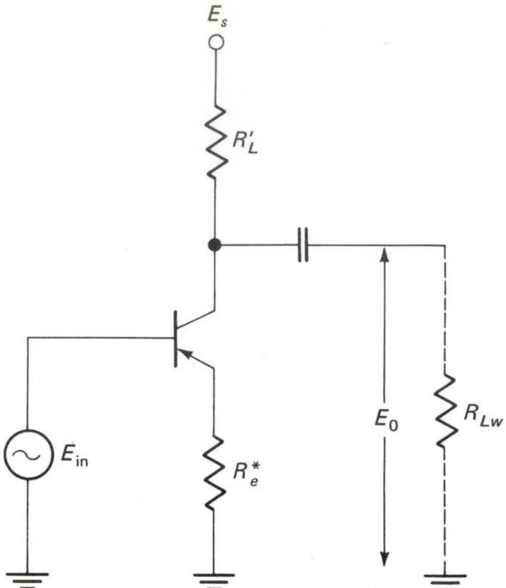

Fig. 3-12 Voltage gain analysis

Figure 3-12 shows the circuit used for analysis.

We shall prove that the voltage gain V_g equals R'_L/R^*_e, where R^*_e consists of the junction resistance R'_e (from the Shockley relationship) plus the resistance of any external emitter (R_e) resistor. See Fig. 3-12. (NOTE: $R'_L = R_{Lac} = R_L \| R_{Lw}$.)

Voltage Gain

To prove that

$$V_g = \frac{R'_L}{R^*_e}$$

assume that $I_c \approx I_e$

1. The emitter current is

   $$I_e = \frac{E_{in}}{R^*_e} \qquad \text{Ohm's law}$$

2. The output voltage is

   $$E_o = I_c R'_L \qquad \text{(Ohm's law)}$$

3. Because $I_c \approx I_e$,

   $$E_o \approx I_e R'_L \qquad \text{(substitution)}$$

4. Because

$$I_e = \frac{E_{in}}{R_e^*}$$

$$E_o \approx \frac{E_{in}}{R_e^*} R_L' \qquad \text{(substitution)}$$

and because

$$V_g = \frac{E_o}{E_{in}} \qquad \text{(by definition)}$$

and we already have

$$E_o \approx \frac{E_{in}}{R_e^*} R_L'$$

if we divide both sides of the equation by E_{in}, we get

$$\frac{E_o}{E_{in}} \approx \frac{R_L'}{R_e^*} \approx V_g$$

Conclusion:

$$V_g \approx \frac{R_L'}{R_e^*}$$

R_L' is the total signal load, including any coupled working load (R_{Lw}); we call the total ac (signal) load R_{Lac}, where $R_{Lac} = R_L \| R_{Lw}$. We would then write the equation as

$$V_g = \frac{R_{Lac}}{R_e^*}$$

Voltage gain is dimensionless; it is simply a factor. If the voltage gain is 100, a 1-mV input signal will produce a 100-mV output signal. The voltage gain is independent of the actual magnitude of the input signal, but there are limits. Any given circuit will have a maximum output voltage, which cannot be exceeded, no matter how large the input signal. For example, an amplifier with a voltage gain of 100 and a *maximum output voltage* of 10 V simply could not accept an input signal of 1 V. One volt × 100 = 100 V, well above the 10-V maximum output. The amplifier is said to be *overdriven* in such cases, and the output waveform would be badly distorted. The amplifier would be overdriven for any input voltage greater than 0.1 V (0.1 V × 100 = 10 V). The distortion caused by overdriving the amplifier is called clipping or flat-top distortion. Figure 3-13 shows the output waveform of an overdriven amplifier. The clipping may occur equally on both peaks (symmetrical clipping) or unequally on the two peaks (asymmetrical).

The maximum output voltage (E_o p-p) is the maximum output voltage

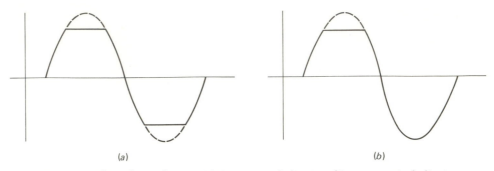

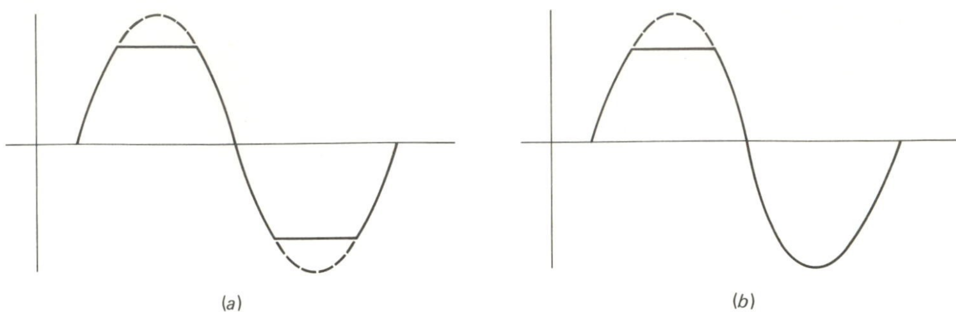

Fig. 3-13 Clipped waveforms. (a) *Symmetrical clipping;* (b) *asymmetrical clipping*

just prior to the start of clipping. The lower limit to the signal input voltage the amplifier will accept is called *sensitivity;* generally it is small enough so that noise problems arise before it is reached. When the signal level gets down to the same order of magnitude as the noise, hum, etc., in the system, the amplifier ceases to be of much use. Even though the signal still gets amplified by the factor V_g, so does the noise, hum, etc.

3-10 POWER AMPLIFICATION IN THE COMMON EMITTER

Every transistor amplifier provides some power gain. In many cases we do not take direct advantage of the power gain available. Where our chief concern is power amplification and the delivery of power to the load, we encounter some special problems and many of our usual problems become greatly exaggerated.

Power gain is the ratio of power out to power in (P_o/P_{in}). The power *in* may be found by multiplying the input voltage by the input current. Power *out* may be found by multiplying the output voltage by the output current.

When power gain is the order of the day, we strive for the largest possible voltage gain–current gain product.

Power gain is the product of voltage gain times the current gain, and it would seem that, in a power amplifier, we would try to get as much of both as possible. This is not generally the case because we must nearly always sacrifice some gain (current, voltage, or some of each) in the interest of circuit stability.

Because of the special problems involved, I will leave further discussion of power gain to a later chapter where we can examine those problems in detail.

3-11 THE COMMON-BASE CIRCUIT

Current Amplification in the Common-base Circuit

The current gain in the common-base configuration is called alpha (α) and is approximately equal to unity in all circuits. In the common-base configuration the input current is the emitter current and the output current is the collector current.

$$\alpha = \frac{I_c}{I_e}$$

Because I_e is composed of the collector current plus the base current, I_e is just *slightly* larger than I_c and α is just *slightly* less than unity. The common-base circuit is not used as a current amplifier, and the figure α is of limited practical importance.

The symbol α appears in equations far more often than it should, mostly for historical reasons. α is often used in equations when β would be more appropriate, but the two are related by the equation

$$\beta = \frac{\alpha}{1 - \alpha}$$

and the term $\alpha/(1 - \alpha)$ can be, and too often is, used in place of β. If you are used to thinking in terms of β, equations written in terms of α can be confusing; and many excellent texts use α frequently. To make it easier for you to read some of these texts, which include some of my references, I will give you a little hint that has stood me in good stead for some time now.

$$\frac{\alpha}{1 - \alpha} \approx \frac{1}{1 - \alpha} \approx \beta$$

Equations written in terms of α will often contain the term $1 - \alpha$ in the denominator; whenever that is the case, you can substitute a β in the numerator for it and change any α that appears in the numerator to a 1 with very little loss in accuracy, but with a significant increase in comprehension.

In this text I will not use α when β is more appropriate; in fact, I will hardly use α at all, but other texts do and you should be aware of the relationship.

Voltage Amplification in the Common-base Configuration

The voltage gain for the common-base circuit is the same as for the common-emitter circuit:

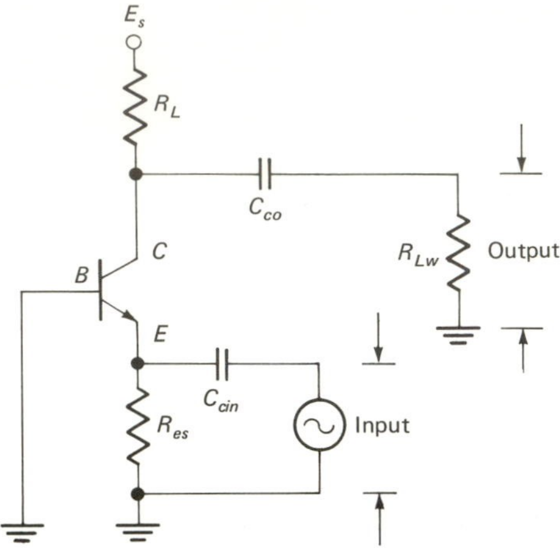

Fig. 3-14 The common-base circuit

$V_g \approx \frac{R_{Lac}}{R'_e}$

NOTE: The common-base circuit has no effective R_e, only R'_e.

where R_{Lac} is the total signal load (usually ignoring the transistor's intrinsic output impedance R_o) and R'_e is the base-emitter junction resistance ($25/I_c$, the Shockley relationship). See Fig. 3-14.

Output Impedance of the Common-base Circuit

$R_{co} \approx R_{Lac}$

See the discussion for the common-emitter circuit.

Input Impedance of the Common-base Circuit

As you can see in Fig. 3-14, the input signal is applied across the resistor R_{es}, which is generally there for stability reasons, and will range from a few hundred to several thousand ohms. If you will look again at Fig. 3-14, you can see that the base-emitter junction is, in effect, in parallel with R_{es}, so that the generator looks into R_{es} in parallel with the junction resistance R'_e.

$Z_{in} = R_{es} \| R'_e$

However, R_{es} is generally so much larger than R'_e (1 kΩ compared to 25 Ω is more or less typical) that Z_{in}, in practical circuits, is about equal to R'_e.

$Z_{in} \approx R'_e$

where $R'_e \approx 25/I_c$ (mA).

3-12 THE COMMON-COLLECTOR CIRCUIT

Current Gain of a Common-collector Circuit

The current gain I_g of a common-collector amplifier is the same as for the common-emitter circuit, β for the transistor alone and generally less than β in practical circuits. See the discussion in Sec. 3-6.

Voltage Gain of a Common-collector Circuit

The voltage gain of a common-collector circuit is slightly less than, but usually very close to, unity. If you will examine Fig. 3-15, you can see that all the signal going in appears across R_{es} (and R_{Lw}, which is in parallel with it) except for that signal voltage which is dropped across the base-emitter junction resistance. Although we would not use the common-collector circuit as a voltage amplifier, because it is more a voltage attenuator than a voltage amplifier, we will, later on, use an equation for finding the voltage gain (actually the voltage attenuation), because for very small values of R_{Lw} that loss can be significant. For now it is sufficient to know that the voltage gain of a common-collector amplifier is less than, but usually close to, unity.

$V_g \approx 1$

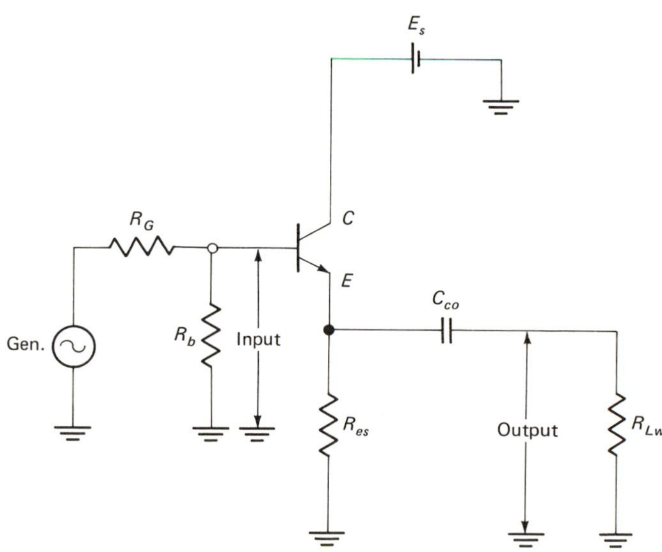

Fig. 3-15 The common-collector circuit

Input Impedance of a Common-collector Circuit

As in the common-emitter circuit, the input impedance of a common-collector circuit is composed of any base-to-ground resistance in parallel with β times the *effective* emitter impedance. In the case of the common-collector circuit, however, the working load R_{Lw} appears in parallel with R_{es}, the external emitter resistor, which makes the *effective* emitter impedance equal to $R_{es} \| R_{Lw}$.

Technically the emitter resistance should be $(R_{es} + R_e') \| R_{Lw}$, but as we shall see later on, R_{es} must be very much larger than R_e' if the circuit is to be stable in the face of temperature variations. We can leave R_e' out of the equation with negligible error. Please remember that, at best, R_e' is never known with any great accuracy. The input impedance of the common-collector circuit is closely approximated by the equation

$$Z_{in} \approx R_b \| \beta(R_{es} \| R_{Lw})$$

(Assume R_G to be insignificant.) (R_G = internal generator impedance.)

Output Impedance of the Common-collector Circuit, R_{co}

I have demonstrated that a resistance in the emitter circuit (of the common-emitter amplifier) is reflected into the base circuit β times as large as its actual ohmic value. If we look into the emitter, it is reasonable to expect base-to-ground resistances to appear β times smaller than they actually are. And, sure enough, that is actually the case. This parameter (R_{co}) is not overwhelmingly important, and I will not take the time to go through the loop equation with you, but you can use the input impedance analysis as a guide if you care to do it for yourself. In Fig. 3-15 R_b is in parallel with R_G, the generator's internal resistance, so that the total equivalent base-to-ground resistance is $R_G \| R_b$ (in some circuits there is no R_b). If we replace R_{Lw}, in Fig. 3-15, with an imaginary eyeball, it will look into an impedance of

$$R_{co} \approx R_{es} \| \left(\frac{R_b \| R_G}{\beta} \right)$$

The equation gives us the output impedance of the common-collector circuit.

It might be well to point out here that the common-collector circuit is often called an emitter-follower circuit. The logic in this terminology is minimal, but the term is an accepted part of our electronics jargon, and you will hear it often.

3-13 THE HYBRID PARAMETERS

The system of hybrid parameters is a mixed bag of input impedances, output admittances, forward current gains, and reverse voltage transfers. Although the hybrid parameters are primarily a way in which the manufacturer can describe his wares, they are important to us as designers and technologists, because these basic parameters provide a starting point for our designs. The hybrid-parameter system is the result of experience by the industry in trying to measure parameters which can be measured on a practical basis, and which can be of some use to the designer. Over the years the hybrid parameters have taken on a most sacred and almost spiritual quality that they don't deserve. The interrelationships among the hybrid parameters are mathematically beautiful, but beyond such esoterica lies the real world, and in the real world the hybrid parameters are no better than any others.

The basic problem is, of course, the variability of parameters with temperature, with operating point, and from unit to unit. The basic design philosophy is to swamp device parameters with external circuitry so that such device parameters and their variations will have but little influence upon the circuit as a whole. In light of this philosophy all we really need to know about the device is maximum operating values, leakage current, the current gain of the transistor, and the emitter resistance R_e'. These things and a few others of less interest are given by the manufacturer, some of them in terms of the hybrid parameters. Let us briefly examine the important hybrid parameters and compare them to the parameters that we have already discussed.

Current Gain (h_{21})

h_{fe}: The common-emitter equivalent of our old friend β
h_{fc}: The common-collector equivalent of our friend β
$h_{fe} \approx h_{fc}$
h_{fb}: The current gain in the common-base configuration
$h_{fb} \approx \alpha$

Input Impedance (h_{11})

h_{ie}: The input impedance in the common-emitter circuit
$h_{ie} \approx \beta R_e' \qquad R_e' \approx 25/I_c$
h_{ic}: The input impedance in the common-collector circuit
$h_{ic} \approx \beta R_e'$ the collector load resistance is 0 Ω
h_{ib}: The input impedance in the common-base configuration
$h_{ib} \approx h_{ie}/h_{fe} \approx h_{ie}/\beta \approx R_e'$
where $R_e' \approx 25/I_c$ (the Shockley relationship)

Output Admittance (h_{22})

h_{oe}: The output admittance of the common-emitter circuit
$h_{oe} \approx 1/R_o$
h_{oc}: The output admittance of the common-collector circuit; the reciprocal of the output impedance

h_{ob}: The output admittance of the common-base circuit
$h_{ob} \approx 1/R_o$

Reverse Transfer Voltage Ratio (h_{12})

$h_{12} = E_{in}/E_o$ and is a measure of how changes in output voltage cause changes in input voltage. For h_{re} (common emitter) and h_{rb} (common base), $h_{12} \approx 5 \times 10^{-4}$ (dimensionless). This is a very small numerical value and is only occasionally important. In the common-collector circuit $h_{12} = h_{rc} \approx 1$. Here the output circuit is part of the input circuit loop, and it is reasonable to expect that changes in the output circuit voltage would have a large effect on the input circuit voltage. All that this parameter means is that there is excellent, but not perfect, isolation between input and output circuits in the common-base and common-emitter configuration, and nearly none at all in the common-collector circuit.

If you care to dig deeper into hybrid parameters, I recommend G. Deboo and C. Burrous, *Integrated Circuits and Semiconductor Devices*, McGraw-Hill Book Company, New York, 1971.

SUMMARY

Please refer to Table 3-1 for this discussion.

Current Gains

In the common-emitter and common-collector circuits, the *device* current gain is measured in a test circuit which eliminates base-to-ground resistances and other circuit influences. The *device* current gain is called β, and β is equal to I_c/I_b. The closest hybrid parameter symbol to β is h_{fe}; β and h_{fe} are identical except at fairly high frequencies.

In most practical circuits there are some *necessary* base-to-ground resistances, which bypass some of the input signal current and make the *circuit* current gain something less than β. This reduction in current gain for the circuit is not necessarily bad, and in fact holding the *circuit* current gain down to 10 or so is often essential to stable operation.

In the common-base circuit, the current gain is less than, but very nearly equal to, unity. This applies to both device and circuit current gains in nearly all cases. The equation for the common-base current gain is

$$\alpha = \frac{I_c}{I_e}$$

where I_c is the collector current and I_e is the emitter current. The nearest hybrid parameter equivalent of α is h_{fb}.

TABLE 3-1. TRANSISTOR PARAMETER SUMMARY

	Current Gain	Voltage Gain	Input Impedance	Output Impedance
		Device Parameter		
Common emitter	$\beta = \frac{I_c}{I_b}$	$V_g = R_L/R'_e$	$Z_{in} = \beta R'_e$	$R_o = \Delta E_c/\Delta I_c$
Common collector	β	$V_g \approx 1$	$Z_{in} = \beta R'_e$	$R_o = R'_e + (R_G/\beta)$
Common base	$\alpha \approx 1$	$V_g = R_L/R'_e$	$Z_{in} = R'_e$	$R_o = \Delta E_c/\Delta I_c$
		Nearest Hybrid Parameter		
Common emitter	h_{fe}	n.a.	h_{ie}	$h_{oe} = 1/R_o$
Common collector	h_{fc}	n.a.	h_{ic}	$h_{oc} = 1/R_o$
Common base	h_{fb}	n.a.	h_{ib}	$h_{ob} = 1/R_o$
		Practical Circuit Parameters		
Common emitter	$I_g < \beta$	$V_g = R_{Lac}/R_e^*$	$Z_{in} = R_l \| \beta R_e^*$	$R_{co} \approx R_L \| R_{Lw}$
Common collector	$I_g < \beta$	$V_g \approx 1$	$Z_{in} = R_b \| \beta R_{Lac}$	$R_{co} \approx R_{es} \| [(R_G \| R_b)/\beta]$
Common base	$\alpha \approx 1$	$V_g = R_{Lac}/R_e^*$	$Z_{in} \approx R'_e$	$R_{co} \approx R_L \| R_{Lw}$

NOTE: R_e^* includes R'_e and any external emitter resistance. Device parameters are measured with no external emitter resistance, and R'_e is the entire emitter resistance for device parameter measurements. $R_{Lac} = R_e$ (external)$\|R_{Lw}$ for common-collector circuits. $R_{Lac} = R_L \| R_{Lw}$ for common-emitter circuits. R_b is any base-to-ground resistance which is external to the transistor.

Many textbooks use the term $\alpha/(1-\alpha)$ instead of β in equations. When this term appears in the denominator of an equation, it can be replaced by a β in the numerator, and an α (if there is one) can be replaced by a 1, without any significant error. Remember, α is almost equal to 1. $(\alpha/1-\alpha) \approx \beta$.

Voltage Gain

Voltage gain is not properly a device parameter at all, for it requires that there be an R_L (collector load) in the circuit. Measurement of device parameters, such as the hybrid parameters, is done without any R_L. For that reason I have used "n.a." in the table to indicate that there is no hybrid parameter which is even remotely equivalent to voltage gain.

The only practical circuit that actually has 0 Ω of collector load resistance is the common-collector circuit, and it essentially produces no voltage gain ($V_g \approx 1$). Every practical common-emitter and common-base circuit *must* have a collector load, and will produce a voltage gain, generally much larger than unity. I have therefore taken the liberty of including voltage gain as a *device* parameter for the simplest practical circuit form. The assumption is that there is no additional working load and no external emitter resistor.

Circuits based on this assumption are not generally very practical because of the need for operating stability, as we shall see in the next chapter. The inclusion of voltage gain as a device parameter is not as artificial as it may seem, and certainly not as artificial as leaving it out altogether, because the voltage gain is always the ratio of the total equivalent collector load impedance to the effective emitter resistance, whether or not there is a working load or external emitter resistance.

$$V_g \approx \frac{\text{effective collector resistance}}{\text{effective emitter resistance}}$$

The effective collector resistance consists of $R_o \| R_L \| R_{Lw}$, where R_o is the collector-base reverse-biased junction resistance, R_L is the collector load resistor, and R_{Lw} is the driven, or working, load. The emitter circuit resistance consists of the base-emitter junction resistance R_e' plus any external emitter resistance. For device parameters we assume that R_e' is the only emitter resistance in the circuit and that there is no R_{Lw}. Table 3-1 also assumes that R_o, which is effectively in parallel with R_L, is so large compared to external collector resistances that it need not be included for either the device parameter or the circuit parameter. The equation for the *circuit* voltage gain will include both working load and any external emitter resistor. The equation for circuit voltage gain is

$$V_g \approx \frac{R_{Lac}}{R_e' + R_{es}}$$

where R_{Lac} consists of R_L in parallel with the working load R_{Lw}, R_e' is the junction resistance, and R_{es} is the external emitter resistance.

Output Resistance

Output resistance (impedance) is not a particularly important parameter in most situations. In the case of the common-base and common-emitter circuits it consists of the reverse-biased collector-base junction dynamic resistance, by itself for device parameters, and in parallel with R_L in all practical circuits. There is a lot of fallacious reasoning about impedance matching and maximum power transfer in transistor circuits, but the truth is that maximum power transfer in transistor circuits is seldom necessary and often of no value at all. Even when transformer coupling is used, which provides the best possibility of maximum *power* transfer and matched impedances, *there is seldom any need for maximum power transfer,* and even when there is a need, it seldom turns out to be very practical. The only excuse for matched impedances is maximum *power* transfer, and maximum power transfer is highly overrated. It is seldom important, and when it is, it is not always possible to obtain. In practical common-base and common-emitter circuits the value of the collector load resistor is the overwhelmingly determinant factor in the output impedance, and if a specific output impedance is required, we can simply select an appropriate value for R_L.

In the common-collector circuit things are a bit more complicated, because any base-to-ground impedances, including the generator impedance, are reflected into the output (emitter) circuit reduced by the factor β in such a way that the output impedance consists of R_{es} (if there is one) in parallel with the equivalent base-to-ground resistance divided by β. The output impedance equation for the common-collector circuit is

$$R_{co} = \frac{R_G \| R_b}{\beta} \| R_{es} \qquad \text{See Fig. 3-15}$$

Input Impedance

The input impedance of a circuit is a very important parameter in both analysis and design. For the common-emitter and common-collector circuits the input impedance consists of any base-to-ground impedance in parallel with β times the effective emitter resistance. I have used the symbol R_e^* in Table 3-1 to stand for "effective" emitter resistance and to illustrate the similarity in equations with and without external emitter resistances. In the simplest case of the common-emitter circuit, the emitter is returned directly to ground, so that the only emitter resistance in the circuit is R_e'. For this case, the input impedance equation is

$$Z_{in} = R_b \| \beta R_e'$$

In the slightly more complex case where there is an external emitter resistor, the input impedance equation reads as

$$Z_{in} = R_b \| \beta (R_e' + R_{es})$$

where R_{es} is the external emitter resistor. Because of certain stability requirements (which we will discuss in the next chapter), R_{es} must nearly always be very large compared to R_e', and so the equation is most often written as

$$Z_{in} = R_b \| (\beta R_{es})$$

In the common-collector circuit, the input impedance still consists of any base-to-ground resistance in parallel with β times the effective emitter resistance (impedance). The difference between the common-emitter and the common-collector circuits is that, because the load is taken off in parallel with R_{es}, the effective emitter resistance is $R_{es} \| R_{Lw}$. For stability reasons R_{es} is much larger than R_e', and R_e' is best left out of the equation.

The equation for the input impedance of the common-collector circuit is

$$Z_{in} = R_b \| \beta(R_{es} \| R_{Lw})$$

In the common-base circuit the input signal is applied across R_{es}, and again for stability reasons R_{es} is much larger than R_e' in practical circuits. Because R_{es} is effectively in parallel with R_e', the input impedance of a common-base circuit is

$$Z_{in} = R_e'$$

Now go back to the chapter preview questions at the beginning of the chapter and see how well you can do.

PROBLEMS

Given the schematic diagram in Fig. 3-16, compute the following parameters:

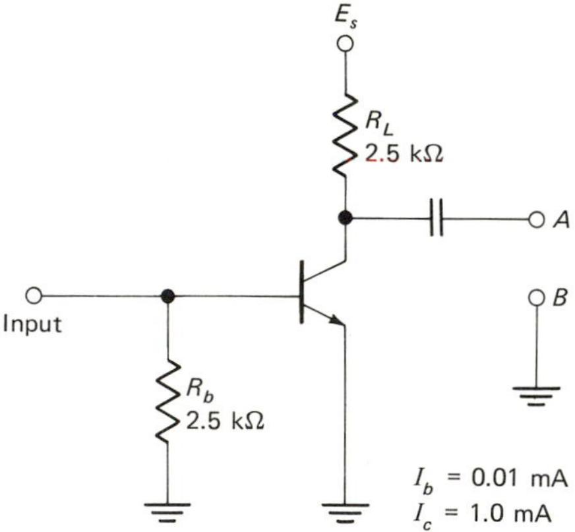

Fig. 3-16 Circuit for Probs. 3-1 to 3-5

3-1 Compute β.
3-2 Compute α.
3-3 Compute R_e' (the Shockley relationship).
3-4 Compute the voltage gain V_g. HINT: Use the value of R_e' from Prob. 3-3.
3-5 Compute the input impedance. Include R_b. HINT: You will need R_e' again.

4

bias and stability

OBJECTIVES:

(Things you should be able to do upon completion of this chapter.)

1. Define *bias* and *quiescent collector current.*
2. List the three causes of collector-current drift and the amount each parameter varies with temperature.
3. Define *swamping.*
4. List the two kinds of negative feedback used to control quiescent collector-current drift.
5. List the typical junction voltages for silicon and germanium junctions.
6. Make a sketch of β vs. quiescent collector current in a transistor.
7. List the effects of voltage- and current-mode feedback on the following circuit parameters:
   a. Voltage gain
   b. Current gain
   c. Distortion
   d. Collector-current drift
   e. Input impedance
   f. Output impedance
   g. Predictability of circuit parameters
8. Solve feedback problems concerning the feedback factor and voltage gain (see Prob. 4-10).
9. Draw the schematic diagram of a single-stage common-emitter circuit with current-mode feedback.
10. Draw the schematic diagram of a single-stage common-emitter circuit with voltage-mode feedback.
11. Define *stability.*
12. List the methods of controlling V_{BE} drift due to temperature changes.
13. List the methods of controlling β drift due to temperature changes.
14. Write the equation for the value of the emitter-resistor voltage drop in a current-mode circuit required to limit quiescent collector-current drift to some specific value at some specific temperature.
15. Define *stability factor.*
16. Define the range of possible stability factors.
17. Define the *practical* range of stability factors for a stable circuit.
18. Be able to solve the following kinds of problems (see the end of the chapter for specific problems):

a. Computing the value of the emitter-resistor voltage drop required to limit Q-point drift (due to V_{BE} drift) to some prescribed value at some maximum expected temperature.
b. Computing the change in the quiescent operating point when I_{co} increases because of an increase in temperature.
c. Computing the change in quiescent collector current due to a large change in β.
d. Given circuit resistor values, computing the stability factor **S**.
e. Given the value **S**, computing the approximate current gain of a circuit.

SELF-TEST

1. What is bias?
2. What is quiescent collector current?
3. How does the transistor "know" the difference between leakage current, signal current, and base bias current?
4. List the three principal causes of quiescent collector-current drift.
5. Which of these causes (from question 4) are controlled by the addition of an added emitter resistor?
6. Which of the three causes of collector-current drift are controlled by reducing the circuit current gain?
7. Define the feedback factor.
8. What effect does negative feedback have upon distortion?
9. Is collector-current drift a form of distortion?
10. What effect does current-mode feedback have on voltage gain? on current gain?
11. How does voltage-mode (negative) feedback affect voltage gain? current gain?
12. What kind of gain, voltage or current, does voltage-mode feedback stabilize? What kind does current-mode feedback stabilize?
13. How are temperature-induced variations in V_{BE} controlled?
14. How does I_{co} vary with temperature in silicon? in germanium?
15. If what you say is true, why is silicon generally preferred to germanium for higher-temperature operation?
16. List the three stability factors.
17. Of the three stability factors, which is convenient and practical to use?
18. Why are practical stability factor equations generally simplified?
19. Define *Miller effect.*
20. In a voltage-mode feedback amplifier, what effect does the Miller effect have on input impedance? on circuit current gain? on voltage gain?
21. Draw the schematic diagram of a basic single-stage transistor amplifier with voltage-mode feedback.

22. In a current-mode feedback amplifier, what effect does the introduction of negative feedback have on input impedance? on voltage gain? on current gain?
23. Draw the schematic diagram of a basic single-stage transistor amplifier with current-mode feedback.
24. What is the importance of the voltage divider base bias scheme?
25. Draw a single-stage transistor amplifier with diode compensation, and explain how the diode helps keep the Q point stable.

INTRODUCTION

In this chapter we will examine and analyze the techniques used for establishing and maintaining the quiescent collector current at a desired value. We call this quiescent current bias current.

The amount of quiescent collector current required for a given application is dictated primarily by the circuit loading. In Chap. 5, we will encounter a table which will make the selection of the proper quiescent collector current an easy task. This chapter, however, is concerned with the problems involved in obtaining and controlling that operating point once an appropriate selection has been made.

There are three important parameters that are very temperature-sensitive, and variations in them result in rather large changes in collector current because the transistor is a current amplifier. In fact, one of the temperature-sensitive parameters is β itself, the transistor's current gain. A second important temperature-dependent parameter is the base-emitter junction voltage, the barrier potential. Because the barrier voltage is part of the base-emitter loop, any change in it will cause a change in the base-emitter current which will appear as β times as large a change in the quiescent collector current.

The third important temperature-controlled variable is I_{co}, the collector-base junction leakage current. I_{co} is primarily a germanium-device problem. It is not very important in silicon devices. This leakage current is generated in the collector-base junction and is injected into the base-emitter circuit to cause β times as large a collector current to flow. All three of these temperature problems are *additive;* but the best practice in silicon devices is to design the circuit in such a way as to minimize quiescent collector-current variations due to variations in β and the base-emitter potential, and let the variations in I_{co} go by. The techniques used to stabilize against variations in β also tend to stabilize against I_{co} problems. In the case of silicon devices, if we properly stabilize against β variations, I_{co} variations become a negligible problem.

Stabilization of β reduces the I_{co} problem in germanium devices as well, but not as completely as in silicon devices. Even so, if we try too hard to take I_{co} variations into account, we are almost certain to end up with a circuit which is overstabilized at the expense of input impedance considerations. It is simply not possible to stabilize germanium circuits as completely as silicon circuits, without unacceptable compromises in signal parameters. The technique for stabilizing against variations in β (and I_{co}) is to lower the *circuit current gain* to somewhere between 1 and 20, so that, even though the transistor may have an intrinsic current gain (β) of 100 or more, the circuit is deliberately designed to have a very small current gain. With the *circuit* current gain reduced to such a low value, the actual β of the transistor becomes relatively insignificant and can vary over a wide range without having much effect on the overall *circuit* current gain.

High β transistors are still desirable, even though such high β contributes very little to the current gain of the circuit. High β elevates the input impedance of the circuit, contributes to stability, and improves the voltage gain in cascades of stages. There are few cases where relatively high *circuit* current gains are required, and special techniques must be used if they are to be attained without a sacrifice in stability. We will look at those techniques when it is time to study power amplifiers. The majority of amplifiers can be designed from a voltage gain standpoint, providing we ensure appropriate impedance levels. Designing for voltage gain leaves us free to use low enough circuit current gains to provide good stability, and that is the approach we will use in most of this book. Current-gain–oriented design techniques are possible, and they are espoused by some designers; but the method is far less convenient, and direct measurements of current gain are very difficult. Voltage gain measurements, on the other hand, are conveniently made with any calibrated oscilloscope (or VTVM).

I suppose what I am saying here is that restricted circuit current gains are essential to the stabilization of β and I_{co} effects, and that these restrictions make current-gain–oriented design procedures less attractive than voltage-gain–oriented methods. In reality we are always concerned with a combination of either voltage gain and impedances or current gain and impedances.

As a part of this process of lowering current gain, we will examine a numerical device known as a stability factor. The stability factor helps to make decisions about how much the current gain should be lowered for each practical situation. It will also make the computations much simpler.

The method used for stabilizing against variations in the junction voltage involves the addition of a much larger voltage in series with the junction voltage, so that even large variations in the junction voltage constitute only a small part of the *total* voltage. This *swamping*[1] voltage can be derived from a separate power supply, and we will study that possibility later, or it can be derived from the voltage drop across a resistor.

There are two distinct cases where we use the voltage drop across a resistor. In one case we use the voltage drop across a resistor in series with the transistor's emitter, and in the other we utilize the voltage drop across the collector load resistor. We call this voltage drop across the emitter (or col-

[1] This term will be explained in detail later in this chapter.

lector) resistor a feedback voltage. It is the collector (output) current which causes the voltage drop, and when a voltage in the output circuit (collector circuit) is connected back to the input circuit (base-emitter circuit), it is called a feedback voltage. By convention, we call the feedback voltage derived from an emitter-resistor voltage drop *current-mode feedback* and the feedback voltage derived from the collector, *voltage-mode feedback*.

We then classify circuits according to the feedback mode. We will take a little time in this chapter to discuss the nature of feedback and considerably more time to discuss practical junction voltage stabilization methods in some detail. Toward the end of the chapter we will examine some important nonfeedback stabilizing methods.

4-1 THE CONCEPT OF TRANSISTOR BIAS

If a transistor is to be a practical amplifying device, certain conditions must be met. They are:

1. The collector-base junction must be reverse-biased. The voltage between collector and emitter must be 2 V or more. This is not an absolute rule, but less than 2 V is seldom practical.
2. The base-emitter junction must be forward-biased. The voltage across this junction is primarily dependent upon the characteristic junction barrier potential and temperature.

This voltage will be in the neighborhood of from 0.1 to 0.25 V for a germanium transistor, and from 0.6 to 0.7 V for a silicon transistor. There is a certain amount of variation in junction voltage with changes in junction current.

In order for the transistor to operate properly, it must be biased to an appropriate idling current. The amount of bias current depends upon the kind of job the circuit is required to do. This is much like the idle speed of an automobile engine. It is normally biased to about 500 rpm for a city car. If the engine were biased much higher, you would not be able to stay within city speed limits in high gear. And, of course, fuel consumption would be greater, owing to the amount of time spent at idle waiting for signals, slow traffic, etc. A car intended for racing might well have its engine biased to a higher idle rpm because it would never be necessary to limit the speed to 30 mi/hr in high gear and because fuel consumption is of little importance.

It probably has never occurred to you that the internal-combustion engine is an amplifier. An amplifier is a device which allows a small amount of power to control a larger power. The force applied to the accelerator pedal is far less than adequate to move the car; nevertheless, the harder you push the accelerator, the more power the engine develops, thousands of times more power than that produced by your foot.

In the automobile example, variation in force applied to the accelerator pedal is the input signal. The accelerator pedal controls the amount of gasoline being delivered to the combustion chambers, and so does the idle screw. There is really no basic difference between the signal (accelerator) and bias (idle), except that one is a fixed adjustment and the other (the signal) communicates the desired variations in power to the system.

This is also the case in the transistor. The bias is a fixed base current which determines the idle collector current. However, in the transistor, there is a temperature problem. As the temperature increases, the idle current increases. Suppose the internal-combustion engine had a similar problem, and as it heated up on a trip, the idle rpm increased until the idle speed was up to 6,000 rpm. You could then either let off or push down on the accelerator pedal without having much, if any, control over the car's speed.

This is exactly what happens in a transistor unless we do something to compensate for the problem. The idle collector current inches up as the temperature increases until the signal has little or no control. The result is either distortion or no response at all. Therefore, our bias system must not only establish the desired idle current, but it must also *regulate* the bias current to keep the idle current at that value despite variations in temperature. Leakage current entering the base-emitter junction is one culprit. As the temperature increases, so does the leakage current. Since the transistor cannot tell the difference between signal current, bias current, and leakage current, it amplifies them all indiscriminately.

The problem of establishing the appropriate bias simply involves the determination of what bias is *appropriate* for a given circuit and a little Ohm's law. The problem of maintaining that bias current constant while the temperature varies is a more difficult problem, and the primary subject of this chapter.

When we discuss bias, we are actually concerned with a steady-state base-emitter current; but we normally think and talk in terms of the resulting collector-emitter current. We do this for the very simple reason that collector current is easy to measure, directly or indirectly, whereas base current is usually difficult to measure, either directly or indirectly. We are justified in taking this position because base current and collector current are directly related by the factor β. The steady-state collector current, which results from the base-emitter bias current, is called the quiescent collector current. The synonymous term *Q point* is also used, and it signifies a particular point on a curve of collector current vs. some other quantity.

4-2 THE CAUSES OF QUIESCENT COLLECTOR-CURRENT DRIFT

There are three transistor parameters whose variations must be controlled or compensated for if any transistor circuit design is to be successful. These are:

1. *Variations in the base-emitter junction voltage drop* (V_{BE}). The typical value of V_{BE} at 20°C is 0.10 to 0.25 V for germanium and about 0.6 to 0.7 V for silicon.

Current-induced variations in V_{BE} are not very predictable, but for most purposes an assumption of a V_{BE} of 0.2 V for germanium and 0.6 V for silicon will cause no serious error. Once the circuit is properly designed, the quiescent base current will remain fairly constant and V_{BE} variations due to base current will cause no appreciable problem. Unfortunately, the base-emitter voltage also varies with temperature, and this is our big problem with V_{BE} variations. Temperature variations in V_{BE} are not quite linear, nor are they independent of base-emitter current. Such complex behavior makes precise analysis very difficult, but experience has yielded some very satisfactory guidelines for the designer. Table 4-1 shows the (empirical) variations in V_{BE} due to temperature variations for various collector-current ranges. Estimates of these *variations* in V_{BE} are quite valid, even though it is nearly impossible to estimate the *absolute* value of V_{BE} accurately at any given temperature.

2. *Variations in the current gain* (β). The β of a transistor varies from transistor to transistor of a given type number with a spread of 200 percent or more. Beta also increases with temperature at the rate of about 1 percent/C°, and β also varies with the operating point in a less predictable fashion. Typically, maximum β occurs somewhere between 0.5 and 2 mA of quiescent collector current, for transistors with a maximum collector current of 200 mA or less. Figure 4-1 shows a typical β vs. I_c curve. As the curve shows, β falls off rapidly at collector currents much below 10 percent, and more gradually but substantially at higher currents. Beta variations constitute a problem of about the same magnitude as V_{BE} variations.

Here again, once the quiescent collector current has been established and the circuit stabilized, we need not be concerned about β variations due to variations in collector current. And again our chief concern will center on those variations in β due to temperature increases.

3. *Variations in collector leakage current* (I_{co}). Variations in collector leakage current due to the temperature may be considered a problem only for germanium devices. When feedback techniques are used to stabilize a silicon

TABLE 4-1. TEMPERATURE-INDUCED VARIATIONS IN BASE-EMITTER VOLTAGE
1. Assume 0.2 V V_{BE} at 20°C for germanium.
2. Assume 0.6 V V_{BE} at 20°C for silicon.

Quiesent Collector Current (QI_c)	**Change in Base-emitter Voltage** (V_{BE})
0–0.1 mA	−3.0 mV/C°
0.1–1.0 mA	−2.75 mV/C°
1.0–10.0 mA	−2.5 mV/C°
10.0–100.0 mA	−2 mV/C°

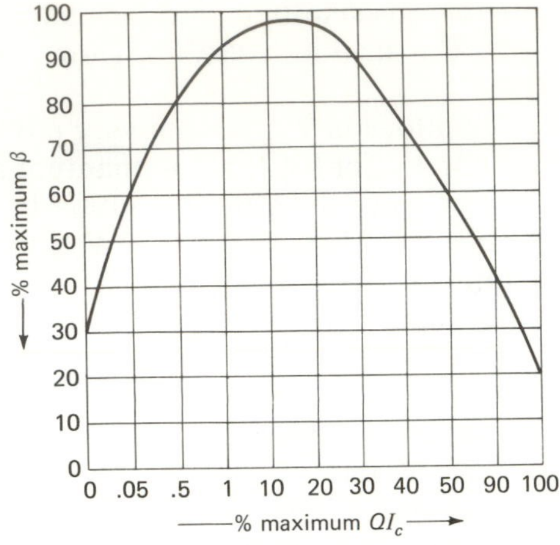

Fig. 4-1 Typical β vs. I_c curve

device against V_{BE} and β variations, adequate I_{co} stabilization follows automatically, owing to the inherently small I_{co} in silicon devices.

In germanium devices, however, it is a different story. In germanium devices, the leakage current problem causes as much quiescent collector-current drift as V_{BE} and β variations put together.

4-3 AN INTRODUCTION TO THE PRINCIPLES OF FEEDBACK STABILIZATION

Throughout this book we will be using the terms *current-mode* and *voltage-mode feedback*. Although most of our design and analysis procedures depend upon the existence of this negative feedback, we will be saying very little about it. We will, in each design, make certain that feedback is involved in sufficient quantities to make the procedures and associated equations valid, but we will not mention it explicitly to any extent. Instead of feedback equations we will more often use simple loop equations which are much less complex to apply and which will do the same job.

Feedback is implicit in the loop equations but it is often so well hidden that we lose sight of its implications altogether. For the most part this is no great disadvantage if we keep the basic consequences of feedback application in the back roads of our minds, where we can draw upon them to provide us with at least an intuitive explanation for some of the things that our loop equations do not account for directly. For example, we will design a circuit called the fully bypassed current-mode (feedback classification) common-emitter circuit, and when we try the circuit in the lab, we will notice a more distorted output signal than we encountered in previous experiments. At that point it is important to understand that the addition of a bypass capacitor across the feedback resistor has nearly *eliminated* the feedback for signal frequencies. The type of feedback involved has the property of reducing, considerably, distortion caused by the nonlinear curve of the transistor, and if we eliminate the feedback for signal frequencies, we should *expect* a more distorted signal. None of the design equations will alert us to this directly; we have to simply know, in advance, the consequences of our action in eliminating the signal feedback, and be prepared to accept them. Are the design procedures deficient because they failed to alert us to this problem? Not really, because we would not have elected to design this particular circuit unless we were in a position where we simply *had to have* a very high voltage gain, and we cannot have that with large amounts of (negative) feedback. It's a little like buying a car. If you *must* have a car, then you *must* expect to include monthly payments in your budget. Nobody likes monthly payments, but if you must buy a car, you must be aware that they are part of the deal. Nobody is too crazy about distortion either, but if you must have high voltage gains, distortion is part of the deal, and you must be aware that high voltage gains cannot be had with low distortion. Eliminating the feedback for signal frequencies provides higher voltage gain, but without (negative) feedback and its distortion-reducing properties, more distortion is inevitable. "You pays your money and takes your choice."

The voltage gain equations that we will use throughout are derived directly from feedback theory, and one of the purposes of this section is to show where these most important equations come from.

The feedback theory in this section falls into much the same category as a *left-handed dutchman* (a special curved pair of shears for cutting circles in sheet metal in a left-handed direction).

During World War II my mother took a job in an aircraft factory; at that time workers were needed and experience was not required. My mother's contribution to the war effort is still somewhat in doubt, but her experience with the left-handed dutchman is a classic story. On her first day on the job her foreman sent her to the tool crib to get—you guessed it—a left-handed dutchman. Now my mother was no rube, she had heard of left-handed monkey wrenches, buckets of steam, and aluminum magnets, and she was just sure she was being put on. She went directly to the coffee shop and killed an hour or so, and after enough time had elapsed, she returned to tell the foreman, "There's no left-handed Dutchman in the tool crib, but there's a left-handed Irishman in Quality Control if you want him." Unfortunately this foreman was a little short in the sense-of-humor department.

Feedback theory, like the left-handed dutchman, is not a universal tool, but when it is needed there is no adequate substitute for it.

In some multistage circuits where feedback involves two or more stages, we shall have to resort to the use of feedback equations, because nothing else is adequate. "Simple" loop equations become too involved to use in some situations, and we must rely, in special cases, on feedback equations.

Feedback involves returning some *fraction* of the output voltage to the input, where it is either added to or subtracted from the input signal. If the output fraction is *added* to the input signal, it is called regenerative or *positive* feedback. Oscillators use positive feedback, and both voltage gain and distortion generally increase.

If the output fraction is *subtracted* from the input signal (that is, combined 180° out of phase for a sinusoidal signal), it is called degenerative or *negative* feedback. Negative feedback reduces the voltage gain, but it makes it very stable, predictable, and independent of transistor characteristics, and reduces distortion.

For the most part, in this book, our concern with feedback will be directed toward *negative* feedback because it is essential to nearly all kinds of transistor circuit stability. Nearly all transistor circuits depend upon negative feedback to maintain desired operating points, and without control over the quiescent operating point most transistor circuits would be useless over any normal range of environmental temperatures.

The *big* problem in transistor circuits is the tendency for the quiescent collector current to drift upward as the temperature increases. We call this instability, but in reality it is a form of distortion, and in the case of the transistor it is by far the most serious and troublesome kind of distortion we will encounter. In the majority of transistor circuits we add negative feedback to reduce Q-point drift and we *always do it at the expense of gain.*

If we use current-mode feedback, we exchange *voltage gain* for stability (reduction of this kind of distortion), and if we use voltage-mode feedback, we trade *current gain* for stability. There are other modes of feedback, but only these two are of special interest to the transistor circuit designer. Dr. Millman and Dr. Halkias provide an excellent treatment of feedback principles in general as well as their specific application to transistors, and I heartily recommend this reference if you desire to delve more deeply into the subject.[2] We must have stability, and if it costs us some gain, we must have it anyway, or forget about most transistor technology. As it happens, the trade-off is not usually a hard one to live with, and we will find that feedback, properly applied, quite often makes the difference between the transistor's being a laboratory curiosity and the foundation of an entire technology. There are other ways in which to deal with the transistor's inherent temperature instability and large manufacturing tolerances, but none of these other methods are even slightly as satisfactory as negative feedback for most applications. I stress this point because, at first, the cost in gain for the required stability will seem to be very high, and you may feel it is not worth it, but later we will develop some clever ways of (almost) having our cake and eating it too, as the not too modern saying goes.

One of the possibilities, and one that is commonly used, is to use large

[2] Jacob Millman and Christos Halkias, *Electronic Circuits and Devices*, McGraw-Hill Book Company, New York, 1967.

amounts of negative feedback for reducing Q-point drift, while using smaller amounts (or none at all) for signal frequencies.

The separation is easily made with capacitors because Q-point drift is essentially a dc phenomenon whereas most signals are ac. Stated another way, the signal rate of change is so much higher than the rate of change of Q-point drift that a chosen capacitor will exhibit nearly 0 Ω of X_c for signal frequencies but will appear to be an open circuit for the very slow-changing Q-point drift.

Since we are talking about various kinds of *distortion,* it might be well to provide a working definition of the word. Distortion is anything that comes out of an amplifier which is unlike the signal that goes into the amplifier. The magnitudes may be different, of course, but nothing else. The signal *in* does not contain any Q-point drift, and if the output does, it is *distortion.* If a given waveform comes out different in shape from the waveform that went in, that is distortion too. Even such things as hum and noise, which are added to the signal by the amplifier, are forms of distortion because they were not part of the original signal. Negative feedback reduces all these kinds of distortion. Let us take a quantitative look at feedback.

Voltage Gain and Current-mode Feedback

Now let us examine the effect of negative feedback upon the circuit voltage gain. The following discussion is based on Fig. 4-2.

A fraction of the output signal, called the feedback fraction, is returned

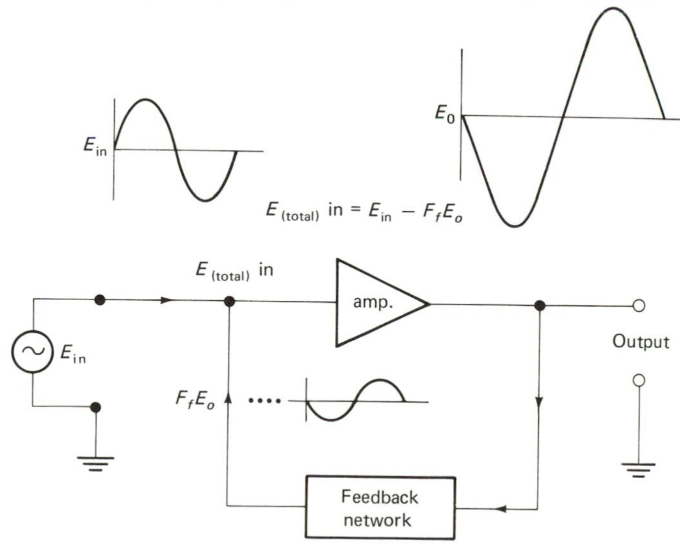

F_f is the feedback fraction

Fig. 4-2 *The general negative feedback loop*

to the input 180° out of phase with the input signal. The two signals are added algebraically at the input, canceling part of the input signal. Any distortion generated in the amplifier is also fed to the input, where it adds a kind of predistortion to the signal which is mixed in the amplifier and tends to cancel the distortion being generated there. The actual signal reaching the input of the amplifier (see Fig. 4-2) is the algebraic sum of the signal from the generator and the feedback-fraction signal being returned from the amplifier output. $E_{\text{total in}} = E_{in} - F_f E_o$, where F_f is the feedback fraction, E_{in} is the signal from the generator, and E_o is the amplifier output voltage. In most literature on feedback the feedback fraction is called β, but to avoid confusion here I will call it F_f. *There is no relationship between the feedback fraction and the β of a transistor.*

Voltage Gain Relationships

Voltage gain without feedback:

$$V_g = \frac{E_o}{E_{in}} \tag{4-1}$$

Voltage gain with feedback (V_g'); the total input signal as seen at the amplifier input:

$$E_{\text{total in}} = E_{in} + F_f E_o \tag{4-2}$$

$$V_g' = \frac{E_o}{E_{in} + F_f E_o} \tag{4-3}$$

Note that for negative feedback F_f carries a negative sign and Eq. (4-3) could have $E_{in} - F_f E_o$ as its denominator. The equation as it is written is applicable to both positive and negative feedback.

The output voltage equation is:

$$E_o = V_g E_{\text{total in}} \qquad \text{(with feedback)} \tag{4-4}$$

Substituting Eq. (4-2) for $E_{\text{total in}}$,

$$E_o = V_g(E_{in} + F_f E_o) \qquad \text{(with feedback)} \tag{4-5}$$

Multiplying through by V_g,

$$E_o = V_g E_{in} + V_g F_f E_o \qquad \text{(with feedback)} \tag{4-6}$$

Transposing ($V_g F_f E_o$) to the left side of the equal-sign, we get $E_o - (V_g F_f E_o)$ on the left side of the equal-sign, and

$$E_o - (V_g F_f E_o) = V_g E_{in} \qquad \text{(with feedback)} \tag{4-7}$$

Factoring out E_o in Eq. (4-7), we get

$$E_o(1 - V_g F_f) = V_g E_{in} \qquad \text{(with feedback)} \tag{4-8}$$

Transposing Eq. (4-8),

$$\frac{E_o}{E_{in}} = V_g' = \frac{V_g}{1 - V_g F_f} \qquad \text{(with feedback)} \tag{4-9}$$

The voltage gain of an amplifier *without* feedback is called the *open-loop gain*.

The equation for the voltage gain with feedback is

$$V_g' = \frac{V_g}{1 - V_g F_f}$$

There are times when the open-loop gain may not be accurately known. If the product of $V_g F_f$ is much greater than 1 (say 10 or so), the following simplified equation can be used to find the voltage gain *with feedback*. When $V_g F_f \gg 1$,

$$V_g' \approx \frac{1}{F_f} \tag{4-10}$$

Example (See Also Table 4-2)

An amplifier has an open-loop gain of 35. What is its voltage gain with 10 percent feedback? ($F_f = 0.1$)

1. $V_g' = \dfrac{V_g}{1 - V_g F_f} = \dfrac{35}{1 - 35\ (0.10)} = 14$

2. $V_g' \approx \dfrac{1}{F_f} = \dfrac{1}{0.1} = 10$

Feedback and Distortion

The distortion and hum generated by an amplifier are also reduced by the same amount that the voltage gain is reduced when negative feedback is used.

$$D' = \frac{D}{1 - F_f V_g}$$

where D is the distortion without feedback and D' is the distortion after feedback.

TABLE 4-2. TABLE OF OPEN-LOOP AND CLOSED-LOOP VOLTAGE GAINS FOR THE EXAMPLE

Open-loop Gain, V_g	Feedback Factor, F_f	$V'_g = \frac{V_g}{1 - V_g F_f}$ (The Complete Equation)	$V'_g = \frac{1}{F_f}$ (The Approximate Equation)
35	0.1	14	10
40	0.1	13.3	10
100	0.1	11.1	10
200	0.1	11.1	10
400	0.1	10.2	10
800	0.1	10.1	10
1,000	0.1	10.1	10
10,000	0.1	10.01	10

Table 4-2 illustrates the fact that the higher the open-loop gain of the amplifier, the more accurate the approximate equation becomes.

Current-mode Feedback

Figure 4-3*a* shows a *general* amplifier with current-mode feedback. The feedback resistor R_f carries both input and output currents. The input resistance R_{in} and the generator internal resistance are both assumed to be small compared to R_f. R_L is also assumed to be large compared to R_f. (Figure 4-3*b* shows a single transistor with current-mode feedback.)

R_L and R_f form a voltage divider. Most of the output voltage appears across R_L; a smaller part of the output voltage is dropped across R_f.

The fraction of the total output voltage appearing across R_f is in series (opposing) with the generator (E_{in}) voltage.

The feedback fraction F_f can be found by using the classic voltage divider equation:

$$F_f = \frac{R_f}{R_f + R_L} \tag{4-11}$$

Because R_f is always small compared to R_L, R_f can be dropped from the denominator of Eq. (4-11), yielding the more common equation:

$$F_f \approx \frac{R_f}{R_L} \tag{4-12}$$

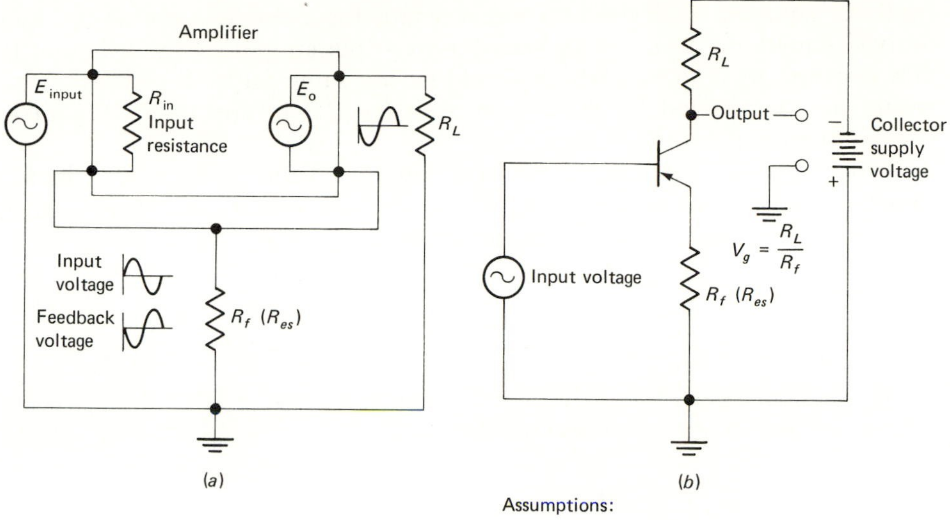

Assumptions:

1. R_e' is small compared to R_f.
2. R_f is small compared to R_L.
3. The internal resistance of the generator is small compared to R_f.

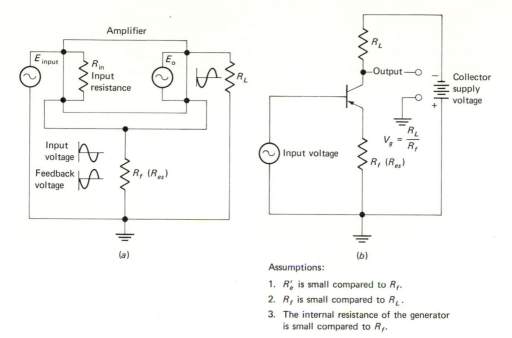

Fig. 4-3 Current-mode feedback. (a) The general amplifier with current-mode feedback; (b) a single transistor with current-mode feedback

From Eq. (4-10) $V_g' \approx 1/F_f$. We can substitute R_f/R_L for F_f and get

$$V_g' = \frac{1}{R_f/R_L} = \frac{R_L}{R_f} \tag{4-13}$$

V_g' is the voltage gain with feedback.

INPUT IMPEDANCE AND CURRENT-MODE FEEDBACK. In Chap. 3 we examined the effect of the emitter resistor on input impedance, through the use of simple loop equations, and we came up with the following equation:

$$Z_{in} = \beta R_f \qquad \text{See Fig. 4-4.} \tag{4-14}$$

(R_f was called R_e in Chap. 3 because we were not talking about the fact that R_e is actually the feedback resistor.)

Equation (4-14) is fine for a single stage, but if we must deal with current-mode feedback involving two (or more) stages, the equation is inadequate, and we must go to the more *general* feedback equation for input impedance. I won't go through the math to show that both equations are valid if correctly interpreted, but I will give you an example to show that the *general* feedback equation will yield the same result, for a single stage, as the loop-derived equation. In order to keep the example simple, we will ignore the bias resistors R_{b1} and R_{b2}, which would be the same for both conditions and would not be affected by the existence of R_f (R_e) in any case. Figure 4-4 shows the circuit for the example.

In Fig. 4-4, according to the loop equation (from Chap. 3), Z_{in}, looking into the base, would be $Z_{in} = \beta R_e = 100 \times 1\ \text{k}\Omega = 100\ \text{k}\Omega$. The feedback equation, taken from classic literature on the subject, is

$$Z_{in} = \beta R'_e(V_g F_f) \tag{4-15}$$

1. $\beta = 100$
2. R'_e (at 1 mA) $= 25\ \Omega$ the Shockley relationship
3. V_g, without R_e (R_f), without feedback

   $$V_g = \frac{R_L}{R'_e} = \frac{10\ \text{k}\Omega}{25} = 400$$

4. $F_f = \dfrac{R_f}{R_L} = \dfrac{1\ \text{k}\Omega}{10\ \text{k}\Omega} = 0.1$
5. $Z_{in} = \beta R'_e(V_g F_f)$
   $Z_{in} = 100 \times 25(400 \times 0.1)$
   $Z_{in} = 100 \times 25 \times 40$
   $Z_{in} = 100\ \text{k}\Omega$

So, you see we get 100 kΩ for Z'_{in} (ignoring R_{b1} and R_{b2}, the bias resistors),

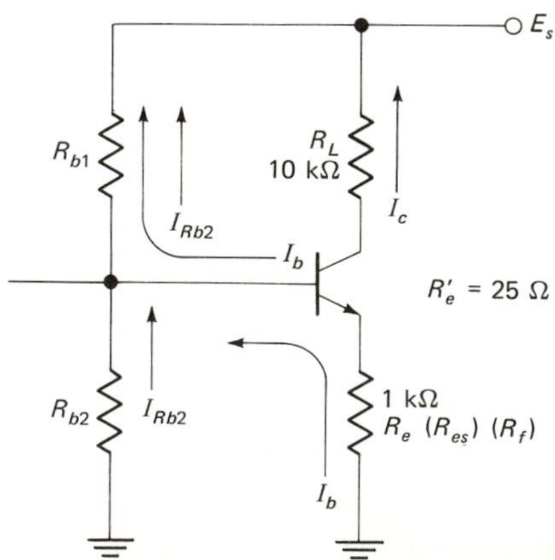

Fig. 4-4 Circuit for the current-mode input Z (impedance) example

whichever equation we use. In both cases, if we include the bias resistors, we would have (see Fig. 4-4):

$$Z_{in} = R_{b1} \| R_{b2} \| 100\ \text{k}\Omega$$

Many textbooks give the input impedance equation for a current-mode feedback amplifier as

$$Z_{in} = \beta R'_e \frac{V_g}{V'_g} \tag{4-16}$$

There is no *real* difference between these two forms, but it might be worthwhile to build a little algebraic bridge between them, so that when you encounter Eq. (4-16), it will cause you no problem.

$$V'_g = \frac{V_g}{1 - V_g F_f} \qquad \text{from Eq. (4-9)}$$

V'_g is the voltage gain with feedback. Rearranging, we get

$$\frac{V'_g}{V_g} = \frac{1}{1 - V_g F_f}$$

Taking the reciprocal of both sides, we get

$$\frac{V_g}{V'_g} = 1 - V_g F_f \tag{4-17}$$

Because the product $V_g F_f$ is much larger than 1, we can drop the 1 from the right-hand member of Eq. (4-17). The minus sign simply tells us that we are dealing with *negative* feedback. So if we drop the 1 and the minus sign, we get

$$\frac{V_g}{V'_g} = V_g F_f$$

We can substitute $V_g F_f$ for V_g/V'_g in Eq. (4-16) and be right back at our first equation (4-15).

Voltage-mode Feedback

Voltage-mode feedback stabilizes the amplifier current gain in the same way that current-mode feedback stabilizes the voltage gain.

The current gain for an amplifier with voltage-mode feedback is

$$\frac{I_o}{I_{in}} = I'_g = \frac{I_g}{1 - I_g F_f} \qquad \begin{array}{l} I_g\text{—without feedback} \\ I'_g\text{—with feedback} \end{array}$$

The derivation of this equation is exactly the same as what we went through for the voltage gain equation for current-mode feedback, except that we substitute I_g for V_g and I for E whenever they appear in Eqs. (4-1) to (4-9).

Again, when the current gain times F_f (product) is reasonably high, the equation can be simplified to

$$I'_g = \frac{1}{F_f} \tag{4-18}$$

Figure 4-5*a* shows the *general* amplifier with voltage-mode feedback, and Fig. 4-5*b* shows a single stage with voltage-mode feedback.

The feedback factor F_f is given by (see Fig. 4-5*a*):

$$F_f = \frac{R_{in}}{R_f + R_{in}} \tag{4-19}$$

R_f is normally quite large compared to R_{in}, and so we can drop R_{in} from Eq. (4-19) to get

$$F_f = \frac{R_{in}}{R_f} \tag{4-20}$$

Combining Eqs. (4-18) and (4-20), we get

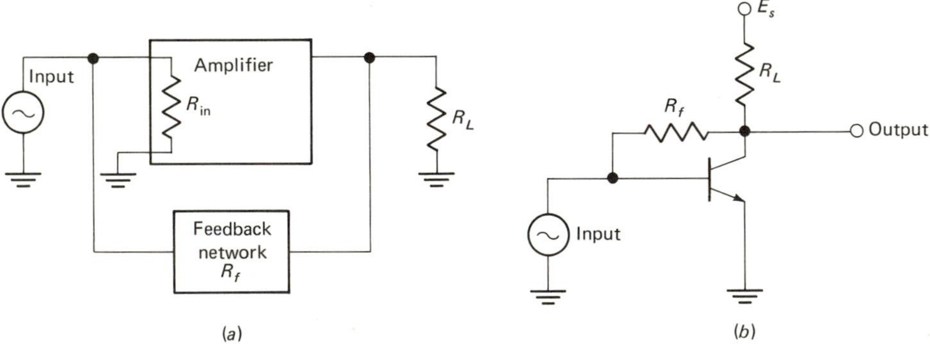

Note: The generator impedance must be high enough not to load the feedback circuit.

Fig. 4-5 Voltage-mode feedback amplifiers. (a) *The general voltage-mode feedback amplifier;* (b) *single stage*

TABLE 4-3. SUMMARY OF THE EFFECTS OF NEGATIVE FEEDBACK ON VARIOUS CIRCUIT PARAMETERS

Feedback Type	Input Impedance	Voltage Gain	Current Gain	Distortion
Current-mode feedback	Increased by factor β or I_g	Decreased Stabilized $V_g' = \frac{1}{F_f}$	Negligible effect	Reduced $D' = \frac{D}{1 - F_f V_g}$
Voltage-mode feedback	Decreased by factor V_g	Negligible effect	Decreased Stabilized $I_g = \frac{1}{F_f}$	Reduced $D' = \frac{D}{1 - F_f V_g}$

$$I_g' = \frac{1}{R_{in}/R_f}$$

$$I_g' \approx \frac{R_f}{R_{in}} \tag{4-21}$$

This last equation may not be valid for a single stage because the $I_g F_f$ product may be too small. Later in this chapter we will use a different equation for finding the current gain of a single stage with voltage-mode feedback.

THE INPUT IMPEDANCE OF A VOLTAGE-MODE FEEDBACK AMPLIFIER. The input impedance with voltage-mode feedback is *lowered* by a factor V_g (voltage gain). This phenomenon is called the Miller effect; we will discuss it in considerable detail later in this chapter (Sec. 4-7).

Table 4-3 is a summary of the effects of negative feedback on various circuit parameters.

Stability

In preparation for the discussions in the next few sections concerning stability, let us write the bias loop equation and look at those factors that influence quiescent current drift with temperature, and also preview the techniques for controlling those factors.

Writing the Kirchhoff loop equation for Fig. 4-6, we get

$$E_{\text{bias}} - I_b R_b - I_e R_e - V_{BE} = 0$$

Assuming that $I_e \approx I_c$, which is a good assumption for most practical transistors, we can write

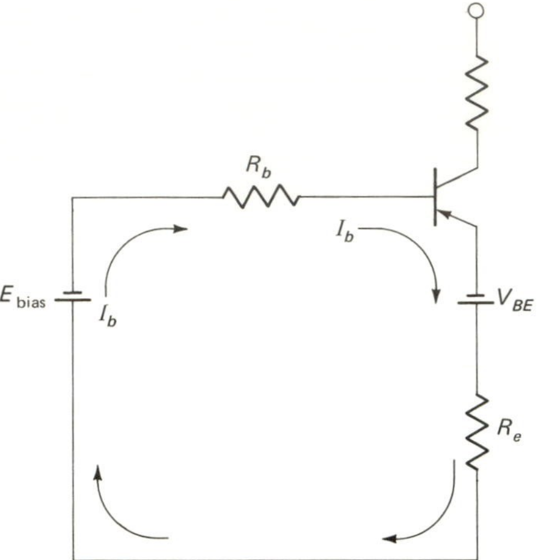

Fig. 4-6 The bias loop

$$E_{bias} - I_b R_b - I_c R_e - V_{BE} = 0$$

Because $I_b = I_c/\beta$, we can substitute

$$E_{bias} - \frac{I_c R_b}{\beta} - I_c R_e - V_{BE} = 0$$

Rearranging,

$$I_c R_e + \frac{I_c R_b}{\beta} = E_{bias} - V_{BE}$$

Factoring out I_c, we get

$$I_c \left(R_e + \frac{R_b}{\beta}\right) = E_{bias} - V_{BE}$$

Solving for I_c yields

$$I_c = \frac{E_{bias} - V_{BE}}{R_e + R_b/\beta}$$

In this equation the two factors that cause I_c to drift with temperature are V_{Be} and β.

We can make the collector current independent of β if we can make R_b/β very small compared to R_e. High values of β would contribute to this, but β is an unpredictable parameter and we cannot depend on it. However, R_b is generally under our control, and we can make it small enough in each case to make the term R_b/β negligible. Obviously, if we could make R_b equal to 0 Ω, β would drop out of the equation altogether and the collector current would no longer be dependent on β at all. We can do this in the common-base configuration, but not in any other configuration because the input signal would be shorted out. We would have a stable but useless circuit. The problem is to select a value for R_b which will provide acceptable immunity from temperature-induced β variations without excessive loading of the signal. In this chapter we will look at quantitative methods for determining an acceptable value for R_b for any circuit we may encounter.

V_{BE} is our other problem, and because we have no control over its value (0.6 V for silicon and 0.25 V for germanium), the only way we can make it negligible compared to E_{bias} is to make E_{bias} larger.

If you will examine the bias loop in Fig. 4-6, you can see that E_{bias} is distributed among R_b, V_{BE}, and R_e. V_{BE} must be only a very small part of E_{bias} if E_{bias} is to be much larger than V_{BE}. And to reduce β effects, R_b must be small compared to R_e, so that we can expect most of E_{bias} to be dropped across R_e. Because of this fact we will approach the problem in terms of the voltage drop across R_e instead of in terms of the ohmic value of R_e. We will find this approach a very practical one for quantitative determination of the proper size for R_e for any given circuit. If we can select values of E_{bias} and R_b which make our two problem terms R_b/β and V_{BE} negligible, then any variations in V_{BE} or β will also become insignificant.

Figure 4-7 shows three bias configurations. Figure 4-7*a* shows simple base bias, which unfortunately cannot be used because R_b (for any practical circuit) cannot be small enough for stability and still supply the proper bias current. Figure 4-7*b* shows a modified version of base bias where an addi-

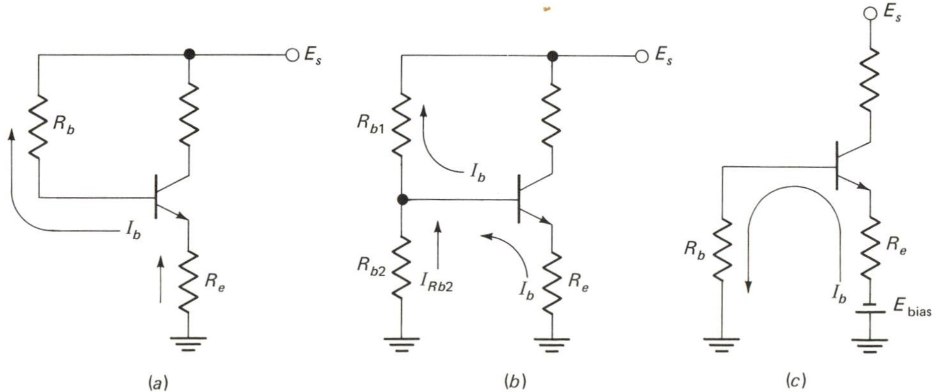

Fig. 4-7 Three bias configurations. (a) Base bias; (b) a practical base bias circuit; (c) emitter bias

tional resistor is placed from base to ground to satisfy the requirement that R_b be small enough.

Figure 4-7*c* shows the emitter bias circuit, which not only is simple but can easily satisfy the requirements for low values of R_b and large values of E_{bias}. If it weren't for the extra bias supply, this might well be the preferred bias method for all occasions. It is the preferred method for some classes of circuits.

For the circuit in Fig. 4-7*c*, $QI_c = E_{\text{bias}}/R_e$ if it is properly designed.

4-4 HOLDING THE QUIESCENT COLLECTOR CURRENT STEADY IN SPITE OF BASE-EMITTER VOLTAGE VARIATIONS

Before we go into a detailed analysis of the method involved in stabilizing the quiescent collector current against variations caused by base-emitter voltage variations, let me provide a simple analogous situation as an illustration. A little intuitive feel for the approach can make the mathematics a little less mysterious.

Let us suppose that we have an unstable battery, one which varies over a range of ±100 percent, and that this unstable battery *must* be a part of the circuit. Let us also assume that the only circuit requirement is a *constant* 1 mA of current. Let us say that our hypothetical battery has a nominal voltage of 0.1 V. Our 1.0-mA circuit would look like Fig. 4-8.

Now if the battery voltage should increase, say to 0.2 V, the current would increase to 2 mA. Where we wanted 1 mA, we now have 2 mA; the current has doubled because the voltage has. A 100 percent increase in the unstable battery voltage has resulted in a 100 percent change in the circuit current. Now suppose we add another battery (a stable one) with a voltage of 1 V and change the resistance, so that we will still have a circuit current of 1 mA. Our circuit now would look like Fig. 4-9.

Now in this case, if our unstable 0.1-V battery increases its voltage to 0.2 V, let us calculate the change in current.

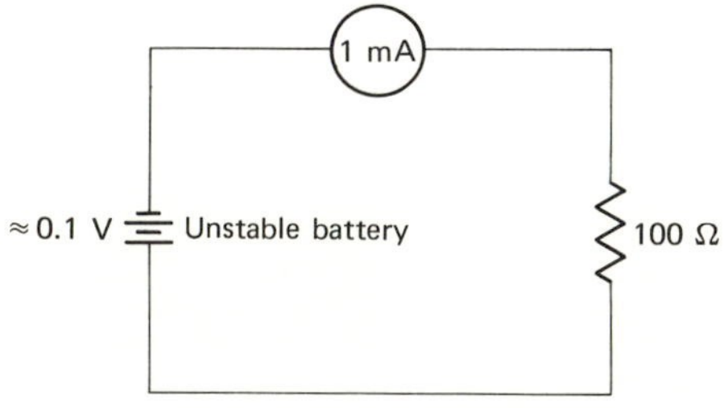

Fig. 4-8 The circuit with its unstable battery

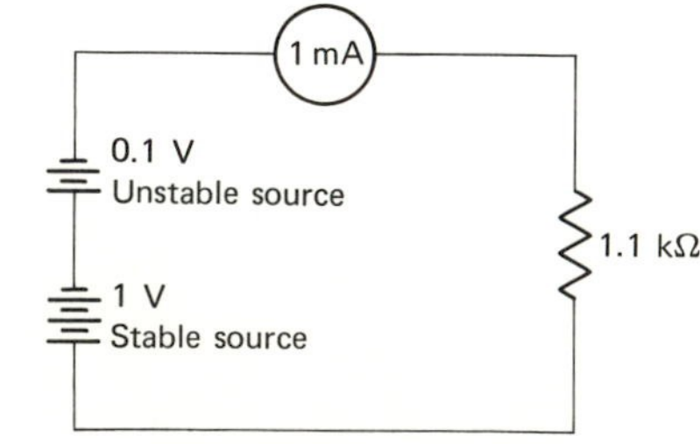

Fig. 4-9 The addition of a large-value stable voltage source

$$I = \frac{E}{R} \qquad \text{Ohm's law}$$

$$I = \frac{1.2\ \text{V}}{1.1\ \text{k}\Omega} \approx 1.09\ \text{mA}$$

The new current is 1.09 mA, which is a change of only 9 percent. In our first circuit (Fig. 4-8), a 100 percent change in the voltage of the unstable battery resulted in a 100 percent change in the circuit current. In the revised circuit in Fig. 4-9, that same 100 percent change in the unstable battery voltage resulted in only 9 percent change in current. The technique of adding larger-value elements to reduce the overall effect of elements that have undesirable characteristics is called *swamping*. Let us take a look at how the swamping technique can be used to minimize the effect of base-emitter voltage variations in the transistor.

Figure 4-10 shows the equivalent transistor input circuit for a *pnp* transistor.

Although collector current must be flowing to make R_e look like $\beta \times R_e$ to the base circuit, the actual value of collector current has no direct influence upon the base-emitter circuit. Therefore, we shall not include the collector circuit in our input circuit model (Fig. 4-10*b*). There is an indirect effect exerted upon the input circuit by the amount of collector current. Because β varies with the quiescent collector current and because β is a part of our input circuit, the input (base) current will be influenced by the quiescent collector current. However, our second problem in stability will be the stabilization of β. For the sake of this discussion, we shall assume that β has already been stabilized and ignore its potential effects.

If we hold the base voltage (E_b) and β constant at some value, the only

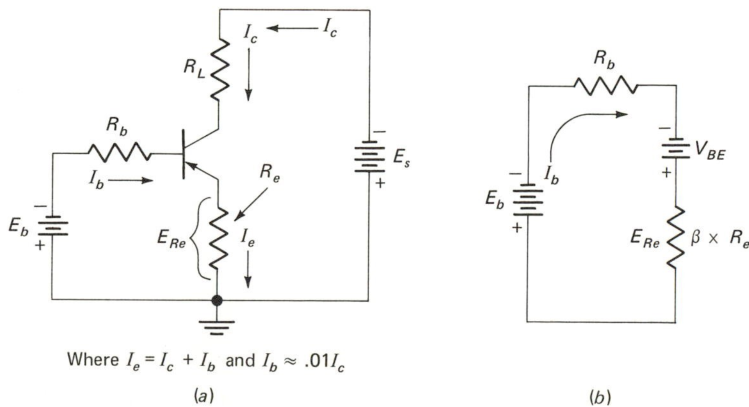

Fig. 4-10 (a) The simplest general bias circuit; (b) the base-emitter loop from (a)

factor which will cause a change in base current due to temperature (ignoring leakage current for the moment) is a change in V_{BE}. A change in V_{BE} results in a change in base current, which in turn results in a larger change in quiescent collector current. The question is: What factors control the variations in base current, and which ones are under our control? What can we do to reduce the effects of temperature-induced changes in V_{BE}? Let us redraw the circuit from Fig. 4-10 without E_b and call it Fig. 4-11.

By Ohm's law, the change in base current is

$$\Delta I_b = \frac{\Delta V_{BE}}{R_b + (\beta \times R_e)} \tag{4-22}$$

As we shall see a little later, the compensation required to stabilize β and leakage current demands that R_b be small compared to $\beta \times R_e$. Thus, R_b will add but little to the denominator of Eq. (4-22).

If, on that assumption, we drop R_b from Eq. (4-22), we get

$$\Delta I_b = \frac{\Delta V_{BE}}{\beta \times R_e} \tag{4-23}$$

The collector-current drift away from the established quiescent value is the actual problem; so let us look at the relationship between Eq. (4-23) and variations in collector current.

Since we know that

$$I_c = \beta \times I_b$$

we can assume that

$$\Delta I_c = \beta \times \Delta I_b$$

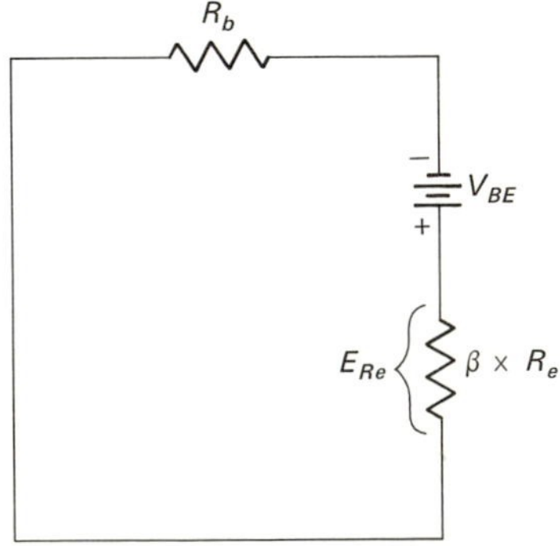

Fig. 4-11 The base-emitter loop with E_b removed

If we multiply both sides of Eq. (4-23) by β, we get

$$\beta \Delta I_b = \beta \frac{\Delta V_{BE}}{\beta \times R_e} \tag{4-24}$$

or

$$\Delta I_c = \frac{\Delta V_{BE}}{R_e}$$

Since we are interested in the fractional amount of variation in QI_c, let us divide both sides of Eq. (4-24) by QI_c.

$$\frac{\Delta I_c}{QI_c} = \frac{\Delta V_{BE}}{QI_c R_e} \tag{4-25}$$

Since the emitter current and collector current are approximately equal, we can substitute I_e for I_c in the right-hand member of Eq. (4-25). We get

$$\frac{\Delta I_c}{QI_c} = \frac{\Delta V_{BE}}{I_e R_e} \tag{4-26}$$

Because $E = IR$, we substitute E_{Re} for the denominator of the right-hand member of Eq. (4-26) to get

$$\frac{\Delta I_c}{QI_c} = \frac{\Delta V_{BE}}{E_{Re}} \tag{4-27}$$

Equation (4-27) tells us that the only control we have over the effect of variations in V_{BE} on the collector current is the voltage drop across the emitter resistor. The larger we make E_{Re} with respect to the *variations* in V_{BE}, the more stable the quiescent point will be. For example, if E_{Re} is 10 times as large as ΔV_{BE}, the quiescent collector-current drift will be reduced to about one-tenth (10 percent) of what it would be without the emitter resistor (and its voltage drop). This is clearly another example of the swamping technique. We make E_{Re} large enough to minimize the effects of ΔV_{BE}.

Equation (4-27) is expressed in terms of the fractional change in QI_c:

$$\frac{\Delta I_c}{QI_c}$$

We can express this fraction as a percentage if we multiply it by 100.

$$\% \Delta QI_c = 100 \frac{\Delta I_c}{QI_c}$$

If we multiply the left member of the equation by 100, we must do the same to the other side. Equation (4-28) is Eq. (4-27) expressed in percentage form:

$$\%\Delta QI_c = \frac{\Delta V_{BE} \times 100}{E_{Re}} \tag{4-28}$$

This form of the equation is useful for analysis purposes, although it is not usually a part of the analysis for troubleshooting situations, unless there is a suspicion that the circuit has been installed in a more hostile environment than the designer intended. Although the equation is rarely needed in practical analysis, we will always use it in design work. In the design process, however, we will normally be interested in the amount of voltage drop required across the emitter resistor to prevent QI_c from drifting more than some specified amount at some *highest expected temperature*. Let us rearrange Eq. (4-28) to solve for E_{Re}:

$$E_{Re} = \frac{\Delta V_{BE} \times 100}{\%\Delta QI_c} \tag{4-29}$$

The upper operating temperature will normally be a part of the design specifications, and it is this upper temperature which will determine the total amount of ΔV_{BE}. The designer will also have to decide upon the maximum percentage increase in QI_c that he finds acceptable at the highest expected temperature. In many cases, the *amount* of QI_c variation allowed by the designer is less important than the fact that the designer *knows* with some certainty how much variation to expect. We will take 20°C as a reference temperature (25°C is also often used). Let us try an example.

Example 4-1

Find the emitter-resistor voltage drop E_{Re} required to limit the increase in quiescent collector current QI_c to 10 percent at an upper temperature of 70°C. The quiescent collector current is to be approximately 1 mA at 20°C. The equation is

$$E_{Re} = \frac{\Delta V_{BE} \times 100}{\%\Delta QI_c}$$

1. Find the total amount of change in the base-emitter voltage.

Consulting Table 4-1, we find that at approximately 1.0 mA V_{BE} decreases at the rate of 2.5 mV/C° increase in temperature.

In this case, the total temperature change is from 20° to 70°C, a total change of 50°. The total $\Delta V_{BE} = 50° \times 2.5$ mV, or 125 mV.

2. Solve for E_{Re}.

$$E_{Re} = \frac{125 \times 10^{-3}\ \text{V} \times 100}{10\ (\%)}$$

$$E_{Re} = 1.25\ \text{V}$$

At 70°C, the quiescent collector current will have increased by 10 percent. If it was 1 mA at 20°C, it will be 1.1 mA at 70°C. This, of course, accounts only for the drift in QI_c due to variations in V_{BE}. Variations in β (and in I_{co} in germanium devices) will occur also. If we know the drift caused by each of the factors, we can find the total drift by finding the sum of the individual drifts. In design, the problem is, "How much deviation in QI_c do I allow?" At this early juncture I would advise an average allowance of 10 percent, because this 10 percent, coupled with variations due to β and I_{co} changes with temperature, will yield a total drift of nearly 20 percent, which is considered to be a normally acceptable error in electronics production.

4-5 HOLDING THE QUIESCENT CURRENT STEADY IN SPITE OF VARIATIONS IN LEAKAGE CURRENT AND β (USING CURRENT-MODE FEEDBACK)

Because of the wide variation in β due to temperature and operating current, there is no justification for holding manufacturing tolerances for β closer than $+100, -50$ percent. The variation in β with temperature, over fairly narrow temperature ranges, amounts to somewhere between 0.5 and 2.0 percent/C°. An average value (normally an increase) of 1 percent/C° is generally used. It might be a good idea, at this juncture, to clarify the selection of a working β. The ultimate rule is: *be conservative.* If the manufacturer specifies *typical* β, divide his figure by 2. If the manufacturer specifies *minimum* β, reduce it by at least 10 percent. Since we cannot normally know the actual value of β without measuring individual transistors, not a very practical procedure, we must compensate for our inexact knowledge.

The extent of our *ignorance* of the true β under specific conditions is so complete that it forces us to assume that the largest possible variation in β is bound to occur. The only way to cope with this problem is to alter the circuit in such a way as to make the collector current as nearly independent of variations in β as possible. Figure 4-12 shows one of the most important bias circuits which does just that. The resistor R_{b2} forms a current divider with the input resistance of the transistor, so that any current delivered to point A in Fig. 4-12 splits into two parts. The larger current passes through R_{b2} to ground and does not get amplified, while a smaller current passes through the base-emitter junction and gets amplified to appear β times as large in the collector circuit. The value of R_{b2} is purposely selected so that its value is

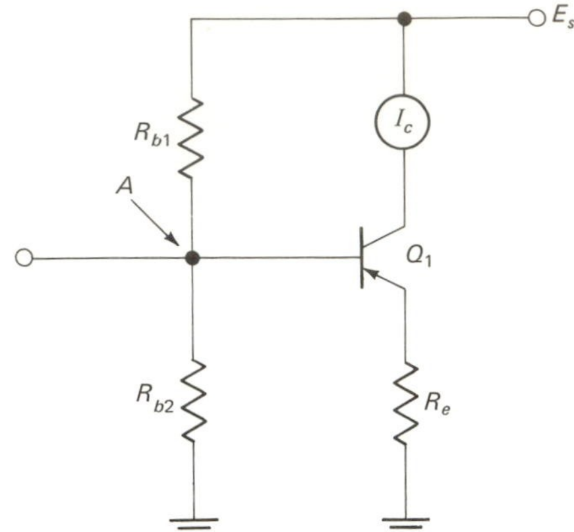

Fig. 4-12 The voltage divider base bias circuit

low compared to the input resistance of the transistor ($\beta \times R_e$). This results in a *circuit* current gain which is much lower than β and almost independent of β. The *circuit* current gain is normally held to a value between 1 and 10. This means that leakage current I_{co} is amplified only by a factor between 1 and 10 instead of by a factor of β, which can be as high as several hundred. The resistor R_{b1} delivers a fairly constant current to supply base-emitter bias current and current for R_{b2}. Suppose we try a numerical example to illustrate the benefits of this kind of bias circuit in the realm of leakage currents. Then we will demonstrate, by another numerical example, how this bias circuit makes the quiescent collector current nearly independent of the transistor's β. Remember that the leakage component I_{co} is independent of the applied voltage and is solely a function of temperature and the kind of semiconductor material involved. Therefore, we can represent I_{co} by a constant-current generator (at a given temperature).

In Fig. 4-13*a*, we have left the bias supply resistor R_{b1} out and indicated the leakage current by a constant-current generator. We will be concerned only with that collector current which results from leakage current injected into the input circuit.

For the circuit in Fig. 4-13 let:

1. $I_{co} = 1\ \mu\text{A}$
2. $\beta = 50$
3. $R_e = 1\ \text{k}\Omega$
4. $R_{b2} = 10\ \text{k}\Omega$

The stability against variations in I_{co} and β is represented by the factor

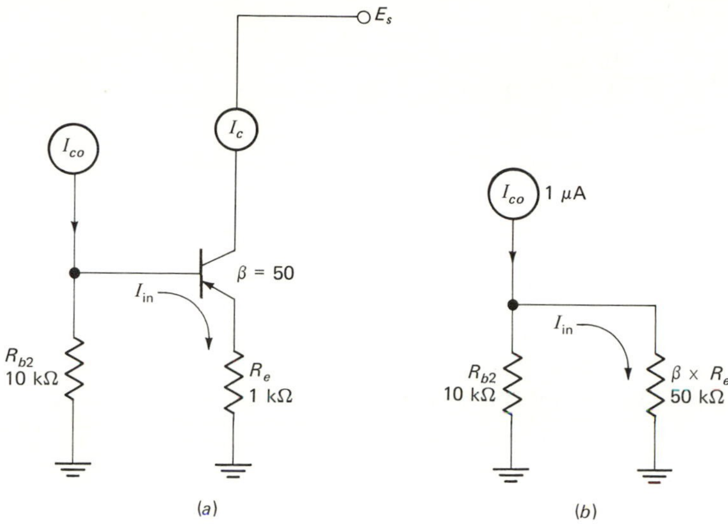

Fig. 4-13 (a) *The circuit model for the stability factor explanation;* (b) *the equivalent circuit of* (a)

S, which is known as the stability factor. The figure **S** is approximately equal to the *circuit* current gain. **S** is also approximately equal to R_{b2} divided by R_e.

1. $\mathbf{S} \approx$ the circuit current gain
2. $\mathbf{S} \approx \dfrac{R_{b2}}{R_e}$

S will prove to be a very useful relationship. Now let's get on with a numerical example (see Fig. 4-13).

Figure 4-13*b* shows a resistive model of Fig. 4-13*a*.

1. Find the current (I_{in}). This is the base-emitter current which appears in the collector circuit, β times as large. The current I_{co} divides between the two branches. The part of the current that passes through R_{b2} *does not* get amplified and *does not* appear at all in the collector circuit.
2. $R_{in} = \beta \times R_e$
3. The current divider equation (finding the input current):

$$I_{in} = I_{\text{total}} \frac{R_{b2}}{R_{b2} + R_{in}}$$

where $I_{\text{total}} = I_{co}$

4. Substituting in values,

$$I_{in} = 1\ \mu\text{A}\ \frac{10\ \text{k}\Omega}{10\ \text{k}\Omega + 50\ \text{k}\Omega}$$

   $I_{in} = 0.166\ \mu\text{A}$
5. Since the collector current is β times the input current,
   $I_c = 0.166\ \mu\text{A} \times \beta$
   $I_c = 0.166 \times 50$
   $I_c = 8.3\ \mu\text{A}$
   If we leave out R_{b2} (only for comparison purposes),
   $I_c = \beta \times I_{in}$
   $I_c = 1\ \mu\text{A} \times 50 = 50\ \mu\text{A}$
   $I_c = 50\ \mu\text{A}$ without R_{b2}
6. The stability factor in this case is

$$\mathbf{S} \approx \frac{R_{b2}}{R_e}$$

$$\mathbf{S} \approx \frac{10\ \text{k}\Omega}{1\ \text{k}\Omega} = 10$$

7. The effective circuit gain for I_{co} is

$$I_g\ \text{(Current gain)} = \frac{I_c}{I_{co}}$$

$$I_g\ \text{(Current gain)} = \frac{8.3\ \mu\text{A}}{1\ \mu\text{A}}$$

   Circuit current gain $= 8.3 \approx \mathbf{S}$ conservative, $\mathbf{S} < 10$

WHAT HAPPENS WHEN β VARIES. Let us look at the *circuit* current gain when the β varies.

Let us take the original circuit in Fig. 4-13, where we compute a circuit current gain of 8.3, increase β to 100, and recalculate the circuit gain, again using the 1-μA leakage current as the input signal.

1. Computing R_{in} with the new β 100,
   $R_{in} = \beta \times R_e$
   $R_{in} = 100 \times 1\ \text{k}\Omega = 100\ \text{k}\Omega$
2. Computing I_{in}, using the current divider equation,

$$I_{in} = I_{\text{total}} \frac{R_{b2}}{R_{b2} + R_{in}} \qquad (I_{\text{total}} = I_{co})$$

$$I_{in} = I\ \mu\text{A}\ \frac{10\ \text{k}\Omega}{10\ \text{k}\Omega + 100\ \text{k}\Omega}$$

   $I_{in} = 0.09\ \mu\text{A}$
3. Computing the collector current with the new β,

$I_c = \beta \times I_{in}$
$I_c = 100 \times 0.09 = 9\ \mu A$

4. And the new *circuit* current gain is

$$\text{Current gain} = \frac{I_c}{I_{co}} = \frac{9\ \mu A}{1\ \mu A}$$

The new current gain is 9.

As you see, β has increased from 50 to 100, but the *circuit* current gain has changed only from 8.3 to 9; what is more important, however, is that the collector current has changed only from 8.3 to 9 μA. This change in collector current is in contrast to a change from 50 μA of collector current to 100 μA without the current division produced by R_{b2}.

For our example we used I_{co} as an input signal, but we could as well have been talking about the bias current delivered by R_{b1}. Without the addition of R_{b2}, a 100 percent change in β would have resulted in a 100 percent shift in the collector current; but with R_{b2} in the circuit a 100 percent shift in β would cause a shift of only a few percent in the collector current. Actually, R_{b1} also tends to lower the *circuit* current gain and add to the stability, and technically it should be included in the stability equation. The equation would then read as

$$\mathbf{S} = \frac{R_{b2} \| R_{b1}}{R_{es}}$$

However, normally R_{b1} is substantially larger than R_{b2} and can be left out. The most drastic result of leaving it out will be a slightly more stable circuit than we intended.

4-6 THE STABILITY FACTOR: ITS MEANING AND IMPLICATIONS

There are three quantities which must be stabilized, V_{BE}, β, and I_{co}. Each of these quantities can be assigned a factor which, in each case, expresses the relationship between variations in the quantity and the resulting collector-current drift.

$$\mathbf{S}_{VBE} = \frac{\Delta I_c}{\Delta V_{BE}} \tag{4-30}$$

$$\mathbf{S}_{\beta} = \frac{\Delta I_c}{\Delta \beta} \tag{4-31}$$

$$\mathbf{S}_{Ico} = \frac{\Delta I_c}{\Delta I_{co}} \tag{4-32}$$

Equations (4-30) to (4-32) define the three stability factors. These definitions seem simple and straightforward, but unfortunately the relationships in each case are actually very complex. For example, the relationship between I_{co} and collector current involves a complex I_{co}-temperature relationship, an additional complex temperature-controlled β variation, and β variations due to operating currents.

In addition to the individual complexity of each stability factor, the three factors are so interdependent that the mathematics becomes almost overwhelming. Professor J. F. Pierce[3] provides a very thorough and rigorous discussion of stability factors. If you take the time to examine this reference, you will be convinced of the truth of the statement I am about to make. "The system of stability factors is too complex to be used easily in practical design situations without computer aid, or an impractical amount of human effort." Even with the aid of the computer, there is the problem of providing values for all the variables involved. Most of the variables will, in all practical situations, include considerable uncertainty about absolute values.

As a result of the complexity involved, *practical approximate* stability factor equations have been devised. Several of these stability factor equations yield similar results when properly applied. In this book we will be interested only in that stability factor which deals with circuit current gain for the control of variations in β and I_{co}.

$$\mathbf{S}_\beta = \frac{\Delta I_c}{\Delta \beta}$$

and

$$\mathbf{S}_{I_{co}} = \frac{\Delta I_c}{\Delta I_{co}}$$

We will not apply the label *stability factor* to our stabilization techniques for base-emitter variations. The equation we came up with in Sec. 4-4 is

$$E_{Re} = \frac{\Delta V_{BE} \times 100}{\%\Delta QI_c}$$

This arrangement suits our practical needs better than the corresponding stability factor, even though this equation is little more than an algebraic rearrangement of the base-emitter stability factor equation:

$$\mathbf{S}_{VBE} = \frac{\Delta I_c}{\Delta V_{BE}} \qquad \text{with all primary variables included}$$

In the case of β variations, an approximate stability factor is most convenient, because we must always assume very large variations from the value

[3] J. F. Pierce, *Transistor Circuit Theory and Design*, Charles E. Merrill Books, Inc., Columbus, Ohio, 1963.

stated in the transistor manual (because of large manufacturing tolerances and temperature-induced changes in β). In germanium devices where I_{co} is an important factor, we can take advantage of the approximate relationship $\mathbf{S} \approx I_g$, where I_g is the circuit current gain. We will examine I_{co} problems in more detail later in this chapter.

The literature gives several different versions of this stability factor, and it might be well to explain how this comes about. **S** is not a linear function; it is an S-shaped curve, and it is asymptotic to $\mathbf{S} = 1$ at one end and $\mathbf{S} = \beta + 1$ at the other.[4] In between these two points lie quite a number of slightly different practical equations for **S**. There are several points of special interest, and the equation used generally corresponds to one of these points. At the center of the curve the equation

$$\mathbf{S} = \frac{R_b}{R_e}$$

applies. In the lower half of the curve,

$$\mathbf{S} = 1 + \frac{R_b}{R_e}$$

is a good approximation.

A *general* equation would involve the use of the calculus and would apply to the entire curve, including those portions of it which represent unstable conditions. In order to avoid complex differential equations, designers have selected short linear (or nearly so) segments of the curve and have written simple algebraic equations for these short segments. Careful selection of certain short segments of the complex S-shaped stability curve results in simple equations with *very little* sacrifice in final circuit results. The upper half of the S-shaped stability curve is not used because it represents conditions of inadequate stability. The two equations ($\mathbf{S} = R_b/R_e$ and $\mathbf{S} = 1 + R_b/R_e$) are most often used because they contain no small-order terms and do represent conditions of satisfactory stability.

In the equation $\mathbf{S} = R_b/R_e$, R_b is the effective resistance from base to ground, and to be strictly accurate, R_{b1} (see Fig. 4-12) must be considered to be in parallel with R_{b2}. I have elected to use a compromise stability formula which ignores R_{b1}, which usually has only a very small numerical effect, and express the stability factor in terms of R_{b2} and R_e.

$$\mathbf{S} = \frac{R_{b2}}{R_e}$$

It is a sloppy compromise in the sense that its exact position on the curve varies, depending upon bias conditions, but it *always* lies somewhere

[4] M. V. Joyce and K. K. Clarke, *Transistor Circuit Analysis*, Addison-Wesley, Reading, Mass., 1961.

between the two commonly preferred values. The particular advantage of using the equation $\mathbf{S} = R_{b2}/R_e$ is that it not only yields values which lie between two acceptable values, but also (and equally important) facilitates the design procedure by not forcing us to predict R_{b1} in advance. The advance prediction of R_{b1} would be a most difficult chore, with no more purpose than the satisfaction of knowing that we are operating at one of two *equally* acceptable limits, and not somewhere in between.

Table 4-4 tabulates the effect on the quiescent collector current of a 100 percent change in β. The table also compares the **S** value with the circuit current gain. As you can see, the correlation between **S** and the circuit current gain is only approximate; but the assumption that they are approximately equal is a very useful one, particularly for predicting the effect of I_{co} on the collector current.

For now let us concentrate on the lefthand half of the table. The first column, headed **S**, lists stability factors based on the approximate equation:

$$\mathbf{S} = \frac{R_{b2}}{R_e}$$

(See Fig. 4-13.) The next column, headed $\beta = 50 \rightarrow \beta = 100$, lists the percent change in the established operating collector current when β increases from 50 to 100 for selected values of **S**. The third column, headed $\beta = 100 \rightarrow \beta = 200$, lists the percent change in the quiescent collector current (for selected values of **S**) when β increases from 100 to 200.

If you will examine and compare these two columns (second and third), you will observe that there is a somewhat smaller variation in the quiescent collector current when β changes from 100 to 200 than when it changes from 50 to 100. This means that a higher β transistor will yield a smaller quiescent collector-current shift for a given stability factor and temperature rise.

Table 4-4 demonstrates that the lowering of the *circuit* current gain makes the circuit relatively independent of actual β values (for low-stability numbers). Not only does this allow the circuit to operate over a wide temperature range, but it also allows the circuit to function properly in spite of the very large manufacturing tolerances in the β value. Any circuit which is well stabilized, with a stability factor of 10 or less and adequate V_{BE} stability, will perform properly with any transistor (of the same type number) which has a β as large or larger than the value used in designing the circuit.

It is not even necessary to use the same type number as long as no maximum ratings are exceeded and the substitute's β is as high as, or higher than, the figure the designer used in computing the circuit. My laboratory assistant put this to the test. He designed a circuit for a 2N2924 transistor with a β of 100, breadboarded it, and measured all the circuit parameters. Then without making any changes in the circuit, he removed the original transistor and replaced it with a quite different (compound) transistor with a β of 2,400. He

TABLE 4-4. PERCENTAGE CHANGE IN QI_c WITH A 100 PERCENT CHANGE IN β FOR VARIOUS STABILITY FACTORS

	% ΔQI_c		Circuit Current Gain		
S	$\beta = 50 \rightarrow \beta = 100$	$\beta = 100 \rightarrow \beta = 200$	$\beta = 50$	$\beta = 100$	$\beta = 200$
50	25	16.6	25	33	40
20	14.3	8.3	14	16.6	18.2
10	8.3	4.5	8.3	9.0	9.5
9	7.6	4.1	7.6	8.2	8.6
8	6.9	3.7	6.9	7.4	7.6
7	6.1	3.3	6.1	6.5	6.8
6	5.4	2.8	5.4	5.6	5.8
5	4.5	2.4	4.5	4.7	4.9
4	3.7	1.9	3.7	3.8	3.9
3	2.8	1.4	2.83	2.91	2.95
2	1.9	0.98	1.92	1.96	1.98
1	0.98	0.49	0.98	0.99	1

$\mathbf{S} = \frac{R_b}{R_e}$

then remeasured all the circuit parameters, and not one of them had changed by more than 8 percent.

Now let us return to Table 4-4. It is a good practice to assume the worst when it comes to β variations, because we have three distinct problems involved in determining its value.

1. There are large manufacturing tolerances.
2. β varies with collector current in a generally unpredictable fashion.
3. β increases at a general rate of from 0.5 percent/C° to 2 percent/C°.

The total amount of uncertainty involved is enough to make it almost impossible to predict the actual β at a given temperature. And so if we must err, it is better to err in the direction of more stability than we need rather than less.

If you will examine Table 4-4 again, you will see that the $\%\Delta QI_c$ is generally less than **S**. We have, for several years, used an arbitrary rule which has yielded very good practical results, even though (as you can see by the table) it is just a made-up one. As a rough but practical estimate of $\%\Delta QI_c$ due to β variations, we can say that it will always be approximately **S** percent.

$\%\Delta QI_c \approx$ **S** percent

Remember that this is not really an equation, it is simply a rule of thumb that does a good job of making a practical accounting for the three uncertainties in β prediction.

WHAT TO DO ABOUT I_{co} VARIATIONS. Before we get under way here, let me take a moment to put this temperature business into perspective. Most of the tables concerned with temperature in this book go up to 100°C, and this is a most uncomfortable temperature indeed; water boils at 100°C, and so does blood. 50°C is approximately 122°F, and so human survival is a bit iffy at a mere 50°C. 60°C (140°F) is quite probably well beyond human endurance—it is certainly beyond mine. And, of course, most transistorized equipment is operated at temperatures that are tolerable to human beings. That takes us up to a little less than 60°C, and beyond that temperature we must consider the environment to be decidedly hostile. (See Table 4-5.)

Now let us go back to Table 4-4 and look at the right half of the table. Although the relationship is far from precise, it is a good approximation to say that the circuit current gain is approximately equal to **S**.

$$I_g \approx \mathbf{S}$$

where **S** is the stability factor and I_g is the circuit current gain. The leakage current I_{co} will be amplified by the *circuit current* gain, which is approximately **S**.

For example, a silicon transistor is biased to a QI_c of 1.0 mA. At a temperature of 20°C, $I_{co} = 0.1\ \mu A$, $\mathbf{S} = 10$. What is the total collector current including I_{co}?

1. The I_{co} will appear in the collector circuit as $\mathbf{S} \times I_{co}$.
   $\mathbf{S} \times I_{co} = I_c$
   $10 \times 0.1 = 1\ \mu A$
2. This 1 μA will add to the 1 mA of quiescent collector current, but adding 1 μA to 1 mA makes no significant change in the Q point.

That example was at room temperature; now let us try the same example with the temperature raised to 50°C.

TABLE 4-5. CENTIGRADE TO FAHRENHEIT CONVERSION TABLE

Centigrade	Fahrenheit	Centigrade	Fahrenheit
10°	50°	55°	131°
15°	59°	60°	140°
20°	68°	65°	149°
25°	77°	70°	158°
30°	86°	75°	167°
35°	95°	80°	176°
40°	104°	85°	185°
45°	113°	90°	194°
50°	122°	95°	203°
		100°	212°

TABLE 4-6. LEAKAGE CURRENT (I_{co}) TEMPERATURE FACTORS

Temperature, C°	Silicon I_{co} Factors	Germanium I_{co} Factors
25–30	3	2
30–40	9	4
40–50	27	8
50–60	81	16
60–70	243	32
70–80	729	64
80–90	2187	128
90–100	6561	256

Table 4-6 provides the necessary factors by which to multiply the 20° I_{co} to get the total amount of I_{co} at higher temperatures. If we assume this is a silicon device, we find that the new I_{co} at 50° can be found by multiplying the 20° I_{co} by the factor 27. Now our 20° I_{co} (from the last example) is 0.1 μA.

1. $27 \times I_{co}$ at 20°
   $27 \times 0.1\ \mu A = 2.7\ \mu A$
2. This 2.7 μA I_c will appear in the collector circuit as $\mathbf{S} \times 2.7\ \mu A$ ($\mathbf{S} \times I_{co}$).
   $\mathbf{S} = 10$
   $I_{co} = 2.7\ \mu A$

   The collector current due to I_{co} is approximately equal to 27 μA.

This 27 μA added to the 1 mA of QI_c still causes an insignificant increase in the quiescent collector current, from 1.0 mA to 1.027 mA. Some manufacturers do not give room temperature I_{co}, but give a rating at 100° or some other high temperature. I_{co} in modern low-power silicon transistors is simply not much of a problem at temperatures which are not hostile to humans. Even many of the modern germanium devices have leakage currents so low as to be no problem.

4-7 VOLTAGE-MODE FEEDBACK

There is another method for obtaining feedback stabilization where the feedback voltage is taken as the voltage drop across R_L, instead of across an emitter resistor. The circuit is shown in Fig. 4-14. The stability obtainable with voltage-mode feedback is every bit as good as the stability that can be had with current-mode feedback, but the configuration has one inherent characteristic that is usually a serious problem. The circuit has an inherently low input impedance. Not only is there no emitter resistor to elevate the input

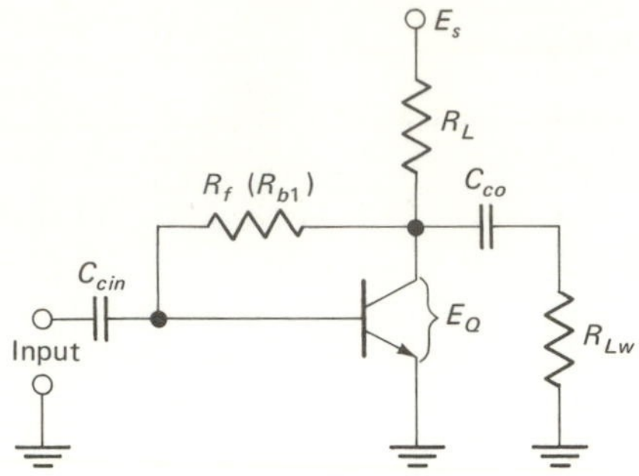

Fig. 4-14 The voltage-mode feedback circuit

impedance, but there is a phenomenon, known as the Miller effect, which makes R_{b1} look like

$$\frac{R_{b1}}{V_g}$$

to the generator. The simple circuit as it is shown in Fig. 4-14 has an additional limitation in that good stability is not possible unless E_Q is small compared to E_{RL}. The stability factor is $\mathbf{S} \approx R_{b1}/R_L$ (see Fig. 4-14). Lowering R_{b1} yields a larger base current and reduces E_Q. Increasing R_L increases the required supply voltage. A reduction in E_Q means a smaller signal output voltage. In the voltage-mode circuit, however, unlike the current-mode circuit, saturation (and its attendant clipping) is a sliding gradual thing, and operation within this sliding quasi-saturation region is permissible and practical. I use the term *quasi-saturation* because the circuit does not actually get biased into saturation until R_{b1} approaches 0 Ω. There are many circuits where only a small output voltage is required and a low value for E_Q is no handicap.

The circuit in Fig. 4-14 is very attractive because of its obvious simplicity, but unfortunately by the time we have added the necessary components to minimize the Miller effect and increase the output swing, the total complexity of the circuit is as great as that of the most complex current-mode circuit.

In addition, it is generally easier to "design in" all the required parameters when current-mode feedback is used. In practice the voltage-mode feedback circuit should be used only when the circuit's basic simplicity can be taken advantage of. This means that an input impedance of (often) less than 500 Ω and signal output voltages of 2 V p-p (with a 10-V supply) must be acceptable. In a later chapter I will show you how to design the basic circuit as well as more complex versions with more desirable properties. Some experienced designers will certainly argue that the voltage-mode feedback circuit can do anything that the current-mode circuit can do, and they are right, but only with more effort.

Stabilizing against Variations in V_{BE}

The mechanism for stabilizing against V_{BE} variations using voltage-mode feedback is exactly the same as that used with current-mode feedback. The only difference is the source of the swamping voltage. In the current-mode feedback case this swamping voltage consists of the voltage drop across R_e, which is in series with the base-emitter circuit, and in the voltage-mode case the bias resistor R_{b1} has been connected to the collector so that R_L is in series with the base-emitter circuit.

Suppose we draw an equivalent circuit for the base-emitter loop with both modes of feedback and analyze the loop as we did for the current-mode circuit. The drawing is shown in Fig. 4-15. To begin the analysis, we must arrange to account for the voltage drop across R_L due to the collector current, because R_L has both base-current and collector-current components flowing through it. Rather than including I_c in our figures, let us simply *simulate* the conditions by "mathematically" elevating R_L by a factor of β. Insofar as the voltage drop across R_L is concerned, it doesn't matter whether β times as much current flows through it or it has β times as many ohms. Then from Ohm's law we can write

$$I_b = \frac{V_{BE}}{R_b + \beta R_L + \beta R_e} \tag{4-33}$$

As in the case of current-mode feedback, β stability (and I_{co} stability) depends upon R_b being small compared to the value of the feedback impedance. In this case R_b must be small compared to $\beta \times R_L$. We can therefore drop R_b from the denominator of Eq. (4-33):

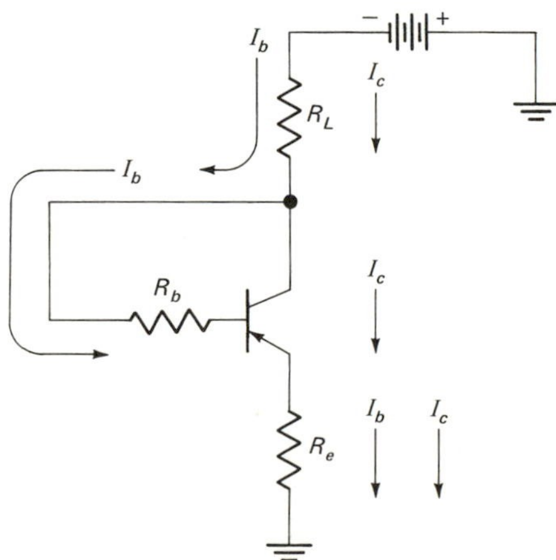

Fig. 4-15 Circuit for combined voltage-mode and current-mode loop analysis

$$I_b = \frac{V_{BE}}{R_b + \beta R_L + \beta R_e} \tag{4-34}$$

Since $I_c = \beta \times I_b$, we can assume that $\Delta I_c = \beta \times \Delta I_b$. If we multiply Eq. (4-34) by β, we get

$$\beta \Delta I_b = \beta \frac{\Delta V_{BE}}{\beta R_L + \beta R_e} \tag{4-35}$$

$$\Delta I_c = \frac{\Delta V_{BE}}{R_L + R_e}$$

Because we are interested in the fractional variation in QI_c, let us divide both sides of Eq. (4-35) by QI_c:

$$\frac{\Delta I_c}{QI_c} = \frac{\Delta V_{BE}}{QI_c(R_L + R_e)} \tag{4-36}$$

Because the emitter current and the collector current are approximately equal, we can write Eq. (4-36) in terms of voltage drops across R_L and R_e.

$$\frac{\Delta I_c}{QI_c} = \frac{\Delta V_{BE}}{I_c R_L + I_e R_e}$$

$$\frac{\Delta I_c}{QI_c} = \frac{\Delta V_{BE}}{E_{RL} + E_{Re}} \tag{4-37}$$

Now, if we express Eq. (4-37) in a percentage form, we get

$$\%\Delta QI_c = \frac{\Delta V_{BE} \times 100}{E_{RL} + E_{Re}} \tag{4-38}$$

Equation (4-38) says that it makes no difference whether we derive the swamping voltage from R_L, R_e, or a combination of the two. A combination of current-mode and voltage-mode feedback is perfectly possible, but it has never been a popular combination, probably because it offers no combination of circuit parameters which cannot be more easily obtained with either mode alone. The equation that we developed here, Eq. (4-38), is simply a general case of the one we developed in Sec. 4-4 for the current-mode circuit Eq. (4-29).

Stabilizing against Variations in β and I_{co} with Voltage-mode Feedback

In the case of current-mode feedback circuits we purposely lowered the *circuit* current gain to minimize the effect of changes in β and I_{co} upon the quiescent collector current. We accomplished the task of lowering the circuit current gain by using a low-value resistance in parallel with the circuit input resistance.

For voltage-mode feedback circuits we also lower the current gain of the circuit by providing a low impedance path to ground in shunt with the transistor input resistance, but we do it a little differently. If you remember, the transistor's current gain (β) made the value of an emitter resistor appear to be β times as large as its actual ohmic value. In the voltage-mode feedback case we return one end of the base bias resistor to the collector. This makes the base bias resistor (looking into the base) seem to be *smaller* by the factor V_g (voltage gain). As far as I know, the phenomenon of current gain elevating the effective value of the emitter resistor has no special name. The opposite case, in voltage-mode feedback, where the resistance between base and collector is effectively reduced by the factor V_g, has had a long history in vacuum-tube technology and is known as the Miller effect.

I had hoped to duck the chore of explaining the Miller effect to you by giving you some references, but it didn't work out. All the references I consulted either provided other references or used a strictly mathematical approach with little or no explanation. The best of the mathematical discussions that I found discussed the Miller effect as it applies to its effect on tuned circuits in vacuum-tube technology. Because of the reactive components involved, the equations are a bit hairy, and it would be difficult to apply them to the comparatively simple case we have to deal with here. Later on we will have to deal with these more complicated situations, but you will be far better prepared if you first gain an understanding of the Miller effect in its simplest form.

The Miller Effect

The circuit in Fig. 4-16*a* shows the simple voltage-mode feedback bias circuit, and Fig. 4-16*b* is a model to be used in the following discussion. I have replaced R_L, E_s, and *Q*1 by a variable battery E_c mechanically linked to E_b in such a way that any change in the voltage of E_b will cause 100 times as large a change in E_c. E_b simulates the input voltage (frequently called E_{in}), and E_c represents the output voltage (generally called E_o).

The 100 to 1 mechanical linkage simulates the voltage gain: E_o/E_{in}, or in our example E_c/E_b. Now, before you react to my mechanical-electrical model, let me state that it is a perfectly legitimate amplifier, as is the wall switch that you use every time you turn on the lights. Little use is made of the kind of hybrid amplifier shown in Fig. 4-16*b* simply because there are very few practical applications for such a scheme. For our purposes it is an excellent model because it leaves us with only those electrical elements involved in our Miller effect discussion, avoiding the confusion of having elements in the circuit which are only incidental to the discussion.

Now let us use a numerical example to demonstrate how the Miller effect works.

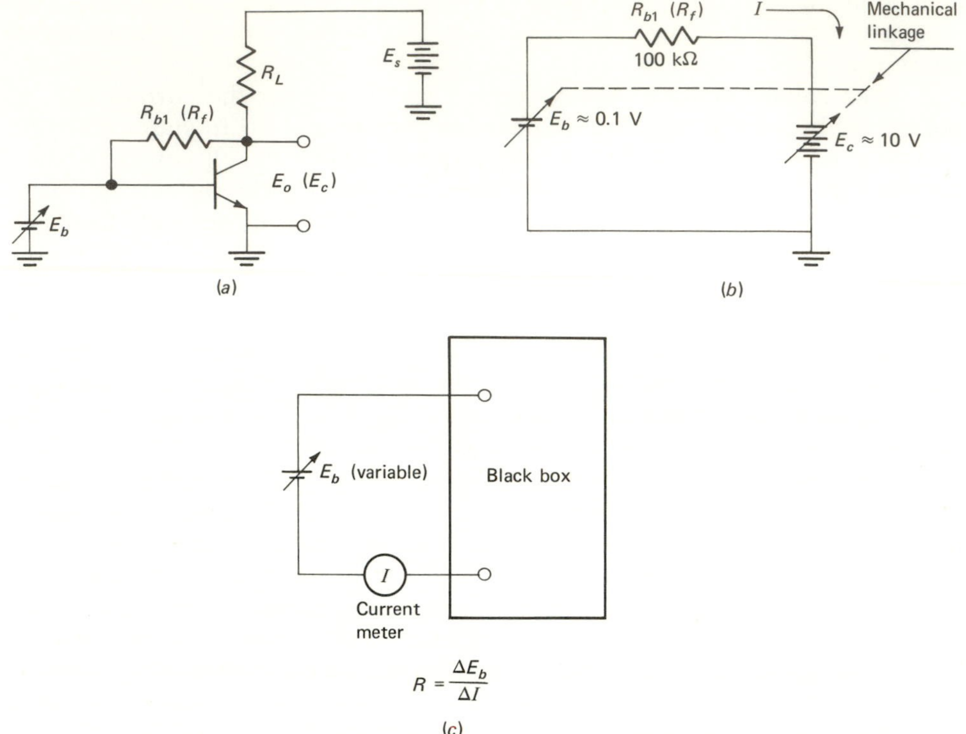

Fig. 4-16 The Miller effect explanation. (a) *The voltage-mode feedback circuit;* (b) *the circuit model for the Miller effect explanation;* (c) *the circuit for measuring the dynamic resistance of a black box*

Example 4-2

For this example we shall refer to Fig. 4-16*b* and make the following assumptions for the purpose of keeping things simple:

1. The voltage gain is to be $V_g = 100$.
2. Any change in the voltage E_b causes E_c to change by 100 times that change in E_b. This is another way of saying that the voltage gain of our model is 100. Some variations:

$$V_g = 100 \qquad V_g = \frac{E_c}{E_b} \qquad 100 = \frac{E_c}{E_b} \qquad E_c = 100E_b$$

3. Now because E_b is only 1 percent of E_c, and any change in E_b will amount to only 1 percent of the change in E_c, we will leave E_b (and ΔE_b) out of computations concerned with the current through R_{b1}. The 1 percent error is negligible, whereas the amount of arithmetic required to include it is not.
4. We will assume that if

$$V_g = \frac{E_c}{E_b}$$

then

$$V_g = \frac{\Delta E_c}{\Delta E_b}$$

I can't really explain anything here without letting the variables change and observing the results. Look at Fig. 4-16*a*. The Miller effect makes R_{b1} appear smaller (by the factor V_g) to the generator which in our case is E_b, a variable-voltage dc source.

Now, let us suppose we have the source E_b and a black box whose dynamic resistance is to be measured. (See Fig. 4-16*c*.) How would we go about it? We would vary E_b and call that variation ΔE_b. That variation in E_b would cause some variation in the circuit current. We would record that change as ΔI. Then using a little Ohm's law, we would compute the effective dynamic resistance.

$$R = \frac{\Delta E}{\Delta I}$$

In our case we are looking for the effective value of R_{b1}.

Suppose we go through this procedure with our model in Fig. 4-16*b*. Suppose we let E_b change by 0.01 V. See Fig. 4-16*b*.

1. If E_b changes by 0.01 V and the voltage gain (V_g) is 100, then $\Delta E_c = V_g \Delta E_b = 100 \times 0.01 = 1$ V. The total change in circuit voltage is 1.01 V, but we won't include the 0.01 V in our calculations.

   $\Delta E_c = 1$ V

2. Now,

   $$I_{Rb1} = \frac{\Delta E_c}{R_{b1}} \qquad \text{or} \qquad I_{Rb1} = \frac{1 \text{ V}}{100 \text{ k}\Omega}$$

   $I_{Rb1} = 1 \times 10^{-5}$ A

So what? Well, we changed E_b by 0.01 V and the circuit current changed by 1×10^{-5} A. This is what actually happened. It happened in our paper model and it would have happened had we constructed the actual circuit; so we know the value ΔE_b because we chose it, and we know what the change in current is with the effect of the voltage gain included. We know ΔE_b and ΔI and can solve for the effective total circuit resistance. If we assume that the internal resistance of the two sources E_b and E_c is negligible, then what we are really solving for is the effective value of R_{b1}.

3. Solving for the effective value of R_{b1}, we get

   $$R_{b1} \text{ (eff)} = \frac{\Delta E_b}{\Delta I} = \frac{1 \times 10^{-2} \text{ V}}{1 \times 10^{-5} \text{ A}} = 10^3 \ \Omega$$

What has happened here is that the voltage gain, coupled with the way in which R_{b1} is connected, lowers the effective value of R_{b1} from 100 kΩ to 1 kΩ. The value of

R_{b1} was effectively lowered by a factor of 100. It is no coincidence that *that* factor and the voltage gain are the same. This is the basic definition of the Miller effect: The effective value of R_{b1} as seen by the source E_b is equal to its ohmic value divided by the voltage gain.
The Miller effect:

$$R_{b1}\ \text{(effective)}\ \frac{R_{b1}\ \text{(ohmic)}}{V_g}$$

Although this example doesn't prove anything in a strict mathematical sense, a careful examination of the model and a little thought will convince you that the example is no fluke. It is also quite easy to prove the Miller effect by using the model and simple loop equations, but I leave that up to you, if you feel like trying it. It grows dark early here in my mountain campground at this wintry time of year, and so I must get along to the next section.

THE STABILITY FACTOR AND THE MILLER EFFECT. If it were not for the Miller effect, voltage-mode feedback would not be very satisfactory. Voltage-mode feedback would provide excellent stability insofar as V_{BE} problems are concerned, but the value of R_{b1} dictated by bias current requirements would be so high that the current gain of the circuit would be almost equal to β. We would find ourselves at the not at all tender mercies of the illusive β.

As it stands, voltage-mode feedback comes out second best, in all but a few applications, when it is compared to the current-mode circuit. The applications where its use is often dictated are those which require the circuit to operate at very low collector voltages or when the absolute maximum signal output voltage for a given collector-to-emitter voltage is required. The latter is often a very important factor in power amplifier driver circuits. Voltage-mode feedback very often yields unsatisfactorily low input impedances, and this rules out its use in a great many cases.

Some designers of radio-frequency circuits use transformer coupling most of the time; at those frequencies transformers are practical and inexpensive, low input impedance is no problem, and consequently they may be more prone to use voltage-mode feedback than most other designers.

The stability factor for the voltage-mode feedback circuits is still based upon the ratio of input resistance to *effective* base-to-ground resistance. The difference in the two cases is the Miller effect.

This stability factor is, as usual, an approximation, but a very satisfactory one. The premise is: if $\mathbf{S} = R_b'/R_e'$, then

$$\mathbf{S} \approx \frac{R_{b1}}{R_L}$$

where R_b' is the *effective* base-to-ground resistance, R_e' is the effective emitter resistance, and R_L is the collector load resistance across which the feedback voltage is developed. See Fig. 4-16.

The stability factor that we developed for current-mode feedback circuits involved the ratio of the emitter-to-ground resistance to the effective base-to-ground resistance, $\mathbf{S} = R_b/R_e$. This has the effect of lowering the circuit current gain.

1. $\mathbf{S} = \dfrac{R_b'}{R_e'}$ (4-39)

2. Because of the Miller effect the effective value of R_b is the actual *resistor* R_{b1} divided by the voltage gain.

$$R_b' = \frac{R_{b1}}{V_g}$$

where $V_g = R_L/R_e'$ (the voltage gain equation developed in Chap. 3).

3. Writing the stability factor equation, including the Miller effect, we get

$$\mathbf{S} \approx \frac{R_{b1}}{V_g} \times \frac{1}{R_e'}$$

Substituting R_L/R_e' for V_g ($V_g = R_L/R_e'$), we get

$$\mathbf{S} \approx \frac{R_{b1}}{1} \times \frac{1}{R_L/R_e'} \times \frac{1}{R_e'}$$

$$\mathbf{S} \approx \frac{R_{b1}}{1} \times \frac{R_e'}{R_L} \times \frac{1}{R_e'}$$

If we cancel the R_e' which appears in both numerator and denominator, we get

$$\mathbf{S} \approx \frac{R_{b1}}{R_L}$$

This is the equation we will use for the stability factor for voltage mode feedback circuits from here on out.

The *circuit* current gain again is approximately equal to **S**, the stability factor. Let me illustrate by an example.

Example (See Fig. 4-17)

$$\mathbf{S} \approx \frac{R_b'}{R_e'} \approx \frac{R_L}{R_e'}$$

1. $R_e' = \dfrac{25}{I_c}$ (the Shockley relationship)

$$R_e' = \frac{25}{1\ \text{mA}} = 25\ \Omega$$

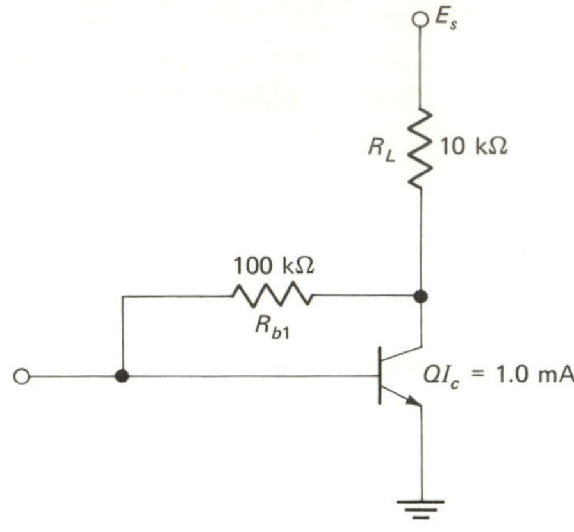

Fig. 4-17 The voltage-mode stability factor example

2. $V_g = \dfrac{R_L}{R_e'} = \dfrac{10 \text{ k}\Omega}{25\ \Omega} = 400$ (voltage gain)

3. $R_b' = \dfrac{R_{b1}}{V_g} = \dfrac{100 \text{ k}\Omega}{400} = 0.25 \text{ k}\Omega = 250\ \Omega$

4. $\mathbf{S} \approx \dfrac{R_b'}{R_e'}$ (derived for the current-mode circuit)

$$\mathbf{S} \approx \frac{250}{25} = 10$$

5. $\mathbf{S} \approx \dfrac{R_{b1}}{R_L} = \dfrac{100 \text{ k}\Omega}{10 \text{ k}\Omega} = 10$

4-8 NONLINEAR COMPENSATION—AN ALTERNATIVE SCHEME

During the early evolution of transistor technology considerable emphasis was placed on nonlinear compensation techniques, where diodes, thermistors, transistor junctions, and silicon resistors were used to steady the Q points in transistor circuits. The purpose of these techniques was to avoid the inevitable loss of gain involved in feedback stabilization methods. In general, these nonlinear stabilization techniques created more problems than they solved, but there are a few which are still useful in certain special cases, particularly in integrated circuitry where good thermal coupling is automatic.

All these techniques involve the addition of a component (or components) in the circuit which tends to reduce the bias current as temperature increases try to raise it. In order for such compensation to work well, near-perfect tracking (in a complex system of variables) between the transistor and the compensating device is required. This is very difficult to achieve in practice, because of large manufacturing tolerances in mass-produced semiconductors, and because of the need for intimate thermal coupling between the device being compensated and the compensating device. In integrated circuitry, closer tolerances are possible among junctions, on a given chip, and fairly good thermal coupling is almost inevitable, so that nonlinear compensation is much more practical. Even in integrated circuitry nonlinear compensation (by itself) is used in only about three forms. Two of these forms involve a working transistor as a resistor with a high dynamic collector resistance, and these we will take up a bit later. The third important form, which we will discuss here, is diode compensation. There are several variations on the basic idea, but we will discuss only the most common one. The circuit is shown in Fig. 4-18. The diode is forward-biased and takes the place of R_{b2} in the voltage divider base-biased circuit. In this case, however, $D1$ is not fixed and its junction voltage varies with temperature in the same fashion as the base-emitter junction voltage of the transistor. Often $D1$ is the base-emitter junction of a transistor of the same type number, to ensure the best possible tracking.

As the temperature increases, V_{BE} of the transistor decreases, allowing a larger base-emitter current to flow. However, at the same time, the junction voltage of $D1$ decreases, drawing additional current through R_{b1}, which reduces the *bias* voltage available to drive current through the base-emitter junction. If these two junction voltages *track exactly*, there will be no change in base-emitter current and, therefore, no change in the quiescent collector current due to changes in V_{BE}. Of course, that is only an ideal situation and never really happens, but it can be a very satisfactory approach in certain cases.

That is how it works; now let's look at some of the difficulties. If we are to expect nearly perfect tracking, we should have about the same current, at quiescence, through both junctions ($D1$ and the base-emitter junction). The impedance of the transistor base-emitter circuit is approximately $\beta \times R_e'$

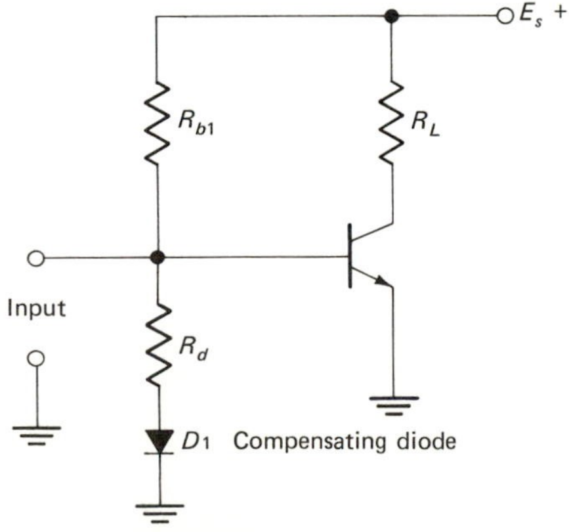

Fig. 4-18 Diode compensation

(because the transistor is an amplifier), while the diode impedance is $25/I_d$, where I_d is the diode junction current, in milliamperes. The result is that about as much input current flows through $D1$ as flows through the base-emitter junction. In short, the compensating diode in Fig. 4-18 yields good, but not perfect, compensation whether we use the added resistor R_d or not. If we do not use the added resistor, the diode's low resistance will shunt most of the input signal current to ground, resulting in a very low current gain, a very low input impedance, and, at the same time, a very low value for **S** (**S** $\approx$ 1). If we are going to put a signal into the base, a resistance of approximately 10 (**S**) times R'_E is in order. This is because, once we get well beyond the hook voltage, R'_e becomes *almost* linear. Thus, by adding a resistor in series with the diode $D1$, we can increase the input impedance to a more or less reasonable value. The resistor in series with $D1$ cannot exceed **S** $\times R_e$ (R_e is the effective emitter resistance) without yielding an unsatisfactory stability factor (for $\Delta\beta$ and ΔI_{co}).

There are some circuits, however, which use a properly biased transistor that has no input signal at all; in that case, the resistor is often left out.

These circuits, which have no input signal, are used as stabilization for circuits which take advantage of the high dynamic collector-base junction resistance to use a transistor in place of an emitter resistor or a base bias resistor. We will examine those two cases later.

SUMMARY

Bias

For normal operation the transistor must have its base-emitter junction forward-biased and its collector-base junction reverse-biased, but in addition some specific *appropriate* amount of base-emitter current must be flowing under zero signal conditions. This base bias current results in β times as large a collector current. This steady-state collector current is called the quiescent or Q-point current.

Stability

For a circuit to perform to specification, the quiescent current must be stable. There are three parameters which vary widely with temperature. They are β, V_{BE}, and I_{co}. Variations in any of these three parameters cause the Q point to change.

Feedback Stabilization

Stabilization of the quiescent collector current is most often accomplished through the use of negative feedback. There are two important modes of negative feedback in transistor circuits: current mode and voltage mode. Both of these feedback modes tend to reduce distortion; quiescent collector-current drift is a form of distortion.

Simpler Equations

Although negative feedback is what actually stabilizes the quiescent point, it is often more convenient to view the system in terms of simple algebraic equations (loop equations mostly), especially for design and analysis purposes. When we use these simpler equations, some of the dynamic character of the feedback system is inevitably lost, but for practical analysis and design this loss is no problem, as long as we set realistic limits on system operation and as long as we understand that our simple equations are essentially static. We must keep in mind that our simple loop equations are valid only within practical limits, and that we use them, in most cases, instead of the dynamic feedback equations because the feedback equations are generally too complex to be practical. Simple Kirchhoff loop analyses are adequate for most design and circuit analysis purposes, if we stay within realistic limits. These limits are not usually easy to define (part of this book will be devoted to quite precise statements of those working limits, in the form of *tables of possibilities*). There are instances where only feedback equations will do. For example, the voltage gain equation,

$$V_g = \frac{R_L}{R_e}$$

Although for practical purposes it seems to be no more than a very simple algebraic equation, it is actually a feedback equation, and there is really no other acceptable way of deriving it except through feedback theory. The point is that even though we can design and analyze circuits with simple loop equations, if, *and only if*, we assign workable limits to these equations, we must realize that it takes both loop equations *and* practical *limits* to approximate adequately what are essentially feedback equations.

Stabilizing V_{BE}

Basically we stabilize base-emitter voltage variations by providing a large feedback voltage drop which *swamps* temperature-induced variations in V_{BE}. In the case of current-mode feedback this swamping voltage is developed across an external emitter resistor. In the case of voltage-mode feedback the swamping voltage is developed across the collector load resistor R_L. It is not important from where this swamping voltage drop is derived; it can be an emitter-resistor voltage drop, a collector load resistor voltage drop, or a combination of the two. What is important is that this swamping voltage is large compared to the temperature-induced variations in V_{BE}.

Stabilizing β and I_{co}

The cure for both β and I_{co} variations is simply to lower the *circuit* current gain to 10 or *less*, no matter what the β of the given transistor is. In the case of current-mode feedback, we lower the *circuit* current gain by providing a shunt path (through R_{b2}, in Fig. 4-12) to ground so that only a *part* of the signal current flows through the base-emitter junction and gets amplified.

In the case of voltage-mode feedback, we depend upon the Miller effect to provide a shunt path so that only part of the input current gets amplified.

In both cases we have effectively reduced the current gain of the circuit, which minimizes the effect of varying β values and reduces the effect of I_{co} on collector current.

Diode Stabilization

Diode stabilization depends upon the fact that the compensating diode behaves much like the transistor base-emitter junction with respect to temperature variations. Nonlinear compensation is not so frequently used as feedback methods, but it does have its place.

PROBLEMS

A transistor amplifier is being designed with an external emitter resistor to limit variations in V_{BE}. Given the following specifications, compute the required voltage drop across the emitter resistor.

4-1 $QI_c = 1$ mA; $\%\Delta QI_c$ to be allowed = 5 percent at 75°C; the reference temperature is 20°C. Compute E_{Re}.

4-2 $QI_c = 10$ mA; $\%\Delta QI_c$ to be allowed = 10 percent at 50°C; the reference temperature is 20°C. Compute E_{Re}.

4-3 A transistor amplifier has a stability factor of 10. The quiescent collector current at 20°C is 1 mA. If the leakage current increases from 5 μA at 20°C to 15 μA at 30°C, what is the quiescent collector current at 30°C?

4-4 What is the quiescent collector current of a transistor amplifier with a stability factor of 5 when β increases from 50 to 100 owing to a temperature rise? The initial collector current is 2 mA. (You may refer to Table 4-4.)

4-5 In a circuit which has a stability factor of 10, a transistor with a β of 100 is replaced by a transistor with a β of 200. What is the new quiescent collector current if it was 1.0 mA with the old transistor (β of 100)? You may consult the appropriate table (Table 4-4).

4-6 Given a current-mode feedback circuit (see Fig. 4-4), compute the value of **S** when $R_{b2} = 10$ kΩ and $R_e = 1$ kΩ.

4-7 Given a voltage-mode feedback circuit (see Fig. 4-14), compute the value of **S** when $R_{b1} = 100$ kΩ and $R_L = 10$ kΩ.

4-8 What is the approximate *circuit* current gain of the circuit in Prob. 4-6?

4-9 What is the approximate *circuit* current gain of the circuit in Prob. 4-7?

4-10 Complete the following table.

Open-loop Voltage Gain, V_g	Feedback Factor, F_f	V' Voltage Gain with Feedback Computed as: $V'_g = \frac{V_g}{1 - V_g F_f}$	Computed as: $V'_g = \frac{1}{F_f}$
25	0.1		
1,000	0.1		
500	0.1		
500	0.15		
500	0.20		
40	0.20		
80	0.20		
60	0.25		
75	0.25		

book two

techniques

5

practical design procedures for amplifiers with current-mode feedback

OBJECTIVES:

(Things you should be able to do upon completion of this chapter)

1. Design the basic current-mode-feedback common-emitter amplifier circuit using the table of possibilities (Table 5-5). (See the specific design problems in the chapter.)

   NOTE: An exercise in using the table of possibilities is included in the chapter.

EVALUATION:

The *final* evaluation will be based upon your ability to design circuits *that work* in the laboratory. See the laboratory manual for details.

INTRODUCTION

This chapter is probably the most important one in the book, for it explains the design procedure in considerable detail and lays the groundwork for all subsequent design and analysis procedures. This chapter is likely to be as difficult as it is important, because there is so much detail. Not only must all the details be learned, but some overall picture of the order and substance of these details must also be formed.

The design process consists of a *definite* sequence of calculations, and order is highly important. Because of the basic difficulty in mastering so much related detail, I will take time to preview the chapter for you so that you will know better what to expect. I will also make a flowchart (Table 5-2) of the entire design process, so that you can see at a glance where things are going and what the all-important order of operations is. Whenever you lose sight of the overall picture in the maze of individual steps, you can refer to the flowchart and regain your bearings. The flowchart is fairly detailed and contains marginal references to sections, figures, etc., which discuss the factors involved in each operation in greater detail.

Because the design procedure consists of a number of well-defined steps in a definite order, I have developed standardized design sheets to help make the learning, and particularly its applications, easier. These design sheets provide a standardized structure that makes cross-referencing among explanations, examples, and problems easier. I will explain the mechanics of this and some other aids later in this introduction.

The beginner always has one fundamental handicap—his lack of experience. To help overcome that problem I have developed some tables of "condensed experience," which I call *tables of possibilities*. These tables will make it possible for you to make intelligent decisions in some most important areas without the benefit of long experience. You will encounter the first table of possibilities in this chapter. Beyond this extra effort on my part in this introduction, it is up to you to struggle, step by step, through each worked example and explanation. Once you have mastered this chapter, the rest of the book is a cinch.

THEORY AND PHILOSOPHY

In this and subsequent chapters, we shall be considering practical design and analysis procedures. The simplicity of the procedures does not really represent a lack of rigor. The procedure depends upon negative feedback, where any deviation from specified operating conditions is sensed and corrected. This correction is quite easily accomplished in several ways, using but few inexpensive components. Although such compensation is not obtained without cost in both components and gain, it must be had, whatever the cost. There are several reasons for including negative feedback in the basic design of even single stages.

First, transistor parameters are difficult to measure and even more difficult to control in production. As a result, tolerances are much larger than can be accommodated within the framework of traditional device-oriented methods of design. And if mass production of transistor circuitry is to be a reality, transistors of the same type must be interchangeable. Such broad tolerances make compensation essential. Secondly, there is little agreement among manufacturers as to how basic parameters should be measured. This is largely because the transistor departs so far from the *ideal* amplifier. The gap between laboratory behavior of the device and its behavior in actual circuits is so great and so variable that the most precise laboratory measurements become only convenient approximations for the circuit designer.

In addition to the broad tolerances, there is the problem of the semiconductor's extreme temperature sensitivity and the variability of basic parameters at various operating currents. Even though closer tolerances may be possible with time and experience, temperature-sensitive parameter variability is an inherent quality of semiconductor devices and is a fact of life with which the designer must live.

Fortunately, proper application of negative feedback can compensate for temperature- and current-induced variations and also make interchangeability possible. The purpose of using negative feedback is simply to minimize the effects of all those parameters that are not thoroughly under the designer's control. These include most of the transistor's fundamental parameters. And if we use feedback properly, we shall find ourselves losing interest in most of the transistor's published parameters. This is the reason that, for the purpose of this text, we shall not devote much space to those overemphasized parameters. We shall become very independent of specific transistor parameters, and this is the essence of good circuit design. We shall find that, having designed a circuit, and having done it properly, we can plug in any transistor which meets certain minimum standards; and the circuit will function as well as it did with the transistor for which it was designed. This is the ultimate in interchangeability and is also characteristic of any good design. It is more traditional to begin with published parameters and end up with a circuit which is independent of those much-labored-over parameters. The methods we will use are not new and astounding. They are simply the reflection of those practical processes used every day by practicing engineers, where results are paramount and the methodology is not published.

The published parameters, of course, must be our starting point for design or analysis, but we must be selective in our choice and interpretation of them. Beta (β) will be our primary parameter, which we will take from the manual with a large grain of salt; that is to say, we'll be very conservative in our use of it. Absolute ratings will be taken from the manual and treated as sacred, but again we will be conservative and stay well below those maximums. The rest of the multitude of device parameters we will not worry about at all or, when the occasion arises, we will use approximations to estimate them to an accuracy which is appropriate to the problems at hand, or to the best degree possible if suitable accuracy is not possible.

This careful application of feedback means transistor interchangeability, and even easy substitutions, and that standard-value 10 or 20 percent resistors can generally be used (without later trimming and diddling). We have already briefly discussed negative feedback and the method by which it reduces collector-current drift and distortion, but one aspect of the concept needs some emphasis here.

A system using negative feedback is automatically a closed-loop system, and any break in the loop causes the entire system to stop functioning (except in a fashion vastly different from its intended function, and in a fashion which is nearly always undesirable).

What is important to us, as circuit designers, is the other side of the coin: the fact that if the loop is functioning, all the individual elements making up the loop will also be performing properly. Because of the operating feedback loop and its self-correcting properties, we are freed from the necessity of using components with precise values. Precise values are, of course, impossible in transistors, with their broad tolerances and acute sensitivity to temperature. Perhaps just as important as being freed from the need to *use* exact values is the fact that we are also freed from the need to *know* exact values, especially exact values of transistor parameters.

In our design procedures, we will frequently compute one value (*a*) based upon another value (*b*) which has not yet been calculated. To compound the felony, when we do get around to computing value *b*, we will use the value *a* (which we computed earlier) in computing value *b*. A lawyer would accuse us of reaching conclusions based upon facts not in evidence.

In reality, we are designing the *loop* in such a way that the *circuit* will work properly, rather than simply designing the circuit. Because of the nature of the closed loop, we can *assume* that if the loop operates properly, each parameter involved (voltages, etc.) will be correct and vice versa. I suppose that philosophically we should refer to all this as feedback loop design rather than transistor circuit design, but that is only philosophically. A really coherent circuit design system based wholly upon feedback loop design principles and procedures would soon become a Frankenstein's monster of unmanageable complexity.

Most authors in the transistor design or analysis field avoid getting too deep into feedback loop concepts, and they do so with good reason. This author will also soft-pedal the concept after a brief comment or two. The design process we will be using could not work if we did not use negative feedback of one sort or another. This is an important thing to remember, but not a thing to worry about.

THE MECHANICS OF THIS CHAPTER

The development of the design procedures we will study here took a number of years, and in the process the system sort of "growed" like Topsy. The result was that each different circuit required a different design procedure, even though the basic equations were valid from circuit to circuit.

This seemed to me to be pedagogically undesirable, and in an attempt to unify those procedures I have resorted to a branching type of program; this is similar to a computer program or to the system often used in programmed learning schemes. There has been no attempt to carry the program concept any further into the realm of computerized design or programmed learning. The branching program was not the only effort to unify the material, but it proved to be the best one tried, because the structure of the system permits a desirable consistency among explanations, examples, and design problems. I have adopted the following terminology:

FIELD: A linear segment (a graphic step-by-step procedure) representing several steps in the design procedure
FRAME: A single step in the design procedure
BRANCH: A short sequence of design steps which departs from the main program

It is these branches which accommodate the individual differences among the various circuits.

There are also some *pseudo*-branches which could very well be branches; however, when treating them as branches seemed to make things more difficult, I did not do it. If the greater uniformity which results appeals to you, these pseudo-branches can be thought of as ordinary branches. The flow diagram officially begins the chapter. It briefly describes each step in the design procedure *in its proper order.* The field-frame-branch terminology is used in the flowchart.

Section 5-2 describes important parameters as they apply to the problems of circuit design, how to select them, and how to use them. This part is not mentioned on the flowchart. Next, each field, frame, and branch is discussed and explained in the same order as it is presented on the flowchart. This section is followed by a worked and explained example using the design sheet format, which parallels the flowchart (Table 5-2) step for step.

Section 5-5 introduces the first table of possibilities (Table 5-5) and tells you how to avoid the trial-and-error (mostly error) element. Such a table can substitute for considerable experience and guarantee a workable practical circuit every time. An example shows how to use the table of possibilities; it follows the same format step for step as it is spelled out on the flowchart. Then there are some circuits for *you* to design, again following the flowchart.

Referring Back

If you are designing a circuit, for example, and you are working on Field 1, Frame 1, and have some doubts, you can refer to Field 1, Frame 1, of the worked example and find the same operation you are working on there. If the example does not help, you can refer to Field 1, Frame 1, of the discussion about the circuit under consideration and find the same topic discussed there. If your problem is a need for a more basic discussion, you can *refer to the flowchart,* Field 1, Frame 1, and find a reference in the margin; this will direct you to Sec. 4-4, which deals with variations in emitter-base voltage and its relation to the calculation you are trying to make. If you were simply reading through the text in normal manner, you would not even need to be aware of this part of its organization.

There are many ways to study a textbook. Some people prefer to master each page before going on to the next. Most books lend themselves well to this approach, and so does this one. And then there are people like me who jump around in an unpredictable fashion, but few books go out of their way to make it easy for us! Those of you who like to skip around or who may have cause to use this book for reference will, I think, find the organization a handy one.

LIMITS AND CONSTRAINTS

A thorny part of any kind of design is designing within practical limits. It is typical for equations to be general and to be as valid for values which are difficult or impossible to attain as they are for values which are common and convenient. For example, the Ohm's law equation $E = IR$ is as valid for a current of 10 A and a resistance of 1 MΩ as it is for a resistance of 1 kΩ and a current of 1 mA. However, the second case requires a voltage of 1 V, which is common and convenient, whereas the first case requires a voltage of 10 million volts; this not only is impractical, but it borders on the ridiculous (and convenient it ain't!).

The example was obvious, because only two interdependent variables were involved; but suppose five or six variables were involved! This is exactly the situation in a transistor circuit and each variable has its limits, some of which are absolute and some of which are convenient or practical. Good design implies that all values be convenient and that none shall be too close to absolute limits.

In most cases, long experience is relied upon to enable the designer to keep out of trouble on this score, and the novice is allowed to make all the mistakes he can manage on his road to becoming an experienced designer. But because this book is not designed for those who intend to make a lifetime career of transistor circuit design, and who have time to gain experience, I have compressed a great deal of experience into several basic tables. These tables detail nearly all the circuit possibilities in a very quantitative fashion. These tables of circuit possibilities will make it possible for you to do a quite sophisticated job of circuit design right away. Because the tables are, in a sense, compressed experience, it is not necessary (or expected) that you understand them completely; it is likely that you will be somewhat dependent on the tables for a while, but that should not bother you too much, since, in time, you can substitute experience for them. I will also explain later on how the tables were designed. Table 5-1 lists and explains the symbols used in this chapter and throughout the remainder of the text. From time to time we will introduce a new one. I would suggest that you take time now to memorize

TABLE 5-1. LIST OF SYMBOLS USED

Resistances

1. R_L Collector load resistor.
   Resistor-*l*oad.
2. R_{Lw} Working load resistance (or impedance). The load being driven "worked" by the stage.
   Resistor-*l*oad-*w*orking.
3. R_{Lac} The total signal load impedance. This is composed of R_L and R_{Lw} in parallel.
   Resistor-*l*oad-*ac*.
4. R_{eac} The unbypassed portion of the emitter resistor.
   Resistor-emitter-*ac*.
5. R_{edc} The bypassed portion of the emitter resistor.
   Resistor-emitter-*dc*.
6. R_{es} The entire emitter resistor ($R_{eac} + R_{edc}$).
   Resistor-emitter-stabilizing.
7. R'_e The transistor's intrinsic emitter resistance.
8. R_b The resistance from base of the transistor to ground (external).
9. R_{b1} The upper base bias resistor. Generally returned to ground through a forward-biasing voltage.
10. R_{b2} The lower base bias resistor. Generally returned from base directly to ground.

Voltages

1. E_{RL} Voltage drop across the collector load (R_L).
2. E_{Res} Voltage drop across R_{es}, the emitter stability resistance. (R_{edc} and R_{eac})
3. E_Q Voltage drop across the transistor, between collector and emitter.
4. E_o Maximum peak-to-peak ac signal output voltage.
5. E_s Dc power supply voltage.
6. E_{Rb1} The voltage drop across the upper base bias resistor R_{b1}.
7. E_{Rb2} The voltage drop across the lower base bias resistor (base stability resistor) R_{b2}.

Capacitors

1. C_{bp} Emitter bypass capacitor.
2. C_{cin} Input coupling capacitor.
3. C_{co} Output coupling capacitor.

Miscellaneous

1. V_{BE} The base-emitter junction voltage drop (see also V_j).
2. ΔV_{BE} Variations in the base-emitter voltage drop (usually due to temperature).
3. β Beta, the common-emitter current gain of the transistor.

   $$\beta = \frac{I_c}{I_b}$$

   where I_c is the collector current and I_b is the base-emitter current
4. f_o The lowest frequency of interest. The low-frequency cutoff value, where the voltage gain begins to roll off.
5. t Temperature in centigrade degrees.
6. Δt The change in temperature $t_1 - t_0$, where t_0 is generally 20°C and t_1 is the highest expected temperature.
7. V_j The junction voltage 0.6 V in silicon and 0.2 V in germanium. Also called V_{BE}.

Current

1. QI_c Quiescent collector current. (I_c is collector current)
2. I_b Base current.

Impedances

1. Z_{in} The total ac signal input impedance.
   $Z_{in} = R_{b1} \| R_{b2} \| \beta \times R_{eac}$
   where $\|$ means in parallel with
2. R_{int} The dc input resistance looking into the base, but not including R_{b1}.
   $R_{int} = R_{b2} \| \beta \times R_{es}$

them, or at least to memorize the page number so that you can refer to it easily. My students have discovered that some of these letter groups can be pronounced and have formed the habit of pronouncing them, which is often easier than saying each letter separately.

5-1 A FLOWCHART OF THE DESIGN PROCESS

Table 5-2 is a flowchart of the entire design process for all common current-mode feedback circuits. The chart is pretty much self-explanatory, but I will call your attention to the boxes in the left margin which make comments about the factors involved in a particular frame. These comments simply *flag* important ideas. I will also remind you that the right margin contains references to *in-depth* information about the contents of each frame. This flowchart and its associated design sheet (Fig. 5-5) provide the design procedure and equations for some nine basic circuits. All these can be designed by following the appropriate branches on chart and design sheet. Figure 5-1 shows all nine variations. All these circuits are classified as *current-mode* feedback circuits. Secondly, each circuit is also classified as common-emitter, common-collector, common-base, or differential. Thirdly, each circuit is classified according to whether it is emitter-biased or base-biased. The common-emitter base-biased circuit is further broken down into its three common variants: the unbypassed circuit, the fully bypassed circuit, and the split emitter circuit. These three variants differ only in the use (or not) of a bypass capacitor across the emitter feedback resistor. Each variant has a different range of available voltage gains and input impedances. In this chapter we will deal in detail with only the *base-biased* common-emitter amplifier. Chapter 6 covers the rest of the *current-mode* common-emitter circuits. The rest of the circuits shown in Fig. 5-1 will be covered in subsequent chapters. The differential amplifier is so important that it will be given a chapter of its own. The design procedure for *voltage-mode* feedback circuits is so different that it requires two special design sheets. We will study this second major classification of circuits in Chap. 7. All this may seem to be frighteningly complicated at this stage of the game, but by the time you have finished this chapter, you will begin to see the overall picture and it will all seem quite simple from there on out.

5-2 HOW TO SELECT AND USE IMPORTANT PARAMETERS

A. Beta (β)

This is the most important design parameter we will have to deal with, and we shall have to make some assumptions to make the published values useful.

TABLE 5-2

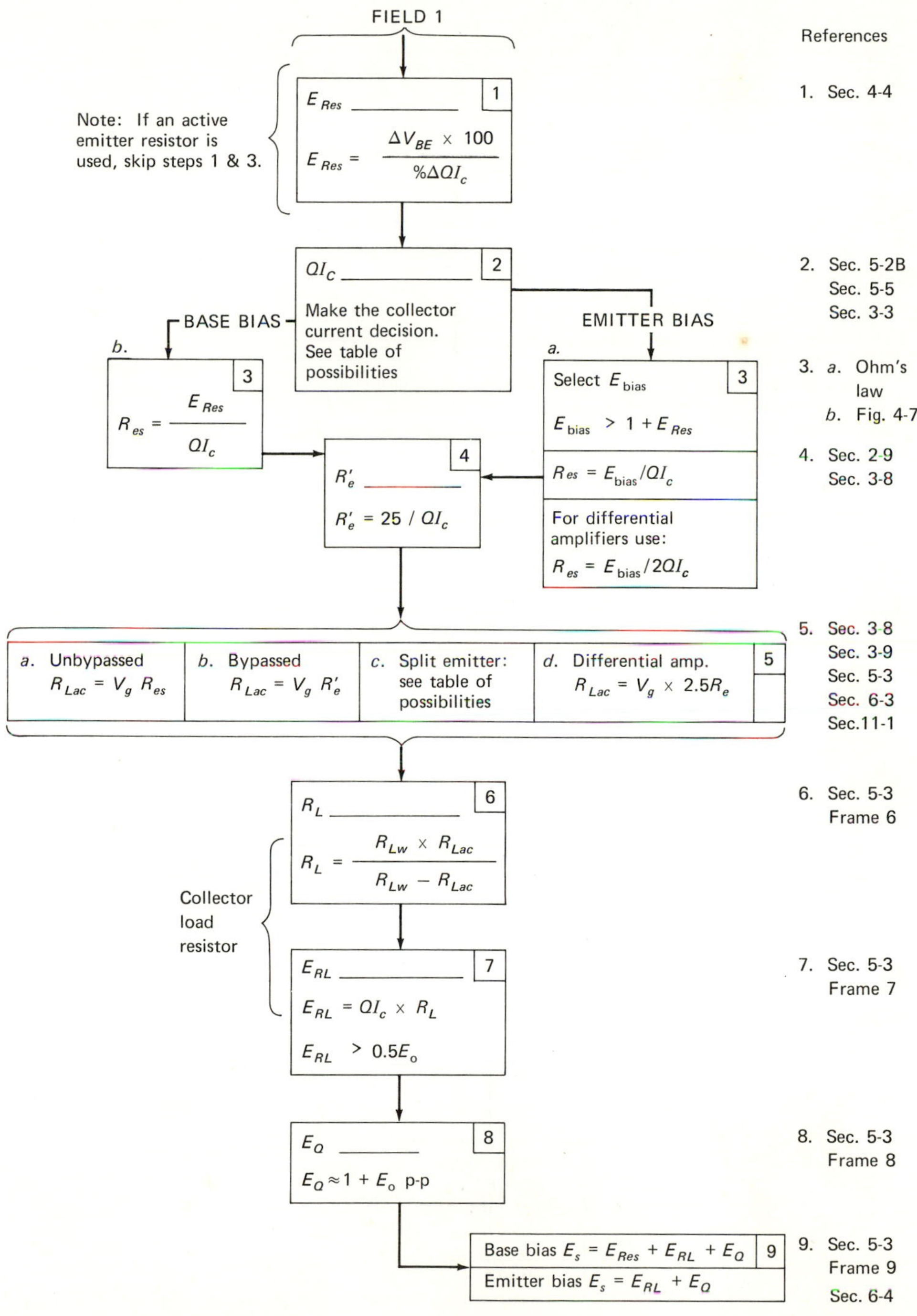

FIELD 1
Note: If an active emitter resistor is used, skip steps 1 & 3.
1
E_{Res} ________
$E_{Res} = \frac{\Delta V_{BE} \times 100}{\%\Delta QI_c}$
2
QI_C ________
Make the collector current decision. See table of possibilities
BASE BIAS
b.
3
$R_{es} = \frac{E_{Res}}{QI_c}$
EMITTER BIAS
a.
3
Select E_{bias}
$E_{bias} > 1 + E_{Res}$
$R_{es} = E_{bias}/QI_c$
For differential amplifiers use:
$R_{es} = E_{bias}/2QI_c$
4
R'_e ________
$R'_e = 25 / QI_c$
5
a. Unbypassed $R_{Lac} = V_g\, R_{es}$
b. Bypassed $R_{Lac} = V_g\, R'_e$
c. Split emitter: see table of possibilities
d. Differential amp. $R_{Lac} = V_g \times 2.5R_e$
6
R_L ________
$R_L = \frac{R_{Lw} \times R_{Lac}}{R_{Lw} - R_{Lac}}$
Collector load resistor
7
E_{RL} ________
$E_{RL} = QI_c \times R_L$
$E_{RL} > 0.5E_o$
8
E_Q ________
$E_Q \approx 1 + E_o$ p-p
9
Base bias $E_s = E_{Res} + E_{RL} + E_Q$
Emitter bias $E_s = E_{RL} + E_Q$

References

1. Sec. 4-4
2. Sec. 5-2B
   Sec. 5-5
   Sec. 3-3
3. *a.* Ohm's law
   b. Fig. 4-7
4. Sec. 2-9
   Sec. 3-8
5. Sec. 3-8
   Sec. 3-9
   Sec. 5-3
   Sec. 6-3
   Sec. 11-1
6. Sec. 5-3
   Frame 6
7. Sec. 5-3
   Frame 7
8. Sec. 5-3
   Frame 8
9. Sec. 5-3
   Frame 9
   Sec. 6-4

TABLE 5-2 (Cont.)

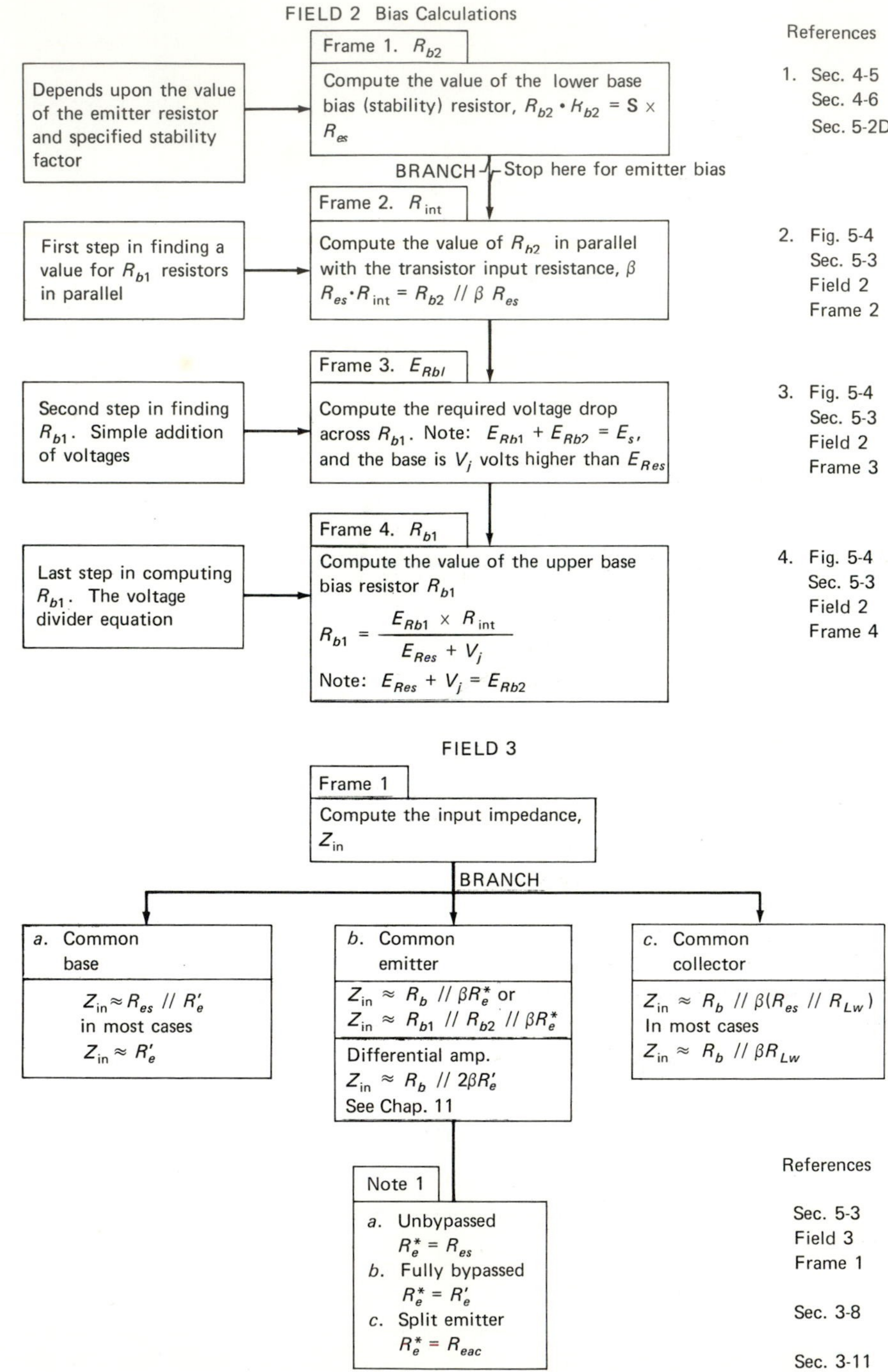

FIELD 2 Bias Calculations
References
Frame 1. R_{b2}
Depends upon the value of the emitter resistor and specified stability factor
Compute the value of the lower base bias (stability) resistor, $R_{b2} \cdot R_{b2} = S \times R_{es}$
1. Sec. 4-5
Sec. 4-6
Sec. 5-2D
BRANCH
Stop here for emitter bias
Frame 2. R_{int}
First step in finding a value for R_{b1} resistors in parallel
Compute the value of R_{b2} in parallel with the transistor input resistance, $\beta R_{es} \cdot R_{int} = R_{b2} // \beta R_{es}$
2. Fig. 5-4
Sec. 5-3
Field 2
Frame 2
Frame 3. E_{Rbl}
Second step in finding R_{b1}. Simple addition of voltages
Compute the required voltage drop across R_{b1}. Note: $E_{Rb1} + E_{Rb2} = E_s$, and the base is V_j volts higher than E_{Res}
3. Fig. 5-4
Sec. 5-3
Field 2
Frame 3
Frame 4. R_{b1}
Last step in computing R_{b1}. The voltage divider equation
Compute the value of the upper base bias resistor R_{b1}
$R_{b1} = \frac{E_{Rb1} \times R_{int}}{E_{Res} + V_j}$
Note: $E_{Res} + V_j = E_{Rb2}$
4. Fig. 5-4
Sec. 5-3
Field 2
Frame 4
FIELD 3
Frame 1
Compute the input impedance, Z_{in}
BRANCH
a. Common base
$Z_{in} \approx R_{es} // R'_e$
in most cases
$Z_{in} \approx R'_e$
b. Common emitter
$Z_{in} \approx R_b // \beta R^*_e$ or
$Z_{in} \approx R_{b1} // R_{b2} // \beta R^*_e$
Differential amp.
$Z_{in} \approx R_b // 2\beta R'_e$
See Chap. 11
c. Common collector
$Z_{in} \approx R_b // \beta(R_{es} // R_{Lw})$
In most cases
$Z_{in} \approx R_b // \beta R_{Lw}$
Note 1
a. Unbypassed
$R^*_e = R_{es}$
b. Fully bypassed
$R^*_e = R'_e$
c. Split emitter
$R^*_e = R_{eac}$
References
Sec. 5-3
Field 3
Frame 1
Sec. 3-8
Sec. 3-11

TABLE 5-2 (*Cont.*)

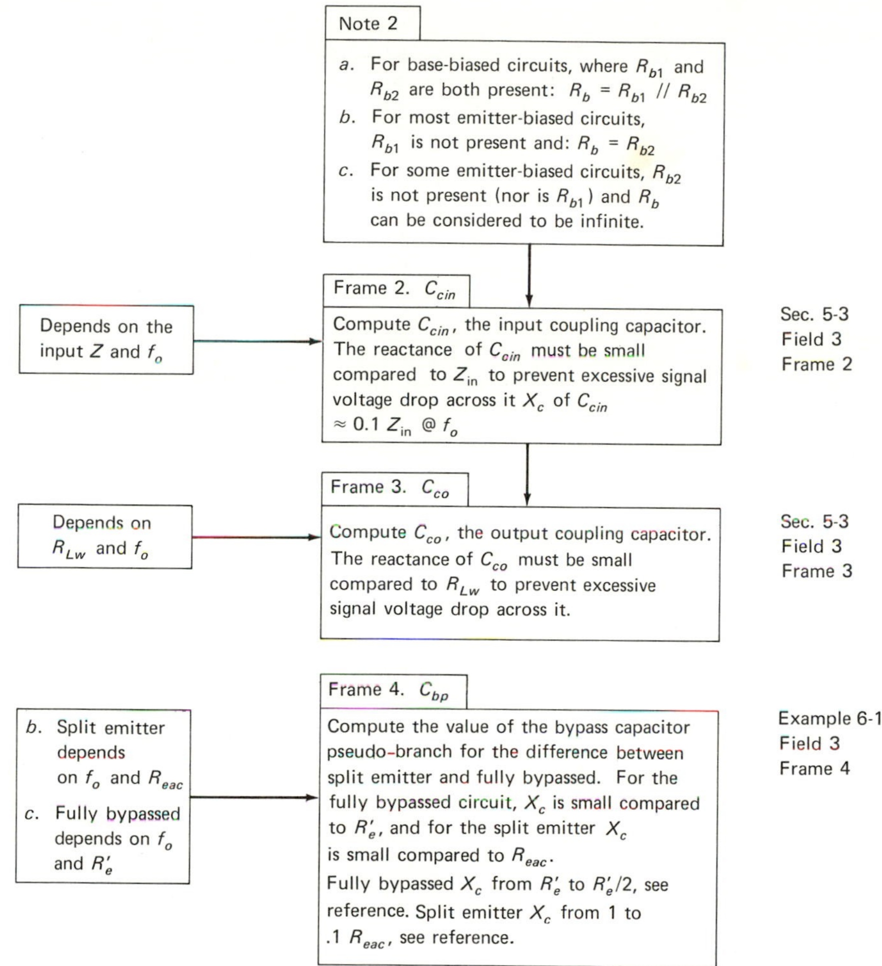

Note 2
a. For base-biased circuits, where R_{b1} and R_{b2} are both present: $R_b = R_{b1} // R_{b2}$
b. For most emitter-biased circuits, R_{b1} is not present and: $R_b = R_{b2}$
c. For some emitter-biased circuits, R_{b2} is not present (nor is R_{b1}) and R_b can be considered to be infinite.
Depends on the input Z and f_o
Frame 2. C_{cin}
Compute C_{cin}, the input coupling capacitor. The reactance of C_{cin} must be small compared to Z_{in} to prevent excessive signal voltage drop across it X_c of C_{cin} $\approx 0.1\ Z_{in}$ @ f_o
Sec. 5-3
Field 3
Frame 2
Depends on R_{Lw} and f_o
Frame 3. C_{co}
Compute C_{co}, the output coupling capacitor. The reactance of C_{co} must be small compared to R_{Lw} to prevent excessive signal voltage drop across it.
Sec. 5-3
Field 3
Frame 3
b. Split emitter depends on f_o and R_{eac}
c. Fully bypassed depends on f_o and R'_e
Frame 4. C_{bp}
Compute the value of the bypass capacitor pseudo-branch for the difference between split emitter and fully bypassed. For the fully bypassed circuit, X_c is small compared to R'_e, and for the split emitter X_c is small compared to R_{eac}.
Fully bypassed X_c from R'_e to $R'_e/2$, see reference. Split emitter X_c from 1 to .1 R_{eac}, see reference.
Example 6-1
Field 3
Frame 4

As you know, β varies with temperature and quiescent collector current. Because the temperature problem is most difficult to deal with, we shall assume, as a starting point, a standard temperature of 20° to 25°C; we shall assume also that the published value is measured at one of those temperatures and that the collector current at which the measured value of β is given represents that current at which β is at its maximum. Figure 5-2 shows a typical β vs. Q-point collector current curve. With this curve and the manufacturer's published value of β, you can make a pretty satisfactory estimate of the expected β at any desired collector current. In nearly every design problem we shall make the quiescent collector current for our circuit one of the very first

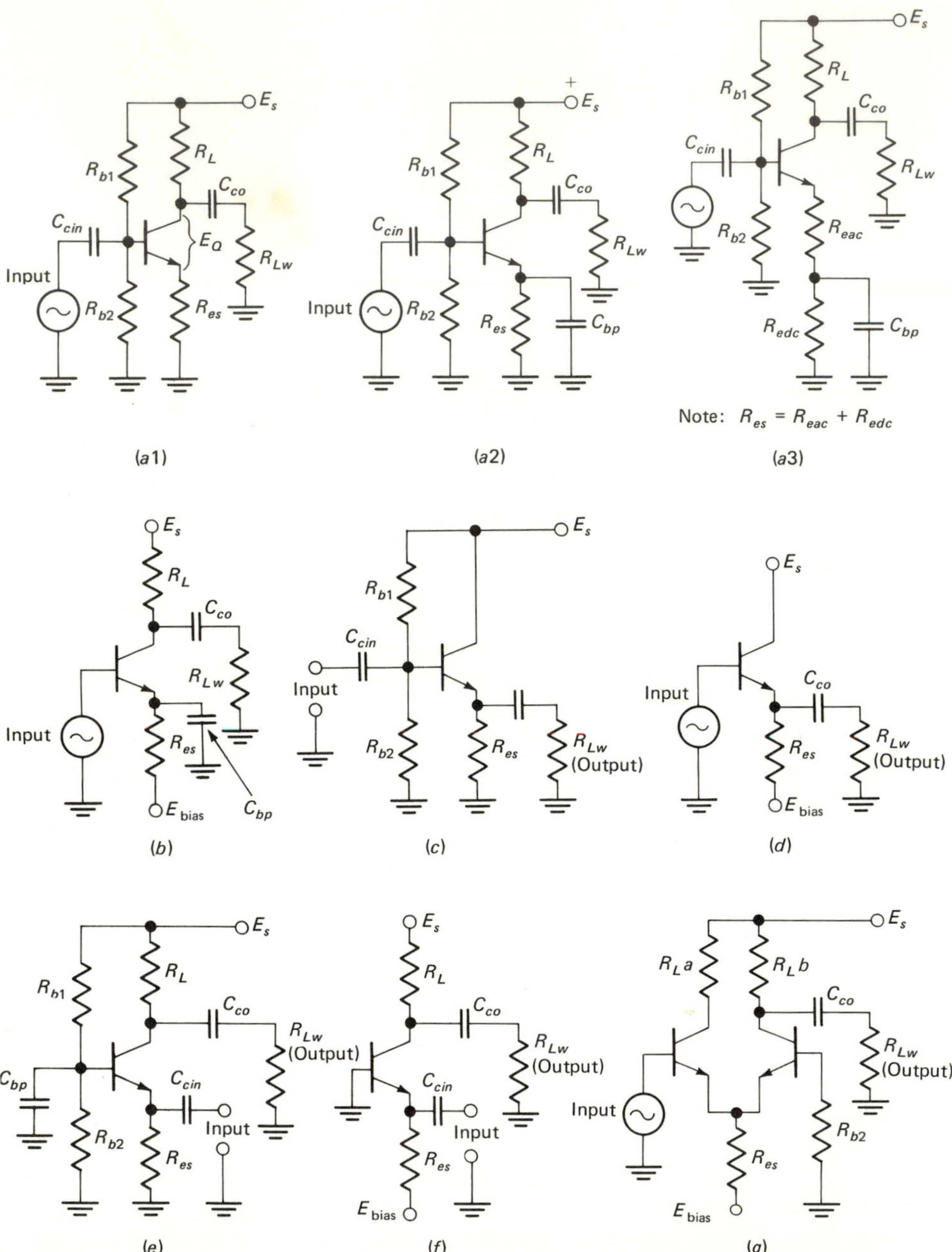

Fig. 5-1 The circuits included in the flowchart, Table 5-2. (a) Three variations of the base-biased common-emitter circuit: 1. the unbypassed circuit, 2. the fully bypassed circuit, 3. the split emitter circuit; (b) the emitter-biased common-emitter circuit (fully bypassed); (c) the base-biased common-collector circuit; (d) the emitter-biased common-collector circuit; (e) the base-biased common-base circuit; (f) the emitter-biased common-base circuit; (g) the differential amplifier-emitter biased

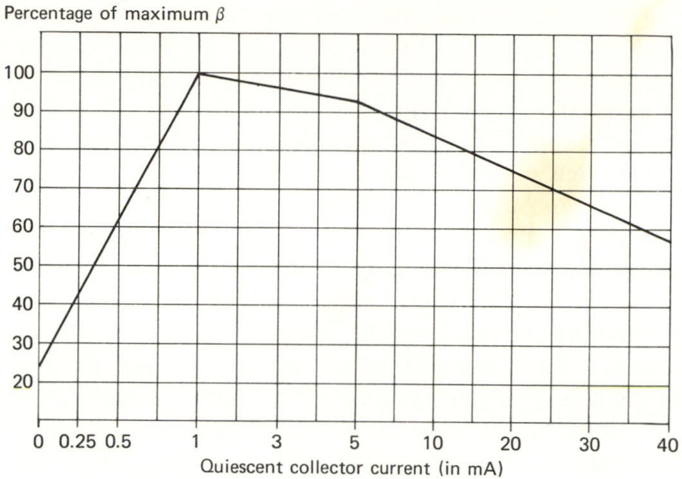

Fig. 5-2 Typical β vs. quiescent collector current curve (based on a transistor with an I_c max of 100 mA)

orders of business. Soon after we have made that decision, we will become interested in the expected β of the transistor we have selected (or will select) at our chosen operating current. If the manufacturer specified *minimum* β, then we need only to reduce this figure by 10 percent or so before applying it to the collector current vs. β graph. If *typical* β is specified, it is proper to reduce the value by at least 25 percent before applying it to the collector current vs. β graph.

B. Maximum Collector Current (I_c max)

In all applications other than power amplifiers, there is no excuse for operating at a quiescent current above 40 percent of the maximum collector current specified by the manufacturer. There is no need for it and no profit in it. Power amplifier circuits are sometimes operated at up to, but never in excess of, 50 percent of the maximum current rating. In general, except for power amplifiers, the lower the quiescent collector current, the better the overall performance will be. There are two primary limits as to how low we can make the Q point. One limit is imposed by the working load under heavy loading situations. The second limit to how low the Q point can be set is dictated by the amount of leakage current. In order that the quiescent collector current can remain under our control, we must make it large (by a factor of 10 or so) with respect to the leakage component, which is for the most part not under our control. A fairly satisfactory rule of thumb is to make the quiescent collector current greater than 100 times the I_{co} value given in the transistor manual. This rule makes the quiescent current about 10 times as high as the leakage component in the *collector* circuit. The rule assumes that I_{co}

will appear in the collector loop as about $10 \times I_{co}$ in a properly designed circuit (it would be βI_{co} in an uncompensated circuit). There are actually a great many aspects to the collector current decision, but they are so complex and often contradictory that finding the optimum quiescent point for any given set of conditions is bound to be a compromise. Many designers simply select that quiescent current which yields the highest β. It is generally good policy to operate somewhere between the lower limit as set by the leakage current and two times the collector current at which maximum β occurs. Some loading conditions will dictate higher Q points and will give the designer no choice in the matter. We will see some examples of this a little later. Experienced designers often select somewhat higher Q points for the sake of stability or lower ones for the sake of higher circuit impedances as the situation requires, but such fine distinctions require experience and experience takes time. You will find that for most practical problems this degree of sophistication is not necessary, and satisfactory performance *can* be had at most Q points below 40 percent of maximum. The tables of *circuit possibilities* will help greatly in making this important decision.

C. The Maximum Collector Voltage

The transistor manual gives the maximum collector-to-base voltage, but the voltage between collector and emitter is more convenient and differs from the collector-to-base voltage by no more than 0.6 V or so. As a further practical consideration, the power supply voltage should not exceed the manufacturer's published collector-base maximum voltage, because it is always possible for a large signal to turn the transistor all the way off. When this happens, the entire supply voltage appears across the transistor and can cause the collector-base junction to zener, with possibly fatal results to the transistor. THE RULE: The power supply voltage E_s should never be higher than the manufacturer's maximum collector voltage rating.

D. The Stability Factor

In Chap. 4 we saw that the current gain of a circuit is nearly always lower than the transistor's β figure, because the base bias resistors bypassed some of the incoming signal current to ground (or because of the Miller effect), making it unavailable for amplification. If we can make the total shunt path from base to ground a low resistance, we can reduce the *circuit* current gain to somewhere between 1 and 10. This contributes to stability on two fronts. First, any leakage current in the base circuit has two paths to follow: a low resistance path to ground which does not get amplified and does not appear in the collector circuit, and a higher impedance path through the base-emitter circuit where it does get amplified and does appear in the collector circuit. Thus, leakage current divides between the two branches, and only a small percentage of it gets amplified to appear as collector current. On the second front, the problem of variations in β, due to temperature, operating point, or differences in individual transistors, is greatly reduced because the *circuit* gain is reduced to approximately **S** (the stability factor) and

$$\mathbf{S} \approx \frac{R_{b2}}{R_{es}}$$

The factor β is not mentioned in the equation, because it is now only slightly significant. Stated another way, the *circuit* gain has become largely independent of β; and so it takes a very large change in β to cause a much smaller change in the *circuit* gain. β is, of course, a factor—otherwise, we would not be concerned with it at all—but by making E_{Res} large and R_{b2} small, we have shifted β's role from that of a completely dominant parameter to that of a very minor parameter in the total system. The voltage divider arrangement for the bias circuit comes about, because the required base current is not compatible with the low-value resistance required for stability. The upper base resistor R_{b1} supplies the bias current for the base-emitter loop; the lower resistor R_{b2} provides the low base-to-ground resistance required for stability. R_{b1}, when there is one, also contributes to the ground return path, but in most cases its contribution is small. R_{b1} returns to ground through the power supply, which, in any practical system, has a very low internal resistance. In Chap. 4 we discussed the stability factor in some mathematical and theoretical detail, and we have reviewed it briefly here; but our problem now is how to decide upon a suitable stability factor for a given design problem. Here again, we can get quite sophisticated in our approach, but it is normally not necessary. We can generally do quite well with a handful of working guidelines.

1. The practical range of stability factors is from about 1 to 10.
2. Stability factors from 1 to 5 are used only for the most critical applications, or those where abnormally large temperature variations can be expected.
3. Stability factors from 1 to 4 are used for direct-coupled circuits, which are to be used in a normal temperature range and are not highly critical circuits.
4. A stability factor of 10 is an excellent compromise for all general-purpose capacitor-coupled circuits.
5. A stability factor of up to 10 can be used for general-purpose capacitor-coupled circuits when silicon transistors are used. We may amend these rules slightly later to accommodate special cases and circumstances.

E. Circuit Specifications

Every design problem begins with a set of *specifications* (parameters). These specifications describe, quantitatively and in detail, the job to be performed. For a while I will provide you with specifications that are realistic and practical; but a little later on you will be able to write your own "specs," and they too will be realistic and practical. Here I will list those specifications which will be adequate for our design needs. There are two parts to the list. The first part lists those transistor parameters that are essential to the task (unessential parameters will not be listed). The second part lists those circuit characteristics that we will wish to design into the circuit.

Transistor Parameters

1. Beta (β)
2. I_c (max) the maximum collector current
3. E_c (max) the maximum collector voltage
4. I_{co} the collector leakage current
5. Device type: *npn, pnp;* silicon, germanium

Circuit Parameters

1. Voltage gain (V_g)
2. Maximum expected operating temperature (usually environmental)
3. Allowable quiescent current variation (due to V_{BE}) at the maximum expected temperature
4. The stability factor (**S**)
5. Input impedance (Z_{in})
6. The working load (R_{Lw}) (This is the load to be driven. It is often the input impedance to another stage but may be any other device.)
7. Low-frequency cutoff (f_o), the lowest frequency of interest
8. Maximum output voltage (E_o) (This is the peak-to-peak amplitude of the output signal before flat-top (clipping) distortion occurs.)

5-3 THE DESIGN PROCEDURE

Let us study an outline of the procedure first, and then try our hand at the first example.

FIELD 1 COLLECTOR SIDE COMPUTATIONS

FRAME 1

Compute E_{Res}, the voltage drop required across the emitter resistor

$$\text{Minimum } E_{Res} = \frac{\Delta V_{BE} \times 100}{\%\Delta QI_c}$$

This is the voltage drop across the emitter resistor, which is necessary to swamp out variations in the base-emitter junction voltage. We derived this equation in Chap. 4, and it tells us what *minimum* voltage drop we must have across R_{es} (resistor-emitter stability) for a given amount of quiescent collector-current drift at the highest expected environmental temperature. ΔV_{BE} is the total amount of *change* in the base-emitter voltage as the temperature moves from 20°C to the maximum expected environmental temperature. The average variation is about 2.5 mV/C°, but this varies a little with different operating (collector) currents. This variation is shown in Table 5-3. $\%\Delta QI_c$ in the formula is the amount that the *collector current* will be allowed to drift at the highest expected temperature. For example, a circuit's quiescent point might be *allowed* to be off by 10 percent at the maximum expected temperature. This would mean that most voltages would also drift by about 10 percent. Just exactly how much drift you, as the designer, choose to allow is not as important as the fact that you do make a decision and know how much to expect.

We will find that our design procedure demands that we compute the voltage drop across the emitter resistor E_{Res} before we know the exact operating current. This might seem to be a problem, because if we don't know QI_c, we don't know how many millivolts per degree variation to expect. If we don't have any idea what the quiescent collector current is going to be, we simply select what we believe to be the lowest reasonable current. This will be the worst-case condition. We can always peek at the table of possibilities, even though the design procedure does not require us to use it until the next step, and get the approximate range from it.

The circuit that we will be discussing in this chapter has rather limited gain capabilities, but it is representative of all the current-mode feedback circuits we shall examine. Minor modifications of the circuit can improve the gain by factors of from 5 to 40 without damaging the circuit's built-in stability. The end point of all the efforts to prevent the quiescent operating point from drifting is to achieve a circuit which will process the signal in the desired manner. We generally treat signal handling and bias stability problems as separate entities, but they are intimately related and it is important to keep this fact in mind. Those things that we do to ensure bias stability influence nearly all the signal parameters of the amplifier, including gain (both current and voltage gain), input impedance, and maximum output voltage. Near-perfect bias stability results in very low gains and unacceptably low input impedances! Therefore, the rules that we establish for making stability decisions

TABLE 5-3. TABLE OF VARIATIONS IN V_{BE} AT VARIOUS CURRENTS

QI_c	ΔV_{BE}
0–0.1 mA	−3.0 mV/C°
0.1–1.0 mA	−2.75 mV/C°
1.0–10.0 mA	−2.5 mV/C°
10.0–100 mA	−2.0 mV/C°

must necessarily be tempered by signal requirements and will represent some essential compromises.

FRAME 2

Make the quiescent collector current decision.

This is discussed in Sec. 5-2A. Normally our chief guide, in lieu of greater experience, will be the appropriate table of circuit possibilities.

FRAME 3

Compute R_{es}, the value of the emitter stability resistor.

This frame has two branches, one for base-biased circuits and one for emitter-biased circuits. There is also a pseudo-branch for emitter-biased differential amplifiers.

Branch a. The base bias branch

$$R_{es} = \frac{E_{Res}}{QI_c}$$

This is simply Ohm's law (E/I). We have made a decision about the quiescent collector current in Frame 2, and so we know I (QI_c); and we have calculated the *voltage* drop required across the emitter resistor to provide us with the desired V_{BE} stability; and so we know E and can easily solve for R.

Branch b. The emitter bias branch (see Fig. 5-1*b*)

The emitter bias arrangement requires an extra bias power supply. In some situations there are enough advantages to emitter bias to justify the extra cost (in this chapter we will not be particularly interested in this branch since we will be concerned only with base bias).

Before R_{es} can be computed, a decision must be made about the bias voltage. The bias voltage may be any *convenient* voltage which is at least 1 V larger than the emitter-resistor voltage drop required by V_{BE} stability (most of the bias voltage will be dropped across R_{es}). Once the bias voltage has been decided upon, the calculation of R_{es} is just more Ohm's law.

$$R_{es} = \frac{E_{\text{bias}}}{QI_c}$$

FRAME 4

Compute the value of the base-emitter junction resistance R'_e.

This is our friend the Shockley relationship. It is only an approximation, but it is the best we can do in most practical situations.

$$R'_e = \frac{25}{QI_c} \qquad QI_c \text{ in mA}$$

The constant 25 is actually millivolts. We will not need to know R_e' in every case, but we will make it part of the routine anyway. (Later, on occasion, you may skip this frame.)

FRAME 5

Compute the value of R_{Lac}, the ac signal load resistance.

The total ac signal load resistance consists of R_L in parallel with the working load R_{Lw} (see Fig. 5-1*a*). The purpose of this frame is to compute the value of R_{Lac} to yield the specified voltage gain. The general voltage gain equation is

$$V_g = \frac{R_{Lac}}{R_e}$$

where R_e may be R_{es}, R_e', or R_{eac}, depending upon the particular circuit configuration (see Fig. 5-1*a*, *b*, *c*). At this stage of the procedure we know V_g because it was one of the original circuit specifications, and in all cases, except for the split emitter circuit, we will also know R_e. We can rearrange the voltage gain equation to solve for R_{Lac} and get

$$R_{Lac} = V_g \times R_e$$

Frame 5 has four branches which cover all various effective emitter circuit resistances that are likely to be encountered.

Branch a. $R_{Lac} = V_g \times R_{es}$ (unbypassed circuit)

Branch b. $R_{Lac} = V_g \times R_e'$ (fully bypassed circuit)

Branch c. $R_{Lac} = V_g \times R_{eac}$ (split emitter circuit) (see table of possibilities—Table 6-2)

Branch d. $R_{Lac} = V_g \times 2.5\ R_e'$ (differential amplifier)

Some of these equations apply to other circuits in addition to those specifically listed; I will call your attention to those cases as they come up. In this chapter we will be concerned with only the first branch, the unbypassed circuit.

R_{Lac} consists of R_L in parallel with R_{Lw}. Two assumptions are made here. First, it is assumed that when we reach that stage in the design process, we will make the coupling capacitor C_{co} a virtual short circuit at signal frequencies. To be a little more specific, we will select a capacitor whose reactance at the lowest frequency of interest is so small compared to R_L and R_{Lw} that it can be considered negligible. The second assumption is that the impedance of the power supply, at signal frequencies, is very low, placing the upper end of R_L (see Fig. 5-1), for all practical purposes, at ground potential insofar as the signal is concerned. The second assumption is generally a valid one, because of the large capacitors used in power supplies, and because low power supply impedance is absolutely essential to multistage circuits. There is a phenomenon called motorboating because of its characteristic sound in audio equipment. Motorboating is an oscillation, and an almost inevitable consequence of too high a power supply impedance.

We often find large capacitors across small batteries to "artificially" lower their impedance to prevent oscillation, and "decoupling" networks (which we will discuss later) in ac-operated supplies to accomplish the same purpose. We can always assume that the power supply impedance is sufficiently low. If, for any reason, this is not true, we can lower the impedance "artificially" and *make* the assumption valid.

This first circuit that we will design has a highly predictable voltage gain, but not very much of it (a range of from 1 to 10). A large capacitor connected across the emitter resistor R_{es} can increase the voltage gain by as much as a factor of 40, more or less (a V_g range of 40 to 400). In exchange for increased voltage gain through bypassing the entire emitter resistor, we lose most of our gain predictability and control. We get a lot of gain, but we don't know exactly how much. The reason is that, once we bypass the emitter resistor, the transistor's internal emitter resistance R'_e becomes the controlling factor, and we *cannot* know that value with any great precision.

A third alternative is to divide the emitter resistor into two parts and to place a large bypass capacitor across only part of the emitter resistance. This is a compromise situation which allows fairly respectable voltage gains (a range of 10 to 70) and good predictability as well.

Which of these options we choose determines which branch in the design program we follow. The decision to design one or another of the three options is something we will take up later when we discuss the problem of selecting an appropriate circuit for a given set of requirements. The fourth option, the differential amplifier, will be discussed in a chapter of its own. For now we will be content with the knowledge that each of the options provides a different set of specifications, and that our design program has a branch for each. The voltage gain equations for the first three cases are:

1. For the unbypassed emitter resistor (see Fig. 5-3*a*):

$$V_g = \frac{R_{Lac}}{R_{es}} \qquad \text{or} \qquad V_g = \frac{R_{Lac}}{R_{es} + R'_e}$$

The equation states that the voltage gain is a function of the collector impedance and the emitter impedance. The first (and preferred) equation is a simplified one, since the transistor's internal emitter resistance R'_e has been omitted because, for all stable current-mode circuits, R'_e is much smaller than R_{es}. In a typical stable circuit, R_{es} may be 100 to 500 times as large as R'_e, so that, at most, leaving R'_e out of the equation results in only a 1 percent error.

Including R_e' in the equation really adds nothing to the accuracy of design results, because we can only estimate its value; we cannot know it exactly, and R_{es} will normally be no better than a 5 percent tolerance resistor anyway. We can make a good deal of extra busywork for ourselves by including R_e'.

2. For the fully bypassed emitter resistor (see Fig. 5-3b):

$$V_g = \frac{R_{Lac}}{R_e'}$$

In this case R_e' becomes the only significant emitter resistance in the circuit. The voltage gain we are concerned with is an ac parameter, and we have pur-

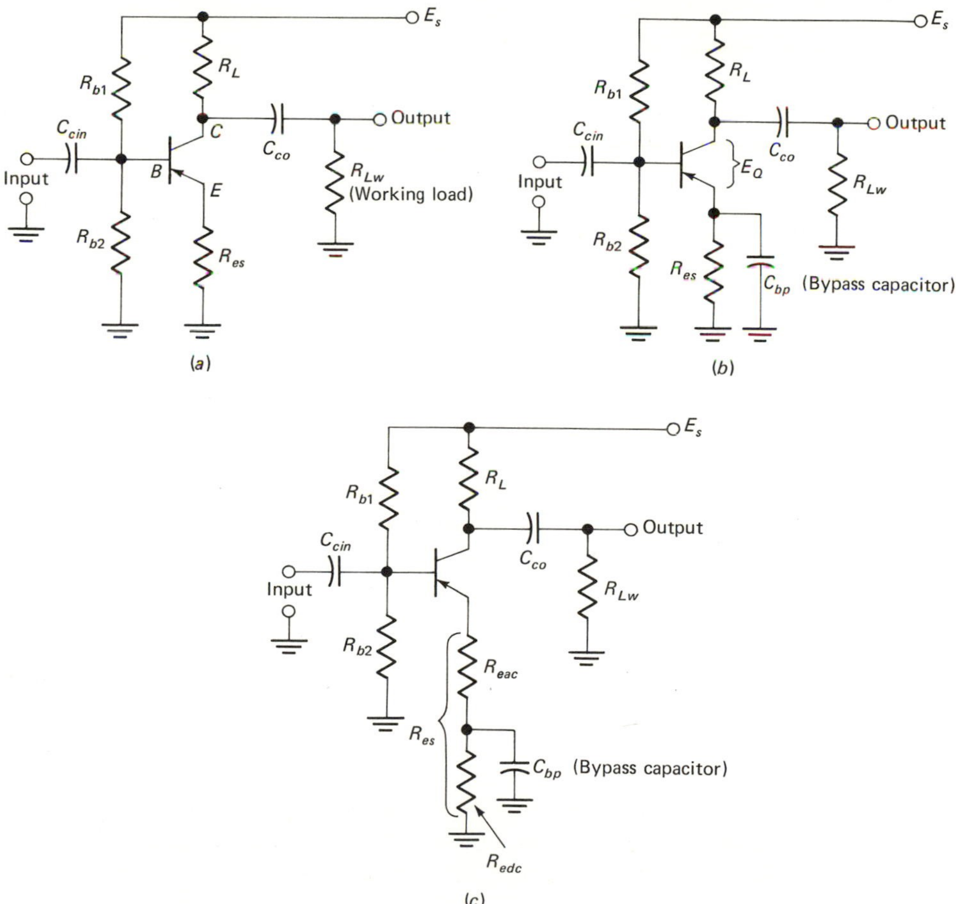

Fig. 5-3 (a) The unbypassed common-emitter circuit; (b) the fully bypassed common-emitter circuit; (c) the split emitter common-emitter circuit

posely selected a value for the bypass capacitor that would appear as a short circuit ($\rightarrow 0\ \Omega$) at all signal frequencies. So as far as the *signal* is concerned, the value of R_{es} has been reduced effectively to 0 Ω. This means that even though R_e' may be only 25 Ω (we can only estimate its value), it is the only emitter resistance in the circuit and it becomes the entire denominator in the voltage gain equation. R_e' can be approximated by the empirical equation

$$R_e' \approx \frac{25}{QI_c}$$

where 25 is an empirical constant and QI_c is the quiescent collector current *in milliamperes*.

3. For the split emitter resistor (see Fig. 5-3*c*):

$$V_g = \frac{R_{Lac}}{R_{eac} + R_e'} \approx \frac{R_{Lac}}{R_{eac}}$$

R_{eac} is the unbypassed portion of the emitter resistor. In this circuit, we normally try to make R_{eac} fairly large (5 to 10 times) with respect to R_e'. If this is possible, we can use the abbreviated equation

$$V_g = \frac{R_{Lac}}{R_{eac}}$$

Because R_e' is always a fairly rough approximation, the abbreviated form of the equation is normally preferred. We will always use it in this text (with one or two rare exceptions) because a very important empirical loading-effect correction curve, the Young-Albertson curve,[1] is based upon the abbreviated form of the equation.

In the unbypassed and the fully bypassed circuits we have no real control over the value of the emitter resistance. In the unbypassed case the emitter resistance R_{es} is dictated by stability requirements, and by the time we become involved in voltage gain calculations, the value of R_{es} has already been determined. In the case of the fully bypassed circuit, the emitter resistance R_e' becomes fixed as soon as we make the collector current decision.

In the split emitter circuit the value of R_{eac} is not computed in any previous step, and we must *select* an *appropriate* value for it. If we try to make an arbitrary decision, we will more often than not make a poor choice. A few ohms one way or the other in the R_{eac} decision can mean the difference between a convenient value for R_L and an impossible one. In order to avoid the pitfalls of guess-and-try, I have provided the split emitter table of possibilities with a superstructure containing appropriate values of R_{Lac} for given val-

[1] An empirical split emitter loading correction curve, discovered by Edward B. Young and Ross Albertson in 1969.

ues of R_{Lw}, V_g, and QI_c. We can look up values for R_{Lac} on the table, then compute R_{eac} from them, and be certain of convenient values all around.

FRAME 6

Compute the value for R_L, the collector load resistor.

At this point we know the value of R_{Lac}, which consists of R_L in parallel with R_{Lw}:

$$R_{Lac} = \frac{R_L \times R_{Lw}}{R_L + R_{Lw}}$$

We also know the value of R_{Lw}, because it was one of the original specifications. What we do not know is R_L. We can manipulate the R_{Lac} equation algebraically to solve for R_L. The result is

$$R_L = \frac{R_{Lw} \times R_{Lac}}{R_{Lw} - R_{Lac}}$$

Here we have a case where we know the equivalent resistance of two resistors in parallel ($R_{Lw} \| R_L$) and the value of only one of the resistors (R_{Lw}), and we must solve for the value of one of the resistors (R_L).

FRAME 7

Compute E_{RL}, the voltage drop across R_L.

$$E_{RL} = QI_c \times R_L$$

This is just a little Ohm's law. We know QI_c and R_L, and we must solve for E_{RL}.

FRAME 8

Decide upon a reasonable voltage drop from collector to emitter of the transistor (E_Q).

This should never be less than 3V and must be large enough to accommodate the required output voltage swing. Rule of thumb: *Minimum* $E_Q = 1 + E_o$ p-p. As the transistor is driven harder by the signal, the voltage drop between collector and emitter decreases. The voltage E_Q can only decrease to 0 V, and any attempt to drive the transistor any harder results in clipping of the top (or bottom) of the waveform. This is a severe form of distortion, and the maximum output voltage E_o is measured just before clipping occurs.

Most textbook equations, and there are several different ones, assume that E_Q can go clear down to 0 V. This is simply not true. Transistor action ceases as soon as (or a little before) the collector-base junction goes into a forward-bias state. At just what voltage (E_Q) this *saturation* begins to occur is quite difficult to predict in most practical situations. It is normally around 0.5 to 0.8 V, but I have seen it as low as 0.2 V and as high as 1.5 V. The rule of thumb is a conservative rule which checks out against the measured value in nearly all circuits. If you prefer a more scientific, but often less satisfactory, equation, I recommend the following as one of the most realistic of those available:

$$E_o \text{ p-p} = 2(E_Q - 0.7)\,\frac{1}{1 + R_{es}/R_{Lac}} \tag{5-1}$$

The 0.7 in the equation is a 0.7-V allowance, because E_Q cannot go to zero.

FRAME 9

Compute the value of the power supply voltage required.

$E_s = E_{RL} + E_Q + E_{Res}$

This is simply the sum of the collector side voltage drops. If this value turns out to be an odd value, something less than the power supply voltage you had in mind, you may increase the value of E_Q to make the supply voltage more convenient.

Perhaps you will realize by the time you complete the first design problem that we have taken an almost frivolous attitude toward the value of E_Q. We found a minimum value for the sake of the desired output swing, then revised it upward in a most arbitrary way to allow for a more convenient power supply voltage. Having assigned an arbitrary value for E_Q, we will not find E_Q mentioned again in the design program; and yet E_Q will end up being almost exactly the voltage assigned to it. Without the feedback loop that I mentioned earlier, the value of E_Q could end up being a most difficult hurdle. Also very important to our flexibility in E_Q is the fact that collector current is *not* a function of collector voltage (except to a very small extent). In base-biased circuits the main power supply voltage (E_s) is distributed among E_{RL}, E_Q, and E_{Res}. The design sheet includes a branch, in this frame, for emitter bias, where $E_s = E_{RL} + E_Q$. In the emitter-biased circuit the voltage drop across R_{es} is supplied by the bias supply, so that E_s is distributed only between E_{RL} and E_Q. See the flowchart (Table 5-2) margin for a reference. In the emitter-biased circuit, E_Q can still be adjusted to allow for more convenient power supply (E_s) voltages.

FIELD 2 BASE SIDE CALCULATIONS

This field involves the computation of resistor values for R_{b1} and R_{b2} (see Fig. 5-1). The computation of R_{b2} is simple and requires only one step.

FRAME 1

Compute R_{b2}.

$R_{b2} = \mathbf{S} \times R_{es}$

where **S** is the stability factor. See Sec. 5-2D for that discussion.

At this point we branch. If the circuit is emitter-biased, we stop here, because there is no R_{b1} in emitter-biased circuits, and proceed to Field 3. If the circuit is base-biased, we must compute a value for R_{b1}, and we go on to Frames 2, 3, and 4. These three frames, in this field, are used for base-biased circuits *only*.

Base Bias Circuit Only Branch

The calculation of R_{b1} involves two preliminary calculations, and so it has been broken up into three frames. The equation for finding R_{b1} is

$$R_{b1} = \frac{E_{Rb1} \times R_{int}}{E_{Res} + V_j} \tag{5-2}$$

Before we can actually plug the values into this equation, we must calculate R_{int} and E_{Rb1}.

FRAME 2

Compute R_{int}.

$$R_{int} = R_{b2} \| \beta R_{es}$$

where R_{es} is the emitter resistor.

R_{int} is the partial input impedance of the base-emitter circuit. It includes R_{b2} and the input resistance due to the emitter resistor (βR_{es}). It does not include R_{b1}, because R_{b1} is not yet in the circuit; and the base-emitter junction resistance is not included because it adds only insignificantly to the total. Any attempt to include the junction resistance for the sake of "rigor" is self-deception because this parameter is impossible to compute with any great accuracy. Later we will encounter circuits where the junction resistance is dominant, and at that time we shall have to make the best of poor approximations of this most elusive parameter; but for now it is better ignored.

FRAME 3

Compute E_{Rb1}.

E_{Rb1} is the voltage across the bias resistor R_{b1}. This voltage is the sum of the voltage drops across E_Q and the load resistor R_L, less the base-emitter junction voltage drop, which is 0.2 V for germanium devices and 0.6 V for silicon. In equation form it looks like this:

$$E_{Rb1} = E_Q + E_{RL} - V_j$$

where $V_j = 0.2$ V for germanium and 0.6 V for silicon.

FRAME 4

Compute R_{b1}.

This is the final step, that of computing the actual resistor value.

$$R_{b1} = \frac{E_{Rb1} \times R_{int}}{E_{Res} + V_j}$$

There is no real mystery to this equation, but let me show you where it came from. Refer to Fig. 5-4 during the following discussion.

1. The part of the input impedance contributed by the transistor and its emitter resistor is $\beta \times R_{es}$. This is a dc parameter in this case, and so the bypass capacitor, if there is one, is not involved. See Fig. 5-4*a*.
2. Adding R_{b2} (see Fig. 5-4*b*), we get a total equivalent base-to-ground resistance of $R_{b2} \| \beta R_{es}$. We call this value R_{int}. The voltage across R_{b2} is about 0.6 V (0.2 for germanium) higher than the voltage drop across R_{es}. We can combine these two into a single equivalent value of $R_{b2} \| \beta R_{es}$ Ω, with a voltage drop of $E_{Res} + V_j$.
3. Now if we add R_{b1}, we have the equivalent of two resistors in series, R_{b1} and $R_{b2} \| \beta R_{es}$. We do not know the current through the equivalent circuit in Fig. 5-4*c*, but we do know it is the same through both resistances. We know the voltage drop across $R_{b2} \| \beta R_{es}$ and its resistance, and the voltage drop across R_{b1}: $E_{Rb1} = E_s - E_{Rb2}$. We can then set up a proportion:

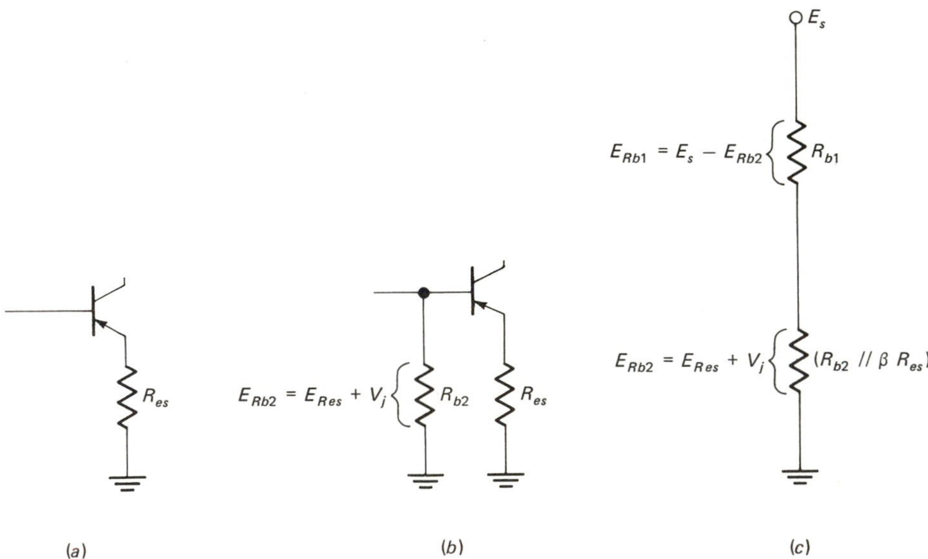

Fig. 5-4 The evolution of the equation for R_{b1}. (a) Looking into the base, we see $\beta \times R_{es}$ Ω; (b) with R_{b2} added, we see $R_{b2} \| \beta \times R_{es}$ Ω; (c) combining R_{b2} and $\beta \times R_{es}$ and adding R_{b1}, we get this result.

$$\frac{E_{Rb1}}{R_{b1}} = \frac{E_{Rb2}}{R_{b2} \| \beta R_{es}}$$

or

$$\frac{E_{Rb1}}{R_{b1}} = \frac{E_{Res} + V_j}{R_{int}}$$

Solving for R_{b1} yields

$$R_{b1} = \frac{E_{Rb1} \times R_{int}}{E_{Res} + V_j}$$

FIELD 3 CAPACITOR AND INPUT IMPEDANCE CALCULATIONS (NOT ON THE DESIGN SHEET)

FRAME 1

Compute the value of the input impedance.

The next step in the procedure is the determination of the coupling capacitor values. The input impedance of the stage and the lowest frequency to be amplified determine the value of the input coupling capacitor, and so we must compute the input impedance first.

Looking into the base, we see the impedance of R_{b1}, in parallel with R_{b2}, in parallel with the transistor's signal input impedance. The transistor's signal input impedance is always β times the effective ac emitter resistance, but because each of the four configurations has a different effective emitter resistance, each has a different total input impedance. The equations are as follows:

For the unbypassed circuit: $Z_{in} = R_{b1} \| R_{b2} \| \beta \times R_{es}$
For the fully bypassed circuit: $Z_{in} = R_{b1} \| R_{b2} \| \beta \times R'_e$
For the split emitter circuit: $Z_{in} = R_{b1} \| R_{b2} \| \beta \times R_{eac}$ (assuming that R_{eac} is 10, or more, times R'_e)
For the differential amplifier: $Z_{in} \approx 2\beta R'_e$

In each case—except the split emitter, where it is a toss-up—one factor dominates, and the input impedance is *approximately* equal to that dominant factor. In the unbypassed circuit, R_{b2} is dominant and, for estimation purposes, $R_{b2} \approx Z_{in}$. In the fully bypassed configuration, the factor $\beta R'_e$ dominates and $Z_{in} \approx \beta R'_e$, again for estimation purposes only. For the differential amplifier, for reasons to be explained later, $Z_{in} \approx 2\beta R'_e$, for both actual calculations and estimation purposes.

FRAME 2

Compute the value of the input coupling capacitor C_{cin}.

We must decide upon a low-frequency cutoff value (this should be a part of the original circuit specifications). We then select a coupling capacitor value, whose reactance at this lowest frequency of interest f_o is approximately one-tenth of the input impedance. In equation form we have

$$X_c \text{ (at } f_o) = 0.1\, Z_{in} \qquad \text{and} \qquad C_{cin} = \frac{1}{2\pi f_o X_c}$$

FRAME 3

Compute the value for the output coupling capacitor C_{co}.

The output capacitor value C_{co} is dependent upon the working load value R_{Lw}. The reactance of C_{co} at the lowest frequency of interest should be about one-tenth the value of R_{Lw}.

Larger capacitor values than those calculated may nearly always be used if physical size and cost are not important, in the laboratory, for example. The use of values smaller than calculated can damage low-frequency response and even reduce the voltage gain at middle frequencies. The output capacitor value C_{co} is dependent upon the working load value R_{Lw}. We will discuss bypass capacitors in Chap. 6. Now let us try our hand at designing a circuit with the following set of specifications.

5-4 THE CURRENT-MODE UNIVERSAL DESIGN SHEET

The design process is necessarily made up of specific calculations made in a definite order. In a sense it is, by nature, a step-by-step process, and any step-by-step procedure carries with it the danger that the student can learn the steps without developing any real insight into, or intuitive feel for, the meaning of those steps and their sequence. I have resorted to several methods and devices in this text to help ensure that you learn the what and why of the design process. One of those devices, which has worked well with my students (after many revisions), is the universal design sheet shown in Fig. 5-5, which has the numerical values filled in for the example. The design sheet is a graphic summary of the design process for all current-mode circuits. I will use the design sheet along with most of the current-mode design examples in this text. I hope that by the time you finish the book, you will have seen the design sheet so many times that you will no longer have any need for it. The design sheet follows the flowchart exactly. There are other design and analysis sheets in the book, and they also are intended as learning tools, and *not* something for you to become dependent upon.

Example 5-1

Design a current-mode common-emitter amplifier with the following specifications.

UNIVERSAL DESIGN SHEET

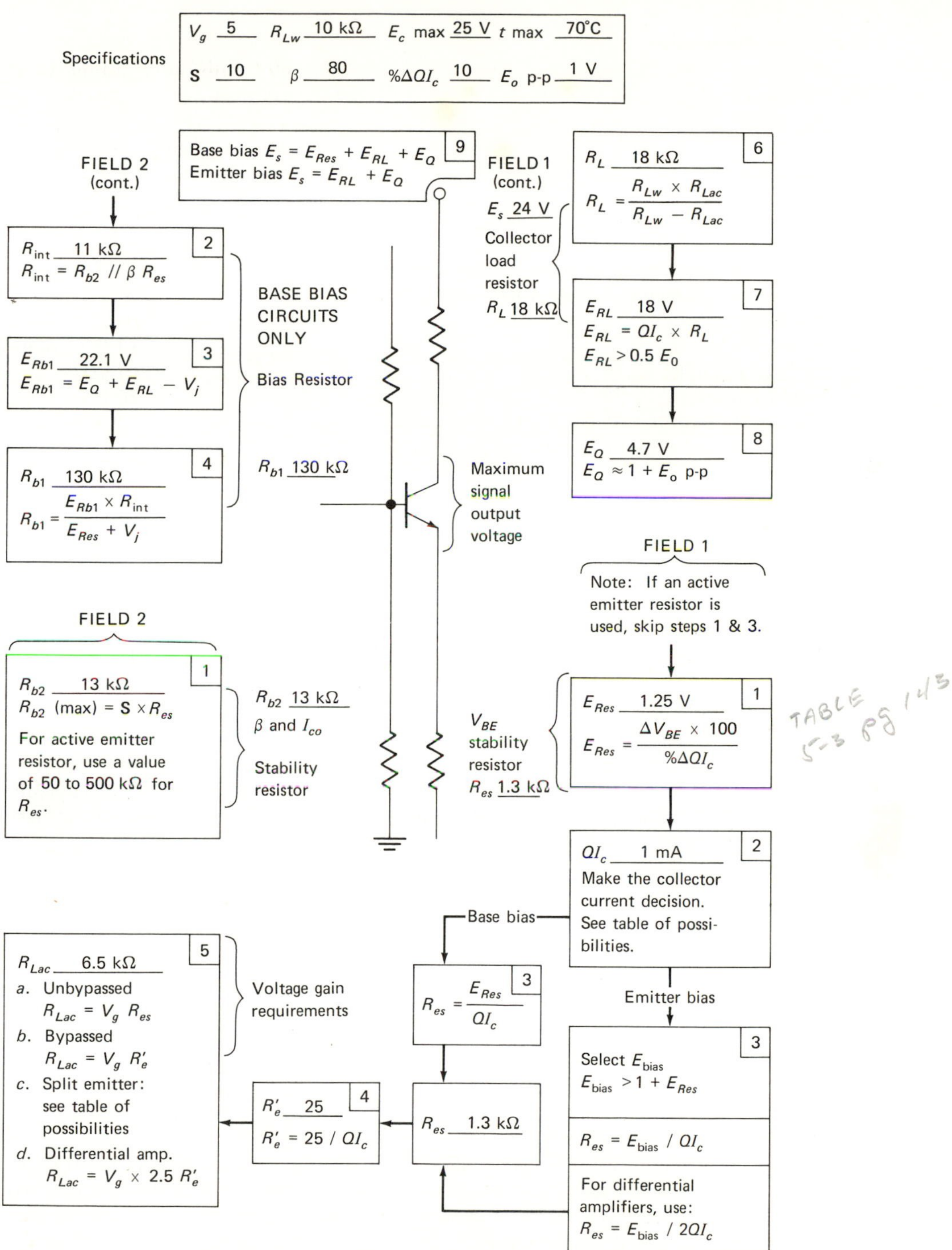

Fig. 5-5 The design sheet for Example 5-1

Transistor Parameters

1. $\beta =$ — 80 (minimum β minus 10%)
2. I_c (max) — 100 mA
3. E_c (max) — 25 V
4. I_{co} — 3.0 μA at 20°C
5. Device type: — *npn* silicon

Circuit Parameters

1. Voltage gain (V_g) — 5
2. Maximum expected operating temperature — 70°C
3. Allowable quiescent current variation at 70°C — 10%
4. Stability factor — 10
5. Input impedance — 5 kΩ or more
6. Working load (R_{Lw}) — 10 kΩ
7. Low-frequency cutoff (f_o) — 20 Hz
8. Maximum output voltage (E_o) peak to peak — 1.0 V

The basic circuit (without values) is shown in Fig. 5-3*a*.

FIELD 1

FRAME 1

Compute the value of E_{Res}, the voltage drop across the emitter resistor.

$$E_{Res} = \frac{\Delta V_{BE} \times 100}{\%\Delta QI_c}$$

Consulting Table 5-3 and assuming 1 mA of QI_c, we find that at a collector current of 1 mA, we can expect the base-emitter junction voltage to decrease by about 2.5 mV/C°. If the highest temperature expected is 70°C and the reference temperature is 20°C, we have a change in temperature of 50°C (70° − 20° = Δt of 50°C). The specifications allow 10 percent variation in the quiescent collector current with an increase in temperature of 50°, and so our equation with values filled in is

$$E_{Res} = \frac{50 \times 2.5 \times 10^{-3} \times 10^{2}}{10\ (\%)} = 1.25\ \text{V}$$

Now look back at the design sheet (Fig. 5-5) to see what you have done and what is coming next. If you do this frequently as you go through each example, it will help you maintain some perspective and keep you from getting so involved in individual steps that you lose sight of the overall picture.

FRAME 2

Make the collector current decision.

For this example, we will use a collector current of 1.0 mA. This is an arbitrary decision, and it is made only to make the arithmetic easy to follow for this first example. We will discuss the collector current decision and the table of possibilities as soon as we have finished this example.

FRAME 3

Compute the value of R_{es}.

$$R_{es} = \frac{E_{Res}}{QI_c}$$

NOTE: Take the base bias branch.

At 1.0 mA of QI_c, this would amount to 1.25 kΩ of resistance. Since 1.3 kΩ is the nearest *standard* value, we shall select 1.3 kΩ for R_{es}. Table 5-4 is a table of *standard* resistor values.

There are two ways to approach the substitution of standard values for resistors in place of computed values, which are often odd.

The first approach is to select the appropriate *standard* value at each step and use that *standard* value in all future steps that involve that particular resistor.

The second approach is to use computed values throughout the design process, and then when it is all finished, select *standard* values for each of the computed values.

TABLE 5-4. STANDARD VALUES FOR RESISTORS

Tolerance		
20%	**10%**	**5%**
10	10	10
		11
	12	12
		13
15	15	15
		16
	18	18
		20
22	22	22
		24
	27	27
		30
33	33	33
		36
	39	39
		43
47	47	47
		51
	56	56
		62
68	68	68
		75
	82	82
		91
100	100	100

NOTE: All numbers are to be multiplied by the desired power of 10. For design work use resistors with highest tolerances when possible.

It would seem logical, at first, to expect different results from the two approaches, but because of the large amount of negative feedback, the differences in final results are negligible. I will demonstrate the first approach in this example. Later I will stick to computed values (and let *you* worry about *standard* values).

FRAME 4

Compute the transistor base-emitter junction resistance R'_e.

$$R'_e \approx \frac{25}{QI_c} = \frac{25}{1} = 25\ \Omega$$

For the circuit in this example this calculation is only for drill; we will not need it. Most of the time this will be a necessary calculation. Frequently we need to know its value even when it is not used in any formal equation.

FRAME 5

Compute the value of the ac collector load R_{Lac}.

This frame sets up the appropriate value for R_{Lac} required to get the specified voltage gain. There are four branches in this frame to account for differences in effective circuit emitter resistances for the various configurations. The circuit in this example is the unbypassed configuration and we will use Branch *a*.

$R_{Lac} = V_g \times R_{es}$ V_g is given in the specifications
$R_{Lac} = 5 \times 1.3\ \text{k}\Omega = 6.5\ \text{k}\Omega$

FRAME 6

Compute the value of the dc load R_L.

$$R_L = \frac{R_{Lw} \times R_{Lac}}{R_{Lw} - R_{Lac}}$$

$$R_L = \frac{10\ \text{k}\Omega \times 6.5\ \text{k}\Omega}{10\ \text{k}\Omega - 6.5\ \text{k}\Omega}$$

$$R_L = 18.57\ \text{k}\Omega$$

The nearest standard resistor value is 18 kΩ.
$R_L = 18\ \text{k}\Omega$ See Table 5-4.

FRAME 7

Compute E_{RL}, the voltage drop across the collector load resistor.

$E_{RL} = QI_c \times R_L$ Ohm's law
$E_{RL} = 1\ \text{mA} \times 18\ \text{k}\Omega$
$E_{RL} = 18\ \text{V}$

The E_o p-p in Frame 7 is simply to remind you that the voltage drop across R_L must be greater than one-half the required p-p output signal voltage. This is only rarely a problem, but it can cause the output swing to be less than you specified. It happens so seldom that you are likely to miss it when it does come up if we don't include it in the design sheet.

FRAME 8

Compute the collector-emitter voltage drop E_Q.

$E_Q \approx 1 + E_o$ p-p

The output swing given in the specification sheet was 1.0 V p-p. Therefore $E_Q \approx 1 + 1 = 2.0$ V. In the next frame, this will yield a supply voltage of 21.3 V; and if we have a 24-V power supply, we can increase E_Q to accommodate it, thus making $E_Q = 4.7$ V.

FRAME 9

Compute the required power supply voltage. NOTE: $E_s < E_{c(\max)}$ (always)

$$E_s = E_Q + E_{RL} + E_{Res} = 4.7 + 18 + 1.3 = 24 \text{ V}$$
$$E_s = 24 \text{ V}$$

Figure 5-6 shows a summary of Field 1 calculations.

FIELD 2 BASE SIDE CALCULATIONS

FRAME 1

Compute the value of the lower base bias resistor R_{b2}.

$$R_{b2} = \mathbf{S} \times R_{es}$$

where **S** is the stability factor, which in this case is 10.

$$R_{b2} = 10 \times 1.3 \text{ k}\Omega$$
$$R_{b2} = 13 \text{ k}\Omega$$

If we were designing an emitter-biased circuit, we would *branch* here to Field 3. Since we are designing a base-biased circuit, we will complete the rest of the frames in this field to get a value for R_{b1}.

FRAME 2

Compute R_{int}, the partial circuit input impedance.

$$R_{int} = R_{b2} \| \beta \times R_{es}$$
$$\beta \times R_{es} = 80 \times 1.3 \text{ k}\Omega = 104 \text{ k}\Omega$$
$$R_{int} = \frac{13 \text{ k}\Omega \times 104 \text{ k}\Omega}{13 \text{ k}\Omega + 104 \text{ k}\Omega}$$
$$R_{int} = 11 \text{ k}\Omega$$

FRAME 3

Compute E_{Rb1}, the upper base bias resistor voltage drop.

$$E_{Rb1} = E_Q + E_{RL} - V_j$$
$$E_{Rb1} = 4.7 + 18 - 0.6$$
$$E_{Rb1} = 22.1 \text{ V}$$

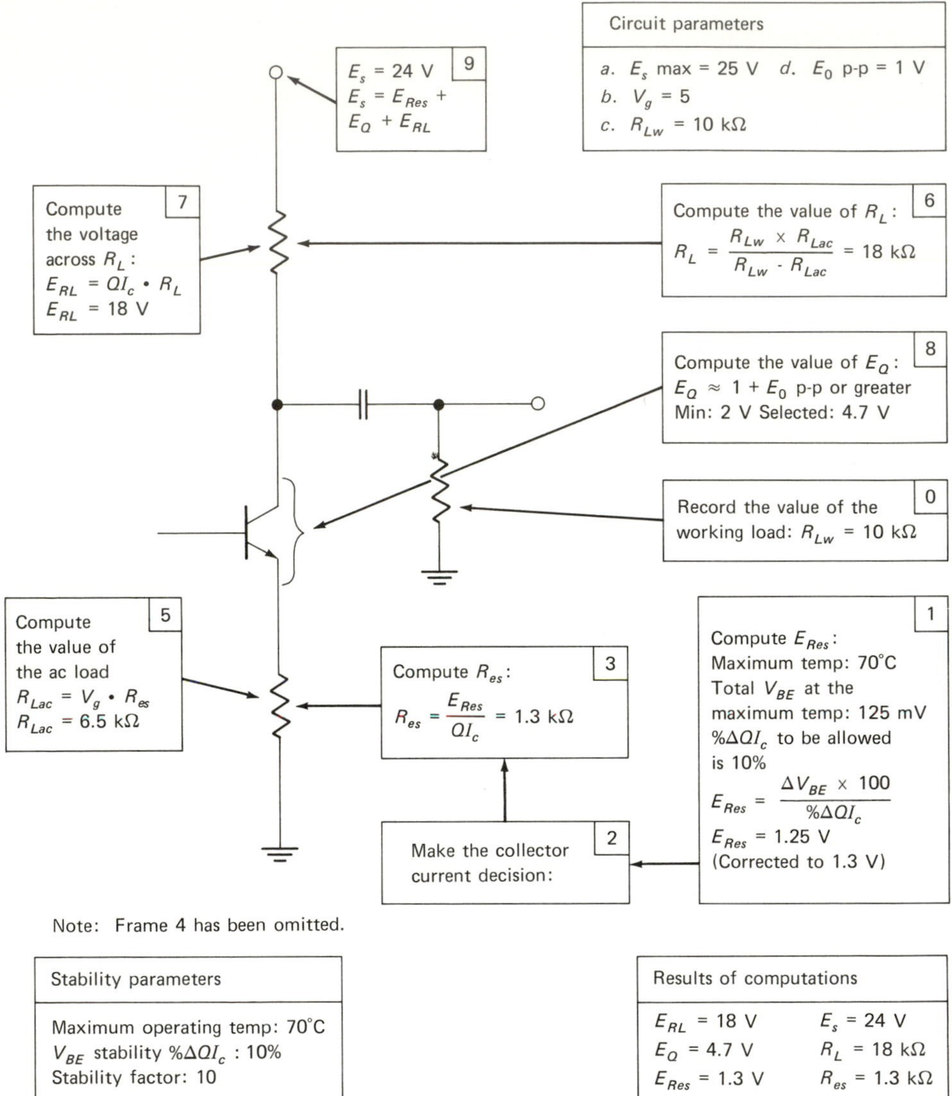

Fig. 5-6 Summary of Field 1 collector side calculations for Example 5-1

FRAME 4

Compute the value of the upper base bias resistor R_{b1}.

$$R_{b1} = \frac{E_{Rb1} \times R_{int}}{E_{Res} + V_j}$$

$$R_{b1} = \frac{22.1 \text{ V} \times 11 \text{ k}\Omega}{1.3 \text{ V} + 0.6 \text{ V}}$$

$$R_{b1} = 130 \text{ k}\Omega$$

Figure 5-7 is a summary of the base side calculations.

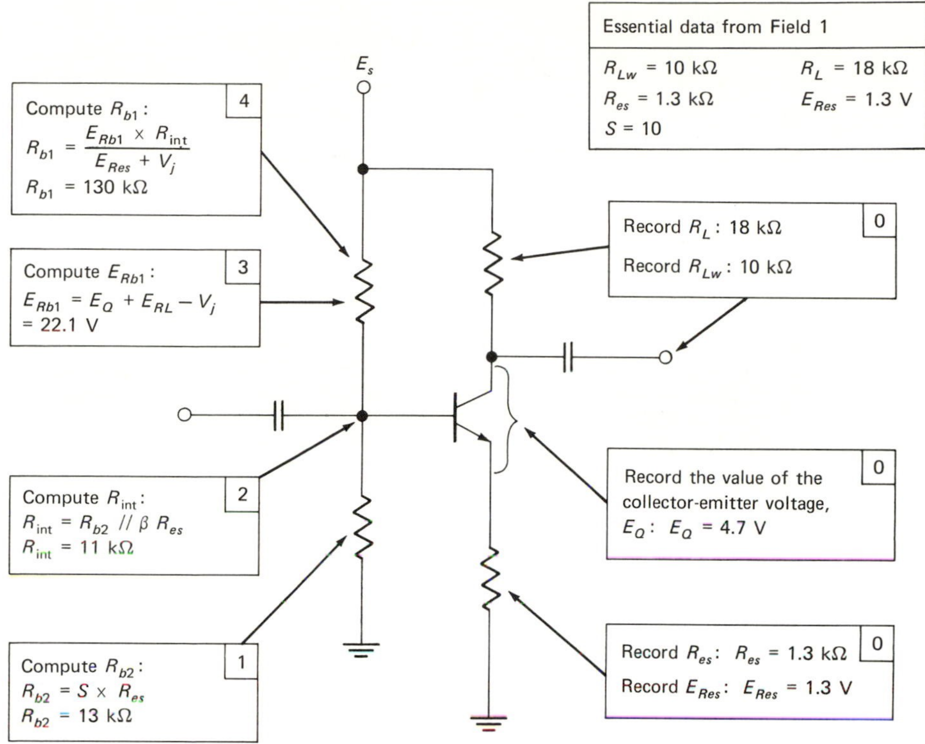

Fig. 5-7 Summary of Field 2 bias calculations for Example 5-1

FIELD 3 INPUT IMPEDANCE AND COUPLING CAPACITOR CALCULATIONS

FRAME 1

Compute the circuit input impedance.

$$Z_{in} = R_{b1} \| R_{b2} \| \beta R_{es}$$

Here we can take a little shortcut. Since we already have computed the value of R_{int}, we can simply find the value of R_{b1} in parallel with R_{int}. Thus,

$$Z_{in} = \frac{R_{int} \times R_{b1}}{R_{int} + R_{b1}} = \frac{11 \times 130 \text{ k}\Omega}{11 \text{ k}\Omega + 130 \text{ k}\Omega}$$

$$Z_{in} \approx 10 \text{ k}\Omega$$

FRAME 2

Compute the value of the input coupling capacitor C_{cin}.

The reactance of C_{cin} should be about 0.1 times the input impedance of the circuit at f_o. This amounts to a reactance of 1,000 Ω.

$$C_{cin} = \frac{1}{2\pi f_o X_c} = \frac{1}{6.28 \times 20 \times 1000}$$

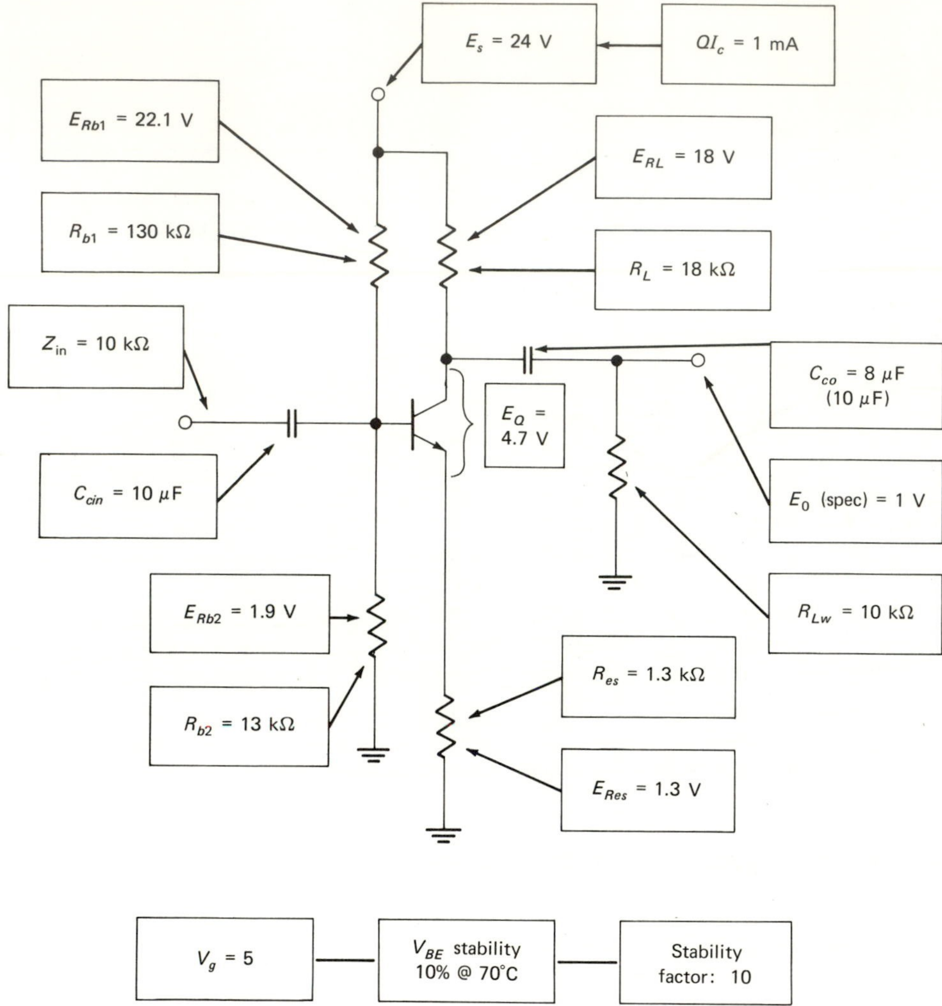

Fig. 5-8 The complete circuit for Example 5-1

The nearest value available is 10 μF.

$C_{cin} = 10\ \mu\text{F}$

FRAME 3

Compute the value of the output coupling capacitor C_{co}.

The specification sheet gives 20 Hz as the lowest frequency of interest and 10 kΩ for R_{Lw}. The output coupling capacitor should have a reactance at 20 Hz (f_o) of about one-tenth of R_{Lw}, or about 1 kΩ.

$$C_{co} = \frac{1}{2\pi f_o X_c}$$

$$C_{co} = \frac{1}{6.28 \times 20 \times 1 \times 10^3}$$

$$C_{co} = 8\ \mu\text{F}$$

Selecting the next common value, we have 10 μF. See Fig. 5-8 for the complete circuit.

5-5 HOW TO AVOID THE IMPOSSIBLE

In the previous example we have apparently picked circuit specifications out of the air and everything has come out all right. This is often the case with textbook examples, but it is important for you to know how the author came up with specifications that did not turn out to be impossible to meet. Here I shall let you in on my secret. Before constructing examples, I took the precaution of designing a number of circuits with a quiescent collector current of 1.0 mA and 1-V drop across the emitter resistor. With these two factors held fixed for all designs, I determined the quiescent voltage drop across the load for quite a number of values of working load at various voltage gains. Then I made a table showing all these conditions. Later on I *scaled* the table to give information on conditions at currents other than 1 mA, and values for E_{Res} other than 1 V. The table now includes most practical, possible conditions for transistors with maximum collector currents up to 100 mA. Those entries marked with an asterisk are impossible. Some other entries may represent conditions which are prohibitive under some circumstances, but they are not impossible. This table is Table 5-5.

How to Use the Table of Possibilities

Table 5-5 is used in the design procedure to make the important, and otherwise difficult, quiescent collector current decision. Later on, when you have become thoroughly familiar with the table by using it for that purpose, you will find it equally useful in setting up your own circuit specifications, which will always be realistic and comfortably workable. These two areas, the collector current decision and the problem of setting realistic specifications, are the most difficult in transistor circuit design. Generally, it takes a good deal of experience to do either of these well, but with the tables even the novice can do a good job of them.

Because the table of possibilities looks so formidable until you know what it is all about, I will try to explain it, provide an example or two in its use, and let you practice using it before we go on. To begin with, we must first know the amount of load the circuit must drive. We call this the working load R_{Lw}, and it, directly or indirectly, is what dictates what we can and cannot do with the circuit. No mechanical engineer would try to design an engine without knowing what size and kind of vehicle it will ultimately drive. A 400-hp engine on the family lawn mower is as absurd as a 3-hp engine in a

TABLE 5-5. TABLE OF POSSIBILITIES FOR THE UNBYPASSED EMITTER CIRCUIT

							Voltage Gains								
						Emitter resistor voltage drops **(E_{Res})**									
						2.0 V	1	2	3	4	5	6	7	8	9
						1.5 V	1.3	2.6	4	5.3	6.6	8	9.3	10.6	12
						1.25 V	1.6	3.2	4.8	6.4	8	10.4	11.6	12.8	14.4
						1.0 V	2	4	6	8	10	12	14	16	18
						0.9 V	2.2	3.4	6.6	8.8	11	13	15.5	17.7	20
						0.8 V	2.5	5	7.5	10	12.5	15	17.5	20	23
						0.7 V	2.8	5.7	8.5	11	14	17	20	22	25.7
						0.6 V	3	6.6	10	13	16.6	20	23	25	30
						0.5 V	4	8	12	16	20	24	28	32	36
R_{Lw} at 0.5 mA	R_{Lw} at 1 mA	R_{Lw} at 2 mA	R_{Lw} at 4 mA	R_{Lw} at 10 mA	R_{Lw} at 20 mA	R_{Lw} at 40 mA	E_{RL}	E_{RL}	E_{RL}	E_{RL}	E_{RL}	E_{RL}	E_{RL}	E_{RL}	E_{RL}
2K	1K	500	250	100	50	25	*	*	*	*	*	*	*	*	*
4K	2K	1K	500	200	100	50	*	*	*	*	*	*	*	*	*
6K	3K	1.5K	750	300	150	75	6	*	*	*	*	*	*	*	*
8K	4K	2K	1K	400	200	100	4	*	*	*	*	*	*	*	*
10K	5K	2.5K	1.25K	500	250	125	3.3	20	*	*	*	*	*	*	*
12K	6K	3K	1.5K	600	300	150	3	12	*	*	*	*	*	*	*
14K	7K	3.5K	1.75K	700	350	175	2.8	9.3	42	*	*	*	*	*	*
16K	8K	4K	2K	800	400	200	2.7	8	24	*	*	*	*	*	*
18K	9K	4.5K	2.25K	900	450	225	2.6	7.2	18	72	*	*	*	*	*
20K	10K	5K	2.5K	1K	500	250	2.5	6.7	15	40	*	*	*	*	*
24K	12K	6K	3K	1.25K	600	300	2.4	6.0	12	24	60	*	*	*	*
30K	15K	7.5K	3.75K	1.5K	700	350	2.3	5.5	10	17	30	60	210	*	*
40K	20K	10K	5K	2K	1K	500	2.2	5.0	8.5	13	20	30	47	80	180
50K	25K	12.5K	6.25K	2.5K	1.25K	625	2.2	4.7	7.9	11	17	23	32	44	64
100K	50K	25K	12.5K	5K	2.5K	1.25K	2	4.3	6.8	9.5	12.5	16	19	24	28
200K	100K	50K	25K	10K	5K	2.5K	2	4	6.4	8.7	11	13.5	17	19	22

* An asterisk signifies an impossible condition.

large truck. By the same measure, we should not attempt to design a transistor amplifier circuit without knowing the load that it must drive and the work it must do. R_{Lw} will often be the input circuit of another stage, and we can compute the value of the input impedance with no trouble at all. At other times it will be some other device, and in that case we must find out what its impedance is. *We must know R_{Lw} in order to design a circuit intelligently.*

For our purposes here I shall, at first, assign values of R_{Lw} which will be selected to demonstrate certain aspects of the design process. Later in the book when we approach multistage circuits, we will find that values for R_{Lw} will be *dictated*, as they nearly always are in the real world of circuit design.

The Importance of E_{RL}

The voltage gain equation is

$$V_g = \frac{R_{Lac}}{R_{es}}$$

where $R_{Lac} = R_{Lw} \| R_L$. The first step in the design procedure is to compute the voltage drop across R_{es}. So, by the second step in the design procedure, we have fixed the voltage drop across R_{es} (first step in the design procedure), the value of R_{Lw} (a specification), and the voltage gain (another specification). The numerator of the voltage gain equation (R_{Lac}) is made up of R_L and R_{Lw}; we can vary only the value of R_L. The denominator of the voltage gain equation is R_{es}, but since the voltage drop across that resistor is already established, and fixed, we can change the resistance only if we also change the current through it. In the case of R_{es}, the current through it is the quiescent collector current.

So it seems we have two things that we can vary, R_L and QI_c. To achieve an increase in voltage gain, we must increase either R_L or QI_c. However, in nearly every case *the optimum circuit uses the lowest practical quiescent collector current.* Practical design demands that we increase R_L first, and only when that ceases to be practical, begin increasing QI_c. Increasing the value of R_L ceases to be practical when the voltage drop across it becomes excessive. For example, a transistor with a maximum collector voltage of 25 V demands that the power supply voltage not exceed 25 V. A 25-V drop across R_L is obviously impossible, because it leaves no voltage available for E_Q (the voltage drop across the transistor) or for E_{Res} (the voltage drop across the emitter resistor). The tables of possibilities are designed so that starting at the lowest collector current value, you can look up the specified value of R_{Lw}, proceed from there to one or more acceptable voltage drop values for E_{RL}, and then look up the table to find out if the *specified* voltage gain can be achieved under those conditions. Some combinations of R_{Lw} and V_g are *impossible* at some quiescent collector currents. The table identifies these impossible conditions by an asterisk. Possible conditions may not always be practical ones, because the value of R_L may have to be increased to a point where the voltage drop across it is excessive. The table is computed directly in terms of the *voltage drop across* R_L, rather than in terms of the resistance of R_L. This not only yields E_{RL} directly (which is what we are interested in), but it also allows the same chart to be used over a wide range of quiescent collector currents.

In order to use the table, we must make some advance estimate of what a reasonable value of E_{RL} is.

1. The sum of $E_{RL} + E_Q + E_{Res}$ *must be less* than the transistor manufacturer's maximum collector voltage rating.
2. E_Q *may never be less than 1 V.* I personally use 3 V as an absolute minimum for E_Q and allow some safety margin. *Let E_Q equal $1 + E_o$ V at least! (And never less than 3 V.)*
3. E_{Res}, the voltage drop across the emitter resistor, is determined as the first step in the design procedure. The table is used in the second step of the design procedure, and so we will know E_{Res}.

All this, taken together, means that an absolute maximum acceptable value for E_{RL} is E_{RL} (max) = E_c (max) − $(1 + E_o)$ − E_{Res}, or E_{RL} (max) = E_c (max) − 3 − E_{Res}. For example, let $E_o = 5$ V, E_c (max) = 25 V, $E_{Res} = 1$ V (computed in Frame 1 of the design procedure). Then E_{RL} (max) = 25 V − (5 + 1) − 1 V, or E_{RL} (max) = 18 V. And, as a second example, let E_o = 500 mV, E_c (max) = 25 V, $E_{Res} = 1$ V. Because the minimum value, in my opinion, for E_Q is 3 V, we will use that value in this example. Because E_o is lower than 3 V, we will use this other "limit."

$$E_{RL}\ (\text{max}) = 25\ \text{V} - 3 - 1$$
$$E_{RL}\ (\text{max}) = 21\ \text{V}$$

In a while, you will be able to perform the arithmetic involved in determining the maximum acceptable E_{RL} in your head while you are using the table. Suppose we try an example of the collector current decision.

Example

Select the most appropriate quiescent collector current for a circuit with the following specifications:

$$R_{Lw} = 6\ \text{k}\Omega$$
$$E_c\ (\text{max}) = 20\ \text{V}$$
$$V_g = 4$$
$$E_o = 3\ \text{V p-p}$$

And let us assume that E_{Res} computed out, in the first design step, at 1.0 V.

1. First find the maximum acceptable E_{RL}.
   E_{RL} (max) $= E_c - (1 + E_o) - E_{Res}$
   E_{RL} (max) $= 20 - (1 + 3) - 1$
   E_{RL} (max) $= 20 - 5 = 15$ V
2. Find the column in Table 5-5 labeled "R_{Lw} at 0.5 mA." We are starting with the *lowest* value of collector current listed in the table. Quiescent currents lower than 0.5 mA are sometimes used, but not very often in this kind of circuit.
3. Look down the column labeled "R_{Lw} at 0.5 mA" until you find the value of R_{Lw} which appeared in the specifications. In this case, we are looking for $R_{Lw} = 6$ kΩ.
4. Look across that *row* ($R_{Lw} = 6$ kΩ). *All* E_{RL} values that we are concerned with appear in that *row* and nowhere else. The only E_{RL} entry in that *row* is a 6 (volts). The rest of the *row* contains only asterisks, indicating that the voltage gains listed in the superstructure above those asterisks are impossible to attain at this particular quiescent current (0.5 mA). Because 6 (volts) is the only entry, let us see what voltage gain is possible with 6 V for E_{RL}.

A Word about the Voltage Gain Entries

All voltage gains must be taken with respect to some value of E_{Res}, which means that we are only interested in the entries in the *row* (in the superstructure) corresponding to the actual value of E_{Res} and no others.

5. Make a left turn at the 6 in the R_{Lw} at 0.5 mA = 6 kΩ *row*, and go up the column until you come to the intersection of this *column* and the $E_{Res} = 1$ V *row*. The entry there is a 2. This means a voltage gain of 2 is all that is possible in this *column*. Normally, our next move would be to move to the right (along the R_{Lw} at 0.5 mA row) to a higher E_{Rl} value, turn left, and go up the column again and check the voltage gain. We would continue moving to the right, to higher values of E_{RL}, until we found a value which would yield the desired voltage gain. In this case, we are stopped by an asterisk. In other cases we will encounter the situation where E_{RL} gets too high by the time we reach the desired voltage gain. In either case, we *must* move to a higher quiescent collector current. Let us do that in this case.
   a. Remember the maximum E_{RL} is 15 V.
   b. Find the *column* in Table 5-5 labeled "R_{Lw} at 1 mA." Look down the column until you find the specified value of R_{Lw}, 6 kΩ.
   c. Move to the right until you come to the first entry under the E_{RL} heading (3 V), turn left, and go up the *column* until it intersects the $E_{Res} = 1$ V *row*. We find there a voltage gain of 2. We want a gain of 4.
   d. Go back to the R_{Lw} at 1 mA = 6 kΩ *row*. Move across the *row*. Pass the 3 and continue moving right to the next entry, which is a 12. Twelve volts is still less than the maximum E_{RL} of 15 V; so turn left and go up the *column* until it intersects the $E_{Res} = 1$ V *row* in the superstructure. Here we find an entry of 4. And there it is, the voltage gain of 4 with an E_{RL} of less than 15 V. So it seems that 1 mA is the lowest quiescent collector current at which this circuit is possible, and 1 mA would be our selection. We have no reason to go to any higher current, although the circuit will be *possible* at any current 1 mA and above.

What If We had Wanted a Gain of 3 or 5?

A voltage gain of 3 lies between the values listed in the $E_{Res} = 1$ V *row*. Values of 2, 4, 8, etc., are listed, but not 3, 5, etc. For a voltage gain of 3, we can interpolate between an E_{RL} of 3 V and an E_{RL} of 12 V. The interpolation is not linear, and the safe way is to shoot for the next higher listed gain. Go for the next higher listed voltage gain is my advice for now. When you have developed some *feel* for the table, then you can consider interpolation. By this rule, a gain of 5 for our example is impossible, because the next higher listed gain of 6 has an asterisk for its entry for E_{RL}. Not interpolating can be wasteful, and I advise you to begin to do it as soon as you have sufficient confidence to do so.

What If the Exact Computed Value of E_{Res} Is not Listed in the Table?

Again, as you become more familiar with the table, you may wish to interpolate, but for now I suggest that you use the next *higher* listed value of E_{Res}.

A Little Bit of Practice

1. Find the lowest collector (Q point) current at which the following circuit is possible.
   Specifications:
   $R_{Lw} = 1.5$ kΩ
   $E_{Res} = 0.8$ V
   $V_g = 5$
   $E_o = 3$ V p-p
   E_c (max) transistor collector voltage 20 V
   a. Find the maximum permissible value for E_{RL}.
   b. Find the smallest QI_c which will yield a voltage gain of 5 with an E_{RL} lower than that computed in part *a*.
2. Given the following sets of specifications, find the lowest value of QI_c at which the circuits are possible.
   a. Specifications:
   $R_{Lw} = 10$ kΩ
   $V_g = 4$
   E_c (max) = 20 V
   $E_o = 1.0$ V p-p
   $E_{Res} = 1$ V
   b. Specifications:
   $R_{Lw} = 4$ kΩ
   $V_g = 4$
   E_c (max) = 15 V
   $E_o = 1$ V
   $E_{Res} = 0.95$ V

Answers

1*a*. 15.2 V
b. 4 mA
2*a*. 1 mA
b. 2 mA

The table of possibilities is valuable for more than simply making the collector current decision, and the more you use it, the more you will find it can be used for. With a little study on your part, how the chart works will become quite obvious; but I will explain its use as best I can as we proceed with the next example.

Example 5-2 A Design Problem

The design problem specifications:

Transistor Parameters

1. Device type:	silicon
2. β	50 (minimum minus 10%)
3. I_c max	100 mA
4. E_c max	50 V

Circuit Parameters

1. Voltage gain (V_g)	10
2. Maximum expected operating temperature	70°C
3. Allowable quiescent current variation at 70°C	10%
4. Stability factor	10
5. Input impedance	5 kΩ or more
6. Working load (R_{Lw})	2 kΩ
7. Low-frequency cutoff (f_o)	20 Hz
8. Maximum output voltage (E_o) peak to peak	1.0 V

The basic circuit (without values) is shown in Fig. 5-9. Figure 5-10 shows the completed design sheet.

FIELD 1 THE COLLECTOR SIDE CALCULATIONS

FRAME 1

Compute the value of the E_{Res}, the voltage drop across the emitter resistor.

$$E_{Res} = \frac{\Delta V_{BE} \times 100}{\% \Delta QI_c} \qquad \text{See Fig. 5-9}$$

Consulting Table 5-3, we find that at a collector current of 1.0 mA we can expect the base-emitter junction voltage to decrease by about 2.5 mV/C°. If the highest temperature expected is 70°C, we have a change in temperature of 50°C (70° − 20° = Δt of 50°C). The specifications allow 10 percent variation in the quiescent collector current with an increase in temperature of 50°, and so our equation with values filled in is

$$E_{Res} = \frac{50 \times 2.5 \times 10^{-3} \times 10^2}{10\ (\%)} = 1.25\ \text{V} \qquad \text{(standard resistor yields 1.3 V)}$$

This frame involves a dilemma, because we must know at least the range of QI_c before we can compute E_{Res}; and we cannot make the QI_c decision without knowing E_{Res}. For the purpose of this example, we will make a conservative *estimate* of QI_c, which will yield a higher value of E_{Res} than necessary; and then select QI_c on the basis

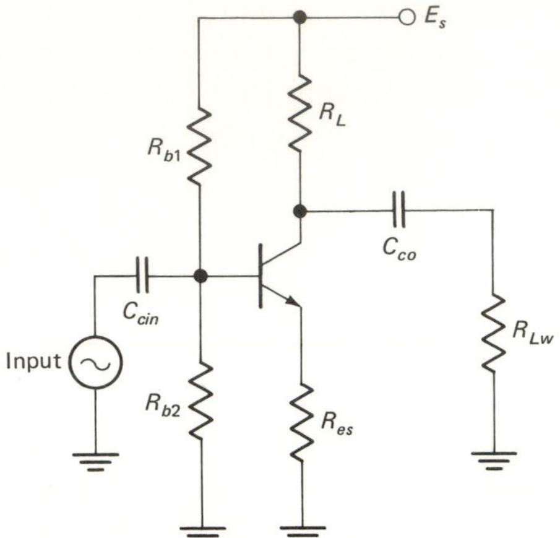

Fig. 5-9 The schematic diagram for Example 5-2

of that value of E_{Res}. This will always yield a workable circuit, but it may eliminate some possibilities at the extreme ends of the possible.

If the conservative estimate proves too limiting, we can, with a little experience, revise E_{Res} until we optimize the circuit. This procedure is normally not necessary. Here we have selected a 1 mA *estimate*, because it is likely to be a conservative value. By conservative, I mean that the required QI_c is not likely to be any lower than 1 mA.

FRAME 2

Make the collector current decision. Now is the time to use the table of possibilities (Table 5-5).

1. Can we use 0.5 mA?
   Find the value for R_{Lw} of 2 kΩ under the heading "R_{Lw} at 0.5 mA." Read across the row until you find a value of E_{RL} that will do. In this case, we find only asterisks and no values at all. An asterisk signifies an impossible condition, and so there is no way we can use 0.5 mA for QI_c with a 2-kΩ load.
2. Can we use 1 mA?
   We find the value 2 kΩ under the 1 mA heading, and again we find a row of asterisks. Therefore, a 1-mA QI_c is impossible with a 2-kΩ load and a V_g of 10.
3. Can we use 2 mA?
   We find an R_{Lw} value of 2 kΩ under the 2 mA heading and read across the row until we come to a number. In this case we find a 4, which means 4 V across the collector load resistor. Four volts for E_{RL} is quite respectable, but if we move vertically up that column (starting at the 4 V E_{Res} entry), we find that no voltage gains as high as 10 are listed. So it appears that 2 mA won't do. As a matter of fact, if we examine the point where our E_{RL} *column* of 4 V intersects the $E_{Res} = 1.25$ V *row*, we will find a voltage gain of only 1.6; the specifications call for a voltage gain of 10.

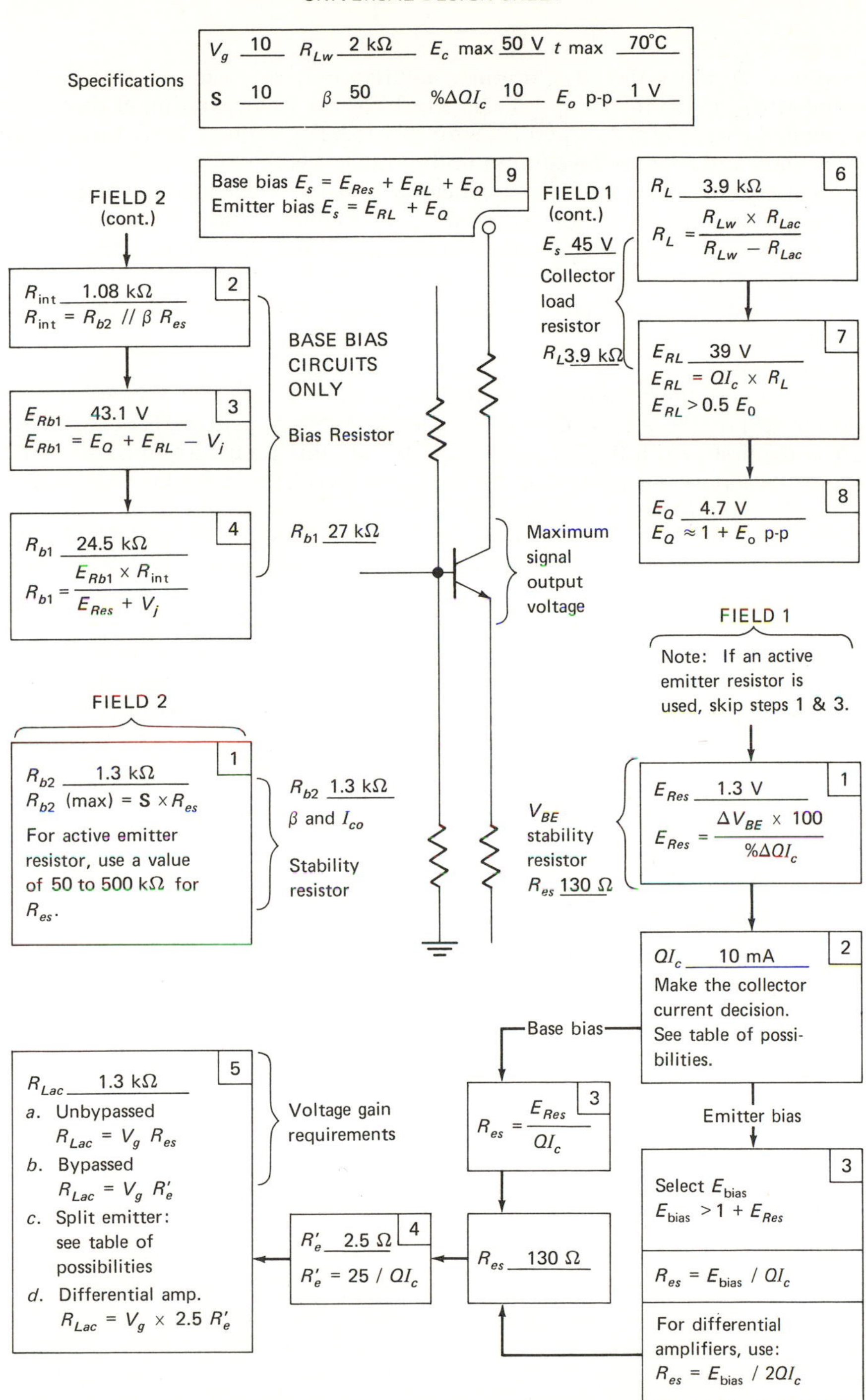

UNIVERSAL DESIGN SHEET
Specifications
V_g 10 R_{Lw} 2 kΩ E_c max 50 V t max 70°C
S 10 β 50 %ΔQI_c 10 E_o p-p 1 V
9 Base bias $E_s = E_{Res} + E_{RL} + E_Q$
Emitter bias $E_s = E_{RL} + E_Q$
FIELD 2 (cont.)
2 R_{int} 1.08 kΩ
$R_{int} = R_{b2} // \beta R_{es}$
3 E_{Rb1} 43.1 V
$E_{Rb1} = E_Q + E_{RL} - V_j$
4 R_{b1} 24.5 kΩ
$R_{b1} = \frac{E_{Rb1} \times R_{int}}{E_{Res} + V_j}$
BASE BIAS CIRCUITS ONLY
Bias Resistor
R_{b1} 27 kΩ
FIELD 1 (cont.)
E_s 45 V
Collector load resistor
R_L 3.9 kΩ
6 R_L 3.9 kΩ
$R_L = \frac{R_{Lw} \times R_{Lac}}{R_{Lw} - R_{Lac}}$
7 E_{RL} 39 V
$E_{RL} = QI_c \times R_L$
$E_{RL} > 0.5\ E_0$
8 E_Q 4.7 V
$E_Q \approx 1 + E_o$ p-p
Maximum signal output voltage
FIELD 1
Note: If an active emitter resistor is used, skip steps 1 & 3.
FIELD 2
1 R_{b2} 1.3 kΩ
R_{b2} (max) = S × R_{es}
For active emitter resistor, use a value of 50 to 500 kΩ for R_{es}.
R_{b2} 1.3 kΩ
β and I_{co}
Stability resistor
V_{BE} stability resistor
R_{es} 130 Ω
1 E_{Res} 1.3 V
$E_{Res} = \frac{\Delta V_{BE} \times 100}{\%\Delta QI_c}$
2 QI_c 10 mA
Make the collector current decision. See table of possibilities.
Base bias
Emitter bias
3 $R_{es} = \frac{E_{Res}}{QI_c}$
3 Select E_{bias}
$E_{bias} > 1 + E_{Res}$
$R_{es} = E_{bias} / QI_c$
For differential amplifiers, use:
$R_{es} = E_{bias} / 2QI_c$
R_{es} 130 Ω
4 R'_e 2.5 Ω
$R'_e = 25 / QI_c$
5 R_{Lac} 1.3 kΩ
a. Unbypassed
$R_{Lac} = V_g\ R_{es}$
b. Bypassed
$R_{Lac} = V_g\ R'_e$
c. Split emitter: see table of possibilities
d. Differential amp.
$R_{Lac} = V_g \times 2.5\ R'_e$
Voltage gain requirements

Fig. 5-10 The completed design sheet for Example 5-2

4. Can we use 4 mA?
Here we read: (*a*) down the "R_{Lw} at 4 mA" *column* until we reach 2 kΩ, which is the value of R_{Lw} given in the specifications; (*b*) to the right, starting at the 2 kΩ entry, we find entries of 2.7, 8, and 24 V for possible E_{RL} values. If we turn left at the 2.7 V entry and read up the *column* until it intersects the $E_{Res} = 1.25$ V *row*, we find a voltage gain of 1.6. We want a voltage gain of 10. If we turn left at $E_{Res} = 8$ V ($QI_c =$ mA and $R_{Lw} = 2$ kΩ) and read up that *column* until it intersects the $E_{Res} = 1.25$ V row, we get a V_g of 3.2. This still won't do. If we turn left at $E_{Res} =$ 24 V and read up that *column* until it intersects the $E_{Res} = 1.25$ V *row*, we find a V_g of 4.8. This is almost half enough, and it appears that a QI_c of 4 mA is not adequate.
5. How about 10 mA?
The intersection of the *column* $E_{RL} = 30$ V and the *row* $E_{Res} = 1.25$ V yields a voltage gain of 10.4. This is very close to the specified voltage gain of 10. So a QI_c of 10 mA is the first possibility we have come to. At this point, we must stop and ask a very important question: *Can the transistor stand the collector voltage dictated here by the table?* In this case, $E_{RL} = 30$ V according to the chart. We have already decided that we would need a voltage drop across R_{es} of 1.25 V. So far, we need a power supply voltage of at least 31.25 V. We must then add the voltage across the transistor (E_Q). E_Q must be greater than $1 + E_o$ p-p. In this case, the specifications call for an E_o of 1 V; thus $E_Q =$ (at least) $1 + 1 = 2$ V. The grand total is $E_{RL} + E_{Res} + E_Q$, or $30 + 1.25 + 2 = 33.25$ V. We must have a power supply voltage of not less than 33.25 V. The transistor must, therefore, have a maximum collector voltage of at least 33.25 V; and some safety margin is preferred. If we go back to the specifications sheet, we find that the transistor we selected is capable of a collector voltage of 50 V; and we are home free. If we had specified a transistor with a maximum collector voltage of, say, 30 V, our only alternative would be to find another transistor, to go to a higher QI_c, or to abandon the whole problem! And so it looks as though 10 mA is the smallest value of QI_c that we can use. Normally, we should not use a collector current higher than necessary. Thus, for this circuit, $QI_c = 10$ mA.

FRAME 3

Compute the value of R_{es}.

$R_{es} = E_{Res}/QI_c$.

At 10 mA of current, this would be 125 Ω of resistance. Since 130 is the nearest standard value, we shall select 130 Ω for R_{es}. (See Table 5-4 for standard resistor values.) Use the base bias branch.

FRAME 4

Compute the value of the transistor's base-emitter junction resistance R'_e (the Shockley relationship).

$$R'_e \approx \frac{25}{QI_c} = \frac{25}{10} = 2.5\ \Omega$$

(Here, again, we will not have need of this figure.)

FRAME 5

Compute the value of R_{Lac}, the ac load.

$R_{Lac} = V_g \times R_{es} = 10 \times 130 = 1.3\ \text{k}\Omega$
$R_{Lac} = 1.3\ \text{k}\Omega$

Use Branch *a* for the unbypassed circuit.

FRAME 6

Compute the value of the dc load.

$$R_L = \frac{R_{Lw} \times R_{Lac}}{R_{Lw} - R_{Lac}} = \frac{2\ \text{k}\Omega \times 1.3\ \text{k}\Omega}{2\ \text{k}\Omega - 1.3\ \text{k}\Omega} = 3.7\ \text{k}\Omega$$

$R_L = 3.9\ \text{k}\Omega$, the nearest standard value

FRAME 7

Compute E_{RL}.

$E_{RL} = QI_c \times R_L$
$E_{RL} = 10 \times 10^{-3} \times 3.9 \times 10^3$
$E_{RL} = 39\ \text{V}$

FRAME 8

Compute E_Q.
$E_Q \approx 1 + E_o$ p-p
$E_Q \approx 1 + 1 = 2.0$ (3 V min preferred)

If, at this point, we add $E_{Res} + E_Q + E_{RL}$, we get $1.3 + 2.0 + 39 = 42.3$. Assuming, for no special reason, that a 45-V power supply is available, we can adjust E_Q to 4.7 V. Thus $1.3 + 4.7 + 39 = 45$ V for E_s.

FRAME 9

Compute the required power supply voltage E_s (see Frame 8).

FIELD 2

FRAME 1

Compute the value of the base stability resistor R_{b2}. The specification sheet gives the stability factor a value of 10.

$R_{b2} = \mathbf{S} \times R_{es} = 10 \times 130 = 1{,}300\ \Omega$
$R_{b2} = 1.3\ \text{k}\Omega$

FRAME 2

Compute R_{int}

$R_{int} = R_{b2} \| \beta R_{es}$ where $\beta = 50$

$$R_{int} = \frac{6.5\ \text{k}\Omega \times 1.3\ \text{k}\Omega}{6.5\ \text{k}\Omega + 1.3\ \text{k}\Omega} = 1.08\ \text{k}\Omega$$

$R_{int} = 1.08\ \text{k}\Omega$

FRAME 3

Compute E_{Rb1}.

$E_{Rb1} = E_Q + E_{RL} - V_j$
$E_{Rb1} = 4.7\ \text{V} + 39\ \text{V} - 0.6\ \text{V}$
$E_{Rb1} = 43.1\ \text{V}$

FRAME 4

Compute R_{b1}.

$$R_{b1} = \frac{E_{Rb1} \times R_{int}}{E_{Res} + V_j}$$

$$R_{b1} = \frac{43.1\ \text{V} \times 1.08\ \text{k}\Omega}{1.9\ \text{V}}$$

$$R_{b1} = \frac{46.548}{1.9\ \text{V}}$$

$R_{b1} = 24.5\ \text{k}\Omega$

The next higher standard value (by Table 5-4) is 27 kΩ.

FIELD 3 CAPACITOR AND INPUT IMPEDANCE COMPUTATIONS

FRAME 1

Compute the circuit input impedance.

$Z_{in} = R_{b1} \| R_{b2} \| \beta R_{es}$

Here we can take a little shortcut. Since we already have computed the value of R_{int}, we can simply find the value of R_{b1} in parallel with R_{int}. Thus,

$$Z_{in} = \frac{R_{int} \times R_{b1}}{R_{int} + R_{b1}} = \frac{1.08\ \text{k}\Omega \times 27\ \text{k}\Omega}{1.08\ \text{k}\Omega + 27\ \text{k}\Omega}$$

$Z_{in} \approx 1\ \text{k}\Omega$

FRAME 2

Compute the value of the input coupling capacitor C_{cin}.

The reactance of C_{cin} should be about 0.1 times the input impedance of the circuit at f_o. This amounts to a reactance of 100 Ω.

$$C_{cin} = \frac{1}{2\pi f_o X_c} = \frac{1}{6.28 \times 20 \times 96}$$

The nearest value available is 100 μF. $C_{cin} = 100$ μF. Figure 5-11 shows the complete circuit and all important voltage and resistance values.

FRAME 3

Compute the value of the output coupling capacitor C_{co}.

The specification sheet gives 20 Hz as the lowest frequency of interest and 2 kΩ for R_{Lw}. The reactance of the output coupling capacitor should have a reactance at 20 Hz (f_o) of about one-tenth of R_{Lw}, or about 200 Ω.

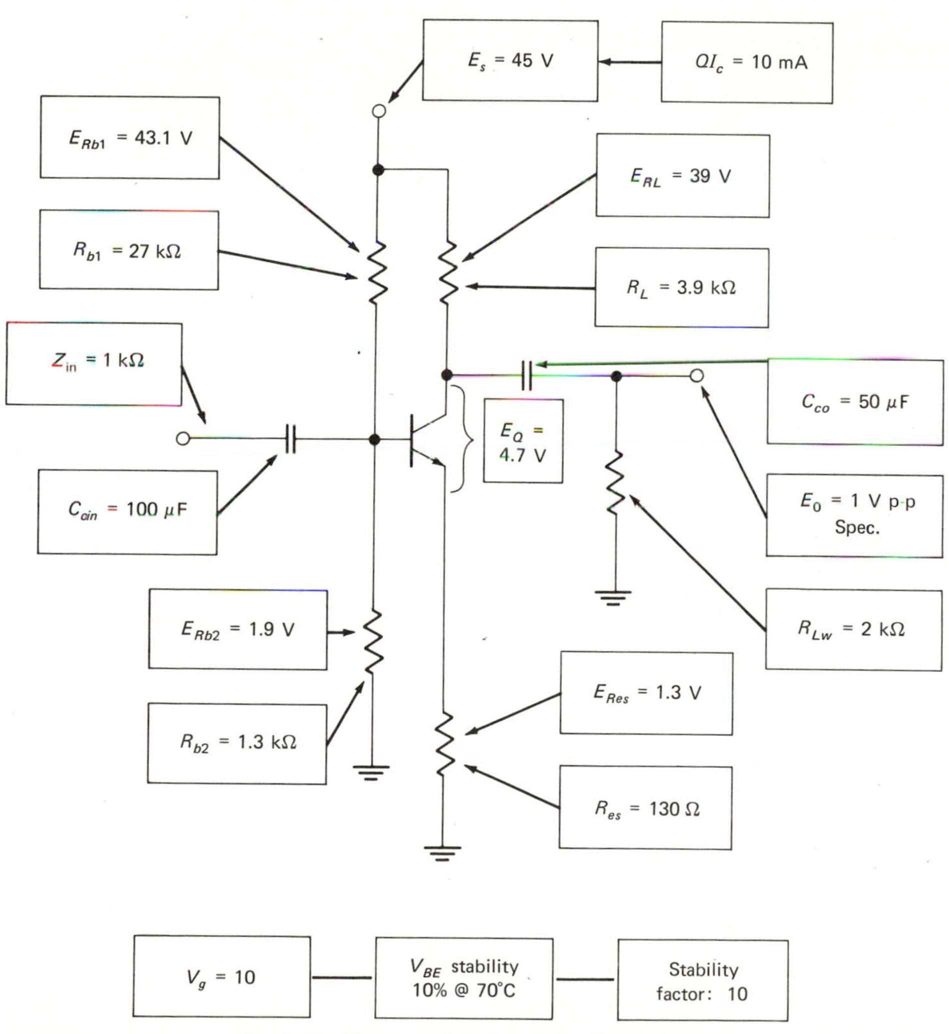

Fig. 5-11 The complete circuit for Example 5-2

$$C_{co} = \frac{1}{2\pi f_o X_c}$$

$$C_{co} = \frac{1}{6.28 \times 20 \times 2 \times 10^2}$$

$$C_{co} \approx 40\ \mu F$$

Selecting the next common value, we have 50 μF.

EXERCISES

Here is some practice in using Table 5-5, the table of possibilities. Use it to make the collector current decision for the sets of conditions in Exercises 5-1 to 5-4.

5-1 $V_g = 10$ E_c max $= 30$ V I_c max $= 100$ mA
$R_{Lw} = 5$ kΩ $E_{Res} = 1.0$ V
ANS. a. QI_c ________ b. E_{RL} ________

5-2 $V_g = 10$ E_c max $= 30$ V I_c max $= 100$ mA
$R_{Lw} = 10$ kΩ $E_{Res} = 0.8$ V
ANS. a. QI_c ________ b. E_{RL} ________

5-3 $V_g = 5$ E_c max $= 30$ V I_c max $= 100$ mA
$R_{Lw} = 10$ kΩ $E_{Res} = 0.8$ V
ANS. a. QI_c ________ b. E_{RL} ________

5-4 $V_g = 6$ E_c max $= 30$ V I_c max $= 100$ mA
$R_{Lw} = 15$ kΩ $E_{Res} = 0.9$ V
ANS. a. QI_c ________ b. E_{RL} ________

5-5 What voltage may be varied in the design process to allow the use of a predetermined power supply voltage?
5-6 Which voltage drop prevents excessive drift in quiescent collector current due to variations in V_{BE}?
5-7 A circuit has a stability factor of 10. What is its approximate current gain?
5-8 What factors limit the maximum possible voltage gain in the simple unbypassed circuit?

Problem 5-1

TITLE: **A Case of Significant Working Load**
PROCEDURE: **Design, construct, and measure the parameters of a current-mode common-emitter amplifier, with the following specifications.**
Note: See Laboratory Manual for Measurement Techniques.

Transistor Parameters

1. β	50 (min)
2. I_c (max)	100 mA
3. E_c (max)	20 V

Circuit Parameters

1. Voltage gain	5
2. E_o p-p	4.0 V
3. R_{Lw}	12 kΩ
4. Stability factor	10
5. $\%\Delta QI_c$ to be allowed:	5
6. Maximum operating temperature:	60°C
7. f_o	25 Hz

Problem 5-2

TITLE: **A Case Where the Load Controls the Selection of the Quiescent Collector Current.**

PROCEDURE: **Design, construct, and measure the parameters of a current-mode emitter amplifier, with the following specifications.**

Transistor Parameters

1. β	80
2. I_c (max)	100 mA
3. E_c (max)	at least 25 V

Circuit Parameters

1. Voltage gain	8
2. E_o p-p	2.0 V
3. R_{Lw}	2.5 kΩ
4. Stability factor	9
5. $\%\Delta QI_c$ to be allowed:	10%
6. Maximum operating temperature:	65°C
7. f_o	25 Hz

5-6 BOOTSTRAPPING FOR HIGHER INPUT IMPEDANCES

Sometimes stability requirements dictate a value for R_{b2}, in a base-biased circuit, which makes the input impedance of the stage uncomfortably low for some applications. When this happens, a technique known as bootstrapping can be used to make R_{b2} appear *very* large to the input signal. See Fig. 5-12. Figure 5-12*a* shows the simplest case, where only R_{b2} is bootstrapped, and we will discuss this case in detail. Figure 5-12*b* shows how this technique can be applied to both R_{b1} and R_{b2}. The operation is simple but effective. The lower base bias (stability) resistor R_{b2} is split into two equal parts, as shown in Fig. 5-12*a*, and a large capacitor is connected between the emitter and the junction of the split R_{b2}. There is nothing sacred about splitting R_{b2} into equal parts, and some designers try to come up with the absolute maximum input impedance by selecting some other ratio of $R_{b2}a$ to $R_{b2}b$, but in practice the improvement is generally too small to make the effort worthwhile.

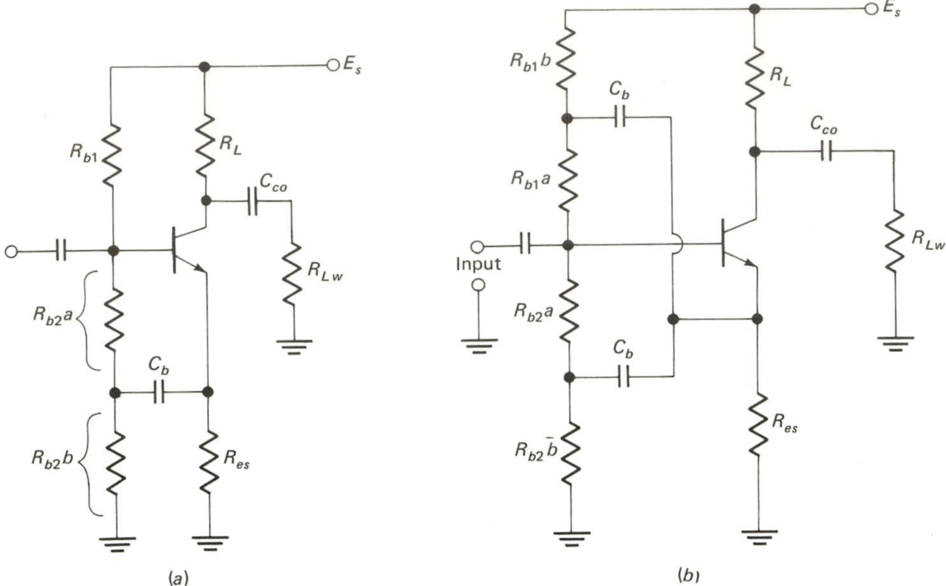

Fig. 5-12 Bootstrapping. (a) Bootstrapping R_{b2}; (b) bootstrapping both R_{b1} and R_{b2}

How It Works

As the base (Fig. 5-12) goes more positive, so does the emitter. The base end of the resistor $R_{b2}a$ goes more positive and, at the same time, the capacitor raises the lower end of $R_{b2}a$ to a more positive value. This is a case of positive feedback, and if 100 percent of the input signal were fed through C_b to the bottom of $R_{b2}a$, the potential difference of signal across $R_{b2}a$ would always be zero. If the potential difference across a resistor is 0 V, there will be no current through that resistor. Now if a voltage (the signal voltage in this case) is applied across a resistor and 0A of current flows through the resistor, the (signal) source *sees* what *appears* to be an infinite impedance. The positive feedback through C_1 is not 100 percent because of the signal voltage drop across the base-emitter junction. The effective value of $R_{b2}a$ is given by the equation

$$R_{b2}a\ (\text{eff}) \approx \frac{R_{b2}a}{1 - V_g} \tag{5-3}$$

In this case V_g is the voltage gain from base to emitter: essentially a common-collector amplifier is driving C_1. The voltage gain of a common-collector circuit is

$$V_g \approx \frac{1}{1 + (R_e'/R_{Lac})} \tag{5-4}$$

where $R_{Lac} \approx R_{es} \| R_{b2}b$. Since we deliberately make R_{es} fairly large compared to R'_e (by making E_{Res} large compared to ΔV_{BE}), we can assume that V_g approaches, but never quite reaches, a value of 1. If you will examine Eq. (5-3), you can see that as V_g approaches 1, the effective value of $R_{b2}a$ approaches infinity. For the circuit in Fig. 5-12*a*, the input impedance is

$$Z_{in} \approx R_{b1} \| \beta \left(R_{es} \middle\| \frac{R_{b2}}{2} \right)$$

(assuming that $R_{b2}a = R_{b2}b$). For Fig. 5-12*b*, where positive feedback is applied to both R_{b1} and R_{b2}, the input impedance is

$$Z_{in} \approx \beta(R_{es} \| \tfrac{1}{2} R_{b2} \| \tfrac{1}{2} R_{b1})$$

or about 250 kΩ, whichever is less. Once the base-emitter circuit impedance gets above 100 kΩ or so, we must start including the base-collector junction impedance as a factor. For practical purposes we can count upon the circuit of Fig. 5-12*b* yielding an input impedance of at least 100 kΩ, but there are such intangibles, as an accurate knowledge of β and V_g, that we cannot *depend* upon a total input impedance of *much more* than 100 kΩ, or $\beta \times R_{es}$, whichever is lower. Still, there is often a considerable advantage in bootstrapping.

To design a bootstrapped circuit, simply complete the design process as demonstrated in this chapter. Then split R_{b1} and R_{b2} into two halves, and add the capacitors C_d. The capacitor feeding R_{b2} should have a reactance at the lowest frequency of interest of 0.1 $R_{b2}b$, and the capacitor feeding R_{b1} should have a reactance of 0.1 $R_{b1}b$.

The flowchart (Table 5-1) and design chart (Fig. 5-5) will serve as a summary for this chapter.

Some Important Practical Information

1. E_q in practical circuits must be at least 3 V regardless of the E_o p-p required.
2. Ideally E_{Res} should never be less than 2 V. This is not always possible in the unbypassed circuit. It is almost always possible in fully bypassed, split emitter, and differential circuits.

WHAT HAPPENS IF YOU VIOLATE THESE TWO RULES?

1. These two factors are interrelated, but it is always the value of E_Q that suffers. It will be lower than predicted, and it may be too low for the transistor to work at all. In most cases a fairly large error in E_Q will not keep you from meeting all the signal specifications, but the collector side quiescent voltages will be off. Table 5-6 shows a statistical summary of this situation.

TABLE 5-6. TABLE OF PERCENTAGE OF ERROR BETWEEN COMPUTED AND MEASURED VALUES FOR BASE-BIASED CIRCUITS

	$E_{Res} = 1$ V Typ.	Max	$E_{Res} = 1.5$ V Typ.	Max	$E_{Res} = 2$ V Typ.	Max
E_{Res}	+21%	±28%	±6.8%	±20%	±6%	±20%
E_{RL}	+20%	±25%	±8.5%	±20%	+8%	±20%
E_Q	−45%	−55%	−18%	−25%	−10.5%	±25%
E_{Rb2}	±15%	±20%	±6.2%	±20%	+4%	±20%
V_g	−10%	±20%	−10%	±20%	−5%	−20%
E_o p-p	+50%	−5% +100%	+50%	−5% +100%	+45%	−5% +100%
Z_{in}	+10%	+20%	+10%	+20%	+9%	+20%
QI_c	+10%	±20%	+8%	±15%	±5%	±15%

S = 10

2. What to do about selecting standard resistance values:
   a. Always make these resistors the nearest higher value—never lower R_{b1} and R_{es}.
   b. Always make R_{b2} and R_L the next lower value—never higher.
   These are practical rules, but they make good sense too. Increasing R_{es} or lowering R_L increases the amount of negative feedback, making the circuit more stable. Decreasing R_{b2} adds to stability and helps reduce the bias and raise E_Q slightly. Increasing R_{b1} also reduces the bias and raises E_Q.
   c. Once in a while you find a properly designed circuit that has a volt or so for E_Q indicating an overbiased condition. This can happen because of accumulated tolerances. Raising R_{b1} to the next higher standard value will nearly always cure the problem.

6

designing for higher voltage gains

OBJECTIVES

(Things you should be able to do upon completion of this chapter)
NOTE: See the end of the chapter for specific design problems.

1. Design the *fully bypassed* common-emitter current-mode single-stage amplifier circuit.
2. Design a *split emitter* common-emitter current-mode single-stage amplifier circuit.
3. Design a fully bypassed emitter-biased common-emitter single-stage circuit.

EVALUATION:

1. There are some self-test questions and design problems at the end of the chapter.
2. The final evaluation will be based upon your ability to design circuits that *work* in the laboratory according to the criteria and specifications stated in the laboratory manual.

INTRODUCTION

If you have investigated the circuits in Chap. 5 in the laboratory, you have come to expect a very close correlation between computed characteristics and measured ones. Quiescent operating voltages and currents, as well as gain parameters, are very predictable in those circuits. Voltage gains for the circuits in Chap. 5, however, were quite low. In this chapter we will trade off some of the predictability (with respect to signal parameters only) for nearly a 40-fold increase in voltage gain.

In addition to a reduction in predictability, we will lose the capability of amplifying direct currents and will introduce a definite low-frequency cutoff value. The circuits we will be exploring in this chapter involve only the minor modification of adding an emitter bypass capacitor to the unbypassed circuit in Chap. 5; all the dc quiescent rules and stability requirements will be the same.

The Fully Bypassed Circuit ($V_g \approx$ 40 to 500)

The first circuit we will study is completely dependent upon transistor parameters, insofar as voltage gain and input impedance are concerned. Our ability to predict, or design in, signal parameters then hinges upon our ability to predict or control the transistor's parameters. The act of modifying the circuit to obtain the maximum voltage gain relinquishes some of our control over signal parameters.

The transistor's input impedance is a very complex function of many factors, and its predictability is not good. There is an empirical (approximate) equation which we will be forced to rely upon because, even though it often yields errors of 30 to 50 percent in practical circuits, it is the best and only really useful one in existence. The voltage gain is also partly a function in this circuit of the input impedance, and so it will also be only approximately predictable.

This uncertainty about signal parameters is not as serious a problem as it may seem; and in the face of the nearly fantastic improvement in voltage gain, we shall find it (like the inconvenience of being rich) quite easy to live with.

The Split Emitter Circuit ($V_g \approx$ 10 to 40)

The second circuit is called the split emitter circuit, and it is an effective compromise between predictability and gain. Gains of from 10 to 40 are possible with the predictability of voltage gain and input impedance nearly as good as that of the unbypassed circuit in Chap. 5.

This circuit is slightly more expensive than the unbypassed circuit or the fully bypassed circuit. It is one of the more difficult circuits to design. The circuit is very useful when predictable and stable signal parameters are required, and it will often be found in high-quality equipment.

The Fully Bypassed Emitter-biased Circuit ($V_g \approx$ 40 to 500)

The third circuit we will study in this chapter is basically a fully bypassed circuit with a modified bias circuit. The circuit requires a separate bias supply, which makes it a little less than overwhelmingly popular. The principal reason for presenting it here is that this circuit is an excellent example of the principles and practices of emitter biasing in its simplest form. Later on we will study a very important circuit, called the differential amplifier, where emitter bias, in a slightly more complex form, will be the preferred bias method.

6-1 THE FULLY BYPASSED CIRCUIT

The fully bypassed circuit is shown in Fig. 6-1. You will notice that the only difference between this and the common-emitter amplifier circuits in Chap. 5 is the large capacitor across the emitter resistor R_{es}. In the circuits in Chap. 5, we did not have to be concerned with a lowest "amplifiable" frequency (except for coupling capacitors), because the circuits had as much gain (small though it was) for 0 Hz as they did for mid-frequencies. In the fully bypassed circuit, we will have much higher gains at mid-frequencies than we had with the unbypassed circuits in Chap. 5, but we will also have a definite low-frequency cutoff.

The voltage gain for frequencies below the cutoff frequency will drop rapidly as the frequency decreases. The mid-frequencies will see gains in the order of 40 to 500, where frequencies much below the low-frequency cutoff

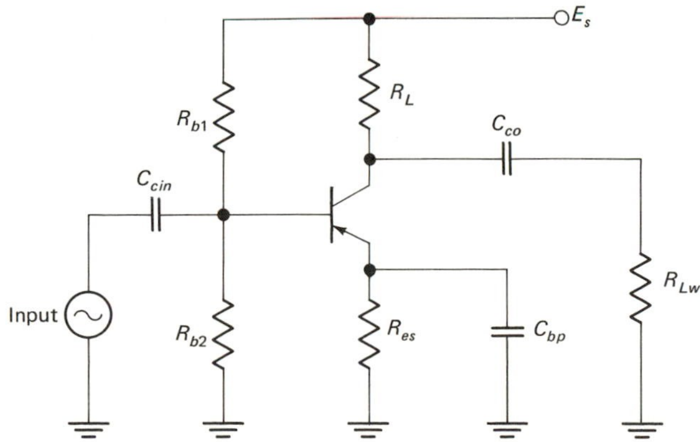

Fig. 6-1 The schematic diagram for Example 6-1, the fully bypassed circuit

value will see gains of from 1 to 10. In practice, this amounts to nearly a total loss of all frequencies much below the low-frequency cutoff value. We can, of course, select any desired value of low-frequency cutoff, but any low-frequency cutoff value below about 10 Hz results in a prohibitively large bypass capacitor (many microfarads, many cubic inches, and much money).

This low-frequency limitation is no real problem, except for certain instrumentation systems where amplification at frequencies approaching 0 Hz is required. For those cases the circuits in Chap. 5 can be used if we are content with the available gains; or we can use more sophisticated circuitry such as the differential amplifier, which will be discussed in detail in Chap. 11.

We also must exchange some of our accuracy in predicting or designing in voltage gain and input impedance because the function of the bypass capacitor is to provide an ac (signal) short across the emitter resistor.

In Chap. 5, we effectively swamped out the transistor's input impedance by developing a large feedback voltage drop across the emitter resistor. This made all circuit parameters independent of the transistor parameters and gave the designer nearly complete control over *all* circuit parameters. The bypass capacitor effectively reduces the signal-frequency impedance from emitter to ground to 0 Ω. This means that we have two sets of parameters for the circuit, the signal parameters and the dc parameters.

Beyond the difficulty in predicting signal parameters, we will also find these parameters varying over a fairly wide range as the temperature varies. We have given up gain stability in exchange for raw gain. We have also placed our signal at the mercy of any nonlinearities in the transistor's transfer curve. This means slightly more distortion than was typical of the unbypassed circuit.

We *can* design the dc circuit (just as we did in Chap. 5), place a capacitor of sufficient size across R_{es}, and have a quite functional fully bypassed circuit with a great deal of gain. *The addition of the bypass capacitor has no effect upon either bias or stability.* However, we will not do it quite that way, because some of our quiescent design decisions in Chap. 5 were, in part, predicated upon signal characteristics and limitations. And since those signal conditions no longer apply once the bypass capacitor is installed, we will alter the quiescent design process slightly to optimize our control over the final signal parameters.

We shall have the same kind of predictable control over quiescent voltages and currents, and over stability, as we did with the circuits in Chap. 5. We will not be so fortunate in our control of signal parameters.

The Shockley Relationship

With the emitter resistor shorted out for signal currents, the emitter circuit impedance is now the transistor's internal emitter resistance R_e'. The signal voltage gain equation is now

$$V_g = \frac{R_{Lac}}{R'_e}$$

This equation is correct enough, but R'_e is a most variable and undependable parameter. As you can see, the accuracy of any voltage gain calculation is dependent upon how accurately we can estimate the value of R'_e.

The best estimate to date was discovered by William Shockley, one of the inventors of the transistor. It is an empirical relationship, which lumps all the many variables involved in R'_e into one simple bulk equation.

The Shockley relationship is

$$R'_e \approx \frac{26}{I_e}$$

where I_e is the quiescent emitter current in mA and 26 is a constant (actually millivolts). The voltage gain equation is

$$V_g = \frac{R_{Lac}}{R'_e} \qquad \text{where } R'_e \approx \frac{26}{I_e} \tag{6-1}$$

I have found that it is easier to see this equation if two slight modifications are made. First, if we take the reciprocal of 25 ($25 \approx 26$) and learn the equation that way, it makes the arithmetic easier. Secondly, since we have become accustomed to thinking in terms of collector current, and since $I_c \approx I_e$, let us substitute QI_c in the equation for I_e. Thus the equation becomes

$$V_g \approx 0.04 QI_c R_{Lac} \tag{6-2}$$

where $0.04 = 1/25$, QI_c is the quiescent collector current in milliamperes, and R_{Lac} is the ac (signal) collector load impedance ($R_{Lac} = R_L \| R_{Lw}$).

It is fairly obvious from Eq. (6-2) that higher quiescent operating points yield higher gains. The maximum value of R_{Lac} is generally controlled by the value of the working load (R_{Lw}); so even though an increase in R_{Lac} would yield an increase in gain, we are normally quite restricted in how much we can raise R_{Lac}. It would seem that operating high on the collector current curve would be a clever thing to do. The truth is that, all things considered, 1 or 2 mA is typical for this circuit as it has been for most previous ones.

Aside from the obvious fact that higher operating currents *can* be a minor hazard to stability, there is another more important limit to how high we can set QI_c in any practical circuit. For example, let us suppose that we have a value for R_L of 10 kΩ (and an R_{Lw} of 100 kΩ). At a collector current of 1 mA the gain would be $V_g \approx 0.04 \times 1 \times 10 \times 10^3$. $V_g \approx 400$ and the voltage drop across R_L would be 10 V (10 kΩ at 1 mA).

It would seem logical to increase the current to 10 mA (still a safe current for a 100-mA transistor) and thereby increase the voltage gain to an astronomical 4,000. The voltage drop across R_L now becomes a problem.

$E_{RL} = 10\ \text{k}\Omega \times 10\ \text{mA}$
$E_{RL} = 100\ \text{V}$

This too, is a bit astronomical for most transistors. In practice, loaded voltage gains generally run from about 50 to 200.

The Input Impedance

In the circuits we have previously designed, the input impedance has been given by the equation $Z_{in} = R_{b1} \| R_{b2} \| \beta R_{es}$. We ignored R'_e in that equation, because we intentionally made R_{es} large compared to R'_e. In this circuit, for ac (signal), we have shunted R_{es} with a capacitor of very low reactance at signal frequencies. With R_{es} effectively reduced to 0 Ω, R'_e becomes the only ac emitter resistance in the circuit, and we can no longer ignore it. The signal input impedance is now

$$Z_{in}\ (\text{ac}) \approx R_{b1} \| R_{b2} \| \beta R'_e \qquad (6\text{-}3)$$

$$\text{where } R'_e \approx \frac{25\ (\text{mV})}{QI_c\ (\text{mA})} \qquad QI_e \approx QI_c$$

Thus, for a transistor with a β of 100 operating at a collector current of 1 mA, we have

TABLE 6-1. COMPARISON OF DC AND SIGNAL PARAMETERS FOR THE FULLY BYPASSED CIRCUIT

Parameter	DC Quiescent Parameters	Signal Parameters
Voltage gain	$V_g = \frac{R_{Lac}}{R_{es}}$ Range ≈ 1 to 10	$R'_e \approx 25/QI_c$ $V_g = \frac{R_{Lac}}{R'_e}$ Range ≈ 40 to 500
Current gain	$I_g \approx$ **S** Range ≈ 1 to 10	I_g Range ≈ 20 to 300
Stability factor	**S** Range ≈ 1 to 10	Range ≈ 20 to 300
Input impedance	$R_{b1}//R_{b2}//\beta R_{es}$ Range ≈ 2 kΩ to 50 kΩ	$R_{b1}//R_{b2}//\beta R'_e$ Range ≈ 500 Ω to 5 kΩ

$Z'_{in}\text{ (ac)} \approx 100 \times \frac{25}{1}$ $\quad Z'_{in}$ does not include the effect of R_{b1} and R_{b2}

$Z'_{in}\text{ (ac)} \approx 2{,}500\ \Omega$, or $2.5\ \text{k}\Omega$ $\quad$ (at a QI_c of 10 mA, $Z'_{in} \approx 250\ \Omega$)

We have now introduced β into the system, and β is another of those highly variable and none too dependable parameters. We now have a case where both R'_e and β are subject to large variations, and we can expect any computation of Z_{in} to be little more than an approximation.

Table 6-1 compares dc parameters and signal parameters for the fully bypassed circuit.

6-2 DESIGNING THE FULLY BYPASSED COMMON-EMITTER CIRCUIT

The first thing we must do before tackling the design of a fully bypassed common-emitter circuit is to examine the capabilities and limitations of such a venture. And to facilitate this examination, I have prepared Table 6-2, the table of possibilities for the fully bypassed circuit. The table required a great deal of effort to prepare, and I would appreciate it if you would refer to it frequently. The introduction to this chapter was perhaps a bit verbose, but it had an important message. That message was, "The best we can ever do here is to make a useful approximation."

Please do not be deceived by expressions in the table, such as 6.2 V, because they only mean *approximately* 6.2 V. The ".2" is really more a devotion to arithmetic than to any devotion to accuracy. In spite of this admonition, you will find that Table 6-2 is accurate enough for most practical design or analysis purposes. It is based upon the Shockley relationship and can be no more accurate than that, but remember that the Shockley relationship is the best we have and we must make that do.

How to Use the Table

The table furnishes the essential information about capabilities and limitations of the circuit. The first thing we must do is set up a list of specifications, but we must make them realistic. We can use Table 6-2 to help in that decision. The following example has been prepared, using Table 6-2.

Example 6-1

Design a fully bypassed common-emitter amplifier with the following specifications. See Fig. 6-1.

Transistor Parameters

1. Device type:	pnp silicon
2. β	100 (minimum β minus 10%)
3. I_c max	100 mA
4. E_c max	25 V

TABLE 6-2. TABLE OF POSSIBILITIES FOR THE FULLY BYPASSED CIRCUIT

Working Loads at Various Values of QI_c							Voltage Drop Across the Collector Load Resistor (R_L) for Various Voltage Gains								
R_{Lw} at 0.5 mA	R_{Lw} at 1.0 mA	R_{Lw} at 2.0 mA	R_{Lw} at 4.0 mA	R_{Lw} at 10 mA	R_{Lw} at 20 mA	R_{Lw} at 40 mA	E_{RL} at $V_g =$ 40	E_{RL} at $V_g =$ 60	E_{RL} at $V_g =$ 80	E_{RL} at $V_g =$ 100	E_{RL} at $V_g =$ 140	E_{RL} at $V_g =$ 160	E_{RL} at $V_g =$ 240	E_{RL} at $V_g =$ 320	E_{RL} at $V_g =$ 400
2K	1K	500	250	100	50	25	*	*	*	*	*	*	*	*	*
4K	2K	1K	500	200	100	50	2.0	6	*	*	*	*	*	*	*
6K	3K	1.5K	750	300	150	75	1.5	3	6	15	*	*	*	*	*
8K	4K	2K	1K	400	200	100	1.3	2.4	4	6.6	28	*	*	*	*
10K	5K	2.5K	1.25K	500	250	125	1.25	2.1	3.3	5.0	11.6	20	*	*	*
12K	6K	3K	1.5K	600	300	150	1.2	2.0	3	4.3	8.4	12	*	*	*
14K	7K	3.5K	1.75K	700	350	175	1.16	1.9	2.8	3.9	7.0	9.3	42	*	*
16K	8K	4K	2K	800	400	200	1.1	1.8	2.7	3.6	6.2	8	24	*	*
18K	9K	4.5K	2.25K	900	450	225	1.1	1.8	2.6	3.5	5.7	7.2	18	72	*
20K	10K	5K	2.5K	1K	500	250	1.1	1.7	2.5	3.3	5.4	6.7	15	40	*
24K	12K	6K	3K	1.2K	600	300	1.1	1.7	2.4	3.1	5.0	6.0	12	24	60
30K	15K	7.5K	3.75K	1.5K	700	350	1.1	1.6	2.3	3.0	4.5	5.5	10	17	30
40K	20K	10K	5K	2K	1K	500	1.1	1.6	2.2	2.8	4.2	5.0	8.5	13	20
50K	25K	12.5K	6.25K	2.5K	1.25K	625	1.1	1.6	2.2	2.7	4.0	4.7	7.9	11	17
100K	50K	25K	12.5K	5K	2.5K	1.25K	1.0	1.6	2	2.6	3.7	4.3	6.8	9.5	12.5
200K	100K	50K	25K	10K	5K	2.5K	1.0	1.6	2	2.5	3.6	4	6.4	8.7	11

* An asterisk signifies an impossible condition.

Circuit Parameters

1.	Voltage gain (V_g)	85–100
2.	Maximum expected operating temperature	70°C
3.	Allowable quiescent current variation at 70°C	10%
4.	Stability factor	10
5.	Input impedance	1 kΩ or more
6.	Working load (R_{Lw})	5 kΩ
7.	Low-frequency cutoff (f_o)	20 Hz
8.	Maximum output voltage (E_o) peak to peak	3.0 V

The basic circuit (without values) is shown in Fig. 6-1.

FIELD 1 THE COLLECTOR SIDE CALCULATIONS

FRAME 1

Compute the value of the voltage drop across the emitter resistor E_{Res}.

$$E_{Res} = \frac{\Delta V_{BE} \times 100}{\%\Delta QI_c}$$

Consulting Table 5-3, we find that at a collector current of 1.0 mA, we can expect the base-emitter junction voltage to decrease by about 2.5 mV/C°. If the highest temperature expected is 70° and the reference temperature is 20°C, we have a change in temperature of 50°C (70° − 20° = Δ*t* of 50°C). The specifications allow 10 percent variation in the quiescent collector current with an increase in temperature of 50°, and so our equation with values filled in is

$$E_{Res} = \frac{50 \times 2.5 \times 10^{-3} \times 10^2}{10(\%)} = 1.25 \text{ V}$$

FRAME 2

Make the collector current decision.

Here, we must use Table 6-2 to determine what quiescent collector current and *approximately* what collector load voltage drop are acceptable. We may vary either of these values a bit later on in the design, but we do need a jumping-off place which will not end in an unsatisfactory or even impossible condition later.

In the previous chapter, I said that it was a good policy to operate at the lowest reasonable collector current. That is still good advice, even though the Shockley relationship promises higher gain at higher currents. When we begin to couple stages together, we will find that the lowered input impedance often more than offsets the increase in gain at higher currents.

The transistor has a maximum collector voltage rating of 25 V with a 1.25 drop across the emitter resistor, and at least a 4-V drop allowed for E_Q; the absolute maximum voltage drop across R_L is 19.75 V. The required voltage gain is 100, and the working load is 5 kΩ. Let us examine Table 6-2 and see what collector currents are possible.

1. Is 0.5 mA possible?
   The table shows that R_{Lw} = 4 kΩ is not possible, 6 kΩ is, and 5 kΩ is not listed.

At 6 kΩ the voltage drop for a gain of 100 is 15 V. Just guessing, 5 kΩ for R_{Lw} and a V_g of 100 would probably require about 30 V drop across R_L. This is well beyond our 25-V limit, and so 0.5 mA does not look practical.

2. How does 1 mA look?
   Well, for an R_{Lw} value of 5 kΩ, a collector current of 1 mA, and a voltage gain of 100, the voltage drop across R_L will be 5.0 V. This looks very promising.
3. How about 2 mA?
   2 mA and an R_{Lw} of 5 kΩ require, according to Table 6-2, a 3.3 voltage drop across R_L for a V_g of 100. 3.3 V is certainly little enough voltage drop, but is it perhaps too little? Remember that we have a specification for peak-to-peak output voltage, in this case 3 V. In this case, assuming a high enough voltage drop for E_Q, there is no problem. We could operate at 2 mA.
4. And perhaps 4 mA?
   Here we would have only 2.8 V across R_L. This is still possible and probably quite workable.

In summary: The figure 0.5 mA is shaky at best; 1 mA yields a nice comfortable 5-V drop across R_L, and it is the lowest *practical* operating current. Now if battery drain or some other factor makes the lowest *possible* current imperative, a design at 0.5 mA might be possible and would be worth a try. However, 1 mA is more practical, and under all *normal* circumstances, for this set of specifications, I would elect to operate at 1.0 mA.

If I seem to have some sort of prejudice in favor of a 1-mA quiescent collector current, I can only admit to it. But please remember that β often peaks at about 1 mA (and that probably means that the input impedance peaks about there also); and if there is no other reason, a 1-mA collector current makes the arithmetic very easy.

In this particular case, it is still about the lowest *practical* operating current according to Table 6-2. So, even though 2 or 4 mA might be quite acceptable, I still recommend 1 mA $QI_c = 1.0$ mA. (At this point you may recognize that you can get more from Table 6-2 than the quiescent operating current, but please don't try to do it yet.)

FRAME 3

Compute the value of R_{es}.

$$R_{es} = E_{Res}/QI_c$$

At 1.0 mA of current this would amount to 1.25 kΩ of resistance. Since 1.3 kΩ is the nearest standard value, we shall select 1.3 kΩ for R_{es} (see Table 5-4 for standard resistor values). Use the base bias branch for this problem.

FRAME 4

Compute the base-emitter junction resistance R_e'.

$$R_e' = \frac{25}{QI_c} = \frac{25}{1} = 25\ \Omega$$

In this case we will need this figure in the next frame and again later in computing capacitor values.

FRAME 5

Compute the value of the ac load R_{Lac}. (Use Branch *b*.)

$$R_{Lac} = V_g \times R_e'$$

$$R_{Lac} = 100 \times R_e' \qquad \text{where } R_e' \approx \frac{25}{QI_c} \approx 25$$

$$R_{Lac} = 2.5\ \text{k}\Omega$$

FRAME 6

Compute the value of the dc load R_L.

$$R_L = \frac{R_{Lw} \times R_{Lac}}{R_{Lw} - R_{Lac}}$$

Figure 6-2 shows the completed design sheet for this example for your reference as you follow it through.

$$R_L = \frac{5\ \text{k}\Omega \times 2.5\ \text{k}\Omega}{5\ \text{k}\Omega - 2.5\ \text{k}\Omega}$$

$$R_L = 5\ \text{k}\Omega$$

The nearest standard resistor value is 4.7 kΩ.

$$R_L = 4.7\ \text{k}\Omega$$

FRAME 7

Compute the voltage drop across R_L.

$$E_{RL} = QI_cR_L = 1 \times 10^{-3} \times 4.7 \times 10^3 = 4.7\ \text{V}$$

FRAME 8

Compute the collector-emitter voltage drop E_Q.

$$E_Q \approx 1 + E_o\ \text{(p-p)}$$

The output swing given in the specification sheet was 3.0 V p-p. Therefore $E_Q \approx 1 + 3 = 4.0$ V. $E_Q = 4.0$ V. (If necessary, this value can be increased to accommodate a given supply voltage E_s.)

FRAME 9

Compute the required power supply voltage.

$$E_s = E_Q + E_{RL} + E_{Res} = 4.0 + 4.7 + 1.3 = 10\ \text{V}$$
$$E_s = 10\ \text{V}$$

UNIVERSAL DESIGN SHEET

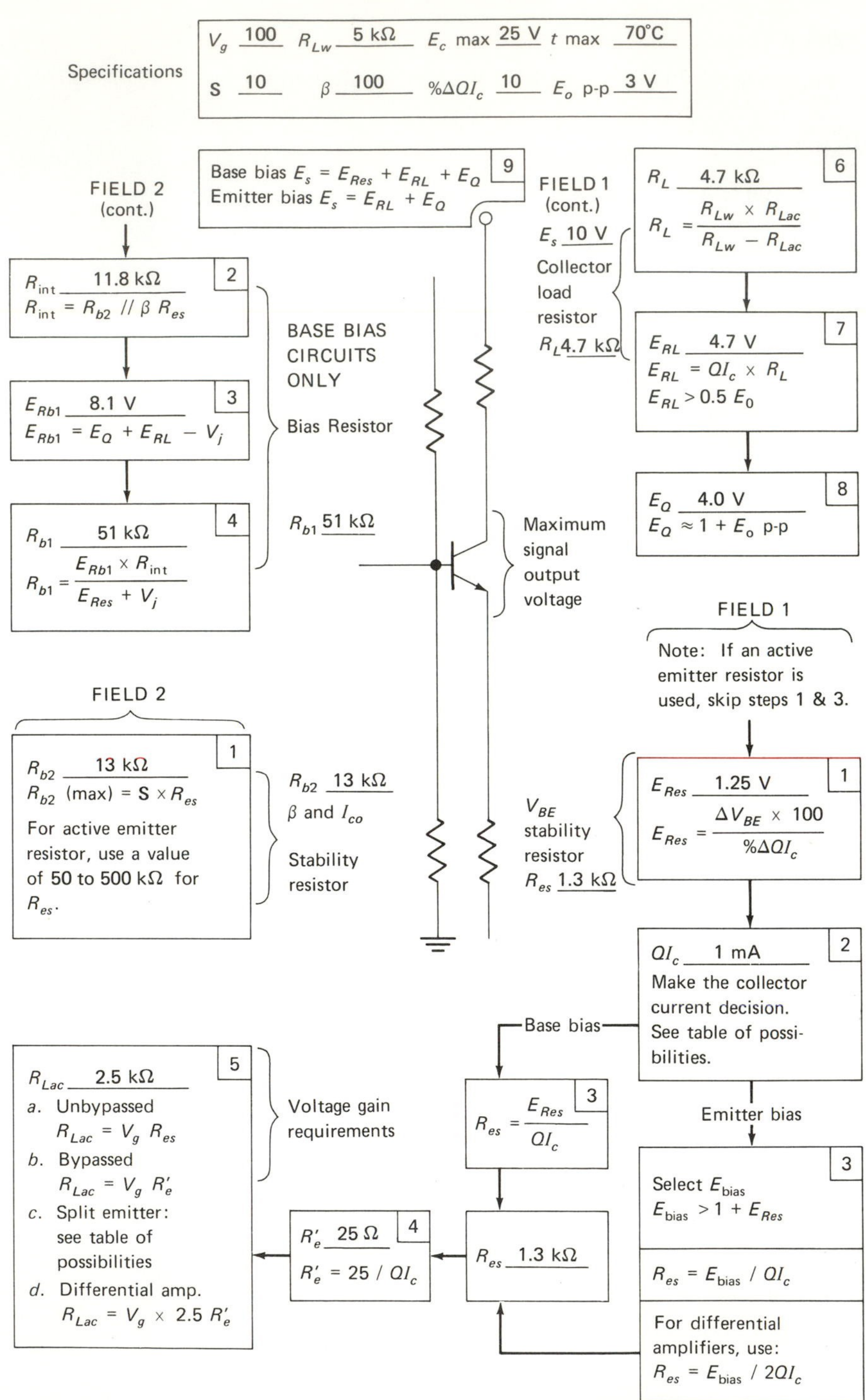

Specifications
Vg 100 RLw 5 kΩ Ec max 25 V t max 70°C
S 10 β 100 %ΔQIc 10 Eo p-p 3 V
Base bias Es = ERes + ERL + EQ
Emitter bias Es = ERL + EQ
9
FIELD 2 (cont.)
Rint 11.8 kΩ
Rint = Rb2 // β Res
2
ERb1 8.1 V
ERb1 = EQ + ERL − Vj
3
Rb1 51 kΩ
Rb1 = ERb1 × Rint / (ERes + Vj)
4
BASE BIAS CIRCUITS ONLY
Bias Resistor
Rb1 51 kΩ
FIELD 1 (cont.)
Es 10 V
Collector load resistor
RL 4.7 kΩ
RL 4.7 kΩ
RL = RLw × RLac / (RLw − RLac)
6
ERL 4.7 V
ERL = QIc × RL
ERL > 0.5 E0
7
EQ 4.0 V
EQ ≈ 1 + Eo p-p
8
Maximum signal output voltage
FIELD 1
Note: If an active emitter resistor is used, skip steps 1 & 3.
FIELD 2
Rb2 13 kΩ
Rb2 (max) = S × Res
For active emitter resistor, use a value of 50 to 500 kΩ for Res.
1
Rb2 13 kΩ
β and Ico
Stability resistor
VBE stability resistor
Res 1.3 kΩ
ERes 1.25 V
ERes = ΔVBE × 100 / %ΔQIc
1
QIc 1 mA
Make the collector current decision. See table of possibilities.
2
Base bias
Emitter bias
Res = ERes / QIc
3
Select Ebias
Ebias > 1 + ERes
3
Res = Ebias / QIc
For differential amplifiers, use:
Res = Ebias / 2QIc
Res 1.3 kΩ
R'e 25 Ω
R'e = 25 / QIc
4
RLac 2.5 kΩ
a. Unbypassed
RLac = Vg Res
b. Bypassed
RLac = Vg R'e
c. Split emitter: see table of possibilities
d. Differential amp.
RLac = Vg × 2.5 R'e
5
Voltage gain requirements

Fig. 6-2 The completed design sheet for Example 6-1

FIELD 2 BIAS CALCULATIONS

FRAME 1

Compute the value of the base stability resistor R_{b2}. The specification sheet gives the stability factor a value of 10.

$$R_{b2} = \mathbf{S} \times R_{es} = 10 \times 1.3\ \text{k}\Omega = 13\ \text{k}\Omega$$
$$R_{b2} = 13\ \text{k}\Omega$$

FRAME 2

Compute the value of the partial input impedance R_{int}.

$$R_{int} = R_{b2} \| \beta R_{es}$$

$$R_{int} = \frac{13\ \text{k}\Omega \times 130\ \text{k}\Omega}{13\ \text{k}\Omega + 130\ \text{k}\Omega}$$

$$R_{int} \approx 11.8\ \text{k}\Omega$$

FRAME 3

Compute the voltage drop across R_{b1}.

$$E_{Rb1} = E_Q + E_{RL} - V_j = 4.0 + 4.7 - 0.6\ \text{V}$$
$$E_{Rb1} = 8.1\ \text{V}$$

FRAME 4

Compute the value of the upper base bias resistor R_{b1}.

$$R_{b1} = \frac{E_{Rb1} \times R_{int}}{E_{Res} + V_j}$$

$$R_{b1} = \frac{8.1\ \text{V} \times 11.8\ \text{k}\Omega}{1.3\ \text{V} + 0.6\ \text{V}}$$

$$R_{b1} \approx 49\ \text{l}\Omega$$
$$R_{b1} = 51\ \text{k}\Omega \qquad \text{nearest standard value}$$

FIELD 3 CAPACITOR AND INPUT IMPEDANCE COMPUTATIONS

FRAME 1

Compute the circuit input impedance.

$$Z_{in} \approx R_{b1} \| R_{b2} \| \beta \times R'_e$$

where

$$R'_e \approx \frac{25}{QI_c}$$

$$R'_e \approx \frac{25}{1} \approx 25\ \Omega$$

$$Z_{in} \approx 47\ \text{k}\Omega \| 13\ \text{k}\Omega \| 2.5\ \text{k}\Omega$$
$$Z_{in} \approx 2\ \text{k}\Omega$$

FRAME 2

Compute the value of the input coupling capacitor C_{cin}.

The reactance of C_{cin} should be about 0.1 times the input impedance of the circuit at f_o. This amounts to a reactance of 200 Ω.

$$C_{cin} = \frac{1}{2\pi f_o X_c} = \frac{1}{6.28 \times 20 \times 200}$$

The nearest value available is 50 μF.

$$C_{cin} = 50\ \mu\text{F}$$

FRAME 3

Compute the value of the output coupling capacitor C_{co}.

The specification sheet gives 20 Hz as the lowest frequency of interest and 5 kΩ for R_{Lw}. The reactance of the output coupling capacitor should have a reactance of 20 Hz (f_o) of about one-tenth of R_{Lw}, or about 500 Ω.

$$C_{co} = \frac{1}{2\pi f_o X_c}$$

$$C_{co} = \frac{1}{6.28 \times 20 \times 5 \times 10^2}$$

$$C_{co} \approx 16\ \mu\text{F}$$

Selecting the next common value, we have 20 μF.

FRAME 4

Compute the value of the emitter bypass capacitor C_{bp}.

The reactance of C_{bp} should be equal (at f_o) to R'_e in parallel with the effective output resistance of the emitter circuit.

$$X_c \approx R'_e \| R_o \tag{6-4}$$

where $R'_e \approx 25/I_c$, and

$$R_o \approx R'_e + \frac{R_{b1} \| R_{b2} \| R_{Gen}}{\beta} \tag{6-5}$$

where R_{Gen} is the impedance of the driving source (the generator).

An X_c for the bypass capacitor C_{bp} of a value described in Eq. (6-5) would yield a voltage gain at f_o about 3 dB (25 percent) less than the gain at 1 kHz. The gain drops off at the rate of about 6 dB per octave for frequencies below f_o.

This seems to be a fairly complex relationship, but in reality it has some very definite limits which will make things easier. If you will examine Eq. (6-4), you will notice that R_o is in parallel with R'_e, so that no matter how large R_o is, the value of the parallel combination can never exceed the value of R'_e. The *upper limit* of the required reactance is R'_e.

Now, examine Eq. (6-5) for the value of R_o. R_o can be no smaller than R'_e even if the value of $(R_{b1} \| R_{b2} \| R_{Gen})/\beta$ approaches zero. If $R_o \approx R'_e$, this represents the lowest possible value for R_o and would yield a reactance of $R'_e \| R'_e$ or $R'_e/2$, so that the range for X_c lies between the values of R'_e and $R'_e/2$.

If we select the lower value ($R'_e/2$), we will always have a conservatively designed circuit and we can use this very well in the laboratory. However, economics being what it is, we would not want to take the simplest way out if we were designing a circuit to sell for profit. For profit we would not want to use a capacitor any larger than necessary; we would want to use Eq. (6-4). Now back to the problem at hand. Let us take the simple way first.

1. $X_c = \dfrac{R'_e}{2} \qquad R'_e = 25\ \Omega$

$$X_c = \frac{25}{2} = 12.5\ \Omega$$

$$c = \frac{1}{2\pi f_o X_c}$$

$$c = \frac{1}{6.28 \times 20 \times 12.5}$$

$$c = 637\ \mu\text{F}$$

2. Now the hard way.

$$X_c \approx R'_e \| R_o$$

$$R_o \approx R'_e + \frac{R_{b1} \| R_{b2}}{\beta}$$

NOTE: Since we have no knowledge about R_{Gen}, we are forced to leave it out this time. This will cause no problems as a rule.

$$R_o \approx 25 + \frac{47\ \text{k}\Omega \| 13\ \text{k}\Omega}{100}$$

$$R_o \approx 25 + \frac{10\ \text{k}\Omega}{100}$$

$$R_o \approx 25 + 100 = 125\ \Omega$$
$$X_c = 25 \| 125$$
$$X_c = 20.8 \approx 21\ \Omega$$

$$c = \frac{1}{2\pi f_o X_c}$$

$$c = \frac{1}{6.28 \times 20 \times 21}$$

TABLE 6-3. TABLE OF SELECTED CAPACITIVE REACTANCE VALUES

f_o = 5 Hz—Range: 16 Ω to 649 Ω

X_C, Ω	16–31.9	32–64.9	65–124.9	125–209	210–319	320–649
C, μF	2,000	1,000	500	250	150	100

f_o = 5 Hz—Range: 650 Ω to 31.9 kΩ

X_C, Ω	650–1.24K	1.25K–3.19K	3.2K–6.49K	6.5–15.9K	16–31.9K
C, μF	50	25	10	2	1

f_o = 10 Hz—Range: 7 Ω to 649.9 Ω

X_C, Ω	7–16	16–25	25–50	50–100	100–160	160–320	320–650
C, μF	2,000	1,000	500	250	150	100	50

f_o = 10 Hz—Range: 650 Ω to 31.9 kΩ

X_C, Ω	650–1.59K	1.6K–3.19K	3.2K–7.99K	8–15.99K	16–31.9K
C, μF	25	10	5	2	1

f_o = 20 Hz—Range: 4 Ω to 799 Ω

X_C, Ω	4–7	8–15	16–31	32–54	55–79	80–159	160–319	320–799
C, μF	2,000	1,000	500	250	150	100	50	25

f_o = 20 Hz—Range: 800 Ω to 15.99 kΩ

X_C, Ω	800–1.59K	1.6K–3.99K	4K–7.99K	8K–15.99K
C, μF	10	5	2	1

f_o = 50 Hz—Range: 1.6 Ω to 64.9 Ω

X_C, Ω	1.6–2.9	3.06–4.9	6.5–11.9	12–20.9	21–31.9	32–64.9
C, μF	2,000	1,000	500	250	150	100

f_o = 50 Hz—Range: 65 Ω to 6 kΩ

X_C, Ω	65–119	120–299	300–649	650–1.59K	1.6K–3.19K	3.2K–6K
C, μF	50	25	10	5	2	1

f_o = 60 HZ—Range: 1.5 Ω to 109.9 Ω

X_C, Ω	1.5–2.5	2.6–4.9	5–10.9	11–12.9	13–25.9	26–54.9	55–109.9
C, μF	2,000	1,000	500	250	150	100	50

TABLE 6-3 (Cont.)

$f_o = 60$ **Hz – Range: 110 Ω to 5 kΩ**

X_C, Ω	110–259	260–499	500–1.29K	1.3K–2.59K	2.6K–5K
C, μF	25	10	5	2	1

$f_o = 100$ **Hz – Range: 0.8 Ω to 64.9 Ω**

X_C, Ω	0.8–1.59	1.6–3.19	3.2–6.49	6.5–9.9	10–15.9	16–31.9	32–64.9
C, μF	2,000	1,000	500	250	150	100	50

$f_o = 100$ **Hz – Range: 65 Ω to 3 kΩ**

X_C, Ω	65–159	160–319	320–799	800–1.59K	1.6K–3K
C, μF	25	10	5	2	1

$$c \approx \frac{1}{2.6 \times 10^3}$$

$$c \approx 380\ \mu\text{F}$$

In this case the hard way was worth the extra effort, if we were designing for profit. Figure 6-2 shows the design sheet computations. Figure 6-3 shows the complete circuit with all values shown.

Table 6-3 can help make easier the selection of bypass and coupling capacitors for laboratory purposes.

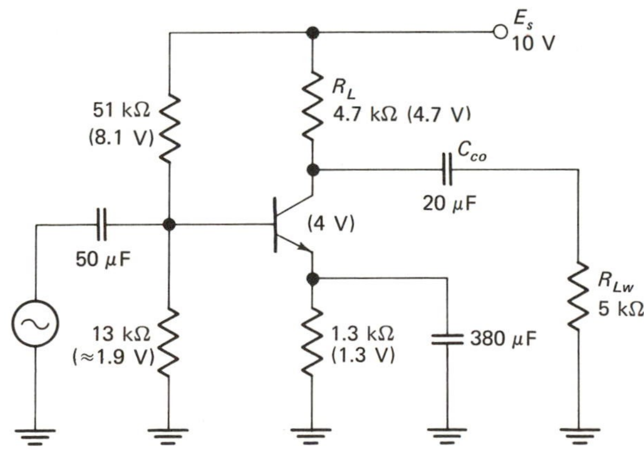

Fig. 6-3 The complete circuit with component values and voltages for Example 6-1

6-3 DESIGNING A SPLIT EMITTER CIRCUIT

The circuit we are about to examine is an excellent compromise between the high voltage gain of the fully bypassed configuration and the excellent gain predictability of the unbypassed circuit. Because this circuit has a set of parameters somewhere between those of the two circuits, I have prepared another table of possibilities for it (Table 6-4). In the previous two circuits the emitter resistance was determined for us, and we had no choice in the matter. In the unbypassed circuit the value of the emitter resistor was determined by the V_{BE} stability formula, and in the fully bypassed circuit the emitter resistance was determined by the collector current according to the Shockley relationship.

In the split emitter circuit (Fig. 6-4), it is up to us to assign a value to the unbypassed portion. We could simply assign an arbitrary value to R_{eac}, but a few ohms in either direction can make the value of R_L compute out to an impractical value. In order to avoid making the R_{eac} decision a trial-and-error process, we will use the table of possibilities. The value of the ac load for each combination listed is implicit in the table, and I have simply listed those values of R_{Lac} in extra rows in the table. Now we can simply look in the table for the value of R_{Lac} and calculate the value of R_{eac} from that. The split emitter circuit offers greater flexibility than either of the two circuits we have studied, but because of that very flexibility it is more difficult to design. We will also encounter a new kind of loading problem, which will also influence the R_{eac} decision. We will use the Young-Albertson curve to make the correction, which will be in the form of reducing the value of R_{eac} when the curve indicates that it is necessary.

If the split emitter circuit is to be the good compromise between the unbypassed and fully bypassed circuits that it should be, R_{eac} should be 10 times (or more) larger than R'_e. It is not hard to meet this condition, but often the Young-Albertson correction can make R_{eac} too low to meet this requirement. Unfortunately, if we use Table 6-4*a*, we will often end up having designed a circuit which is less than ideal, and because the Young-Albertson correction is nearly the last step in the procedure, a lot of work can be done simply to design an inadequate circuit. Professor W. P. Matthews discovered this flaw and communicated the problem to Professor R. T. Burchell, who got together with computer programmer Roy Carr and modified Table 6-4*a* to the form of Table 6-4*b* to ensure against such wasted effort.

Example 6-2

Design a common-emitter amplifier stage with the following specifications.

Transistor Parameters

1. Device type:	*pnp* silicon
2. β	100 (minimum β minus 10%)
3. I_c max	100 mA
4. E_c max	25 V

TABLE 6-4a. TABLE OF POSSIBILITIES FOR THE SPLIT EMITTER CIRCUIT

						QI_c	Values of R_{Lac}								
						40 mA	37.5	50	100	150	200	250	300	350	400
						20 mA	75	100	200	300	400	500	600	700	800
						10 mA	150	200	400	600	800	1K	1.2K	1.4K	1.6K
						4 mA	375	500	1K	1.5K	2K	2.5K	3K	3.5K	4K
						2 mA	750	1K	2K	3K	4K	5K	6K	7K	8K
						1 mA	1.5K	2K	4K	6K	8K	10K	12K	14K	16K
						0.5 mA	3K	4K	8K	12K	16K	20K	24K	28K	32K
						Vg → Range	2.5–15	3–20	6–40	10–60	13–80	16–100	20–120	23–140	26–160
R_{Lac} at 0.5 mA	R_{Lac} at 1 mA	R_{Lac} at 2 mA	R_{Lac} at 4 mA	R_{Lac} at 10 mA	R_{Lac} at 20 mA	R_{Lac} at 40 mA	E_{RL}	E_{RL}	E_{RL}	E_{RL}	E_{RL}	E_{RL}	E_{RL}	E_{RL}	E_{RL}
2K	1K	500	250	100	50	25	*	*	*	*	*	*	*	*	*
4K	2K	1K	500	200	100	50	6	*	*	*	*	*	*	*	*
6K	3K	1.5K	750	300	150	75	3	6	*	*	*	*	*	*	*
8K	4K	2K	1K	400	200	100	2.4	4	*	*	*	*	*	*	*
10K	5K	2.5K	1.25K	500	250	125	2.1	3.3	20	*	*	*	*	*	*
12K	6K	3K	1.5K	600	300	150	2.0	3	12	*	*	*	*	*	*
14K	7K	3.5K	1.75K	700	350	175	1.9	2.8	9.3	42	*	*	*	*	*
16K	8K	4K	2K	800	400	200	1.8	2.7	8	24	*	*	*	*	*
18K	9K	4.5K	2.25K	900	450	225	1.8	2.6	7.2	18	72	*	*	*	*
20K	10K	5K	2.5K	1K	500	250	1.7	2.5	6.7	15	40	*	*	*	*
24K	12K	6K	3K	1.25K	600	300	1.7	2.4	6.0	12	24	60	*	*	*
30K	15K	7.5K	3.75K	1.5K	700	350	1.6	2.3	5.5	10	17	30	60	210	*
40K	20K	10K	5K	2K	1K	500	1.6	2.2	5.0	8.5	13	20	30	47	80
50K	25K	12.5K	6.25K	2.5K	1.25K	625	1.6	2.2	4.7	7.9	11	17	23	32	44
100K	50K	25K	12.5K	5K	2.5K	1.25K	1.5	2	4.3	6.8	9.5	12.5	16	19	24
200K	100K	50K	25K	10K	5K	2.5K	1.5	2	4	6.4	8.7	11	13.5	17	19

* An asterisk signifies an impossible condition.

TABLE 6-4b. MODIFIED SPLIT EMITTER TABLE OF POSSIBILITIES

						QI_c	← R_{Lac} →																	
						40 mA	37.5		50		100		150		200		250		300		350		400	
						20 mA	75		100		200		300		400		500		600		700		800	
						10 mA	150		200		400		600		800		1K		1.2K		1.4K		1.6K	
						4 mA	375		500		1K		1.5K		2K		2.5K		3K		3.5K		4K	
						2 mA	750		1K		2K		3K		4K		5K		6K		7K		8K	
						1 mA	1.5K		2K		4K		6K		8K		10K		12K		14K		16K	
						0.5 mA	3K		4K		8K		12K		16K		20K		24K		28K			
						V_g → range	2.5–6		3–8		6–16		10–24		12–30		20–40		20–40		20–50		20–60	
R_{Lw} at 0.5 mA	**R_{Lw} at 1 mA**	**R_{Lw} at 2 mA**	**R_{Lw} at 4 mA**	**R_{Lw} at 10 mA**	**R_{Lw} at 20 mA**	**R_{Lw} at 40 mA**	E_{RL}	**Mag**	E_{RL}	**Mag**	E_{RL}	**Mag**	E_{RL}	**Mag**	E_{RL}	**Mag**	E_{RL}	**Mag**	E_{RL}	**Mag**	E_{RL}	**Mag**	E_{RL}	**Mag**
2K	1K	500	250	100	50	25	*	*	*	*	*	*	*	*	*	*	*	*	*	*	*	*	*	*
4K	2K	1K	500	200	100	50	6	3.5	*	*	*	*	*	*	*	*	*	*	*	*	*	*	*	*
6K	3K	1.5K	750	300	150	75	3	4	6	5.3	*	*	*	*	*	*	*	*	*	*	*	*	*	*
8K	4K	2K	1K	400	200	100	2.4	5.5	4	5.6	*	*	*	*	*	*	*	*	*	*	*	*	*	*
10K	5K	2.5K	1250	500	250	125	2.1	6	3.3	7.2	20	8	*	*	*	*	*	*	*	*	*	*	*	*
12K	6K	3K	1.5K	600	300	150	2.0	6	3	8.0	12	10	*	*	*	*	*	*	*	*	*	*	*	*
14K	7K	3.5K	1750	700	350	175	1.9	6	2.8	8	9.3	10	42	14	*	*	*	*	*	*	*	*	*	*
16K	8K	4K	2K	800	400	200	1.8	6	2.7	8	8	10	24	14	*	*	*	*	*	*	*	*	*	*
18K	9K	4.5K	2250	900	450	225	1.8	6	2.6	8	7.2	12	18	16	72	15	*	*	*	*	*	*	*	*
20K	10K	5K	2.5K	1K	500	250	1.7	6	2.5	8	6.7	14	15	16	40	15	*	*	*	*	*	*	*	*
24K	12K	6K	3K	1200	600	300	1.7	6	2.4	8	6.0	16	12	16	24	20	60	20	*	*	*	*	*	*
30K	15K	7.5K	3750	1.5K	700	350	1.6	6	2.3	8	5.5	16	10	20	17	20	30	25	60	20	210	30	*	*
40K	20K	10K	5K	2K	1K	500	1.6	6	2.2	8	5.0	16	8.5	24	13	25	20	25	30	30	47	30	80	30
50K	25K	12.5K	6250	2.5K	1250	625	1.6	6	2.2	8	4.7	16	7.9	24	11	30	17	35	23	30	32	30	44	
100K	50K	25K	12.5K	5K	2.5K	1250	1.5	6	2	8	4.3	16	6.8	24	9.5	30	12.5	40	16	40	19	50	24	60
200K	100K	50K	25K	10K	5K	2.5K	1.5	6	2	8	4	16	6.4	24	8.7	30	11	40	13.5	40	17	50	19	60

* An asterisk signifies an impossible condition.

† Practical design limits.

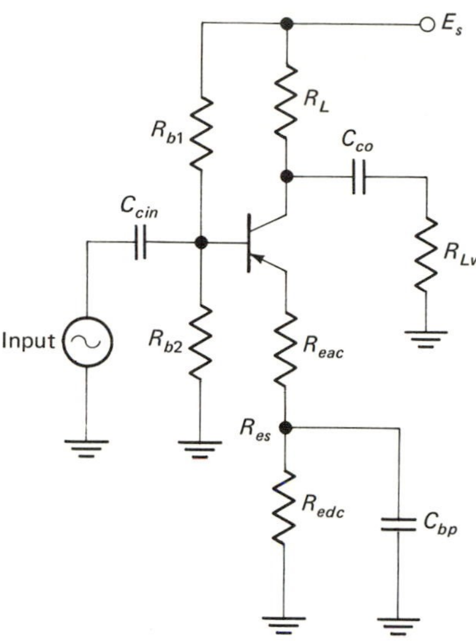

Fig. 6-4 The split emitter circuit

Circuit Parameters

1. Voltage gain (V_g)	10
2. Maximum expected operating temperature	80°C
3. Allowable quiescent current variation at 80°C	5%
4. Stability factor	10
5. Input impedance (minimum)	1.5 kΩ
6. Working load (R_{Lw})	8 kΩ
7. Low-frequency cutoff (f_o)	30 Hz
8. Maximum output voltage (E_o) peak to peak	3.0 V

The basic circuit is shown in Fig. 6-4.

FIELD 1

FRAME 1

Compute the value of the voltage drop across the emitter resistor E_{Res}.

$$E_{Res} = \frac{V_{BE} \times 100}{\%\Delta QI_c}$$

Consulting Table 5-3, we find that at a collector current of 1.0 mA, we can expect the base-emitter junction voltage to decrease by about 2.5 mV/C°. If the highest temperature expected is 80° and the reference temperature is 20°C, we have a change in temperature of 60°C (80° − 20° = Δt of 60°C). The specifications allow 5 percent variation in the quiescent collector current with an increase in temperature of 60°, and so our equation with values filled in is

$$E_{Res} = \frac{60 \times 2.5 \times 10^{-3} \times 10^{2}}{5\ (\%)} = 3.0 \text{ V}$$

FRAME 2

Make the collector current decision.

Table 6-4*a* is a modification of the other tables of possibilities, but it has one defect for the split emitter circuit. Whenever a Young-Albertson correction is required, it means lowering R_{eac}. This correction may yield a value of R_{eac} which is much less than $10 \times R_e'$. Table 6-4*b* is a modified version of Table 6-4*a*; it allows you to avoid this problem. If you use only those entries between the two stair-step lines, you can avoid the problem. "Mag" in the table stands for maximum allowable gain (V_g).

At 0.5 mA we find that a maximum allowable gain (Mag) of 5.6 is the best we can do. So let's try 1 mA with 8 kΩ and 1 mA. We find an entry Mag of 10 with an E_{RL} of 8 V. This looks good. Any circuit where R_e' is greater than one-tenth of R_{eac} is not practical. The reason is simply that there is no point to such a circuit. A circuit operating at 1 mA would have a value for R_e' of 25 Ω, and the modified table will not allow for a corrected R_{eac} value of less than 250 Ω. Once the value of R_{eac} gets lower than this ratio, most of the advantages of the split emitter configuration are virtually lost and it is time to switch to the fully bypassed configuration (or even the unbypassed configuration if the gain requirements can be met).

$QI_c = 1$ mA

FRAME 3

Compute the value of the total emitter resistance R_{es}.

$$R_{es} = \frac{E_{Res}}{QI_c}$$

$$R_{es} = \frac{3.0}{1 \times 10^{-3}}$$

$$R_{es} = 3.0\ \text{k}\Omega$$

In this case there will be no actual 3.0 kΩ resistor involved, but $R_{eac} + R_{edc} =$ 3.0 kΩ (R_{es}). Stated a little differently, R_{es} is the emitter *resistance* which, in the actual circuit, will be split into resistors R_{eac} and R_{edc}.

FRAME 4

Compute the base-emitter junction resistance R_e'.

$$R_e' = \frac{25}{QI_c} = \frac{25}{1} = 25\ \Omega$$

This is a case where we will not actually use the results of this computation for anything but to ensure that other values are reasonable when compared to it. In this case when we compute R_{eac}, R_{eac} must not be less than about 10 times R_e', or gain and input impedance calculations will not agree with circuit performance. If you use the modified Table 6-4*b* correctly, this should never be a problem.

FRAME 5

Find R_{Lac} and R_{eac}.

1. Transfer the value of R_{Lac} from the 1 mA row of the table of possibilities (Table 6-4b). In this case we find 8 kΩ value for R_{Lw} under the 1 mA heading. We read across that row until we come to the 8 (inside the stair-step lines) under the E_{RL} heading, turn left, and read up the column to the 1 mA row, where we find a value of 4.0 kΩ. This is the value of R_{Lac} required for this circuit.
2. Compute the value of R_{eac} required for the desired voltage gain.

$$R_{eac} = \frac{R_{Lac}}{V_g}$$

$$R_{eac} = \frac{4.0\ \text{k}\Omega}{10}$$

$$R_{eac} = 400\ \Omega$$

We could have selected a value for R_{eac} out of thin air and then computed R_{Lac} as $R_{Lac} = V_g \times R_{eac}$. The relationship is correct, but we did not do it that way because a poor selection of R_{eac} invariably yields inconvenient values (or impossible ones) for R_L.

At this point we encounter a new problem. I have mentioned the loading problem and its solution, the Young-Albertson curve (see Fig. 6-5). In this problem we have an R_L value of 8 kΩ and an R_{Lw} of 8 kΩ, and so $R_L = 1R_{Lw}$. An examination of the Young-Albertson curve gives an error value of about 30 percent for $R_L = 1R_{Lw}$. This means that if our gain prediction is to be correct, we must reduce R_{eac} by 30 percent.

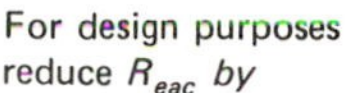

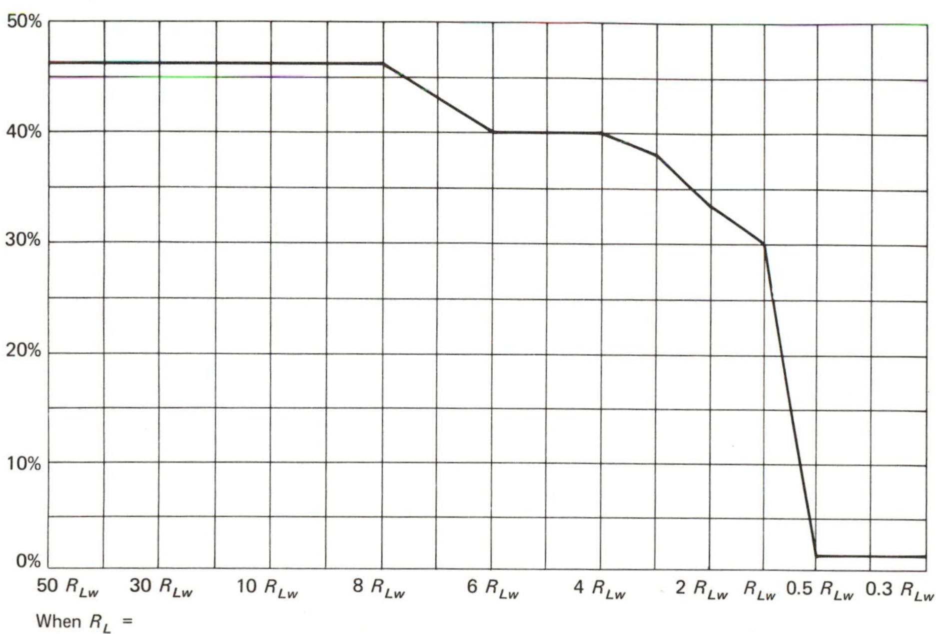

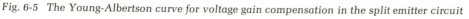

Fig. 6-5 The Young-Albertson curve for voltage gain compensation in the split emitter circuit

$$0.30 \times 400 = 120\ \Omega$$
$$400 - 120 = 280\ \Omega$$
$$R_{eac} \approx 280\ \Omega$$

The nearest standard value is 270 Ω.

3. Later on we will need to know the value of the other half of the emitter resistance R_{edc}.

$$R_{edc} = R_{es} - R_{eac}$$
$$R_{edc} = 3.0\ \text{k}\Omega - 270$$
$$R_{edc} = 2.730\ \text{k}\Omega$$

The nearest standard value is 2.7 kΩ.

FRAME 6

Compute the value of the collector load resistor R_L.

$$R_L = \frac{R_{Lw} \times R_{Lac}}{R_{Lw} - R_{Lac}}$$

$$R_L = \frac{8\ \text{k}\Omega \times 4.0\ \text{k}\Omega}{8\ \text{k}\Omega - 4.0\ \text{k}\Omega}$$

$$R_L = \frac{32\ \text{k}\Omega}{4\ \text{k}\Omega}$$

$$R_L = 8\ \text{k}\Omega$$

FRAME 7

Compute the voltage drop across the collector load resistor E_{RL}.

$$E_{RL} = QI_c \times R_L$$
$$E_{RL} = 1 \times 10^{-3} \times 8 \times 10^3$$
$$E_{RL} = 8\ \text{V}$$

This is just a little exercise in Ohm's law. The required output voltage is 3.0 V p-p. To meet this requirement, the voltage drop across R_L must be greater than 0.5 E_o. $0.5 \times 3 = 1.5$, and since we have an 8 V drop across R_L, there is no problem.

FRAME 8

Compute the collector-to-emitter voltage E_Q.

The minimum value of E_Q is

$$E_Q \approx 1 + E_o$$
$$E_Q \approx 1 + 3 = 4\ \text{V}$$

It is possible to revise this value upward to accommodate a more convenient power supply voltage (E_s). The normal procedure is to compute the minimum value, proceed to the next frame and calculate the minimum value for E_s, adjust E_s to the desired value, and finally return to this frame and make the adjustment in E_Q. In this case we will consider the minimum value in the next frame to be satisfactory and no adjustment will be made.

FRAME 9

Compute the power supply voltage E_s.

$E_s = E_{Res} + E_{RL} + E_Q$
$E_s = 3.0 + 8 + 4$
$E_s = 15.0\ \text{V}$

If we wanted to adjust this value to some *higher*, more convenient value, we could state that desired value and return to Frame 8 and add the additional voltage to E_Q. We would then enter an *adjusted* value for E_Q in Frame 7. In this case we will assume that 15.0 V is perfectly satisfactory and will make no adjustment in E_Q.

FIELD 2 BIAS CALCULATIONS

FRAME 1

Compute the value of the lower base bias resistor R_{b2}.

$R_{b2} = \mathbf{S} \times R_{es}$
$R_{b2} = 10 \times 3.0\ \text{k}\Omega = 30\ \text{k}\Omega$

FRAME 2

Compute the partial input resistance R_{int}.

$$R_{int} = R_{b2} \| \beta \times R_{es}$$
$$\beta \times R_{es} = 100 \times 3.0\ \text{k}\Omega = 300\ \text{k}\Omega$$
$$R_{int} = \frac{30\ \text{k}\Omega \times 300\ \text{k}\Omega}{30\ \text{k}\Omega + 300\ \text{k}\Omega}$$
$$R_{int} = \frac{9{,}000}{330} = 27.2\ \text{k}\Omega$$

FRAME 3

Compute the value of E_{Rb1}, the voltage drop across the upper base bias resistor.

$E_{Rb1} = E_Q + E_{RL} - V_j$
$E_{Rb1} = 4 + 8 - 0.6$
$E_{Rb1} = 11.4\ \text{V}$

FRAME 4

Compute the value of the upper base bias resistor R_{b1}.

$$R_{b1} = \frac{E_{Rb1} \times R_{int}}{E_{Res} + V_j}$$

$$R_{b1} = \frac{11.4 \times 27.2\ \text{k}\Omega}{3.0 + 0.6}$$

$$R_{b1} = \frac{310.08\ \text{k}\Omega}{3.6}$$

$$R_{b1} = 86.13\ \text{k}\Omega$$

Selecting the nearest standard value, we have

$$R_{b1} = 91\ \text{k}\Omega$$

FIELD 3

FRAME 1

Compute the circuit input impedance Z_{in}.

$$Z_{in} = R_{b1} \| R_{b2} \| \beta R_{eac}$$
$$\beta \times R_{eac} = 100 \times 400 = 40\ \text{k}\Omega$$

$$Z_{in} = \frac{1}{1/R_{b1} + 1/R_{b2} + 1/(\beta \times R_{eac})}$$

$$Z_{in} = \frac{1}{1/82\ \text{k}\Omega + 1/30\ \text{k}\Omega + 1/40\ \text{k}\Omega}$$

$$Z_{in} \approx 14\ \text{k}\Omega$$

FRAME 2

Compute the value of the input coupling capacitor C_{cin}.

$$X_c \text{ of } C_{cin} = 0.1\ Z_{in}$$
$$X_c = 0.1 \times 14\ \text{k}\Omega = 1{,}400$$

$$C_{cin} = \frac{1}{2\pi f X_c}$$

$$C_{cin} = \frac{1}{6.28 \times 30 \times 1{,}400}$$

$$C_{cin} = \frac{1}{2.63 \times 10^5}$$

$$C_{cin} \approx 3.7\ \mu\text{F}$$
$$C_{cin} \approx 5\ \mu\text{F}$$

FRAME 3

Compute the value of the output coupling capacitor C_{co}.

$$X_c \text{ of } C_{co} \text{ at 30 Hz} \approx 0.1\ R_{Lw}$$
$$X_c \approx 800$$

$$C_{co} = \frac{1}{2\pi f X_c}$$

$$C_{co} = \frac{1}{6.28 \times 30 \times 800}$$

$$C_{co} = \frac{1}{15.07 \times 10^4}$$

$$C_{co} \approx 10\ \mu F$$

FRAME 4

Compute C_{bp}, the emitter bypass capacitor.

Computing C_{bp} in this case is simply a matter of making its reactance small compared to the resistance of R_{edc}. This is an empirical rule based upon the assumptions that R_{eac} is large compared to R'_e and that R_{edc} is large compared to R_{eac}. These are normally valid assumptions whenever the split emitter circuit is a good choice for the job at hand.

$$X_c \text{ of } C_{bp} \approx 0.1\ R_{edc} \text{ at } f_o$$

$$X_c \approx 0.1 \times 2.7\ k\Omega = 270\ \Omega$$

$$C = \frac{1}{2\pi f X_c} = \frac{1}{6.28 \times 30 \times 270}$$

$$C \approx 19\ \mu F$$

The nearest practical value would probably be 20 μF. Figure 6-6 shows the completed design for the split emitter example.

6-4 THE COMMON-EMITTER CIRCUIT WITH EMITTER BIAS

The circuit in Fig. 6-7 is an example of an emitter-biased common-emitter stage. Base current is supplied by a bias supply connected to the emitter instead of by the collector supply through R_{b1}. The principal advantages of the circuit are that the bias is relatively independent of base circuitry so long as the base resistance to ground is low enough, and that there is a higher circuit current gain in some instances. The circuit is inherently stable because of the very large voltage drop across the emitter resistor. The input impedance would appear to be considerably higher than in the base-biased configuration.

Unfortunately, we cannot take advantage of the potentially high input impedance because of the voltage gain relationship, $V_g = R_{Lac}/R_{es}$, which often yields unbypassed voltage gains of unity or less. We can, of course, bypass R_{es} with a large capacitor to ground and bring the voltage gain up to a quite respectable value, but at the same time the input impedance is drastically reduced. In short, the circuit can provide higher current gain than the base bias configuration when the input generator can be all or part of R_{b2}; it provides more freedom of design in the base circuit; and it allows larger voltage drops across the emitter resistor than are practical with base bias circuits.

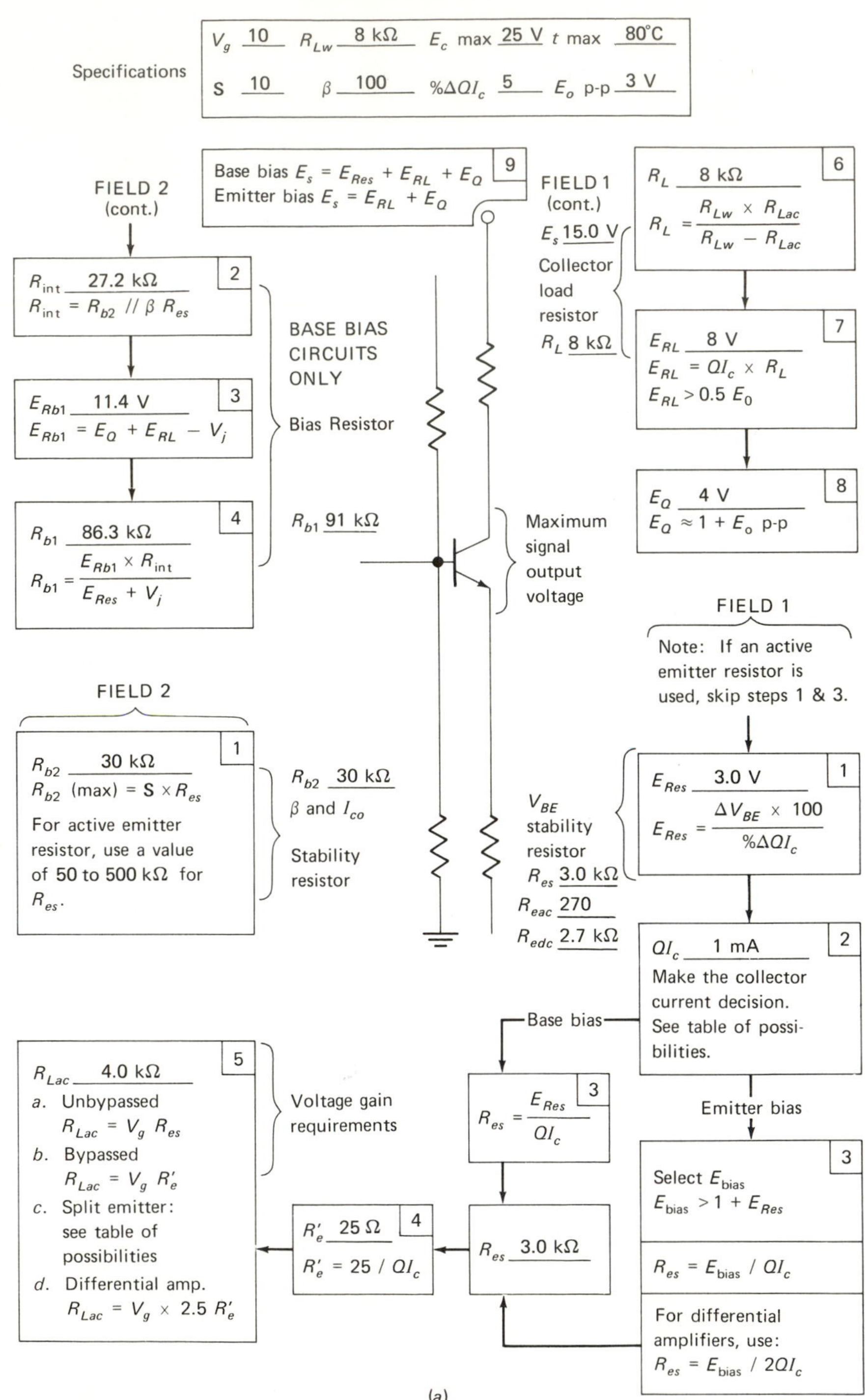

UNIVERSAL DESIGN SHEET
Specifications
V_g 10 R_{Lw} 8 kΩ E_c max 25 V t max 80°C
S 10 β 100 $\%\Delta QI_c$ 5 E_o p-p 3 V
9
Base bias $E_s = E_{Res} + E_{RL} + E_Q$
Emitter bias $E_s = E_{RL} + E_Q$
FIELD 2 (cont.)
2
R_{int} 27.2 kΩ
$R_{int} = R_{b2} \,//\, \beta R_{es}$
3
E_{Rb1} 11.4 V
$E_{Rb1} = E_Q + E_{RL} - V_j$
4
R_{b1} 86.3 kΩ
$R_{b1} = \dfrac{E_{Rb1} \times R_{int}}{E_{Res} + V_j}$
BASE BIAS CIRCUITS ONLY
Bias Resistor
R_{b1} 91 kΩ
FIELD 1 (cont.)
E_s 15.0 V
Collector load resistor
R_L 8 kΩ
6
R_L 8 kΩ
$R_L = \dfrac{R_{Lw} \times R_{Lac}}{R_{Lw} - R_{Lac}}$
7
E_{RL} 8 V
$E_{RL} = QI_c \times R_L$
$E_{RL} > 0.5\, E_0$
8
E_Q 4 V
$E_Q \approx 1 + E_o$ p-p
Maximum signal output voltage
FIELD 1
Note: If an active emitter resistor is used, skip steps 1 & 3.
FIELD 2
1
R_{b2} 30 kΩ
R_{b2} (max) = S × R_{es}
For active emitter resistor, use a value of 50 to 500 kΩ for R_{es}.
R_{b2} 30 kΩ
β and I_{co}
Stability resistor
V_{BE} stability resistor
R_{es} 3.0 kΩ
R_{eac} 270
R_{edc} 2.7 kΩ
1
E_{Res} 3.0 V
$E_{Res} = \dfrac{\Delta V_{BE} \times 100}{\%\Delta QI_c}$
2
QI_c 1 mA
Make the collector current decision. See table of possibilities.
Base bias
Emitter bias
3
$R_{es} = \dfrac{E_{Res}}{QI_c}$
3
Select E_{bias}
$E_{bias} > 1 + E_{Res}$
$R_{es} = E_{bias} \,/\, QI_c$
For differential amplifiers, use:
$R_{es} = E_{bias} \,/\, 2QI_c$
5
R_{Lac} 4.0 kΩ
a. Unbypassed
$R_{Lac} = V_g\ R_{es}$
b. Bypassed
$R_{Lac} = V_g\ R'_e$
c. Split emitter: see table of possibilities
d. Differential amp.
$R_{Lac} = V_g \times 2.5\, R'_e$
Voltage gain requirements
4
R'_e 25 Ω
$R'_e = 25 \,/\, QI_c$
R_{es} 3.0 kΩ
(a)

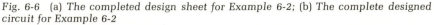

Fig. 6-6 (a) The completed design sheet for Example 6-2; (b) The complete designed circuit for Example 6-2

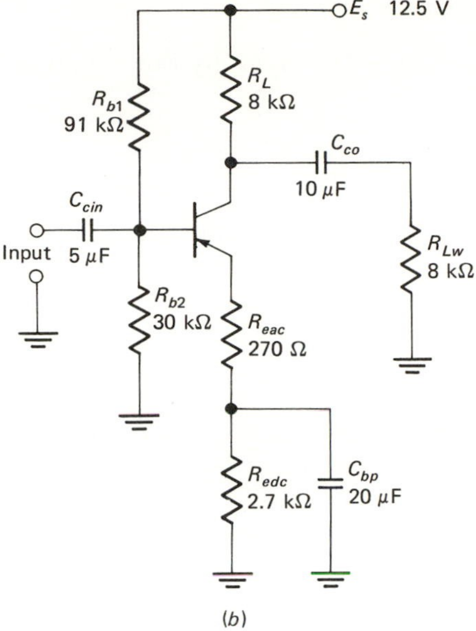

Fig. 6-6 (Cont.)

Don't write this bias method off too soon, however. A little later on we will study the differential amplifier, where this method of biasing offers many advantages. In its basic configuration as we will examine it here, it offers only a little improvement over the circuits we have already studied. I introduce it here primarily to lay the foundation for the differential amplifier, which is probably the most important circuit in linear integrated circuitry, and a very important one in discrete transistor circuitry.

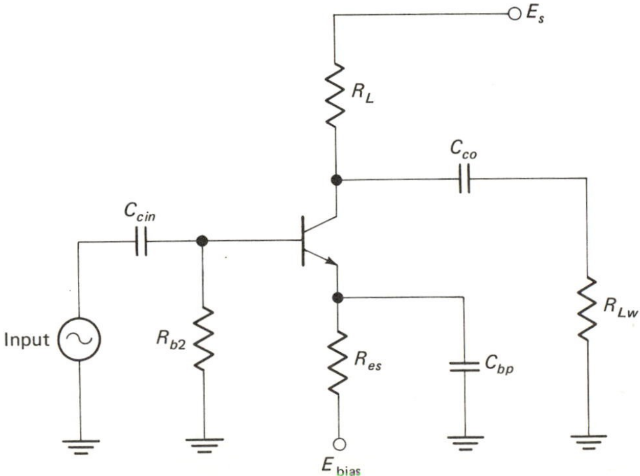

Fig. 6-7 The emitter-biased fully bypassed circuit

Biasing the Circuit

The emitter current in Fig. 6-7 is given by the equation

$$I_e = \frac{E_{\text{bias}} - V_{BE}}{R_{es} + R_b/\beta} \tag{6-6}$$

We wrote the loop equation for the bias circuit in Chap. 4, in the discussion of stability, and derived this equation from the loop equation. If you have forgotten, refer to Fig. 4-6 and the discussion that goes with it. The equation we got from Fig. 4-6 was

$$E_{\text{bias}} - I_b R_b - I_e R_e - V_{BE} = 0$$

We assumed that $I_e \approx I_c$ (QI_c), substituted I_c/β for I_b, and ended up with

$$QI_c \approx \frac{E_{\text{bias}} - V_{BE}}{R_e + R_b/\beta}$$

(NOTE: R_e and R_{es} are the same quantity.)

1. If E_{bias} is much larger than V_{BE}, and we will see that it always is, we can drop V_{BE} from the equation.
2. And if R_b/β is much smaller than R_{es}, and it must be if we are to have good stability.
3. And if $I_e \approx I_c$, which it always is, we can write the equation in the following form:

$$QI_c \approx \frac{E_{\text{bias}}}{R_{es}} \tag{6-7}$$

The circuit is normally a fully bypassed circuit, but a split emitter version is a rare but possible variation. The same signal equation and some of the dc equations apply to both the base-biased circuit and the emitter-biased version. The principal difference, in the collector side voltage distribution, is that the power supply voltage E_s is distributed only between E_{RL} and E_Q, and the *bias* voltage supplies the entire voltage drop across the emitter resistor (E_{Res}). In more concise terms, these conditions always prevail:

1. $E_Q + E_{RL} = E_s$
2. $E_{Res} \approx E_{\text{bias}}$

Because the base-emitter bias is taken care of at the emitter, we only need to provide a return resistance from base to ground which meets the usual stability requirement. The stability requirement states that $R_b = \mathbf{S} \times R_{es}$, where **S** is within a range of from 1 to 10. This circuit frequently has stability factors of less than unity. **S** values of less than unity are perfectly permissible but do not result in any significant improvement in stability. The signal source often provides the base-to-ground return path; but if that is the case, the internal resistance of the source must be low enough to meet the stability factor requirement. In general, bias supply voltages are high enough so that there is quite a large voltage drop across the emitter resistor, resulting in exceptional V_{BE} stability.

In the design example, we will start the process by computing the *minimum* value of E_{Res} as we have done before; but in this case it is only to be certain that we know what the minimum acceptable value is. We will almost always revise the value of E_{Res} upward when we make the bias calculation. Later, you may wish to skip some of the early steps involving the value of E_{Res}. For the time being, the extra steps can be used to provide the perspective necessary to omit them intelligently later.

Example 6-3 The Fully Bypassed Emitter-biased Circuit

Design a fully bypassed emitter-biased amplifier with the following specifications.

Transistor Parameters

1. Device type:	*npn* silicon
2. $\beta =$	100 (minimum β minus 10%)
3. I_c max	100 mA
4. E_c max	25 V

Circuit Parameters

1. Voltage gain	140
2. Maximum expected temperature	75°C
3. Allowable quiescent current variation at 75°C	5%
4. Stability factor	5
5. Working load (R_{Lw})	10 kΩ
6. Low-frequency cutoff (f_o)	25 Hz
7. Maximum output voltage (E_o) p-p	3 V
8. Power supply voltage (E_s)	10 V

The basic circuit is shown in Fig. 6-7. Figure 6-8 shows the completed design sheet.

FIELD 1

FRAME 1

Compute the minimum value of the emitter resistor E_{Res}.

$$E_{Res} = \frac{\Delta V_{BE} \times 100}{\%\Delta QI_c}$$

1. $\Delta V_{BE} = 2.5$ mV/C°
2. $\Delta t = 75° - 20° = 55°$
3. $\Delta V_{BE} = 2.5 \times 55° = 137.5$ mV total

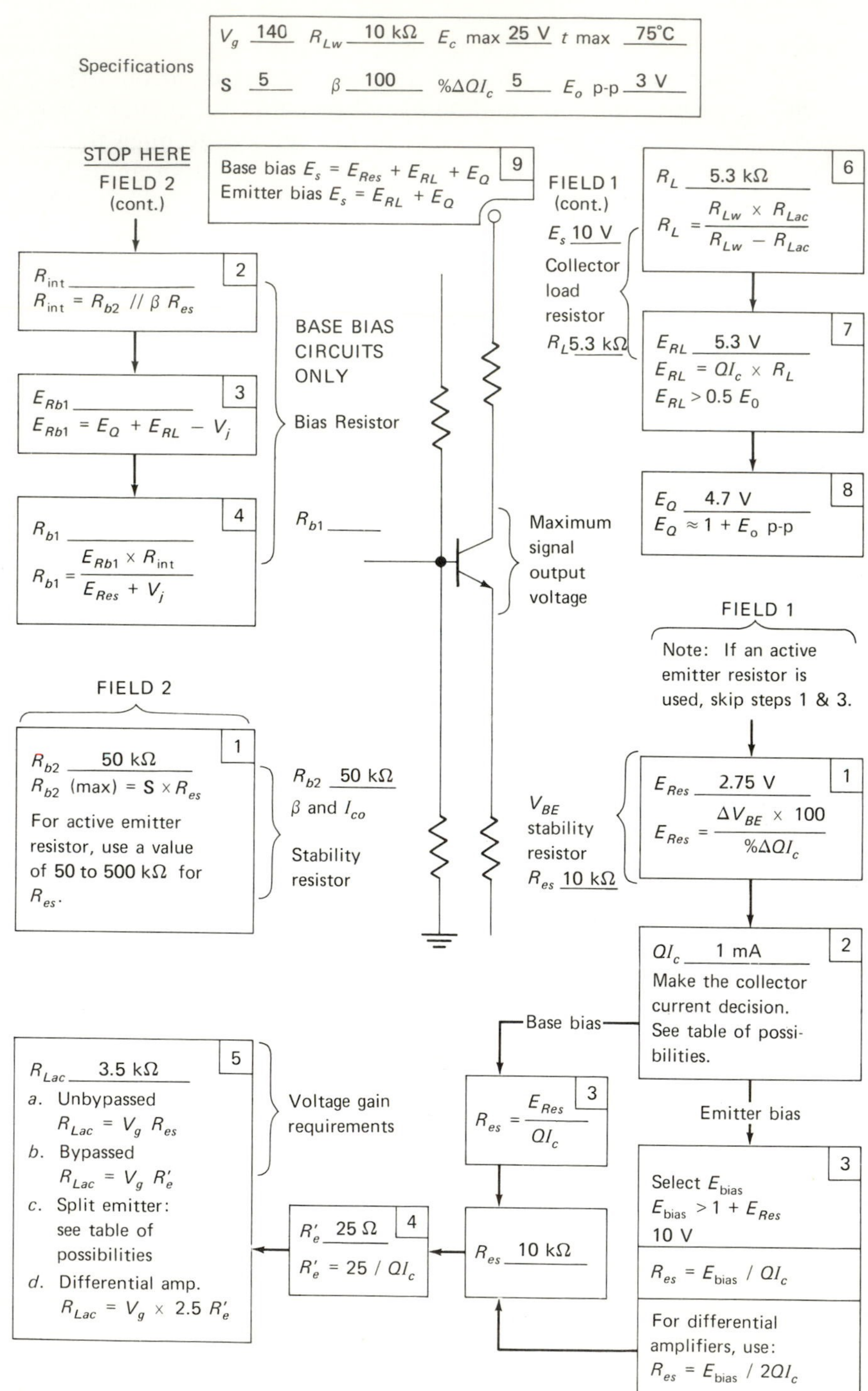

UNIVERSAL DESIGN SHEET
Specifications
V_g 140 R_{Lw} 10 kΩ E_c max 25 V t max 75°C
S 5 β 100 $\%\Delta QI_c$ 5 E_o p-p 3 V
STOP HERE
FIELD 2 (cont.)
Base bias $E_s = E_{Res} + E_{RL} + E_Q$
Emitter bias $E_s = E_{RL} + E_Q$
9
FIELD 1 (cont.)
E_s 10 V
Collector load resistor
R_L 5.3 kΩ
6
R_L 5.3 kΩ
$R_L = \frac{R_{Lw} \times R_{Lac}}{R_{Lw} - R_{Lac}}$
2
R_{int}
$R_{int} = R_{b2} // \beta R_{es}$
BASE BIAS CIRCUITS ONLY
Bias Resistor
7
E_{RL} 5.3 V
$E_{RL} = QI_c \times R_L$
$E_{RL} > 0.5 E_0$
3
E_{Rb1}
$E_{Rb1} = E_Q + E_{RL} - V_j$
4
R_{b1}
$R_{b1} = \frac{E_{Rb1} \times R_{int}}{E_{Res} + V_j}$
R_{b1}
Maximum signal output voltage
8
E_Q 4.7 V
$E_Q \approx 1 + E_o$ p-p
FIELD 1
Note: If an active emitter resistor is used, skip steps 1 & 3.
FIELD 2
1
R_{b2} 50 kΩ
R_{b2} (max) = S × R_{es}
For active emitter resistor, use a value of 50 to 500 kΩ for R_{es}.
R_{b2} 50 kΩ
β and I_{co}
Stability resistor
V_{BE} stability resistor
R_{es} 10 kΩ
1
E_{Res} 2.75 V
$E_{Res} = \frac{\Delta V_{BE} \times 100}{\%\Delta QI_c}$
2
QI_c 1 mA
Make the collector current decision. See table of possi-bilities.
Base bias
Emitter bias
5
R_{Lac} 3.5 kΩ
a. Unbypassed
$R_{Lac} = V_g\ R_{es}$
b. Bypassed
$R_{Lac} = V_g\ R'_e$
c. Split emitter: see table of possibilities
d. Differential amp.
$R_{Lac} = V_g \times 2.5\ R'_e$
Voltage gain requirements
3
$R_{es} = \frac{E_{Res}}{QI_c}$
3
Select E_{bias}
$E_{bias} > 1 + E_{Res}$
10 V
$R_{es} = E_{bias} / QI_c$
For differential amplifiers, use:
$R_{es} = E_{bias} / 2QI_c$
4
R'_e 25 Ω
$R'_e = 25 / QI_c$
R_{es} 10 kΩ

Fig. 6-8 The design sheet for the emitter-biased circuit, Example 6-3

$$E_{Res} = \frac{137.5 \times 10^{-3} \times 10^{2}}{5\ (\%)}$$

$$E_{Res} = 2.75\ \text{V}$$

This step may be interpreted to mean that the bias voltage must be *at least* 3.75 V. In many cases it will be considerably higher, but at least we know the lowest acceptable bias voltage. See Frame 3.

FRAME 2

Make the collector current decision. (See Table 6-2.)

If we examine the table of possibilities, we find that a V_g of 140 and R_{Lw} of 10 kΩ at a collector current of 0.5 mA is a possibility and it would require a voltage drop across R_L of 11.6 V. Ordinarily this would be satisfactory, and 0.5 mA is the lowest current (on the table) that will do the job. However, the original set of specifications has introduced an additional constraint upon the collector current decision in that it states that a maximum value for E_s is 10 V. If we try 1 mA, we find that we can get by with 5.4 V across R_L, leaving 4.6 V for E_Q. Remember that, in this circuit, the bias supply covers the voltage drop across R_{es} and E_s is split between R_L and the transistor (E_Q). QI_c must be 1 mA.

FRAME 3

Compute the value of the emitter resistor R_{es}.

Use the emitter bias branch.

1. *The bias voltage decision*

The bias supply must provide a voltage *at least* equal to the E_{Res} voltage required for V_{BE} stability, plus 1 V. Any convenient voltage which equals or exceeds this requirement will do, but we must make that decision before we can compute a value for R_{es}. In this particular application of emitter bias V_{BE} stability requirements are not very demanding, and convenient power supply voltages for bias will usually be much higher than the minimum required for V_{BE} stability; however, there are high-stability circuits such as the differential amplifier, where stability requirements will demand up to 10 V, and perhaps more, for E_{Res}.

In the case of this example we will need a bias supply of 2.75 V plus an additional volt (or so) for V_{BE} and E_{Rb1}, or 3.75 V. Any voltage larger than this will do. Suppose we assume that a 10-V supply is available for bias use and work from there. Let E_{bias} equal 10 V.

2. *Computing R_{es}*

Now that we have made the bias voltage decision, a little Ohm's law will provide us with a value for R_{es}.

$$R_{es} = \frac{E_{\text{bias}}}{QI_c} = \frac{10\ \text{V}}{1\ \text{mA}} = 10\ \text{k}\Omega$$

Here we have assumed that *all* the bias voltage is dropped across the emitter resistor. This is not quite true, because there will be about 0.6 V (for silicon) drop across the base-emitter junction, and some small voltage drop across R_{b2}. We would expect some error in QI_c if we make this assumption, but because the error is quite small and because of the large amount of negative feedback involved, the error, in practice, is negligible.

FRAME 4

Compute the base-emitter junction resistance R_e'.

$$R_e' = \frac{25}{QI_c} = \frac{25}{1} = 25\ \Omega$$

FRAME 5

Compute the value of the ac load resistance R_{Lac}.

Use Branch *b* on the design sheet.

$R_{Lac} = V_g \times R_e' = 140 \times 25$
$R_{Lac} = 3.5\ \text{k}\Omega$

FRAME 6

Compute the value of the dc load resistor R_L.

$$R_L = \frac{R_{Lw} \times R_{Lac}}{R_{Lw} - R_{Lac}} = \frac{10\ \text{k}\Omega \times 3.5\ \text{k}\Omega}{10\ \text{k}\Omega - 3.5\ \text{k}\Omega}$$

$R_L = 5.3\ \text{k}\Omega$

FRAME 7

Compute the voltage drop across the dc load E_{RL}.

$E_{RL} = QI_c \times R_L = 1 \times 10^{-3} \times 5.3 \times 10^3$
$E_{RL} = 5.3\ \text{V}$

FRAME 8

Compute the voltage drop between collector and emitter, E_Q. (E_o p-p is 3 V.)

$E_Q\ (\text{min}) = 1 + E_o\ \text{p-p} = 1 + 3 = 4\ \text{V}$

Suppose we wish to use a 10-V power supply for E_s.

$E_{RL} + E_Q = 5.3 + 4 = 9.3\ \text{V}$

Remember that in the emitter bias circuit, none of E_s is dropped across R_{es}. The difference between the calculated E_s of 9.3 V and the desired E_s of 10 V is 0.7 V. If we add the leftover 0.7 V to the computed 4-V E_Q, we get a new value for E_Q of 4.7 V.

$E_Q = 4.7\ \text{V}$

FRAME 9

Compute E_s, the power supply voltage.

We have already decided on an E_s of 10 V.

$E_s = 10\ \text{V}$

FIELD 2 BASE SIDE CALCULATIONS

FRAME 1

Compute the base-to-ground resistance R_{b2}.

$R_{b2} = \mathbf{S} \times R_{es} = 5 \times 10\ \text{k}\Omega$
$R_{b2} = 50\ \text{k}\Omega$

Any value 50 kΩ or lower will satisfy the stability requirement.

That takes care of the base side calculations, because there is no R_{b1}, and we branch to Field 3.

FIELD 3 INPUT IMPEDANCE AND CAPACITORS

FRAME 1

Compute the input impedance.

$Z_{in} = R_{b2} \| \beta R_e'$
$Z_{in} = 50\ \text{k}\Omega \| (100 \times 25)$
$Z_{in} \approx 2.5\ \text{k}\Omega$

FRAME 2

Compute the value of the input coupling capacitor C_{cin}.

X_c of $C_{cin} \approx 0.1\ Z_{in}$ at f_o

$$X_c \text{ at } f_o = \frac{2.5\ \text{k}\Omega}{10} = 250\ \Omega$$

$$C = \frac{1}{2\pi f X_c}$$

$$C = \frac{1}{6.28 \times 25 \times 250}$$

$$C \approx 25\ \mu\text{F}$$

FRAME 3

Compute C_{co}, the output coupling capacitor.

X_c of $C_{co} \approx 0.1\ R_{Lw}$ at f_o

$$X_c = \frac{10\ \text{k}\Omega}{10} = 1\ \text{k}\Omega$$

$$C = \frac{1}{2\pi f X_c}$$

$$C = \frac{1}{6.28 \times 25 \times 1 \times 10^3}$$

$$C \approx 6\ \mu\text{F}$$

FRAME 4

Compute the value of the bypass capacitor C_{bp}.

In this case let us be a little cheap and lazy, and use

$X_c \approx R_e'$ at f_o

Because this circuit is a fully bypassed circuit, all the equations and comments made in the last example apply.

X_c at Hz $= 25\ \Omega$ (R_e')

$$C = \frac{1}{2\pi f X_c} = \frac{1}{6.28 \times 25 \times 25}$$

$C \approx 250\ \mu F$

Figure 6-9 shows the complete circuit with all values indicated.

SUMMARY

The addition of a bypass capacitor across the emitter resistor of what is basically an unbypassed common-emitter circuit can increase the voltage gain from 10- to 50-fold (more in very lightly loaded circuits).

The price we pay for raw gains of 40 to 400 (even several thousand in a few instances) is lack of gain stability, considerable uncertainty in predicting

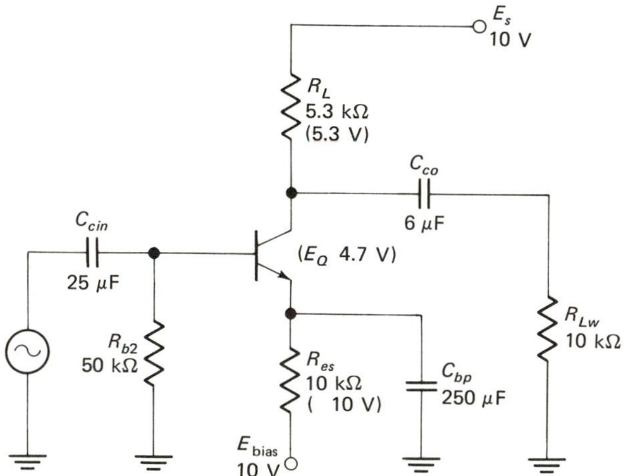

Fig. 6-9 The complete circuit for Example 6-3, with values of resistances and voltages

the voltage gain during the design process, poor input impedance predictability and stability, and more distortion.

None of the stability requirements are affected by the addition of the bypass capacitor. In fact, the added capacitor causes two sets of parameters to exist, stability and signal. Below the lowest frequency of interest V_g, **S**, I_g, Z_{in}, and V_{BE} stability are the same as though the capacitor were not there. For frequencies above this cutoff frequency these parameters are the same as though R_{es} were 0 Ω. Table 6-1 compares these two sets of parameters.

The split emitter circuit is a good compromise between the unbypassed and the fully bypassed configurations. This variation also has a definite low-frequency cutoff value, and below that frequency it has the parameters of the unbypassed circuit, including Q-point stability.

Predictability of signal parameters is almost as good as it is in the unbypassed circuit. Available voltage gains range from about 10 to 50 for moderate loads. The signal input impedance is lower than in the unbypassed circuit and higher than in the fully bypassed circuit.

The emitter-biased circuit provides somewhat better Q-point stability than base bias, and provides considerable base circuit flexibility because of the absence of R_{b1}. It is presented here as a fully bypassed circuit and has signal parameters nearly identical to the base-biased fully bypassed circuit. Later this mode of bias will be used in the differential amplifier, where it is far more practical than base bias.

QUESTIONS

6-1 Why are signal parameters somewhat unpredictable in the fully bypassed circuit?

6-2 Are signal parameters influenced more by temperature in the fully bypassed and split emitter circuits than in the unbypassed circuit? Why?

6-3 Why is the fully bypassed circuit likely to produce more distortion than the unbypassed circuit?

6-4 What effect does the bypass capacitor have on Q-point stability?

6-5 Match the following characteristics with the appropriate circuit configuration.

1. Unbypassed	*a.* The highest voltage gain
	b. The highest input Z
	c. The best ac predictability
	d. The lowest voltage gain
2. Split emitter	*e.* The lowest input Z
	f. The poorest ac predictability
	g. Modest voltage gain
3. Fully bypassed	*h.* Modest input Z
	i. Modest ac predictability

PROBLEMS

Problem 6-1

TITLE: Designing the Fully Bypassed Circuit
PROCEDURE: Design the circuit, using the following specifications. See Fig. 6-1.

Transistor Parameters	
1. (min)	60
2. I_c (max)	100 mA
3. E_c (max)	20 V
Circuit Parameters	
1. Voltage gain	100
2. E_o p-p	3.0 V
3. R_{Lw}	5 kΩ
4. Stability factor	10
5. $\%\Delta QI_c$ to be allowed	5%
6. Maximum operating temperature	85°C
7. f_o	30 Hz

Problem 6-2

TITLE: Designing the Split Emitter Circuit
PROCEDURE: Design the circuit, using the following specifications. See Fig. 6-4.

Transistor Parameters	
1. β (min)	100
2. I_c (max)	100 mA
3. E_c (max)	30 V
Circuit Parameters	
1. Voltage gain	15
2. E_o p-p	3.0 V
3. R_{Lw}	15 kΩ
4. Stability factor	10
5. $\%\Delta QI_c$ to be allowed	5%
6. Maximum operating temperature	90°C
7. f_o	30 Hz

Problem 6-3

TITLE: Designing the Emitter-biased Fully Bypassed Circuit
PROCEDURE: Design the circuit, using the following specifications. See Fig. 6-7.

Transistor Parameters	
1. β (min)	100
2. I_c (max)	50 mA
3. E_c (max)	30 V

Circuit Parameters

1.	Voltage gain	25
2.	E_o p-p	5 V
3.	R_{Lw}	12 kΩ
4.	Stability factor	1
5.	$\%\Delta QI_c$ to be allowed	1%
6.	Maximum operating temperature	85°C
7.	f_o	25 Hz

|7|

potpourri

OBJECTIVES:

(Things you should be able to do upon completion of this chapter)
NOTE: Specific design problems are included in the chapter.

1. Design the simplest case (case 1) of the voltage-mode feedback common-emitter amplifier circuit.
2. Design the more versatile case (case 2) of the voltage-mode feedback common-emitter amplifier circuit.
3. Design the split-load (paraphase) amplifier.
4. Design a common-collector circuit with both emitter and base bias, and including bootstrapping for increased input impedance for ac signals.
5. Design two kinds of common-base circuit, one with base bias and one with emitter bias.

EVALUATION:

1. There are self-test questions and design problems included in this chapter.
2. The *final* evaluation will be based upon your ability to design the circuits in this chapter and have them work in the laboratory according to the criteria and specifications set forth in the laboratory manual.

INTRODUCTION

This chapter is called potpourri for good reason. We will examine a number of circuits and methods, some related and some not.

To begin with, we will examine practical design procedures for circuits using voltage-mode feedback. There are two cases of voltage-mode feedback, a very simple but very limited case, and a more complex but more versatile case. It is sometimes difficult to meet exact specifications with voltage-mode feedback, and in most cases, with a pair of notable exceptions, current-mode feedback is preferred to voltage-mode feedback. The first exception involves the case where very low power supply voltages are involved, say 3 V or less, where current-mode circuits are simply not practical. The second case involves power amplifier drivers, where the absolute maximum E_o p-p is required for a *given* E_Q. Voltage-mode circuits can provide about 50 percent more E_o p-p for a given E_Q than can be had from current-mode circuits.

We will examine a circuit called the split-load (paraphase) amplifier, which is essentially a current-mode common-emitter circuit with one output taken from the collector and one from the emitter. It is really a combination common-emitter and common-collector circuit which provides two output voltages with equal amplitude which are 180° out of phase with each other, but for design purposes it is treated as a special case of the common-emitter circuit. It is treated as a common-emitter circuit because none of the special characteristics of the common-collector part of the circuit can normally be taken advantage of in the instances where this circuit is used.

Next, we will look at the common-collector circuit and its special impedance-elevating characteristics. The common collector, often called the *emitter follower*, is a particularly important *buffer* circuit between low-level stages and power amplifier stages. Often, when we use an emitter-follower (common-collector) circuit, we are interested in getting the highest possible input impedance. When this is the case, we find that bias resistors and, at a little higher impedance level, collector-base junction resistance tend to limit that maximum possible input impedance. We shall see how bootstrapping (for ac signals) can help solve this problem.

There are two common-base circuits, one with base bias and one with emitter bias. The emitter-biased circuit is "true" common-base because the base is returned directly to ground, for both dc and signal currents. The base-biased circuit is really a common-emitter circuit for dc parameters, but a capacitor returns the *signal* current to ground, making it a common-base circuit, insofar as the signal is concerned. Both circuits are designed like their common-emitter counterparts, with one or two minor deviations from the common-emitter design sheet.

7-1 DESIGNING WITH VOLTAGE-MODE FEEDBACK—THE SIMPLEST CASE

The circuit for voltage-mode feedback in its simplest form is shown in Fig. 7-1. The outstanding virtue of this circuit is its simplicity. However, it does have three inherent characteristics which limit its application.

1. Because of the Miller effect its input impedance is generally somewhere on the order of 500 Ω or less, which is about one-quarter of the input impedance of a fully bypassed current-mode circuit.
2. The maximum value for E_Q (for many practical situations) is 3 V or less. This is because of an inherent conflict between acceptable base currents and necessary stability factors.

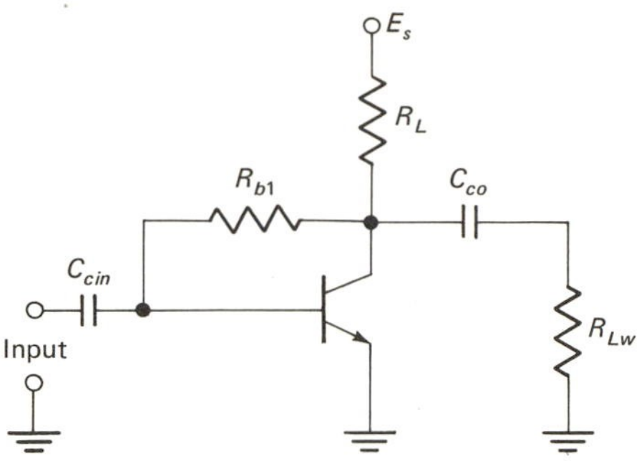

Fig. 7-1 The circuit for the simplest case (case 1) of voltage-mode feedback

Because R_{b1} (See Fig. 7-1) serves the dual purpose of meeting stability requirements and setting the quiescent point, keeping E_Q low compared to E_{RL} is necessary if we are to be able to choose a value for **S** and one for QI_c. Even if we restrict E_Q, it is difficult to get the exact value of **S** required. However, it is not too difficult to get the desired QI_c and some value of **S** which is equal to or smaller than the one specified.

In spite of the basic simplicity (in fact, because of it), it is not generally possible to meet all specifications exactly. In fact, it is very difficult to come up with a really good compromise in all cases. To avoid this difficulty, I have prepared Table 7-1, the voltage-mode stability table, which makes it quite easy to come up with an acceptable compromise circuit for any application for which this simple configuration is suitable.

Applications for Voltage-mode Feedback

In the majority of applications voltage-mode circuits run a poor second to current-mode circuits, at least from a designer's point of view. There are only three instances when voltage mode has the edge on current-mode designs.

WHEN THE VOLTAGE-MODE CIRCUIT'S BASIC SIMPLICITY CAN BE TAKEN ADVANTAGE OF. This generally means that a very low input impedance must be acceptable. The voltage gain, even with this simplest circuit, is comparable to that of the current-mode fully bypassed circuit. We will use the fully bypassed (current-mode) table (Table 6-2) for design purposes.

Stability against V_{BE} variations is inherently very good, and it depends upon the voltage drop across R_L instead of upon the voltage drop across an extra emitter resistor. The voltage drop across R_L is nearly always adequate for V_{BE} stability. However, β and I_{co} stability depends upon a large ratio of E_{RL} to E_Q, which in many cases means a relatively low value for E_Q. Fortunately a low E_Q in a voltage-mode circuit is not as serious a problem as it is in current-mode circuits.

TABLE 7-1. VOLTAGE-MODE STABILITY TABLE

$E_Q = 1.6$ V				$E_Q = 2$ V				$E_Q = 2.6$ V			
E_{RL}	S at $\beta = 50$	S at $\beta = 100$	S at $\beta = 200$	E_{RL}	S at $\beta = 50$	S at $\beta = 100$	S at $\beta = 200$	E_{RL}	S at $\beta = 50$	S at $\beta = 100$	S at $\beta = 200$
2	25	50	100	4	20	40	80	2	50	100	200
4	12.5	25	50	6	13.3	26.6	53.2	4	25	50	100
6	8.3	16.6	33.2	8	10	20	40	6	16.6	33.2	66.4
8	6.25	12.5	25	10	8	16	32	8	12.5	25	50
10	5	10	20	12	6.6	13.3	26.6	10	10	20	30
12	4.15	8.3	16.6	14	5.7	11.4	22.8	12	8.3	16.6	33.2
14	3.5	7.1	14.2	16	5	10	20	14	7.1	14.2	28.4
16	3.1	6.2	12.4	18	4.4	8.8	17.6	16	6.2	12.4	24.8
18	2.75	5.5	11	20	4	8	16	18	5.5	11	22
20	2.5	5.0	10	22	3.85	7.7	15.4	20	5.0	10	20
22	2.25	4.5	9	24	3.3	6.6	13.2	22	4.1	8.6	16.4
24	2.05	4.1	8.2								

In current-mode circuits the maximum p-p output voltage is about $E_Q - 1$ V, whereas in the case of voltage-mode feedback the maximum p-p output voltage is about 1.5 E_Q.

WHEN THE CIRCUIT MUST OPERATE FROM A VERY LOW VOLTAGE POWER SUPPLY. Voltage-mode circuits can operate with collector-emitter voltages (E_Q) of 1 V or less; this is something that cannot be done, reliably, with current-mode circuits.

WHEN THE ABSOLUTE MAXIMUM OUTPUT VOLTAGE (E_o p-p) FOR A GIVEN E_Q IS REQUIRED. Because saturation does not occur in voltage-mode circuits in the abrupt fashion that is common in current-mode circuits, the maximum output voltage, for a given E_Q, is greater for voltage-mode circuits than it is for current-mode circuits.

This is particularly important in certain power amplifier *driver* applications, where the value of E_Q is dictated by required voltage distributions in the power amplifier.

As a general rule such power amplifiers require fairly large values for E_Q, precluding the use of the *simplest case* for these applications. We will examine more elaborate variations of the simplest case of voltage-mode feedback which will be suitable for these applications.

Designing the Simplest-case Voltage-mode Circuit (Refer to Fig. 7-1)

Before I give you the design procedure and a design example, let me take a moment to summarize the important equations for the simplest case.

1. Voltage gain (V_g):

   $$V_g = \frac{R_{Lac}}{R'_e}$$

   You will notice that this is the same equation as that used for the fully bypassed current-mode circuit, which means that we can use the *fully bypassed* (current-mode) *table of possibilities*, Table 6-2, to aid our design process.
2. V_{BE} stability:

   $$E_{RL} = \frac{\Delta V_{BE} \times 100}{\% \Delta QI_c}$$

   This equation is also familiar except that the voltage drop we are interested in is the drop across R_L and not the voltage across an extra emitter resistor. Rare indeed are the cases where it is necessary to bother with this equation for the simplest case. The only instance I can think of would be a circuit designed to operate with a very low E_s, perhaps 3 V or less. This equation is omitted from the design sheet for the simplest case because it is so seldom necessary to use it.
3. **S**, the stability factor:

   $$\mathbf{S} \approx \frac{R_{b1}}{R_L}$$

   This equation was derived in Chap. 4.
4. I_g, the circuit current gain:

   $$I_g \approx \mathbf{S} \approx \frac{R_{b1}}{R_L}$$

   As in previous circuits, the circuit current gain is approximately equal to the stability factor.
5. E_o p-p, the maximum p-p output voltage:

   $$E_o \text{ p-p} \approx 1.5\ E_Q$$

   This is a fairly satisfactory figure for 5 to 20 V of E_Q. The maximum E_o will be a little less for lower E_Q voltages, and slightly higher for higher values of E_Q.
6. The input impedance Z_{in}:

   $$Z_{in} \approx \frac{R_{b1}}{V_g} \| \beta R'_e$$

NOTE: R_{b1} should be divided by V_g and the multiplication of $\beta \times R_e'$ should be accomplished before the paralleling operation

$$\frac{R_1 \times R_2}{R_1 + R_2}$$

is performed. The left side of the $\|$ sign is R_{b1}/V_g. This is the Miller effect as discussed in Chap. 4. On the righthand side of the $\|$ sign is the input resistance of the transistor; this should already be familiar to you.

Example 7-1 Designing the Voltage-mode Circuit, Simplest Case

Design a voltage-mode feedback circuit with the following specifications. See Fig. 7-1.

Transistor Parameters

1. β	50
2. E_c (max)	15 V
3. I_c (max)	100 mA
4. Type	*npn* silicon

Circuit Parameters

1. Voltage gain	100
2. Stability factor	10
3. E_o p-p	500 mV
4. R_{Lw}	4 kΩ

FIELD 1

FRAME 1

Select QI_c.

Consult the *fully bypassed table of possibilities,* Table 6-2, for E_{RL} and QI_c. Here, as usual, we cannot have a supply voltage greater than the transistor maximum collector voltage. If we allow 3 V or so for E_Q, we cannot have an E_{RL} of more than 12 V.

If we examine Table 6-2, we find a QI_c of 1 mA, and 6.6 V for E_{RL} looks favorable. We could shoot for a higher E_{RL} in order to get a larger E_o p-p. If you will examine Table 7-1, the voltage-mode stability table, you will see that stability involves a large ratio of E_{RL} to E_Q: if we need a large E_o p-p, we also need a large E_Q, and if we have a large E_Q, we must have a larger E_{RL} in order to have a stable circuit. For 15 V maximum E_s, 6.6 V for E_{RL} is fairly reasonable, and so we shall use that figure.

$QI_c = 1$ mA $\qquad E_{RL} = 6.6$ V

FRAME 2

Find E_Q.

Consult the voltage-mode stability table, Table 7-1, for an appropriate value for E_Q. According to the values under $\beta = 50$, we can get a stability factor of 8.3 with an E_Q of 1.6 V and an E_{RL} of 6 V. 8.3 is a perfectly satisfactory stability factor, because *it is less than* the **S** of 10 specified. An E_Q of 1.6 V is also adequate for the specified p-p output voltage of 0.5 V. We could use a higher value of E_Q if for some reason it seemed desirable.

$E_Q = 1.6$ V

FRAME 3

Compute R_e'.

$$R_e' \approx \frac{25}{QI_c}$$

$$R_e' = \frac{25}{1 \text{ mA}} = 25\ \Omega$$

FRAME 4

Compute R_{Lac}.

$$R_{Lac} = V_g \times R_e'$$
$$R_{Lac} = 100 \times 25 = 2{,}500\ \Omega$$

FRAME 5

Compute R_L.

$$R_L = \frac{R_{LW} \times R_{Lac}}{R_{LW} - R_{Lac}}$$

$$R_L = \frac{4 \text{ k}\Omega \times 2.5 \text{ k}\Omega}{4 \text{ k}\Omega - 2.5 \text{ k}\Omega}$$

$$R_L = \frac{10 \text{ k}\Omega}{1.5 \text{ k}\Omega} = 6.6 \text{ k}\Omega$$

FRAME 6

Compute E_{RL}.

$$E_{RL} = QI_cR_L$$
$$E_{RL} = 1 \text{ mA} \times 6.6 \text{ k}\Omega$$
$$E_{RL} = 1 \times 10^{-3} \times 6.6 \times 10^{3}$$
$$E_{RL} = 6.6 \text{ V}$$

FRAME 7

Compute the required power supply voltage.

$$E_s = E_Q + E_{RL}$$
$$E_s = 1.6 + 6.6 = 8.2 \text{ V}$$

At this point the designer is faced with a lack of flexibility in the power supply voltage selection. In the case of previous current-mode designs we could conveniently elevate

E_Q to obtain a comfortable value for E_s. We cannot do that here without a sacrifice in stability; in fact, we cannot do anything without altering circuit parameters. The most convenient, and at the same time the simplest, approach to adjusting E_s to a convenient value is to *elevate* E_{RL} and to recalculate R_L. The results of such an adjustment are all in a favorable direction. The voltage gain and stability factor will be altered, with the stability factor being made smaller and the voltage gain increasing. Suppose, in this case, we adjust E_s to 9 V.

1. The calculated value for E_s is 8.2 V.
2. The desired value for E_s is 9 V.
3. The difference is 0.8 V.
4. The original $E_{RL} = 6.6$ V plus 0.8 V, yielding a new value for E_{RL} of 7.4 V.
5. Recalculating R_L,

$$R_L = \frac{E_{RL}}{QI_c} = \frac{7.4\ \text{V}}{1\ \text{mA}} = 7.4\ \text{k}\Omega$$

The recalculated value for R_L (7.4 kΩ) is not a standard value, but this is no problem because the difference in circuit parameters due to the use of the nearest standard value will be a negligible one.

E_s (adjusted) = 9 V
E_{RL} (adjusted) = 7.4 V
R_L (adjusted) = 7.4 kΩ

The new voltage gain after the adjustment of R_L, for this example, is only about 103 compared to the originally specified 100. And, as a matter of fact, we could have adjusted E_s to 20 V or so without any very significant change in voltage gain. If you are in doubt, you can always recalculate the voltage gain using the adjusted value of R_L: $V_g = R_{Lac}/R_e'$, where R_{Lac} equals $R_{Lw} \| R_L$ (adjusted).

FIELD 2

FRAME 1

Compute the base current required for the desired QI_c.

$$I_b = \frac{1\ \text{mA}}{50} = 2 \times 10^{-5}\ \text{A}$$

It is good practice always to use a design value of about 50 for β even though the transistor manual specifies a considerably higher value.

FRAME 2

Compute the value of R_{b1}.

$$R_{b1} = \frac{E_Q - V_J}{I_b}$$

VOLTAGE MODE DESIGN SHEET
Case 1

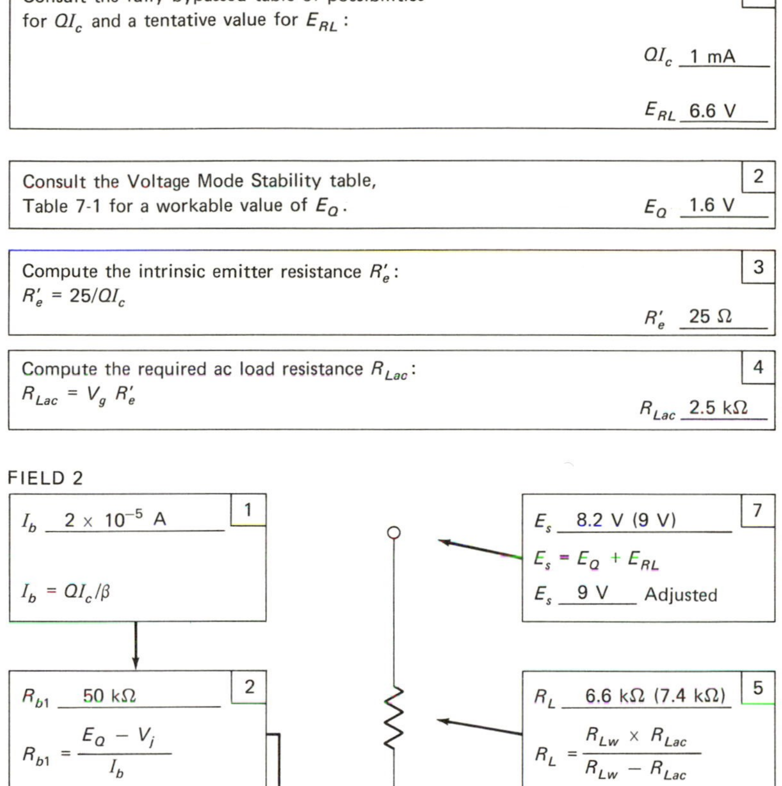

1
Consult the fully bypassed table of possibilities for QI_c and a tentative value for E_{RL}:
QI_c 1 mA
E_{RL} 6.6 V
2
Consult the Voltage Mode Stability table, Table 7-1 for a workable value of E_Q.
E_Q 1.6 V
3
Compute the intrinsic emitter resistance R'_e:
$R'_e = 25/QI_c$
R'_e 25 Ω
4
Compute the required ac load resistance R_{Lac}:
$R_{Lac} = V_g\ R'_e$
R_{Lac} 2.5 kΩ
FIELD 2
1
I_b 2×10^{-5} A
$I_b = QI_c/\beta$
7
E_s 8.2 V (9 V)
$E_s = E_Q + E_{RL}$
E_s 9 V Adjusted
2
R_{b1} 50 kΩ
$R_{b1} = \dfrac{E_Q - V_j}{I_b}$
5
R_L 6.6 kΩ (7.4 kΩ)
$R_L = \dfrac{R_{Lw} \times R_{Lac}}{R_{Lw} - R_{Lac}}$

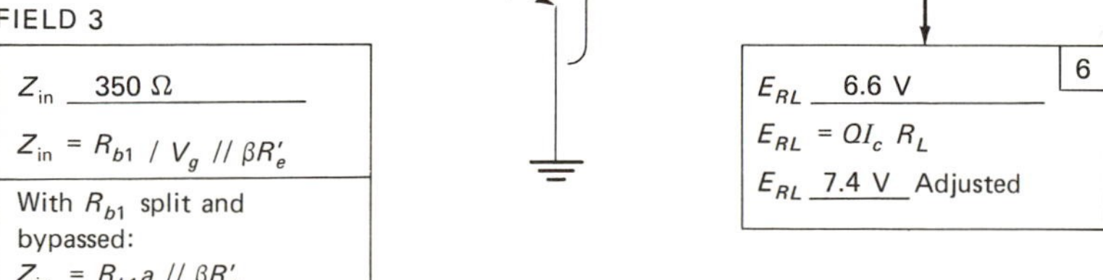

Fig. 7-2 The completed design sheet for Example 7-1

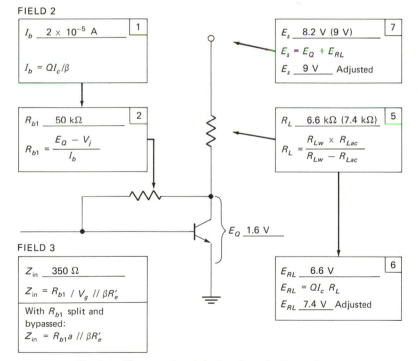

$$R_{b1} = \frac{1.6 - 0.6 \text{ V}}{2 \times 10^{-5}} = \frac{1 \text{ V}}{2 \times 10^{-5}} = 5 \times 10^4 \ \Omega$$

$$R_{b1} = 50 \text{ k}\Omega$$

FIELD 3

FRAME 1

Compute the input impedance.

$$Z_{in} \approx \frac{R_{b1}}{V_g} \| \beta R_e'$$

$$Z_{in} \approx \frac{50 \text{ k}\Omega}{100} \| 50 \times 25$$

$$Z_{in} \approx 500 \| 1{,}250$$

$$Z_{in} \approx 350 \ \Omega$$

FRAME 2

Compute the value of the input coupling capacitor.

X_c of C_{cin} at $f_0 \approx 0.1 Z_{in}$

This is the same rule of thumb we have used consistently for the determination of C_{cin}.

$$C_{cin} \approx 200 \ \mu\text{F}$$

Figure 7-2 shows the completed design sheet for the example.

Increasing the Input Impedance of the Simplest Case

The input impedance (at signal frequencies) in the simplest-case circuit can be increased to almost exactly the same level as that of a current-mode fully bypassed circuit, (operating at the same QI_c) by splitting R_{b1} and filtering out the ac Miller effect feedback with a capacitor. The modified circuit is shown in Fig. 7-3. At this point the circuit complexity[1] is nearly the same as that of the fully bypassed current-mode circuit. The resistor R_{b1} should normally be split near its center so that $R_{b1}a \approx R_{b1}b$. Other splits are possible, but a 50-50 split is generally considered to be the best compromise. Moving the tap too far toward the collector increases the collector loading and reduces the voltage gain; moving it toward the base tends to lower the input impedance.

The equation for the input impedance of the modified circuit in Fig. 7-3 is $Z_{in} \approx R_{b1}a \| \beta \times R_e'$.

The modified circuit is designed by the same procedure as the simplest case in the previous example. After the basic design is completed, R_{b1} is

[1] Component count and approximate cost.

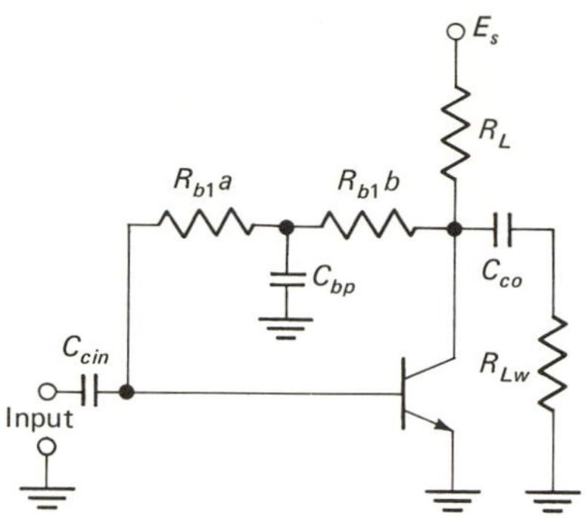

Fig. 7-3 The modified voltage-mode feedback circuit for higher input impedances

split in two, and a capacitor value computed which has an X_c at f_o of about $0.1R_{b1}a$. The input capacitor C_{cin} will be smaller in the modified circuit because the input impedance of the modified circuit will be higher ($\approx$ 1,190 Ω).

Problem 7-1

TITLE: **Designing the Simplest Case (Case 1) of Voltage-mode Feedback**
PROCEDURE: **Design the circuit, using the following specifications. See Fig. 7-1.**

Transistor Parameters

1. β (min)	100 (use 50)
2. I_c (max)	100 mA
3. E_c (max)	25 V
4. Transistor type	*npn* silicon

Circuit Parameters

1. Voltage gain	140
2. E_0 p-p	1 V
3. R_{Lw}	5 kΩ
4. Stability factor	10
5. $\%\Delta QI_c$ to be allowed	—
6. Maximum operating temperature	—
7. f_o	30 Hz

7-2 DESIGNING THE MORE VERSATILE CASE OF THE VOLTAGE-MODE FEEDBACK CIRCUIT

The principal deficiency of the simplest case is that R_{b1} must provide the desired base current and an appropriate stability factor at the same time. This dual requirement forces us to have a large ratio of E_{RL} to E_Q. The net effect is an inadequate degree of flexibility in the selection of values for E_Q. There are some applications where the ratio of E_{RL} to E_Q is dictated by other circuit requirements and where the ratio of E_{RL} to E_Q cannot be high enough for acceptable stability. In such cases an additional resistor can be added from base to ground; this allows us to select a value of R_{b1} which will provide a desired stability factor, without producing excessive base bias current, for any desired value of E_Q. The addition of this resistor forms a current divider which shunts any excess R_{b1} current to ground. The circuit is shown in Fig. 7-4.

Because of the Miller effect, R_{b1} will provide the necessary stability, and although R_{b2} will contribute to stability, its contribution will be relatively small. In general, for large values of E_Q, when R_{b1} is selected for a reasonable stability factor ($\mathbf{S} \approx R_{b1}/R_L$), the value of R_{b1} will be so low that the transistor will be biased nearly into saturation and normal transistor action will cease. The resistor R_{b2} can be computed to shunt any amount of excess current to ground, allowing just the proper amount of base current for any desired quiescent collector current.

The input impedance of this variation is somewhat lower than it is in the simplest case. The input impedance can be raised in this circuit by the same method used in the simplest case, by splitting R_{b1} and adding a large capacitor. This modification is shown in Fig. 7-5. The basic performance equations for the unmodified, case 2 (without R_{b1} being split and the capacitor omitted), form of the more versatile circuit are the same as for the unmodified simplest case, case 1.

The design procedure is different because of the addition of R_{b2} and because of the greater design flexibility this addition allows.

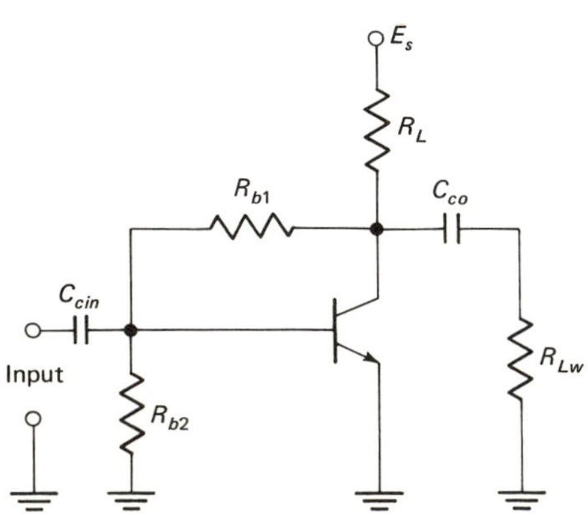

Fig. 7-4 The more versatile case of the voltage-mode feedback circuit (case 2)

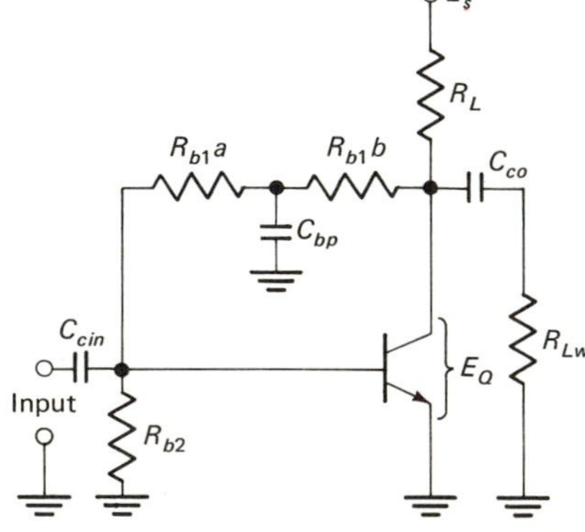

Fig. 7-5 The more versatile case of voltage-mode feedback, modified for higher input impedances (modified case 2)

Summary of the Design Procedure

FIELD 1 COLLECTOR SIDE COMPUTATIONS

FRAME 1

Determine the minimum collector load voltage drop E_{RL} for adequate V_{BE} stability.

$$E_{RL}\ (\text{min}) = \frac{\Delta V_{BE} \times 100}{\%\Delta QI_c}$$

This step can often be bypassed because E_{RL}, more often than not, is more than adequate. For the time being I suggest that you include the step. Later with a bit of experience you can leave it out routinely and go back to it if E_{RL} from the table of possibilities seems suspiciously low.

FRAME 2

Make the collector current decision.

Consult the current-mode fully bypassed table of possibilities, Table 6-2.

The voltage gain equation for this circuit is $V_g \approx R_{Lac}/R'_e$. This is the same equation as that used for the fully bypassed current-mode circuit and the simplest-case voltage-mode circuit, and the same table of possibilities is used for all three circuits.

FRAME 3

Compute R'_e (the Shockley relationship).

$$R'_e \approx \frac{25}{QI_c}$$

FRAME 4

Compute R_{Lac}, the equivalent (signal) ac load.

$$R_{Lac} = V_g \times R'_e$$

Nothing new here.

FRAME 5

Compute R_L, the collector load resistor value.

$$R_L = \frac{R_{Lw} \times R_{Lac}}{R_{Lw} - R_{Lac}}$$

This is the same equation we have used in all previous designs.

FRAME 6

Compute E_{RL} (the voltage drop across the collector load resistor).

$E_{RL} = QI_c \times R_l$ (Ohm's law)

FRAME 7

Compute E_Q (min).

E_Q (min) $\approx 0.66\ E_o$ p-p

This is different from the equation we used in current-mode designs because of the voltage-mode circuit's different saturation characteristics.

The maximum output voltage (E_o p-p) equation is

E_o p-p $\approx 1.5E_Q$

If we solve this equation for E_Q, we get

$$E_Q \approx \frac{1}{1.5} E_o \text{ p-p} = 0.66\ E_o \text{ p-p}$$

FRAME 8

Compute E_s, the supply voltage.

$E_s = E_{RL} + E_Q$ (the sum of the voltage drops)

At this point we *may* adjust E_Q, just as we did in current-mode circuits, to yield a more suitable supply voltage. In some cases E_Q will be dictated by voltage distribution requirements of a following stage. This more versatile configuration makes this kind of adjustment in E_Q possible also.

FIELD 2 BIAS AND STABILITY CALCULATIONS

FRAME 1

Compute the value of R_{b1}.

$R_{b1} = \mathbf{S} \times R_L$

This resistor provides for the stability factor, but it will normally provide more current than is appropriate for base current. It is the function of R_{b2} to divide the current and drain off the excess. R_{b1} is of course involved in the bias current, but because of the Miller effect, it is overwhelmingly dominant in controlling the stability factor; it is best thought of as a stability resistor rather than as a bias resistor. R_{b2} also contributes something, but not much, to the stability of the circuit. Because of R_{b2} the actual **S** of the designed circuit will be *slightly* lower than the specifications call for. Obviously a slight increase in overall stability is no great inconvenience.

This particular design approach treats R_{b1} as a stability resistor *only* and deals with R_{b2} as a bias control resistor *only*. The amount of error in this approach is negligible, but even more important is the fact that this approach almost always yields acceptable input impedance levels.

We could write an equation involving all three variables, bias current, **S**, and input impedance, for each resistor, but the gain in accuracy would be far from worth the effort.

FRAME 2

Compute the voltage drop across R_{b1}.

$$E_{Rb1} = E_Q - V_j$$

The voltage applied across the resistors R_{b1} and R_{b2} in series is the voltage E_Q. R_{b2} is in parallel with the base-emitter junction of the transistor and, because of the nature of the diode forward bias conduction curve, R_{b2} will *always* have the junction voltage (0.6 V for silicon) V_j across it. The balance of the voltage E_Q must therefore appear across R_{b1}.

FRAME 3

Compute I_{Rb1}, the current through R_{b1}.

$$I_{Rb1} = \frac{E_{Rb1}}{R_{b1}} \qquad \text{(Ohm's law)}$$

FRAME 4

Compute I_b, the base current.

$$I_b = \frac{QI_c}{\beta}$$

This is simply an algebraic variation of the definition of β.

$$\beta = \frac{QI_c}{I_b}$$

We are simply computing the required base current to yield the desired collector current.

FRAME 5

Compute I_{Rb2}, the current through R_{b2}.

$$I_{Rb2} = I_{Rb1} - I_b \qquad \text{(simple subtraction)}$$

At this point we know how much current will be flowing down R_{b1}, and how much of that current we need as base-emitter current. What we must determine here is how much excess current we must siphon off through R_{b2}, that is, the current through R_{b2}.

FRAME 6

Compute R_{b2}.

$$R_{b2} = \frac{V_j}{I_{Rb2}} \qquad \text{(Ohm's law)}$$

We know that the voltage drop across R_{b2} is always V_j, and we computed the current through R_{b2} in Frame 5.

FIELD 3

FRAME 1

Compute the input impedance Z_{in}.

$$Z_{in} \approx \frac{R_{b1}}{V_g} \| \beta R'_e \| R_{b2}$$

The only difference between this and the equation for the input impedance of the simplest case is the addition of R_{b2} in parallel with the rest of the input circuit.

FRAME 2

Compute the value of the input capacitor C_{cin}.

$$X_c \text{ at } f_o \text{ of } C_{cin} \approx 0.1Z_{in}$$

Increasing the Input Impedance of the More Versatile Case of Voltage-mode Feedback

The procedure here is exactly the same as that used for the simplest case. R_{b1} is split into two equal-value resistors (each one half of the computed value for R_{b1}), and a bypass capacitor is placed between the junction of the two resistors and ground. See Fig. 7-5.

The input impedance now becomes

$$Z_{in} \approx R_{b1}a \| \beta R'_e \| R_{b2}$$

The input capacitor is X_c at $f_o \approx 0.1Z_{in}$. The reactance of the bypass capacitor C_{bp} is (again) approximately $0.1R_{b1}b$.

Example 7-2 Designing the More Versatile Voltage-mode Circuit (Case 2)

Design a voltage-mode feedback circuit with the following specifications. See Fig. 7-5.

Transistor Parameters

1. β	100
2. E_c (max)	25 V
3. I_c (max)	100 mA
4. Type	*npn* silicon

Circuit Parameters

1. Voltage gain	100
2. Stability factor	10
3. $\%\Delta QI_c$	10%
4. E_o p-p	6 V
5. R_{Lw}	6 kΩ
6. Maximum operating temperature	80°C

FIELD 1 COLLECTOR SIDE CALCULATIONS

FRAME 1

Compute E_{RL} (min).

$$E_{RL}\text{ (min)} = \frac{\Delta V_{BE} \times 100}{\%\Delta QI_c}$$

$t = 80° - 20° = 60°\text{C}$
$V_{BE} = 60 \times 2.5 \text{ mV} = 150 \text{ mV}$

$$E_{RL}\text{ (min)} = \frac{1.5 \times 10^{-1} \times 10^{2}}{10\ (\%)}$$

$$E_{RL}\text{ (min)} = 1.5 \text{ V}$$

FRAME 2

Make the collector current decision (see Table 6-2).

Here we want values for QI_c and an approximate E_{RL}. An examination of Table 6-2 shows that a 1-mA QI_c and 4.3 V for E_{RL} will yield a voltage gain of 100. Frame 1 required at least 1.5 V of E_{RL} to satisfy V_{BE} stability requirements. Since the 4.3 V indicated in the table of possibilities is greater than the 1.5 V required by V_{BE} stability, our selection of 1.0 mA with an E_{RL} of 4.3 V is satisfactory.

FRAME 3

Compute R_e' (the Shockley relationship).

$$R_e' = \frac{25}{QI_c} = \frac{25}{1} = 25\ \Omega$$

FRAME 4

Compute R_{Lac}.

$R_{Lac} = V_g \times R'_e$
$R_{Lac} = 100 \times 25 = 2.5\ \text{k}\Omega$

FRAME 5

Compute R_L

$$R_L = \frac{R_{Lw} \times R_{Lac}}{R_{Lw} - R_{Lac}}$$

$$R_L = \frac{6\ \text{k}\Omega \times 2.5\ \text{k}\Omega}{6\ \text{k}\Omega - 2.5\ \text{k}\Omega}$$

$R_L \approx 4.3\ \text{k}\Omega$

FRAME 6

Compute E_{RL}.

$E_{RL} = QI_c \times R_L = 1 \times 10^{-3}\ \text{A} \times 4.3 \times 10^3 = 4.3\ \text{V}$

FRAME 7

Compute E_Q (min).

$E_Q \approx 0.66E_o$ p-p
$E_Q \approx 0.66 \times 6\ \text{V} = 3.96\ \text{V}$ min
Adjusted $E_Q = 5.7$ V (see next frame)

FRAME 8

Compute E_s.

$E_s = E_{RL} \times E_Q = 4.3\ \text{V} + 3.96\ \text{V}$
$E_s = 8.26\ \text{V}$

Now, since 8.26 V is not a particularly convenient power supply voltage, let us adjust E_s *upward* to 10 V. This means that we must add the difference between 8.26 V and 10 V to E_Q. The difference:

$10 - 8.26 = 1.74\ \text{V}$

The minimum E_Q computed in Frame 7 was 3.96 V. Adding 1.74 V to 3.96 V gives us an adjusted E_Q of 5.7 V.

$E_s = 10\ \text{V}$

FIELD 2

FRAME 1

Compute R_{b1}.

$R_{b1} = \mathbf{S} \times R_L$
$R_{b1} = 10 \times 4.3\ \text{k}\Omega = 43\ \text{k}\Omega$

FRAME 2

Compute E_{Rb1}.

$$E_{Rb1} = E_Q - V_j = 5.7 - 0.6 = 5.1 \text{ V}$$

FRAME 3

Compute I_{Rb1}.

$$I_{Rb1} = \frac{E_{Rb1}}{R_{b1}} = \frac{5.1 \text{ V}}{43 \text{ k}\Omega}$$

$$I_{Rb1} = 1.18 \times 10^{-4} \text{ A}$$

FRAME 4

Compute I_b.

$$I_b = \frac{QI_c}{\beta} = \frac{1 \times 10^{-3}}{1 \times 10^{2}} = 1 \times 10^{-5} \text{ A}$$

FRAME 5

Compute I_{Rb2}.

$$I_{Rb2} = I_{Rb1} - I_b$$

$$I_{Rb2} = (1.18 \times 10^{-4}) - (1 \times 10^{-5})$$

$$\begin{array}{r} 1.18 \times 10^{-4} \\ -0.10 \times 10^{-4} \\ \hline 1.08 \times 10^{-4} \end{array}$$

FRAME 6

Compute R_{b2}.

$$R_{b2} = \frac{V_j}{I_{Rb2}} = \frac{6 \times 10^{-1} \text{ V}}{1.08 \times 10^{-4}}$$

$$R_{b2} \approx 6 \text{ k}\Omega$$

Figure 7-6 shows the completed design sheet for Example 7-2.

FIELD 3 INPUT IMPEDANCE AND COUPLING CAPACITOR CALCULATIONS

FRAME 1

Compute the input impedance Z_{in}.

$$Z_{in} \approx \frac{R_{b1}}{V_g} \| \beta R'_e \| R_{b2}$$

Suppose we examine each factor independently before we combine them, in order to study the relative importance of each.

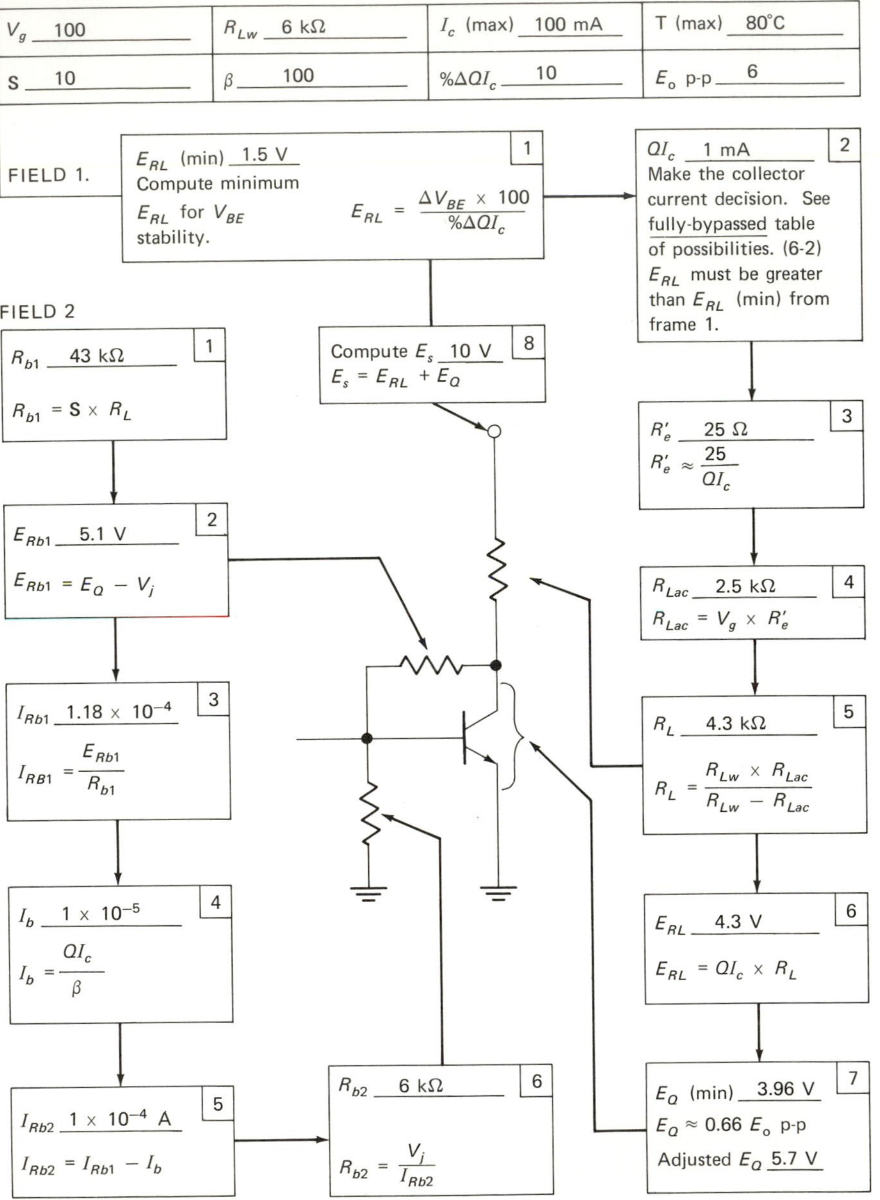

VOLTAGE MODE DESIGN SHEET (case two)
SPECIFICATIONS
V_g 100
R_{Lw} 6 kΩ
I_c (max) 100 mA
T (max) 80°C
S 10
β 100
$\%\Delta QI_c$ 10
E_o p-p 6
FIELD 1.
1
E_{RL} (min) 1.5 V
Compute minimum E_{RL} for V_{BE} stability.
$E_{RL} = \frac{\Delta V_{BE} \times 100}{\%\Delta QI_c}$
2
QI_c 1 mA
Make the collector current decision. See fully-bypassed table of possibilities. (6-2)
E_{RL} must be greater than E_{RL} (min) from frame 1.
3
R'_e 25 Ω
$R'_e \approx \frac{25}{QI_c}$
4
R_{Lac} 2.5 kΩ
$R_{Lac} = V_g \times R'_e$
5
R_L 4.3 kΩ
$R_L = \frac{R_{Lw} \times R_{Lac}}{R_{Lw} - R_{Lac}}$
6
E_{RL} 4.3 V
$E_{RL} = QI_c \times R_L$
7
E_Q (min) 3.96 V
$E_Q \approx 0.66\ E_o$ p-p
Adjusted E_Q 5.7 V
8
Compute E_s 10 V
$E_s = E_{RL} + E_Q$
FIELD 2
1
R_{b1} 43 kΩ
$R_{b1} = S \times R_L$
2
E_{Rb1} 5.1 V
$E_{Rb1} = E_Q - V_j$
3
I_{Rb1} 1.18 × 10^{-4}
$I_{RB1} = \frac{E_{Rb1}}{R_{b1}}$
4
I_b 1 × 10^{-5}
$I_b = \frac{QI_c}{\beta}$
5
I_{Rb2} 1 × 10^{-4} A
$I_{Rb2} = I_{Rb1} - I_b$
6
R_{b2} 6 kΩ
$R_{b2} = \frac{V_j}{I_{Rb2}}$

Fig. 7-6 The completed design sheet for Example 7-2

1. $\frac{R_{b1}}{V_g} = \frac{43\ \text{k}\Omega}{100} = 430\ \Omega$

2. $\beta R_e = 100 \times 25 = 2.5\ \text{k}\Omega$

3. $R_{b2} = 6\ \text{k}\Omega$

Expression 1 is the Miller effect, and as you can see, it dominates the scene to such an extent that factors 2 and 3 will have but little effect on the total input impedance. So for this circuit we will have an input impedance of about 400 Ω. If we were to try to cascade two of these stages, we would find it difficult, to say the least. The input impedance of the second stage would be R_{Lw} for the first stage, and an examination of the table of possibilities (Table 6-2) indicates that the first stage would have to run at a quiescent collector current in the neighborhood of 10 mA. Its resistor values would be quite low, and the first stage would have an input impedance of 200 Ω or less.

In order to cascade this circuit, it is almost essential that we split R_{b1} and add a bypass capacitor as shown in Fig. 7-5.

If we make that modification, the input impedance for Example 7-2 becomes

$$Z_{in} \approx \frac{R_{b1}}{2} \| \beta R'_e \| R_{b2} \qquad \text{(assuming a 50-50 split in } R_{b1}\text{)}$$

Let us examine the individual factors in this case:

1. $\frac{R_{b1}}{2} = \frac{43\ \text{k}\Omega}{2} = 21.5\ \text{k}\Omega$

This is much greater than 400 Ω.

2. $R'_e = 100 \times 25 = 2.5\ \text{k}\Omega$

This is the same as before, but it now is the dominant factor.

3. $R_{b2} = 6\ \text{k}\Omega$

This is also the same as before, but it has a greater total influence.

The net result of splitting R_{b1} and adding the bypass capacitor is an input impedance near 1.5 kΩ instead of near 400 Ω. This is an increase of about a factor of 5, and is almost the same as the input impedance of a fully bypassed current-mode circuit. With this modification cascading is possible.

PROBLEMS

Problem 7-2

TITLE: **Designing the More Versatile Case (Case 2) of Voltage-mode Feedback**
PROCEDURE: **Design the circuit, using the following specifications. See Fig. 7-4.**

Transistor Parameters

1. β (min)	100
2. I_c (max)	100 mA
3. E_c (max)	30 V
4. Transistor type	*npn* silicon

Circuit Parameters

1.	Voltage gain	140
2.	E_o p-p	5 V
3.	R_{Lw}	5 kΩ
4.	Stability factor	10
5.	$\%\Delta QI_c$ to be allowed	10%
6.	Maximum operating temperature	65°C
7.	f_o	30 Hz

7-3 THE SPLIT-LOAD (PARAPHASE) AMPLIFIER

This circuit uses equal values of emitter and collector resistances. The circuit is normally used to feed two loads simultaneously 180° out of phase. This is a common requirement in push-pull circuits and a number of other phase-sensitive circuits. The circuit is shown in Fig. 7-7.

The two loads R_{L1} and R_{L2} are nearly always equal where the output impedances of the two outputs are inherently different. The collector output has an output impedance of $Z_o \approx R_L$ while the output impedance of the emitter output conforms to emitter-follower (common-collector) output impedance, which is somewhat lower, as we will see in Secs. 7-4 and 7-5.

Because the working loads are generally the same, we must design the circuit around the loading in the collector circuit because it is the least tolerant of heavy loading. Other than being careful not to load the collector circuit excessively, there are almost no problems in designing the circuit. The gain is about unity for both outputs, and the voltage drop across E_{Res} is fairly high, ensuring good V_{BE} stability. We should treat the stability factor according to the rules for the common-emitter circuit (see Chap. 5).

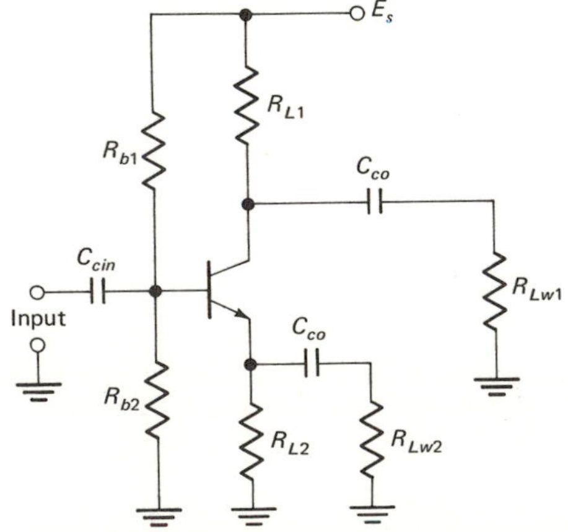

Fig. 7-7 The paraphase amplifier

The one caution in designing the paraphase amplifier is "don't load the collector too heavily." Since this is the only real hazard, a good starting point is to take some pains to avoid the problem.

We will start the design by establishing a collector load resistor value which will not be too heavily loaded by the expected working load. We would like to do this without doing anything that would prevent us from optimizing other parameters. One of the more important of these parameters is input impedance. The best overall results come about *when*

$R_L = K \times R_{Lw}$, where K is between 1 and 5.

The value of K will determine the final input impedance of the circuit, and so I have prepared Table 7-2 to enable you to select a reasonable value of K for an acceptable input impedance. The table can be easily scaled up or down for different values of E_{RL} (and E_{Res}). It is based upon a 5-V drop across R_L. If your circuit has a 10-V E_{RL}, the input impedance will be approximately twice the value shown in the table, a 2.5-Volt E_{RL} will yield an input impedance of about one-half that shown in the table, etc. *Please* note that the values of input impedance shown are only *estimates* for the purpose of helping you make a decision on the value of K.

The design procedure is simple, and the design sheet for current-mode base bias circuits can be used if we ignore Field 1 and use a different approach

TABLE 7-2. TABLE OF APPROXIMATE INPUT IMPEDANCES FOR THE PARAPHASE AMPLIFIER

	Approximate Input Impedances				
R_{Lw}	**K = 1**	**2**	**3**	**4**	**5**
200	1 kΩ	2 kΩ	3 kΩ	4 kΩ	5 kΩ
400	2.5 kΩ	5 kΩ	7.5 kΩ	10 kΩ	12.5 kΩ
500	3 kΩ	6 kΩ	9 kΩ	12 kΩ	15 kΩ
800	5 kΩ	10 kΩ	15 kΩ	20 kΩ	25 kΩ
1,000	6.5 kΩ	13 kΩ	19 kΩ	26 kΩ	32 kΩ
2,000	13 kΩ	26 kΩ	39 kΩ	52 kΩ	65 kΩ
4,000	25 kΩ	50 kΩ	75 kΩ	100 kΩ	125 kΩ
8,000	50 kΩ	100 kΩ	*	*	*
10,000	65 kΩ	130 kΩ	*	*	*

NOTE: The table is based on the following conditions:
1. **S** $\approx$ 10
2. $E_{RL} = 5$ V, $E_Q = 5$ V, $E_{Res} = 5$ V
3. $R_L = R_{es} = K \times R_{Lw}$
4. * indicates input impedances which are high enough for the collector junction impedance to become involved.

to the collector side calculations. Let me outline the collector side operating conditions for you and then proceed to a design example.

Collector Side Operating Conditions

1. $E_{RL} = E_{Res}$
2. $R_L = R_{es}$
3. E_{RL} (min) $= 0.5\ E_o$ p-p
4. E_Q (min) $= 1 + E_o$ p-p
5. $K = 1$ to 5 where $R_L = K \times R_{Lw}$
6. Since there is very little reason to use any stability factor but 10, let us make an **S** of 10 our standard.

Example 7-3 Designing the Paraphase Amplifier

Design a paraphase amplifier to the following requirements.

Transistor Parameters

1. β	100
2. I_c (max)	100 mA
3. E_c (max)	25 V
4. Type	*npn* silicon

Circuit Parameters

1. Stability factor	10
2. R_{Lw}	5 kΩ
3. E_o p-p	3 V
4. Input impedance	25 kΩ or greater

ALTERNATIVE FIELD 1 COLLECTOR SIDE CALCULATIONS

FRAME 1

Select a value for K. ($R_L = K \times R_{Lw}$)

Using Table 7-2, find a value of K which will meet our input impedance requirement with a 5-kΩ working load (R_{Lw}). An examination of Table 7-2 reveals that 5 kΩ is not listed but 4 kΩ and 8 kΩ are, and 5 kΩ lies somewhere between. Under the $K = 1$ column in the table we find an input impedance of 25 kΩ for a 4-kΩ load and a 50-kΩ Z_{in} for an 8-kΩ load. Since a 5-kΩ load is somewhere between 4 kΩ and 8 kΩ, our input impedance for a 5-kΩ load will be between 25 kΩ and 50 kΩ, which meets the input impedance specification. For this problem let us select a K of 1. In some cases we may get to the stage where we compute QI_c and wish we had used a higher value of K which would have yielded a lower value for QI_c. This happens only occasionally, and any guidelines I might give you would probably be more trouble than they are worth. A little experience will guide you quite satisfactorily.

$K = 1$

FRAME 2

Compute R_L.

$R_L = K \times R_{Lw}$
$R_L = 1 \times 5\ \text{k}\Omega$
$R_L = 5\ \text{k}\Omega$

FRAME 3

Compute R_{es}.

$R_{es} = R_L = 5\ \text{k}\Omega$

FRAME 4

Compute E_{RL} (min).

$E_{RL}\ (\text{min}) = 0.5\ E_o$ p-p
$E_{RL}\ (\text{min}) = 0.5 \times 3 = 1.5\ \text{V}$

FRAME 5

Compute E_{Res} (min).

$E_{Res} = E_{RL} = 1.5\ \text{V}$

FRAME 6

Compute E_Q (min).

$E_Q\ (\text{min}) = 1 + E_o$ p-p $= 4\ \text{V}$

FRAME 7

Compute E_s (min).

$E_s\ (\text{min}) = E_{Res} + E_{RL} + E_Q$
$E_s\ (\text{min}) = 1.5 + 1.5 + 4 = 7\ \text{V}$

FRAME 8

Select a convenient E_s.

$E_s > E_s\ (\text{min})$

For this example let us assume we have a 15-V power supply to work with. The voltage distribution must conform to the following rules:

1. $E_Q > E_Q\ (\text{min})$
2. $E_{RL} = E_{Res}$

In the case of this particular example one-third of 15 V is greater than the 3 V required for E_Q, and so we can set up the following voltage distribution:

$E_{RL} = 5$ V
$E_Q = 5$ V
$E_{Res} = 5$ V

Beyond having enough E_Q and enough E_{RL} to satisfy the output voltage requirement, the voltage distribution is completely arbitrary. The voltage drops across R_L and R_{es} must be equal, but E_Q can have any reasonable value above the minimum.

FRAME 9

Compute QI_c.

$$QI_c = \frac{E_{RL}}{R_L} = \frac{5\text{ V}}{5\text{ k}\Omega} = 1\text{ mA}$$

FIELD 2 BASE SIDE CALCULATIONS

From here on the procedure is identical to that of designing a current-mode common-emitter circuit (with base bias).

FRAME 1

Compute R_{b2}.

$R_{b2} = \mathbf{S} \times R_{es}$
$R_{b2} = 10 \times 5\text{ k}\Omega$
$R_{b2} = 50\text{ k}\Omega$

FRAME 2

Compute R_{int}.

$$R_{int} = R_{b2} \| \beta R_{es}$$

$$R_{int} = \frac{50\text{ k}\Omega \times 500\text{ k}\Omega}{50\text{ k}\Omega + 500\text{ k}\Omega}$$

$$R_{int} \approx 45\text{ k}\Omega$$

FRAME 3

Compute E_{Rb1}.

$E_{Rb1} = E_Q + E_{RL} - V_j$
$E_{Rb1} = 5 + 5 - 0.6 = 9.4$ V

FRAME 4

Compute R_{b1}.

$$R_{b1} = \frac{E_{Rb1} \times R_{int}}{E_{Res} + V_j}$$

$$R_{b1} = \frac{9.4 \times 45\text{ k}\Omega}{5 + 0.6}$$

$R_{b1} = 75\text{ k}\Omega$

The completed design sheet is shown in Fig. 7-8.

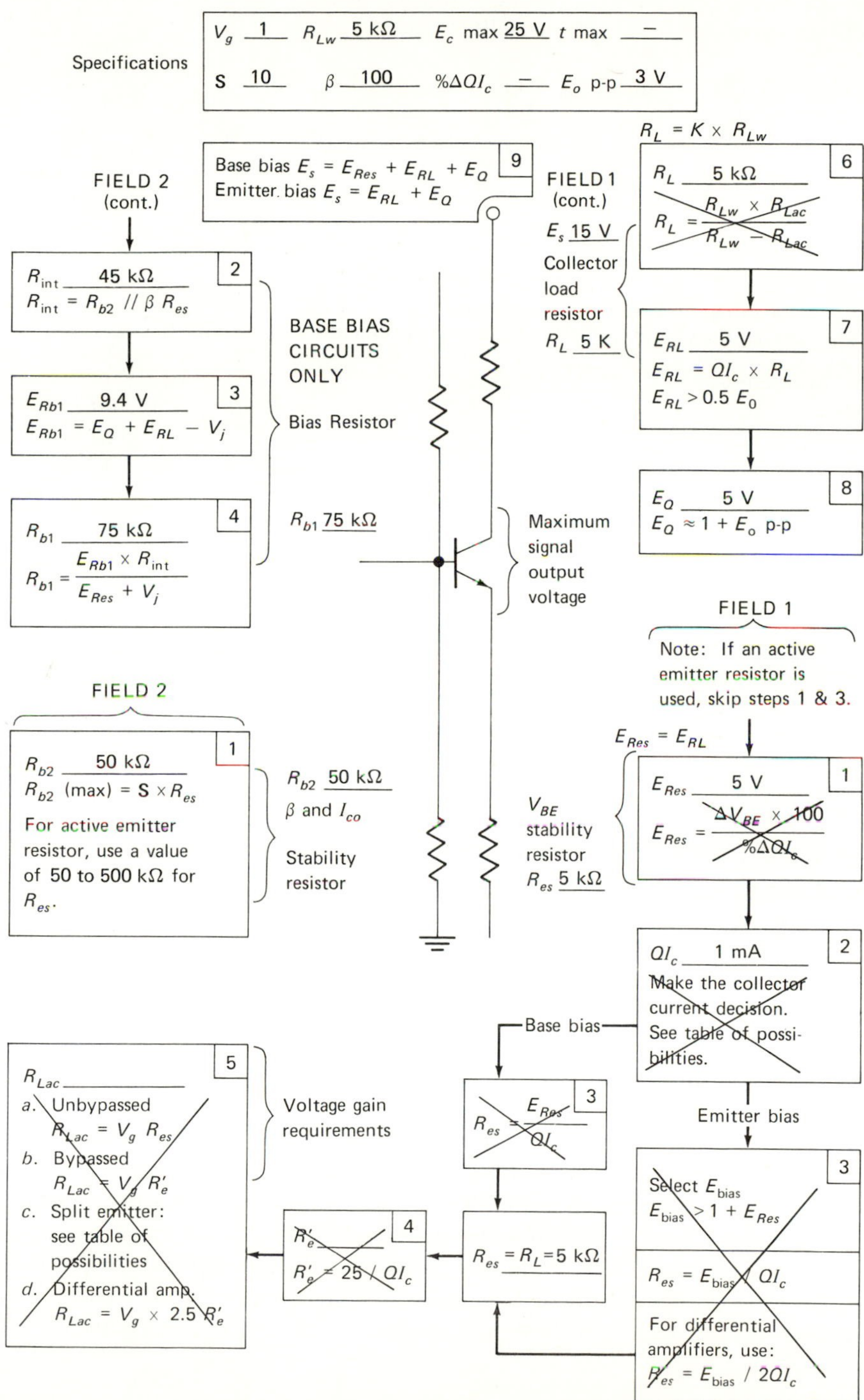

UNIVERSAL DESIGN SHEET
Specifications
Vg 1 RLw 5 kΩ Ec max 25 V t max –
S 10 β 100 %ΔQIc – Eo p-p 3 V
Base bias Es = ERes + ERL + EQ
Emitter bias Es = ERL + EQ
9
FIELD 2 (cont.)
Rint 45 kΩ
Rint = Rb2 // β Res
2
ERb1 9.4 V
ERb1 = EQ + ERL − Vj
3
Rb1 75 kΩ
Rb1 = ERb1 × Rint / ERes + Vj
4
BASE BIAS CIRCUITS ONLY
Bias Resistor
Rb1 75 kΩ
FIELD 1 (cont.)
Es 15 V
Collector load resistor
RL 5 K
RL = K × RLw
RL 5 kΩ
RL = RLw × RLac / RLw − RLac
6
ERL 5 V
ERL = QIc × RL
ERL > 0.5 Eo
7
EQ 5 V
EQ ≈ 1 + Eo p-p
8
Maximum signal output voltage
FIELD 1
Note: If an active emitter resistor is used, skip steps 1 & 3.
ERes = ERL
ERes 5 V
ERes = ΔVBE × 100 / %ΔQIc
1
VBE stability resistor
Res 5 kΩ
FIELD 2
Rb2 50 kΩ
Rb2 (max) = S × Res
For active emitter resistor, use a value of 50 to 500 kΩ for Res.
1
Rb2 50 kΩ
β and Ico
Stability resistor
QIc 1 mA
Make the collector current decision. See table of possibilities.
2
Base bias
Emitter bias
Res = ERes / QIc
3
Select Ebias
Ebias > 1 + ERes
Res = Ebias / QIc
For differential amplifiers, use:
Res = Ebias / 2QIc
3
Res = RL = 5 kΩ
R′e
R′e = 25 / QIc
4
RLac
a. Unbypassed
RLac = Vg Res
b. Bypassed
RLac = Vg R′e
c. Split emitter: see table of possibilities
d. Differential amp.
RLac = Vg × 2.5 R′e
5
Voltage gain requirements

Fig. 7-8 The completed design sheet for Example 7-3

FIELD 3 INPUT IMPEDANCE AND COUPLING CAPACITOR CALCULATIONS

Input impedance and capacitors C_{cin} and C_{co} are handled the same way as they were for the ordinary current-mode common-emitter circuit.

$$X_c \text{ at } f_o \text{ of } C_{cin} \approx 0.1Z_{in} \qquad \text{and} \qquad X_c \text{ at } f_o \text{ of } C_{co} \approx 0.1R_{Lw}$$

PROBLEMS

Problem 7-3

TITLE: **Designing the Paraphase Amplifier**
PROCEDURE: **Design the circuit, using the following specifications. See Fig. 7-7.**

Transistor Parameters

1. β (min)	100
2. I_c (max)	100 mA
3. E_c (max)	25 V
4. Transistor type	*npn* silicon

Circuit Parameters

1. Voltage gain	1
2. E_o p-p	6 V
3. R_{Lw}	10 kΩ
4. Stability factor	10
5. $\%\Delta QI_c$ to be allowed	10%
6. Maximum operating temperature	70°C
7. f_o	25 Hz

7-4 DESIGNING THE COMMON-COLLECTOR (THE EMITTER-FOLLOWER) CIRCUIT WITH BASE BIAS

The emitter follower has a low output impedance and a high input impedance. This pair of characteristics make it an ideal circuit for elevating the effective impedance of a load by a factor as great as β. So far in this book we have been working with circuits that have been primarily voltage amplifiers, operating at relatively low quiescent collector currents, and with working loads R_{Lw} of no less than 1 kΩ or so. A bit later on we will be working with circuits designed to deliver high currents or high powers into a very low impedance load. This kind of operation demands high quiescent collector currents and usually the use of power transistors. The input impedance of a power transistor may be from less than 1 Ω to a few hundred ohms. In an amplifier with many stages and a heavy final load we have two design options. The first involves the gradual reduction of QI_c from output toward input, using low collector currents for only the first few stages. This route is expensive for several reasons: the cost of a transistor increases in proportion to its current-handling ability; resistors must dissipate more power and are therefore more expensive; more power is wasted, and this makes power supply costs increase; mechanical costs increase because of the need to mount the transistors on some kind of heat-radiating body; and finally design becomes more difficult and costly.

The second and almost universally used approach is to drive the power stage with an emitter-follower (common-collector) stage which can raise the impedance level of the load by a factor of 10 to 100 (in some cases even more), up to a level where we can use low-level voltage amplifiers for all the stages ahead of the emitter-follower buffer. The common-collector circuit has a voltage gain of approximately unity, a power gain of V_g (1) $\times I_g$, and a current gain I_g of $\mathbf{S}_{ac}$, where $\mathbf{S}_{ac} \approx R_{b2}/R_{Lac}$, which is generally larger than the dc value of $\mathbf{S}$.

The emitter follower is much like an impedance-matching transformer except that it has some distinct advantages over transformers for low-frequency operation. It is cheaper than an iron-core transformer, it presents no special physical mounting problems, it has a very wide bandwidth (starting at 0 Hz), and it is not subject to core saturation distortion. Exact impedance matching is usually not practical with an emitter follower, but it is also very rare to achieve exact impedance matching with conventional transformers, as we shall see later.

Designing the emitter follower is very much like designing the current-mode common-emitter circuit, at least in so far as the bare mechanics are concerned. The common-collector circuit has a voltage gain which is always very close to, but always less than, unity except when the ac load R_{Lac} approaches the value of R_e', where it may be considerably less than unity. Voltage gain is not one of our primary concerns as it has been in the common-emitter circuit; although we will calculate it as a routine part of the design procedure, it will only occasionally be of any real importance.

Instead, our attention will shift to the problem of getting the desired p-p output voltage. The p-p output voltage is a function of R_{Lac}, the total ac load, and a *valid selection of the quiescent collector current* for the load at hand. Here we are again, back to the old problem of selecting a suitable QI_c to make the circuit do the particular job we have in mind. We are concerned with the p-p output voltage equation instead of the voltage gain equation, but QI_c is still the key to proper design. In order to help you make the proper QI_c selection for each set of specifications you may encounter, I have built a table of possibilities for the common-collector circuit, Table 7-3. The table is quite different from the other *tables of possibilities*, because the problem here is a relationship between load impedance and E_o p-p and not between load impedance and V_g, which was the problem in previous cases. The table of possibilities for the emitter follower provides values of QI_c for p-p output voltage vs. R_{Lac}. It is simply a tabulation of the equation E_o p-p $= 2QI_cR_{Lac}$.

Many of the figures have been rounded off to make them fit better, physically, into the table. All QI_c values are in milliamperes, unless followed by "A," which denotes amperes.

Suppose we combine the design explanation with a design example. We will use the *current-mode universal design sheet* (Fig. 7-10) with minor deviations.

Example 7-4

Design a common-collector (emitter-follower) circuit with the following specifications. See Fig. 7-9.

Transistor Parameters

1. β	100
2. I_c (max)	100 mA
3. E_c (max)	25 V

Circuit Parameters

1. Voltage gain	1 (not too important)
2. E_o p-p	800 mV
3. R_{Lw}	500 Ω
4. Stability factor	10
5. $\%\Delta QI_c$ to be allowed	10%
6. Maximum operating temperature	65°C
7. f_o	25 Hz
8. Z_{in}	20 kΩ or greater

FIELD 1

FRAME 1

Compute the minimum voltage drop across the emitter resistor

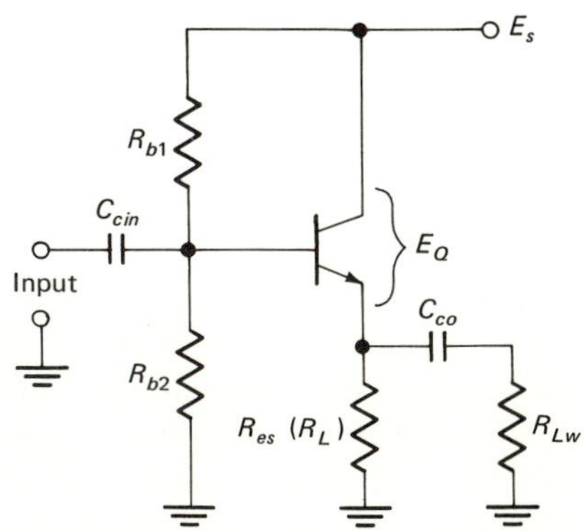

Fig. 7-9 The base-biased common-collector circuit

$$E_{Res} = \frac{112.5 \times 10^{-3} \times 10^2}{10}$$

$$E_{Res} = 1.125 \text{ V}$$

The purpose of making this calculation is simply to make certain that when we make a choice for R_{es}, it will always have a voltage drop sufficient for the specified V_{BE} stability. After you have had a little experience, you may wish to omit this frame except for very special cases.

FRAME 2

Make the QI_c decision.

Use the common-collector table of possibilities, Table 7-3.

The specifications call for a working load value R_{Lw} of 500 Ω and a peak-to-peak output voltage of 800 mV. Here, as in all previous QI_c decisions, we will select the *lowest practical* QI_c. If you will look at Table 7-3, you will see that a p-p output voltage of 800 mV with an R_{Lac} of 500 Ω requires a quiescent collector current of 0.8 mA. If you are wondering why I first referred to the working load R_{Lw} and then looked up R_{Lac} in the table using the same ohmic value, I am proud of you. You are thinking. In order to get away with such a thing, I had to make the assumption that $R_{Lw} \approx R_{Lac}$. We are going to *make* R_{es} (R_L) at least 10 times R_{Lw}; this will *make* that assumption valid. However, 0.8 mA for QI_c is not the most comfortable figure to work with, and because the assumption that $R_{Lw} \approx R_{Lac}$ is only an approximation, we will increase the value of QI_c *slightly* to 1.0 mA. This is not the lowest *possible* value for QI_c, but it is close and we will call it the lowest *practical* value. This is in line with our general policy of not working too close to any limit and of keeping the arithmetic simple whenever there is no compelling reason to do otherwise.

$$QI_c = 1 \text{ mA}$$

FRAME 3

Compute the value for the emitter resistor R_{es} (R_L). Here we will use the usual equation, $R_{es} = E_{Res}/QI_c$, but only to determine a minimum acceptable value for V_{BE} stability. Again with a bit of experience you can avoid this minor computation except in unusual cases. But for now go along with it.

$$R_{es} \text{ (min)} = \frac{1.12 \text{ V}}{1 \text{ mA}} = 1.1 \text{ k}\Omega$$

OK, 1.1 kΩ is our minimum, but it is not what we need for the common-collector circuit. In almost all cases which are capacitor-coupled to the load we will want to make R_{es} (R_L) *at least* 10 times R_{Lw}. Remember that the object of this circuit is to get the highest possible input impedance for a given load impedance. Theoretically we can transform the load impedance up to a value β times the load value. We are interested in transforming R_{Lw}, but we must also have the resistor R_{es} in the circuit. See Fig. 7-9. If we were to make R_{es} 500 Ω, the theoretical input impedance (leaving R_{b1} and R_{b2} out of the picture for now) would be

$$Z_{in} \approx \beta \times (R_{es} \| R_{Lw})$$
$$Z_{in} \approx 100 \times 250 = 25 \text{ k}\Omega$$

TABLE 7-3. TABLE OF POSSIBILITIES FOR THE COMMON-COLLECTOR CIRCUIT

Minimum values of QI_c, mA

	E_o p-p Millivolts					E_o p-p Volts									
R_{Lac}, Ω	50	100	200	400	800	1.6	2	4	8	12	16	20	24	28	32
0.25	100	200	400	800	1.6 A	3.2 A	4 A	8 A	16 A	48 A	*	*	*	*	*
0.3	80	166	333	666	1.4 A	2.7 A	3 A	7 A	13 A	20 A	27 A	34 A	*	*	*
0.6	40	80	160	320	640	1.4 A	2 A	3 A	7 A	10 A	13 A	17 A	20 A	*	*
1.0	25	50	100	200	400	800	1 A	2 A	4 A	6 A	8 A	10 A	12 A	14 A	16 A
2.5	10	20	40	80	160	320	400	800	2 A	2 A	3 A	4 A	5 A	6 A	6 A
5.0	5	10	20	40	80	160	200	400	800	1 A	2 A	2 A	2 A	3 A	3 A
8.0	3	6	12	25	50	100	125	250	500	750	1 A	1 A	2 A	2 A	2 A
10	2.5	5	10	20	40	80	100	200	400	600	800	1 A	1 A	1 A	2 A
16	1.6	3	6	12.5	25	50	63	125	250	380	500	700	750	900	1 A
50	0.5	1.2	2	4	8	16	20	40	80	120	160	200	240	280	320
80	0.3	0.6	1.2	2.5	5.0	10	12.5	25	50	75	100	130	150	175	200
100	0.25	0.5	1	2	4	8	10	20	40	60	80	100	120	140	160
160	0.16	0.3	0.6	1.2	2.5	5	6.3	13	25	38	50	70	75	90	100
250	0.1	0.2	0.4	0.8	1.6	3.2	4	8	16	24	32	40	48	56	64
500	0.05	0.12	0.2	0.4	0.8	1.6	2	4	8	12	16	20	24	28	32
1,000	0.025	0.05	0.1	0.2	0.4	0.8	1	2	4	6	8	10	12	14	16
5,000	0.005	0.012	0.02	0.04	0.08	0.16	0.2	0.4	0.8	1.2	1.6	2	2.4	2.8	3.2
10,000	0.003	0.005	0.01	0.02	0.04	0.08	0.1	0.2	0.4	0.6	0.8	1	1.2	1.4	1.6

* Impossible design conditions

If R_{es} were not there,

$$Z_{in} \approx \beta \times R_{Lw} = 100 \times 500 = 50\ \text{k}\Omega$$

We can't leave R_{es} out of the circuit, but we can make its value high enough so that R_{Lac}, the parallel combination of R_{es} and R_{Lw}, is almost equal to R_{Lw} alone.

$$R_{Lac} \approx R_{Lw}$$

For example, if an R_{es} of 10 kΩ is put in parallel with the 500-Ω R_{Lw}, we get an equivalent resistance of about 470 Ω.

$$470 \approx 500$$

We will normally set R_{es} at a value of somewhere between $10R_{Lw}$ and $50R_{Lw}$, not only for the reason just cited, but also because $R_{b2} = \mathbf{S} \times R_{es}$. The higher we make R_{es}, the higher the value of R_{b2} will be. The actual input impedance of a base-biased common-collector circuit is given by the formula

$$Z_{in} \approx R_{b1} \| R_{b2} \| \beta \times R_{Lac} \qquad R_{Lac} = R_{es} \| R_{Lw}$$

We can maximize the input impedance for a given load R_{Lw} by making R_{es} large compared to R_{Lw} and by making R_{b1} and R_{b2} as high as possible.

Table 7-4 provides a tabulation of estimated input impedance values. The table is not intended to yield precise values but only to provide a guide for selecting a value for R_{es}.

In this example the specifications call for an input impedance of 20 kΩ or greater, $R_{Lw} = 500\ \Omega$, and $\beta = 100$.

If we make $R_{es} = 10 \times R_{Lw}$ with our β of 100, we find (in Table 7-4) an *estimated* input impedance of $33R_{Lw}$. $R_{Lw} = 500$, $33 \times 500 = 16.5$ kΩ, a little lower than the 20 kΩ specified. Suppose we see what an R_{es} of $20 \times R_{Lw}$ will do. The table tells us

TABLE 7-4. APPROXIMATE INPUT IMPEDANCES FOR BASE-BIASED EMITTER FOLLOWERS

	Estimated Input Impedances			
β	$R_{es} =$ **10**R_{Lw}	$R_{es} =$ **20**R_{Lw}	$R_{es} =$ **25**R_{Lw}	$R_{es} =$ **50**R_{Lw}
25	$16R_{Lw}$	$20R_{Lw}$	$21R_{Lw}$	$23R_{Lw}$
50	$25R_{Lw}$	$33R_{Lw}$	$35R_{Lw}$	$40R_{Lw}$
100	$33R_{Lw}$	$50R_{Lw}$	$55R_{Lw}$	$70R_{Lw}$
150	$37R_{Lw}$	$60R_{Lw}$	$68R_{Lw}$	$90R_{Lw}$
200	$40R_{Lw}$	$66R_{Lw}$	$76R_{Lw}$	$100R_{Lw}$

$Z_{in} \approx \beta R_{Lw} \| (\mathbf{S}/2) R R_{Lw}$
$R = R_{es} | R_{Lw}$
Note: Table based on an **S** of 10. Formula and table both assume $\mathbf{E}_{Res} \approx \mathbf{E}_Q$.

that for a β of 100 and $R_{es} = 20 \times R_{Lw}$, our estimated input impedance is $50R_{Lw}$. $50 \times 500 = 25$ kΩ. So it looks as if a choice of $20 \times R_{Lw}$ is a good one.

$R_{es} = 20 \times 500 = 10$ kΩ
$R_{es} = 10$ kΩ

which is much larger than the minimum value for V_{BE} stability. The figure we have written down in Frame 1 is *not* the actual voltage drop across R_{es}.

We had better take a moment to compute the actual voltage drop across the 10 kΩ, R_{es}. We will need this figure later. By Ohm's law,

$E_{Res} = QI_cR_{es}$
$E_{Res} = 1 \times 10^{-3} \times 1 \times 10^4\ \Omega$
$E_{Res} = 10$ V

FRAME 4

Compute the base-emitter junction resistance R_e'.

$$R_e' = \frac{25}{QI_c} = 25\ \Omega$$

FRAME 5

Compute R_{Lac}.

Skip this frame if R_{es} (R_L) is $10 \times R_{Lw}$ or more.

FRAME 6

Compute the value of the collector load resistor.

Since there is no collector load resistor, skip this step. In a later example we will add a collector resistor for bootstrapping purposes, but it will *not* be a *load* resistor, and we will call it R_d instead of R_L.

FRAME 7

Compute E_{RL}.

Obviously if there is no collector load resistor, there is no voltage drop across it. Skip this frame (or enter 0 V).

FRAME 8

Compute the collector-to-emitter voltage drop E_Q.

E_Q (min) $\approx 1 + E_o$ p-p
E_Q (min) $\approx 1 + 0.8$ V (800 mV)
E_Q (min) ≈ 1.8 V

This is another step which will prove unnecessary in many cases, but it is quite important that we don't overlook it on those occasions when it is important. Such a small E_Q voltage would yield a very low value for R_{b1} which would drastically reduce the input impedance. One of the best compromises is to set $E_Q \approx E_{Res}$. A higher value of E_Q would yield a larger value for R_{b1}, but because R_{b1} is in parallel with the rest of the input resistance, it takes a fairly large increase in E_Q to make much of a difference in the total input impedance, and because of the nature of parallel resistances an E_Q much less than E_{Res} can easily defeat the basic purpose of the emitter follower circuit. Rule of thumb: Make E_Q equal to or greater than E_{Res}.

Let $E_Q = 10$ V.

FRAME 9

Compute the power supply voltage E_s.

$E_s = E_{Res} + E_{RL} + E_Q$
$E_s = 10 + 0 + 10 = 20$ V
$E_s = 20$ V

FIELD 2

FRAME 1

Compute R_{b2}, the stability resistor.

$R_{b2} = \mathbf{S} \times R_{es}$
$R_{b2} = 10 \times 10$ k$\Omega = 100$ kΩ

FRAME 2

Compute the partial input resistance R_{int}.

$R_{int} = R_{b2} \| \beta R_{es}$
$R_{int} = 100$ k$\Omega \| 100 \times 10$ kΩ
$R_{int} = 91$ kΩ

This is a dc calculation, and because of the coupling capacitor C_{co}, R_{Lw} is not a part of R_{int}.

FRAME 3

Compute E_{Rb1}, the voltage drop across R_{b1}.

$E_{Rb1} = E_Q + E_{RL} - V_j$
$E_{Rb1} = 10 + 0 - 0.6$
$E_{Rb1} = 9.4$ V

FRAME 4

Compute the value of R_{b1}.

$$R_{b1} = \frac{E_{Rb1} \times R_{int}}{E_{Res} + V_j}$$

$$R_{b1} = \frac{9.4 \times 91 \text{ k}\Omega}{10 + 0.6}$$

$R_{b1} \approx 91$ kΩ

Figure 7-10 shows the completed design sheet for Example 7-4.

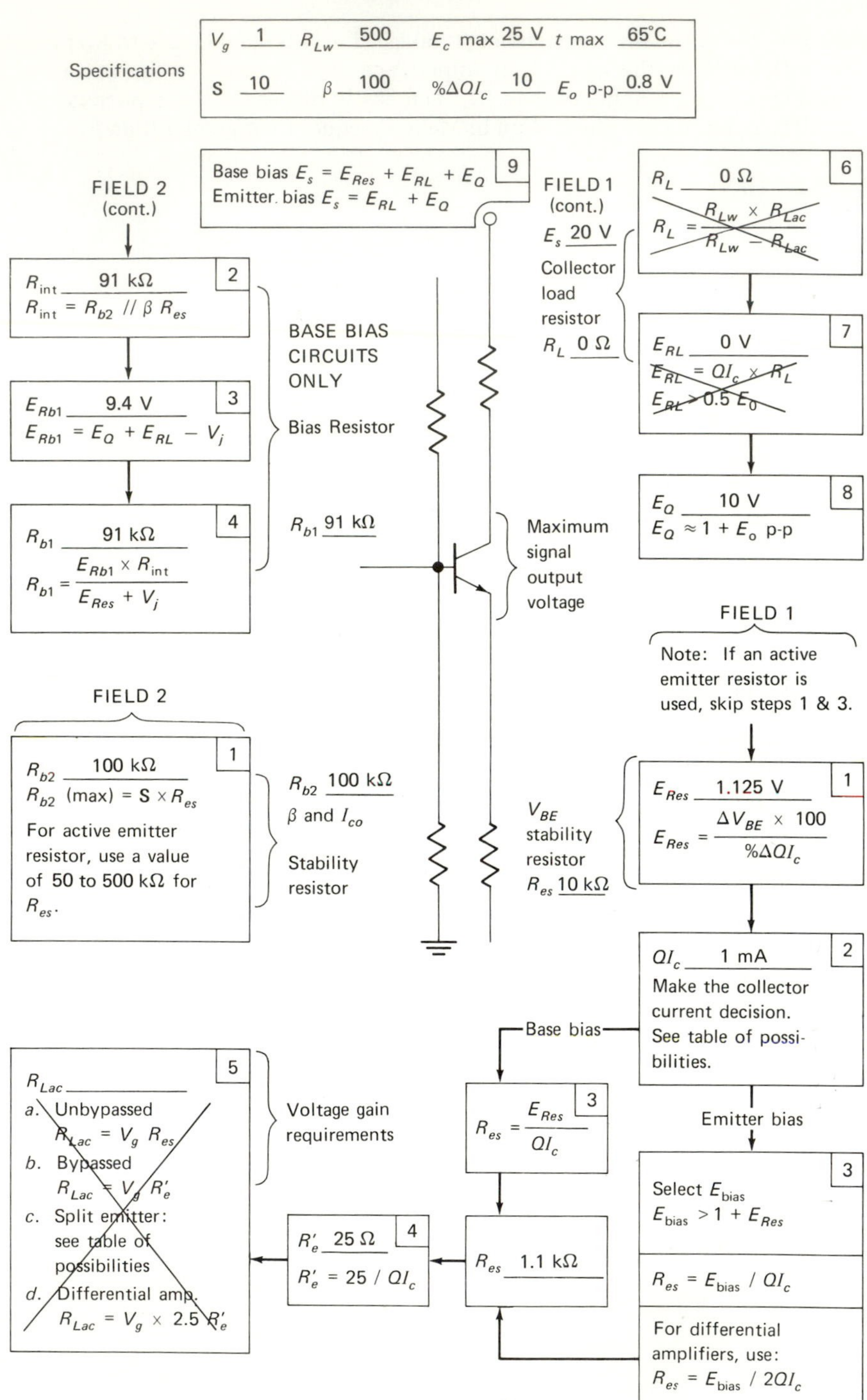

UNIVERSAL DESIGN SHEET
Specifications
V_g 1 R_{Lw} 500 E_c max 25 V t max 65°C
S 10 β 100 %ΔQI_c 10 E_o p-p 0.8 V
Base bias $E_s = E_{Res} + E_{RL} + E_Q$
Emitter bias $E_s = E_{RL} + E_Q$
9
FIELD 2 (cont.)
R_{int} 91 kΩ
$R_{int} = R_{b2} // \beta R_{es}$
2
E_{Rb1} 9.4 V
$E_{Rb1} = E_Q + E_{RL} - V_j$
3
R_{b1} 91 kΩ
$R_{b1} = \frac{E_{Rb1} \times R_{int}}{E_{Res} + V_j}$
4
BASE BIAS CIRCUITS ONLY
Bias Resistor
R_{b1} 91 kΩ
FIELD 1 (cont.)
E_s 20 V
Collector load resistor
R_L 0 Ω
R_L 0 Ω
$R_L = \frac{R_{Lw} \times R_{Lac}}{R_{Lw} - R_{Lac}}$
6
E_{RL} 0 V
$E_{RL} = QI_c \times R_L$
$E_{RL} > 0.5\ E_o$
7
E_Q 10 V
$E_Q \approx 1 + E_o$ p-p
8
Maximum signal output voltage
FIELD 1
Note: If an active emitter resistor is used, skip steps 1 & 3.
FIELD 2
R_{b2} 100 kΩ
R_{b2} (max) = S × R_{es}
For active emitter resistor, use a value of 50 to 500 kΩ for R_{es}.
1
R_{b2} 100 kΩ
β and I_{co}
Stability resistor
V_{BE} stability resistor
R_{es} 10 kΩ
E_{Res} 1.125 V
$E_{Res} = \frac{\Delta V_{BE} \times 100}{\%\Delta QI_c}$
1
QI_c 1 mA
Make the collector current decision. See table of possibilities.
2
Base bias
Emitter bias
R_{Lac}
a. Unbypassed
$R_{Lac} = V_g\ R_{es}$
b. Bypassed
$R_{Lac} = V_g\ R'_e$
c. Split emitter: see table of possibilities
d. Differential amp.
$R_{Lac} = V_g \times 2.5\ R'_e$
5
Voltage gain requirements
$R_{es} = \frac{E_{Res}}{QI_c}$
3
R'_e 25 Ω
$R'_e = 25 / QI_c$
4
R_{es} 1.1 kΩ
Select E_{bias}
$E_{bias} > 1 + E_{Res}$
3
$R_{es} = E_{bias} / QI_c$
For differential amplifiers, use:
$R_{es} = E_{bias} / 2QI_c$

Fig. 7-10 The completed design sheet for Example 7-4

FIELD 3

FRAME 1

Compute the circuit input impedance.

$Z_{in} = R_{b1} \| R_{b2} \| \beta R_{Lac}$

We have gone to some effort to make R_{es} greater than 10 times as large as R_{Lw}, so that $R_{Lac} \approx R_{Lw}$. Now we can rewrite the Z_{in} equation as follows:

$$Z_{in} \approx R_{b1} \| R_{b2} \| \beta R_{Lw}$$
$$R_{b1} = 91 \text{ k}\Omega$$
$$R_{b2} = 100 \text{ k}\Omega$$
$$\beta R_{Lw} = 50 \text{ k}\Omega$$
$$Z_{in} \approx 25 \text{ k}\Omega$$

Here we have managed to elevate a load from its actual value of 500 Ω to an *effective* value of 25,000 Ω, a 50-fold increase.

A 500-Ω load, for a common-emitter amplifier, would either demand a fairly large QI_c or keep the voltage gain down to a low value. If we elect a high QI_c for the common-emitter stage, it will present a heavy load to the stage driving it, and so on. On the other hand, if we had placed the emitter follower we just designed between a common-emitter stage and the 500-Ω load, the common-emitter stage would have seen a load of 25 kΩ, which would allow comfortable operation at 1 mA or so with a very satisfactory voltage gain. Furthermore, all prior stages could be operated at low quiescent currents with high voltage gains.

PROBLEMS

Problem 7-4

TITLE: **Designing the Base-biased Emitter Follower**
PROCEDURE: **Design the circuit, using the following specifications. See Fig. 7-9. Try for the highest practical input impedance.**

Transistor Parameters

1. β (min)	100
2. I_c (max)	100 mA
3. E_c (max)	25 V
4. Transistor type	*npn* silicon

Circuit Parameters

1. Voltage gain	1
2. E_o p-p	1.6 V
3. R_{Lw}	100 Ω
4. Stability factor	10
5. $\%\Delta QI_c$ to be allowed	10%
6. Maximum operating temperature	65°C
7. f_o	30 Hz

7-5 DESIGNING THE EMITTER-FOLLOWER CIRCUIT WITH EMITTER BIAS

Somewhat higher input impedances can be obtained from the emitter-biased version of the common-collector circuit. If the generator is direct-coupled as in Fig. 7-11*a*, the input impedance can be very nearly $\beta \times R_{Lac}$. (In most cases $R_{Lac} \approx R_{Lw}$ in the common-collector circuit.) The circuit in Fig. 7-11*b* has a capacitor-coupled generator and an R_{b2}. The input impedance of the circuit in Fig. 7-11*b* is almost exactly twice that of a similar base-biased circuit. The input impedance can be estimated by doubling values taken from Table 7-4 or by using the following estimation formula: $Z_{in} = \beta R_{Lw} \| \mathbf{S} R R_{Lw}$ where **S** is the stability factor and $R = R_{es}/R_{Lw}$.

We will concentrate our efforts on the circuit in Fig. 7-11*a*. The circuit is so simple that we could very well do away with the design sheet, but we will follow the current-mode design sheet anyway, eliminating any unnecessary steps, for the purpose of maintaining a uniform procedure. If later, when you have had some experience, you elect to ignore the design sheet, I shall not be offended.

Example 7-5

Given the circuit in Fig. 7-11*a*, design a common-collector circuit to the following specifications.

Transistor Parameters

1. β	100
2. I_c (max)	100 mA
3. E_c (max)	25 V

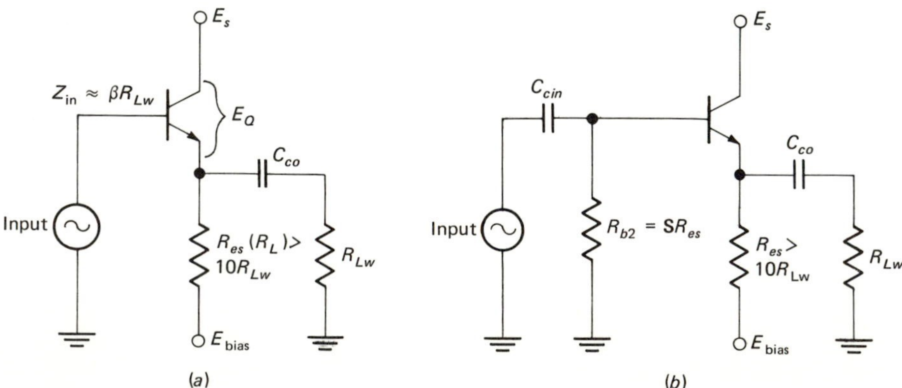

Fig. 7-11 The common-collector circuit, with emitter bias. (a) The circuit with a direct-coupled generator; (b) the circuit with a capacitor-coupled generator

Circuit Parameters

1. Voltage gain	1
2. E_o p-p	4 V
3. R_{Lw}	500 Ω
4. Stability factor	10
5. $\%\Delta QI_c$ to be allowed	10%
6. Maximum operating temperature	65°C
7. f_o	25 Hz
8. Z_{in}	40 kΩ or greater

FIELD 1

FRAME 1

Compute the minimum E_{Res} for V_{BE} stability.

$$E_{Res} = \frac{\Delta V_{BE} \times 100}{\%\Delta QI_c}$$

$$E_{Res} = \frac{112.5 \times 10^{-3} \times 10^2}{10}$$

$$E_{Res} = 1.125 \text{ V}$$

Any voltage higher than this will be acceptable.

FRAME 2

Make the quiescent collector current decision, QI_c.

See the common-collector table of possibilities, Table 7-3.

$$QI_c = 4 \text{ mA}$$

FRAME 3

Find the value of R_{es}.

Here again we want R_{es} to be somewhere between $10R_{Lw}$ and $50R_{Lw}$. In that case the *estimated* input impedance for this circuit is approximately equal to βR_{Lw} and does not vary much (because there is no R_{b2}) between $10R_{Lw}$ and $50R_{Lw}$. The thing that does become important here is the dc voltage drop across R_{es}. In this example, we are talking about a QI_c of 4 mA. If we make R_{es} equal to $50R_{Lw}$, we get an R_{es} of 25 kΩ. 25 kΩ at a QI_c of 4 mA yields a voltage drop of

$$E = IR \text{ (Ohm's law)} = 4 \times 10^{-3} \times 2.5 \times 10^4 = 100 \text{ V}$$

In most cases a 100-V bias supply would be a bit too much. Our real concern here is the bias supply voltage; so let us try the minimum of $R_{es} = 10R_{Lw}$: $R_{es} = 10 \times 500 = 5$ kΩ. This yields a voltage drop of 5 kΩ × 4 mA = 20 V. Twenty volts is much more reasonable. Let $R_{es} = 5$ kΩ.

FRAME 4

Compute R_e'.

$$R'_e = \frac{25}{QI_c}$$

$$R'_e = \frac{25}{4} = 6.25\ \Omega$$

FRAME 5

Compute R_{Lac}.
Omit this frame.

FRAME 6

Compute the value of the collector load resistor R_L.
Omit this frame.

FRAME 7

Compute the voltage drop across R_L, E_{RL}.

$E_{RL} = 0$ V because $R_L = 0\ \Omega$

FRAME 8

Compute E_Q (min).

$$E_Q \approx 1 + E_o \text{ p-p}$$
$$E_Q \text{ (min)} = 1 + 4 = 5 \text{ V}$$

FRAME 9

Compute E_s, the collector supply voltage.
For the emitter bias branch,

$$E_s = E_{RL} + E_Q$$
$$E_s = 0 \text{ V} + 5 \text{ V} = 5 \text{ V}$$

Because there is no collector resistor in this circuit, $E_s = E_Q$. E_s can be *increased* up to E_c maximum to accommodate available power supply voltages.

$E_s = 5$ V

FIELD 2

FRAME 1

$R_{b2} = \mathbf{S} \times R_{es}$

The generator must have a dc resistance equal to or less than the computed value of R_{b2} if β stability is to be adequate. $R_{b2} = 10 \times 5\ \text{k}\Omega = 50\ \text{k}\Omega$.

FIELD 3

FRAME 1

Compute the input impedance.

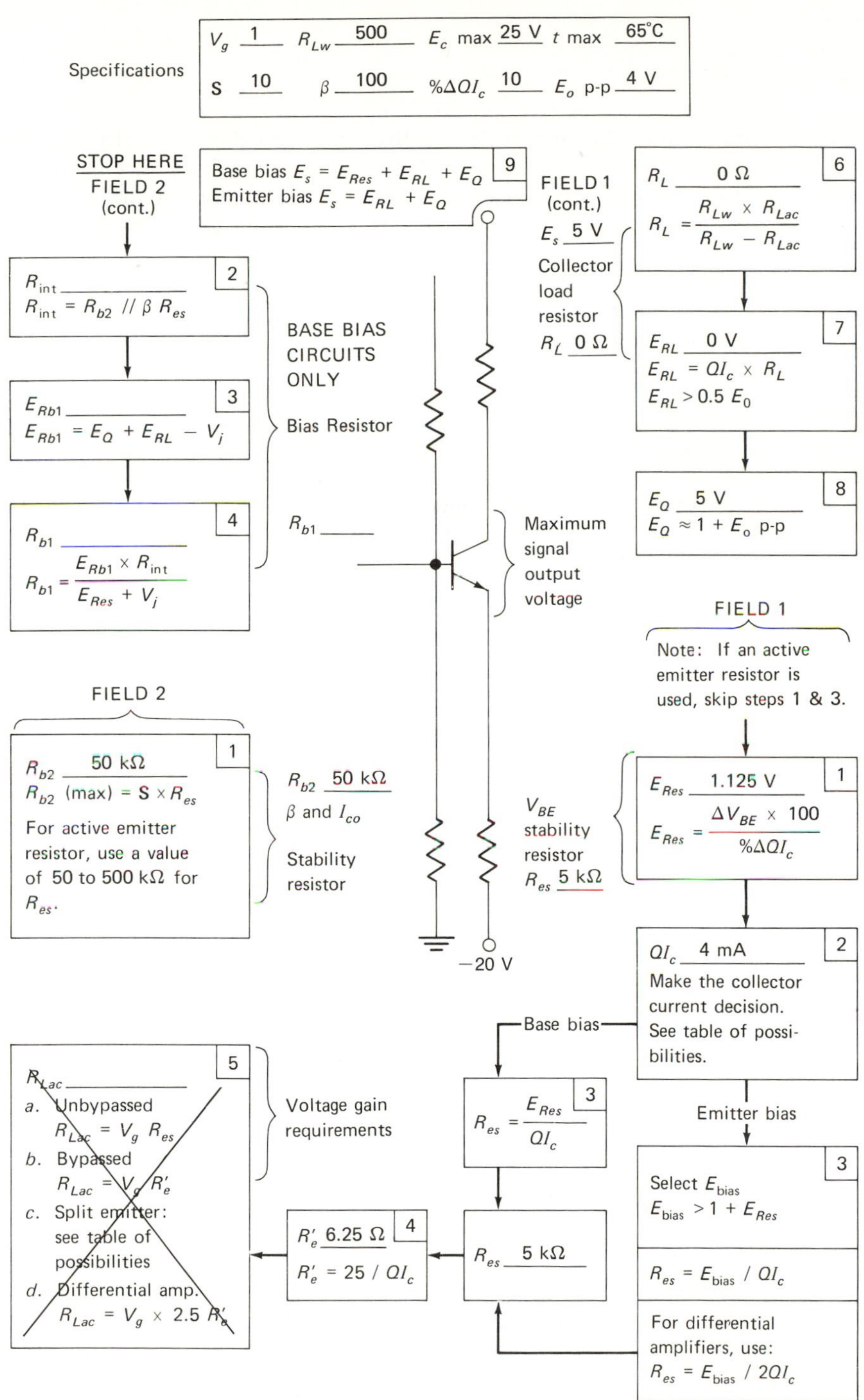

Fig. 7-12 The completed design sheet for Example 7-5

$$Z_{in} \approx \beta \times (R_{Lac} \| R_{Lw})$$
$$Z_{in} \approx 100 \times (500 \| 50\ \text{k}\Omega)$$

Because we decided to make R_{es} equal to $10R_{Lw}$, the input impedance should be about 10 percent lower than the 50 kΩ we calculated here. In practice, if we have been conservative enough in our selection of a working β, the chances are that the input impedance will actually be a bit higher than calculated; this is all to the good. Whenever we get involved with that elusive entity β, we must expect some error. The best we can do is to try to tilt the balance in the most favorable direction by *always* assuming a *conservative* value for β.

Figure 7-12 shows the completed design sheet for Example 7-5.

PROBLEMS

Problem 7-5

TITLE: **Designing the Emitter-biased Common-collector Circuit**
PROCEDURE: **Design the circuit, using the following specifications. See Fig. 7-11. Try for the highest practical input impedance.**

Transistor Parameters

1. β (min)	100
2. I_c (max)	100 mA
3. E_c (max)	25 V
4. Transistor type	*npn* silicon

Circuit Parameters

1. Voltage gain	1
2. E_o p-p	4 V
3. R_{Lw}	500 Ω
4. Stability factor	1
5. $\%\Delta QI_c$ to be allowed	5%
6. Maximum operating temperature	60°C
7. f_o	25 Hz

7-6 BOOTSTRAPPING THE EMITTER FOLLOWER

Figure 7-13 shows a modified base-biased emitter-follower circuit where both R_{b1} and R_{b2} are bootstrapped to increase the input impedance. Input impedances of about the same order of magnitude as those for emitter-biased circuits are possible using this variation.

The circuit is designed exactly as though the bootstrapping was not there, except that, after values for R_{b1} and R_{b2} have been calculated, each one is split into two parts to form the junction for the bootstrap capacitors C_1 and C_2. $R_{b1}a$ and $R_{b2}a$ should be somewhere between 20 and 50 percent of the whole resistance ($R_{b1}a$ = 20 to 50 percent R_{b1} total). The effective resistance of R_{b1}, because of the feedback from the emitter, is

$$R_{b1} \approx \frac{R_{b1}a}{1 - V_g}$$

The effective resistance of R_{b2} is

$$R_{b2}\text{ eff} \approx \frac{R_{b2}a}{1 - V_g}$$

The voltage gain approaches 1. The total input impedance is

$$Z_{in} \approx \frac{R_{b1}a}{1 - V_g} \| \frac{R_{b2}b}{1 - V_g} \| \beta R_{Lac}$$

where R_{Lac} is $R_{es} \| R_{Lw}$.

The capacitors should have a reactance of

$x_c c_2 = 0.1 R_{b2} b$ at f_o and $x_c c_2 = 0.1 R_{b1} b$ at f_o

The voltage gain of an emitter follower is

$$V_g = \frac{1}{1 - R'_e/R_{Lac}}$$

For those conditions where V_g is very close to unity and the values of R_{b1} and R_{b2} are large, $Z_{in} \approx \beta R_{Lac}$.

Obviously high input impedances are much more easily obtained with the emitter-biased circuit, but there are times when the second power supply

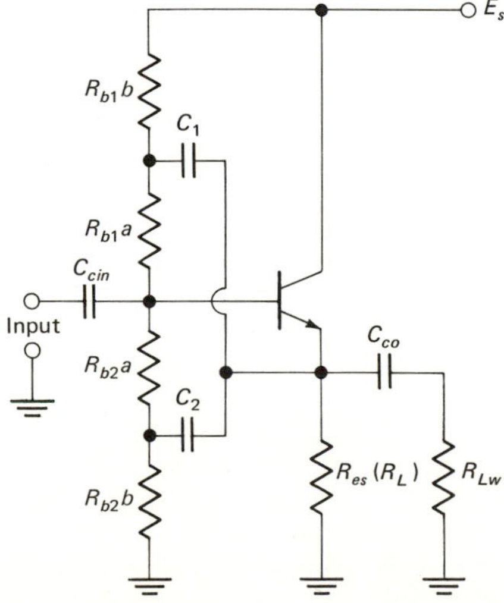

Fig. 7-13 Bootstrapping R_{b1} and R_{b2} in the emitter-follower circuit

for bias is simply not available. Under those circumstances bootstrapping R_{b1} and R_{b2} can be a practical answer to the input impedance problem.

7-7 THE COLLECTOR-BASE JUNCTION IMPEDANCE PROBLEM

Up until now the collector-base junction impedance has not been taken into account in input impedance equations, because the impedance of that reverse-bias junction has generally been 10 to 100 times as high as the combination of base-to-ground resistance and βR_{Lac} values. The collector-base junction impedance is usually somewhere between 500 kΩ and 2 MΩ, and so no matter what we do with the rest of the input circuit, we cannot realize an input impedance any higher than the collector-base junction impedance. There are cases that require input impedances higher than the collector-base junction impedance, and again bootstrapping can come to the rescue. Figure 7-14 shows an emitter follower with the collector bootstrapped.

The resistor R_d must be added to the circuit to prevent the output signal from being shorted to ground through the low impedance of the power supply. R_L should be 10 times (or more) greater than R_{Lw} to prevent loading of the output circuit. The reactance of C_1 should be approximately equal to 0.1 R_d at f_0. The effective collector-base junction impedance with bootstrapping is

$$R_j \text{ eff} \approx \frac{R_J}{1 - V_g}$$

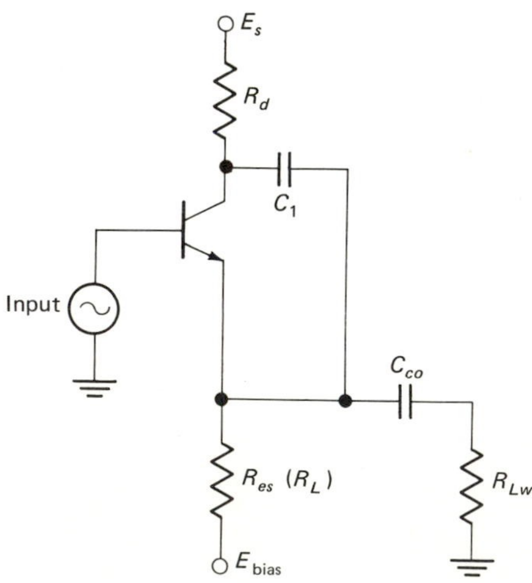

Fig. 7-14 Bootstrapping the collector-base junction impedance

where R_J is the collector-base junction dynamic impedance without bootstrapping.

Example

Find R_J effective when $R_J = 500\ \text{k}\Omega$ and $V_g = 0.9$.

$$R_J \text{ eff} \approx \frac{5 \times 10^5}{1 - 0.9} \approx 5\ \text{M}\Omega$$

Here the *effective* junction impedance has been increased by a factor of 10. Had V_g been 0.99, it would have been raised by about a factor of 100.

Actually we never really know what the collector-base junction value is, and all we can do is to raise its *effective* value so high that, again, we no longer need to include it in our input impedance equations.

7-8 DESIGNING THE COMMON-BASE CIRCUIT

There are two practical versions of the common-base circuit. One is common-base for both bias and signal, with the base returned directly to ground and emitter bias. The emitter-biased common-base circuit is shown in Fig. 7-15.

The circuit here is designed exactly as if it were an emitter-biased common-emitter circuit, except that R_{b2} is set to 0 Ω; that is, the base is returned directly to ground. The input impedance is $Z_{in} \approx R_e'$.

Now let's take a look at the second practical common-base configuration. Figure 7-16 shows this circuit.

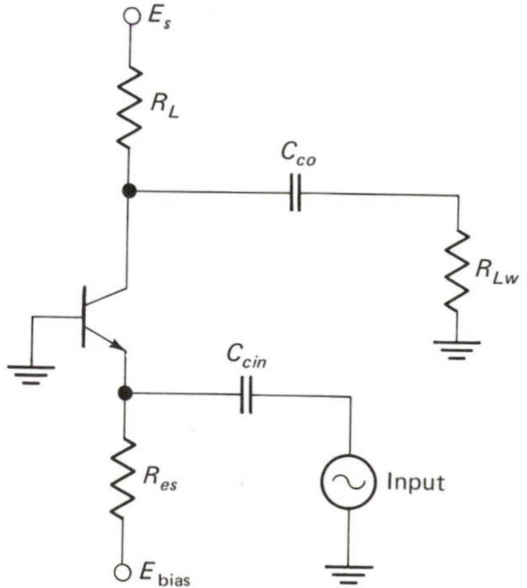

Fig. 7-15 The emitter-biased common-base circuit

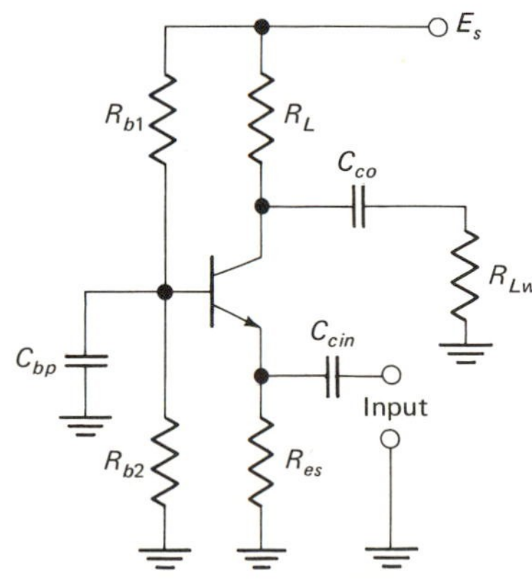

Fig. 7-16 The base-biased common-base circuit

This circuit is basically a common-emitter circuit insofar as the bias circuitry is concerned. For signal frequencies the base is grounded through a large capacitor, making it a grounded base circuit for ac. The circuit is designed exactly as if it were a current-mode common-emitter circuit, with the exception that the input impedance is $Z_{in} \approx R'_e$ and an R_{b2} bypass capacitor C_{bp} is added.

Example 7-6

Design a base-biased common-base circuit, using the following specifications. See Fig. 7-16.

Transistor Parameters	
1. β (min)	100
2. I_c (max)	100 mA
3. E_c (max)	30 V

Circuit Parameters	
1. Voltage gain	200
2. E_o p-p	5 V p-p
3. R_{Lw}	10 kΩ
4. Stability factor	5
5. $\%\Delta QI_c$ to be allowed	10%
6. Maximum operating temperature	75°C
7. f_o	50 Hz

FIELD 1

FRAME 1

Compute E_{Res}.

$$E_{Res} = \frac{V_{BE} \times 100}{\%\Delta QI_c} = \frac{137 \times 10^{-3} \times 10^2}{10} = 1.37 \text{ V}$$

$$E_{Res} \approx 1.4 \text{ V}$$

FRAME 2

Make the QI_c decision.

See the fully bypassed table of possibilities (Table 6-2).

$$QI_c = 1 \text{ mA}$$

FRAME 3

Compute R_{es} (base bias branch).

$$R_{es} = \frac{E_{Res}}{QI_c} = \frac{1.4}{1 \times 10^{-3}} = 1.4 \text{ k}\Omega$$

FRAME 4

Compute R'_e.

$$R'_e = \frac{25}{QI_c} = \frac{25}{1} = 25\ \Omega$$

NOTE: This is Z_{in} for the common-base circuit.

$Z_{in} \approx 25\ \Omega \approx R'_e$

FRAME 5

Compute R_{Lac}.

$R_{Lac} = V_g R'_e = 200 \times 25 = 5\ \text{k}\Omega$

FRAME 6

Compute R_L.

$$R_L = \frac{R_{Lw} \times R_{Lac}}{R_{Lw} - R_{Lac}} = \frac{10\ \text{k}\Omega \times 5\ \text{k}\Omega}{10\ \text{k}\Omega - 5\ \text{k}\Omega} = 10\ \text{k}\Omega$$

FRAME 7

Compute E_{RL}.

$E_{RL} = QI_c R_L = 1 \times 10^{-3}\ \text{A} \times 10 \times 10^3\ \Omega = 10\ \text{V}$

FRAME 8

Compute minimum E_Q.

$E_Q\ (\text{min}) = 1 + E_o\ \text{p-p} = 1 + 5 = 6\ \text{V min}$

FRAME 9

Compute E_s. Assume an E_s of 22 V.

$E_s = 22\ \text{V} \qquad E_{Res} = 1.4\ \text{V} \qquad E_{RL} = 10\ \text{V}$

and E_Q must then be 10.6 V, which is well above the minimum.

FIELD 2

FRAME 1

Compute R_{b1}.

$R_{b2} = \mathbf{S} \times R_{es} = 5 \times 1.4\ \text{k}\Omega = 7\ \text{k}\Omega$

We can use **S** values down to unity in this circuit without adding any input impedance problems. However, low values of R_{b2} will increase the voltage divider current drain, which may be important, particularly in battery-operated circuits.

FRAME 2

Compute R_{int}.

$R_{int} = R_{b2} \| \beta R_{es} = 7\ \text{k}\Omega \| 100 \times 1.4\ \text{k}\Omega$
$R_{int} = 6.6\ \text{k}\Omega \approx 7\ \text{k}\Omega$

FRAME 3

Compute E_{Rb1}.

$E_{Rb1} = E_Q + E_{RL} - V_j = 10.6 + 10 - 0.6$
$E_{Rb1} = 20\ \text{V}$

FRAME 4

Compute R_{b1}.

$$R_{b1} = \frac{E_{Rb1} \times R_{int}}{E_{Res} + V_j} = \frac{20\ \text{V} \times 7\ \text{k}\Omega}{2\ \text{V}} \approx 70\ \text{k}\Omega$$

FIELD 3

FRAME 1

Calculate the input impedance.

$Z_{in} \approx R'_e = 25\ \Omega$

FRAME 2

Compute C_{cin}

x_c of C_{cin} at $f_o \approx 0.1Z_{in}$
x_c at 25 Hz $\approx 2.5\ \Omega$

$$c = \frac{1}{2\pi f x_c} = \frac{1}{6.28 \times 50 \times 2.5} \approx 2{,}000\ \mu\text{F}$$

FRAME 3

Compute C_{bp}.

x_c of C_{bp} at $f_o \approx 0.1R_{b2}$
$x_c \approx 0.7\ \text{k}\Omega$
$C_{bp} \approx 5\ \mu\text{F}$

Figure 7-17 shows the completed design sheet to provide a summary of the results of our calculations.

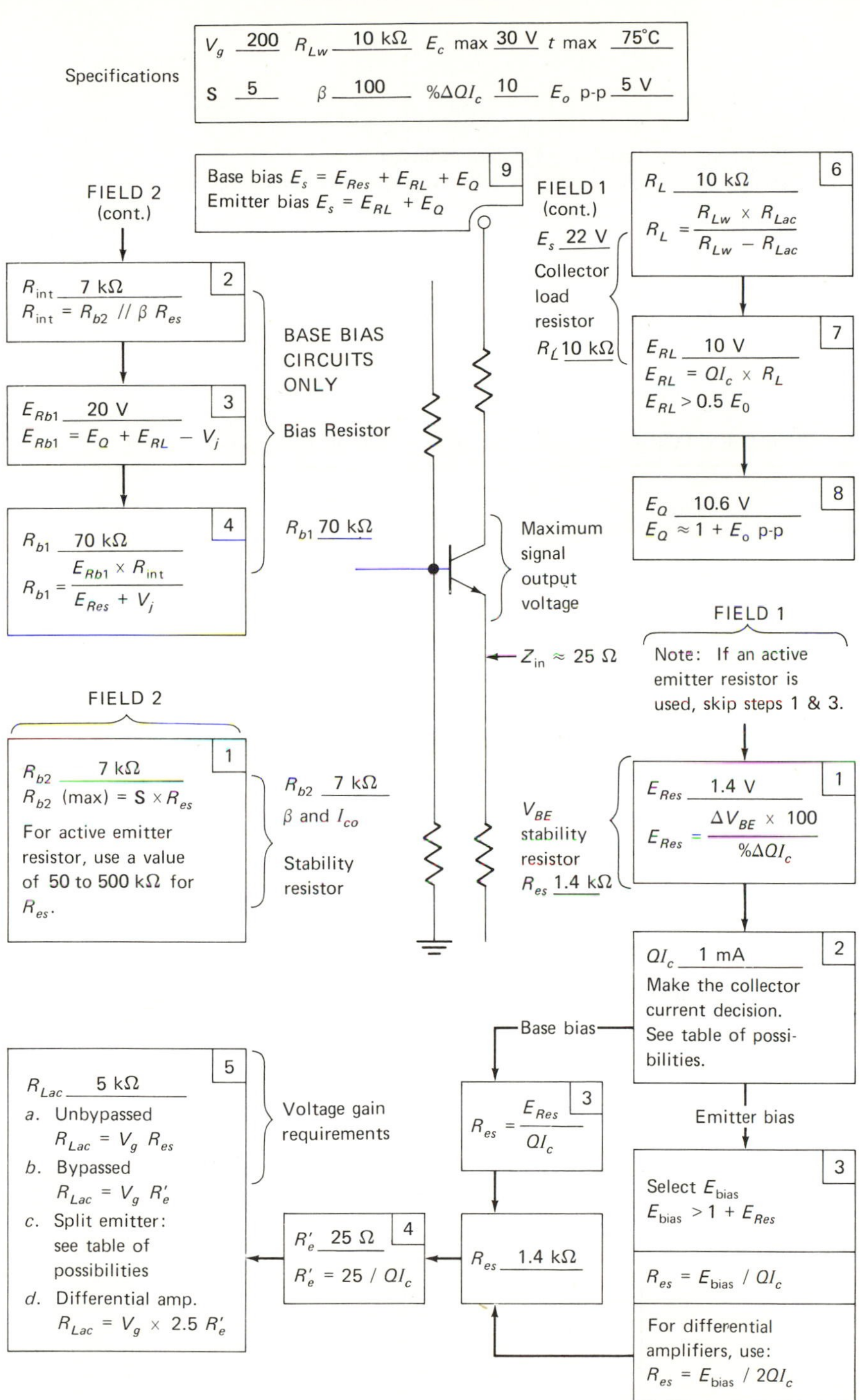

UNIVERSAL DESIGN SHEET
Specifications
V_g 200 R_{Lw} 10 kΩ E_c max 30 V t max 75°C
S 5 β 100 %ΔQI_c 10 E_o p-p 5 V
Base bias $E_s = E_{Res} + E_{RL} + E_Q$
Emitter bias $E_s = E_{RL} + E_Q$
9
FIELD 2 (cont.)
R_{int} 7 kΩ
$R_{int} = R_{b2} \,//\, \beta R_{es}$
2
E_{Rb1} 20 V
$E_{Rb1} = E_Q + E_{RL} - V_j$
3
R_{b1} 70 kΩ
$R_{b1} = \frac{E_{Rb1} \times R_{int}}{E_{Res} + V_j}$
4
BASE BIAS CIRCUITS ONLY
Bias Resistor
R_{b1} 70 kΩ
FIELD 1 (cont.)
E_s 22 V
Collector load resistor
R_L 10 kΩ
R_L 10 kΩ
$R_L = \frac{R_{Lw} \times R_{Lac}}{R_{Lw} - R_{Lac}}$
6
E_{RL} 10 V
$E_{RL} = QI_c \times R_L$
$E_{RL} > 0.5\ E_0$
7
E_Q 10.6 V
$E_Q \approx 1 + E_o$ p-p
8
Maximum signal output voltage
$Z_{in} \approx 25\ \Omega$
FIELD 1
Note: If an active emitter resistor is used, skip steps 1 & 3.
FIELD 2
R_{b2} 7 kΩ
R_{b2} (max) = S × R_{es}
For active emitter resistor, use a value of 50 to 500 kΩ for R_{es}.
1
R_{b2} 7 kΩ
β and I_{co}
Stability resistor
V_{BE} stability resistor
R_{es} 1.4 kΩ
E_{Res} 1.4 V
$E_{Res} = \frac{\Delta V_{BE} \times 100}{\%\Delta QI_c}$
1
QI_c 1 mA
Make the collector current decision. See table of possibilities.
2
Base bias
Emitter bias
$R_{es} = \frac{E_{Res}}{QI_c}$
3
Select E_{bias}
$E_{bias} > 1 + E_{Res}$
3
$R_{es} = E_{bias} \,/\, QI_c$
For differential amplifiers, use:
$R_{es} = E_{bias} \,/\, 2QI_c$
R_{Lac} 5 kΩ
a. Unbypassed
$R_{Lac} = V_g\ R_{es}$
b. Bypassed
$R_{Lac} = V_g\ R'_e$
c. Split emitter: see table of possibilities
d. Differential amp.
$R_{Lac} = V_g \times 2.5\ R'_e$
5
Voltage gain requirements
R'_e 25 Ω
$R'_e = 25 \,/\, QI_c$
4
R_{es} 1.4 kΩ

Fig. 7-17 The completed design sheet for the common-base example (Example 7-6)

PROBLEMS

Problem 7-6

TITLE: **Designing the Base-biased Common-base Circuit**
PROCEDURE: **Design the circuit, using the following specifications. See Fig. 7-16.**

Transistor Parameters

1. β (min)	100
2. I_c (max)	100 mA
3. E_c (max)	30 V

TABLE 7-5. SUMMARY FOR CHAPTER 7—CURRENT-MODE CIRCUITS

Appropriate Table of Possibilities	Circuit Classification		Designed as:		Differences
a Unbypassed, Table 5-5 b Fully bypassed, Table 6-2	Common emitter	Base bias	Common emitter	Base bias	None
c Split emitter, Table 6-4	Common emitter	Emitter bias	Common emitter	Emitter bias	None
Fully bypassed, Table 6-2	Common base	Base bias	Common emitter	Base bias	$Z_{in} \approx R_e'$
Fully bypassed, Table 6-2	Common base	Emitter bias	Common emitter	Emitter bias	$Z_{in} \approx R_e'$
Common collector, Table 7-3	Common collector	Base bias	Common emitter	Base bias	1. $R_{Lac} = R_{es} \| R_{Lw}$ 2. $R_{es} > 10R_{Lw}$ $\therefore R_{Lac} \approx R_{Lw}$ 3. $Z_{in} \approx R_{b1} \| R_{b2} \| \beta R_{Lw}$ 4. $R_L = 0$ 5. $E_{RL} = 0$ V 6. $V_g \approx 1$
Common collector, Table 7-3	Common collector	Emitter bias	Common emitter	Emitter bias	1. $R_{Lac} = R_{es} \| R_{Lw}$ 2. $R_{es} > 10R_{Lw}$ $\therefore R_{Lac} \approx R_{Lw}$ 3. $Z_{in} \approx \beta R_{Lw}$ or $Z_{in} \approx \beta R_{Lw} \| R_{b2}$ 4. $R_L = 0\ \Omega$ 5. $E_{RL} = 0$ V 6. $V_g \approx 1$

Circuit Parameters

1.	Voltage gain	140
2.	E_o p-p	3.0 V
3.	R_{Lw}	5 kΩ
4.	Stability factor	10
5.	$\%\Delta QI_c$ to be allowed	10%
6.	Maximum operating temperature	70°C
7.	f_o	20 Hz

SUMMARY

Although this chapter is concerned with quite a number of circuit variations, there are actually very few new ideas or methods involved.

The voltage-mode circuit design is most different from what we have encountered before, but even here there was only a little new. Voltage-mode circuits come in two classes: case 1, the simplest case, and case 2, a slightly more complex but much more versatile version. The simplest case is limited to situations where E_Q's of 2.6 V or less are acceptable. The case 2 circuit can be used when greater versatility is required. Both circuits have very low input impedances unless the resistor R_{b1} is split and a bypass capacitor is added to eliminate the Miller effect problem for signal frequencies.

Table 7-5 is a summary of the equations involved in this chapter (excluding voltage-mode circuits) and compares them to the current-mode common-emitter circuit equations from previous chapters.

QUESTIONS

7-1 Describe the limitations of the case 1 voltage-mode feedback circuit.

7-2 What causes the very low input impedance of the voltage-mode feedback circuit, and what can be done about it?

7-3 Compare the maximum E_o p-p for a voltage-mode circuit with the E_o p-p of a current-mode circuit.

7-4 What is the purpose of the paraphase amplifier?

7-5 Describe the voltage distributions on the collector side of the paraphase amplifier.

7-6 What are the primary reasons for using a common-collector circuit?

7-7 Why is the emitter follower table of possibilities (Table 7-3) so very different from the other tables of possibilities?

7-8 In the emitter-follower circuit, why is R_{es} (Fig. 7-9) made 10 times (or more) larger than R_{Lw}?

7-9 What are the advantages of an emitter-biased common-collector circuit over a base-biased common-collector circuit?

7-10 What is the purpose of bootstrapping the emitter-follower bias circuit?

7-11 Describe the collector-base junction impedance problem for very high input impedance emitter followers.

7-12 What kind of feedback (negative or positive) is involved in bootstrapping?

7-13 Compare the circuit characteristics of the common-base circuit to those of the common-emitter circuit.

7-14 What are the similarities in the design procedures for common-base and common-emitter circuits?

8

the design process as a troubleshooting tool

OBJECTIVES:

(Things you should be able to do upon completion of this chapter)

1. Analyze the base-biased common-emitter circuit in all three variations: unbypassed, fully bypassed, and split emitter.
2. Analyze the emitter-biased version of the common-emitter fully bypassed circuit.
3. Analyze the voltage-mode (feedback) common-emitter circuit.
4. Analyze the paraphase amplifier circuit.
5. Analyze the common-collector circuit (both emitter and base bias versions).
6. Analyze both emitter and base bias versions of the common-base circuit.

EVALUATION:

1. There are questions and analysis problems in the chapter.
2. The *final* evaluation rests upon the ability of the student to analyze circuits, build them, and have his analysis figures come significantly close to measured values in the laboratory. The laboratory manual contains specific circuits to be analyzed and operating criteria.

INTRODUCTION

This chapter is devoted to the fine art of analyzing circuits designed by someone else. In many ways this is a more difficult art than that of design, because we are not privy to the designer's philosophy, methodology, or decision-making process.

The method presented here is dependent upon an understanding of the design methods you have been studying and is an outgrowth of those methods. The distinct advantage of the method we shall learn in this chapter is that we need not have any knowledge of the designer's thinking, and in most cases we do not even need to know the transistor *type* number to analyze the circuit successfully.

The fact that the analysis process presented here does not require any knowledge of transistor parameters has far-reaching implications in the field of integrated circuit technology. Data on the parameters of individual transistors on an IC chip are nearly *absolutely unavailable*. The method we shall study is ideally suited to this new technology, as well as being a simple and effective tool for the analysis of discrete circuits.

The technician is all too often faced with the problem of troubleshooting a circuit when such things as stage gain, quiescent voltages, and maximum output swing are not known. Schematics are often lacking in some or all of this vital information, and frequently no schematic is available. The schematic can, of course, be drawn and component values obtained from the (discrete) equipment being serviced; but it is often essential to know what voltages and other circuit parameters should be, so that measured values can be compared with what those values should be in the properly operating circuit.

By turning the design procedures inside out, we can determine with a good degree of accuracy what these circuit parameters should be. With a little practice, good estimates of parameters can be made with very little actual computation. These easily obtained estimates will serve for all but the most subtle malfunctions.

The analysis process can be quite simple or nearly as complex as the design process, depending upon the degree of accuracy required. In perhaps 99 percent of the usual routine troubleshooting problems, a first approximation is adequate, and this can be made with only a little calculation and with *no* knowledge of the parameters of the transistor involved.

We shall begin by seeing how this first approximation is made and then follow it up with more careful calculations. I warn you now that the increase in the effort required to achieve that greater accuracy will be greater than the improvement in that accuracy. The basic philosophy of good design is making the circuit parameters as independent as possible of the transistor parameters. Then, by implication, we can analyze a well-designed circuit to at least a fair approximation without including transistor parameters in our analysis. Frequently, the first approximation yields accuracies within 20 percent. The second approximation includes the transistor's β, either as an educated guess or taken from the manual. It makes very little difference which, since either approach involves considerable possibility of error. The second approximation, even with a poor estimation of β, can yield 2 to 5 percent improvement in the accuracy of quiescent voltage calculations. It is difficult to tell exactly how accurate your calculations will be, because so much depends upon how well the circuit is designed in the first place. Some designs are compromised by economic considerations; but as a rule these are not critical circuits, and the error involved is not apt to be troublesome.

The third approximation requires an exact knowledge of β for the particular transistor involved and yields very accurate results. Because β is such a shifty character, the third approximation is seldom used and seldom needed.

In spite of some uncertainty in predicting the amount of error, the method has yielded surprisingly good and consistent results for my students during the six years it has been in use in my laboratory. Integrated circuitry is part of the current technical scene, and *device*-oriented analysis techniques are almost certain to fall short as tools for the analysis of integrated circuits. It is almost a certainty that parameters of individual transistors on a chip will *not* be available to the technician or designer.

And if history is any guide, even overall integrated circuit performance data will often be lacking when the IC is a part of a commercial system. If a schematic diagram of the IC is available, its important performance data can be analyzed out of the schematic. Device-oriented methods could be used by making educated guesses about parameters and plugging those into the appropriate equations. At best this is a makeshift procedure. What I offer you here is a tool which is appropriate to the new technology, and one which is far faster and simpler to use than any of the device-oriented approaches.

The methods of analysis presented in the book are tools for today's technology, equally useful for integrated circuits and discrete systems.

With this circuit-oriented approach it is sufficient, for most design purposes, to know all maximum ratings and to have a rough idea of the transistor's β.

Now, before you get the impression that the approach is somehow less scientific than the more traditional device-oriented methods, let me present the following ideas for your consideration.

Critical transistor parameters vary so greatly both in manufacturing tolerances and with changes in temperature that knowledge of the parameters of a typical transistor of a given type number is of dubious value. Secondly, any design process based on device parameters must ultimately *design out* the transistor's parameters, if the circuit is to function and be stable with changes in temperature. The desired end result is always a *circuit* which performs a particular job as the designer intended, and whatever the design process (and there are several), it eventually, often after much effort, becomes circuit-oriented.

A circuit-oriented approach does not require precise knowledge of transistor parameters; this simplifies things considerably. The methods of analysis in this book are intimately related to circuit-oriented methods of design (a logical extension of them), and as such *do not* require a knowledge of the parameters of the transistor involved to analyze a transistor circuit *quantitatively*. This is of great practical advantage, because circuits encountered in the field, which must be analyzed, often contain transistors about which nothing at all is known.

8-1 THE ANALYSIS OF THE CURRENT-MODE COMMON-EMITTER CIRCUIT

Figure 8-1 shows a universal analysis sheet for all circuits using current-mode feedback.

Field 1 is used *only* for base-biased circuits, but it applies to common-emitter, common-collector, and common-base circuits so long as they are base-biased.

Field 2 starts with the computation of the quiescent collector current QI_c. There are two branches, one for base-biased and one for emitter-biased circuits. If we were analyzing an emitter-biased circuit, the emitter bias branch of Frame 1, Field 2, would be the start of the analysis since we would have skipped Field 1 altogether. Frames 2, 3, and 4 are common to all current-mode circuits. Frame 5 involves the calculation of E_Q. Here there are two branches, one for base-biased circuits and one for emitter-biased circuits. The difference is that the voltage drop across R_{es} in base-biased circuits is supplied by E_s, whereas in the emitter-biased circuit that voltage drop is provided by the emitter bias supply and is not a part of E_s.

Field 3 covers the signal parameters p-p output voltage, voltage gain, and input impedance. Frame 1, the p-p output voltage, is common to all current-mode circuits. Frame 2 involves the voltage gain. There are two branches, one for *all* current-mode circuits except the common-collector circuit, and one for the common-collector circuit. The general equation for the voltage gain of all common-emitter and common-base circuits is $V_g = R_{Lac}/R_e^*$, where R_{Lac} is the effective ac load impedance and R_e^* is the effective ac emitter impedance. The note marked *NOTE: specifies the effective emitter impedance for all the various configurations. Frame 3 contains three branches. Branch *a* is valid for computing the input impedance of all common-emitter and common-collector circuits. Again the NOTE is used to get the effective emitter resistance value for the particular circuit being analyzed. If the circuit you are dealing with in this frame does not have an R_{b1} resistor, simply expunge R_{b1} from the equation without altering the equation otherwise. Similarly if the circuit has no R_{b2}, it may be dropped from the equation also. Branch *c* of Frame 3 is used for the common-base circuit only, both emitter-biased and base-biased versions. Branch *b* is used for differential amplifiers *only*. We will examine the differential amplifier in another chapter. For the time being, ignore all references to the differential amplifier. I have included it on the flowchart and analysis sheets so that when we do get to it, we can look back and see how it fits in with the rest of the circuits. The usual marginal references are provided, and the flowchart and analysis sheet correspond step by step.

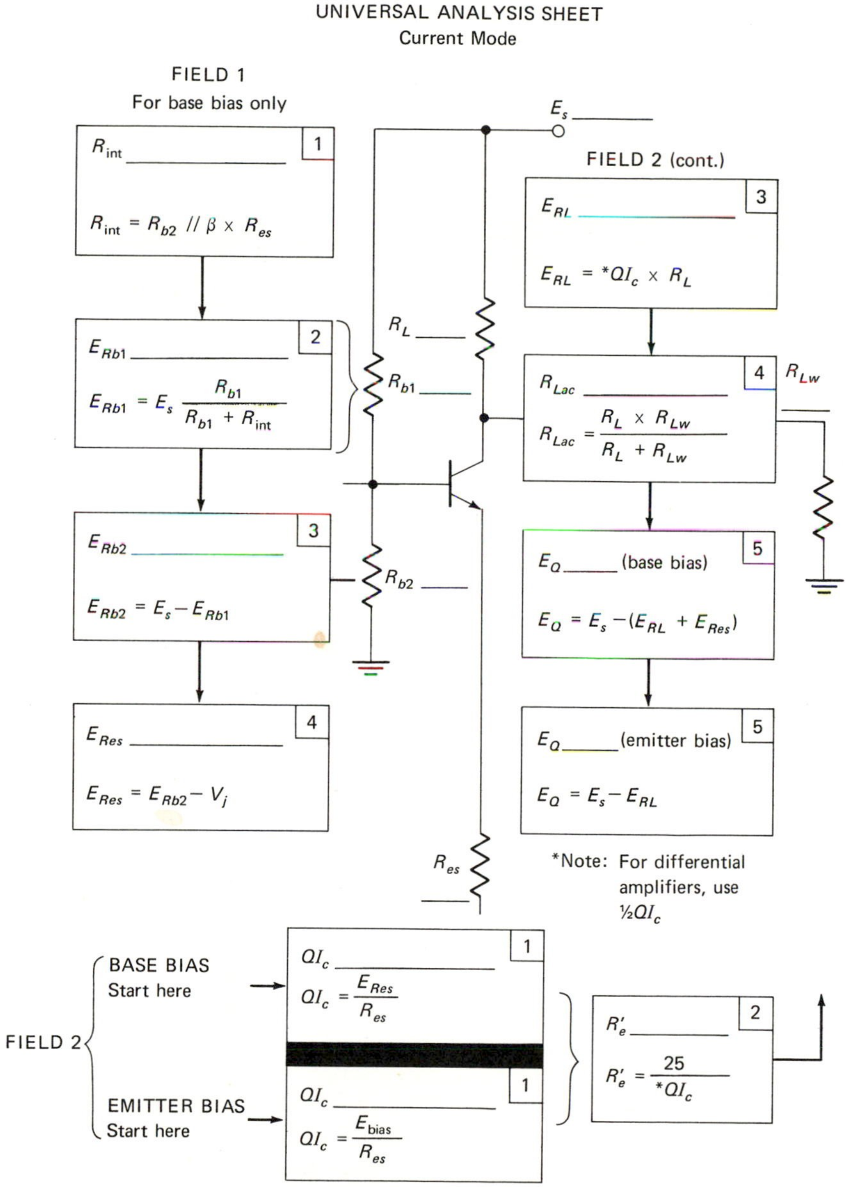

UNIVERSAL ANALYSIS SHEET
Current Mode

FIELD 1
For base bias only

1. R_{int} ______
$R_{int} = R_{b2} \,//\, \beta \times R_{es}$

2. E_{Rb1} ______
$E_{Rb1} = E_s \dfrac{R_{b1}}{R_{b1} + R_{int}}$

3. E_{Rb2} ______
$E_{Rb2} = E_s - E_{Rb1}$

4. E_{Res} ______
$E_{Res} = E_{Rb2} - V_j$

E_s ______

R_L ______

R_{b1} ______

R_{b2} ______

R_{es} ______

R_{Lw} ______

FIELD 2 (cont.)

3. E_{RL} ______
$E_{RL} = {}^*QI_c \times R_L$

4. R_{Lac} ______
$R_{Lac} = \dfrac{R_L \times R_{Lw}}{R_L + R_{Lw}}$

5. E_Q ______ (base bias)
$E_Q = E_s - (E_{RL} + E_{Res})$

5. E_Q ______ (emitter bias)
$E_Q = E_s - E_{RL}$

*Note: For differential amplifiers, use ½QI_c

FIELD 2

BASE BIAS
Start here

1. QI_c ______
$QI_c = \dfrac{E_{Res}}{R_{es}}$

EMITTER BIAS
Start here

1. QI_c ______
$QI_c = \dfrac{E_{bias}}{R_{es}}$

2. R'_e ______
$R'_e = \dfrac{25}{{}^*QI_c}$

Fig. 8-1 *Current-mode analysis sheet*

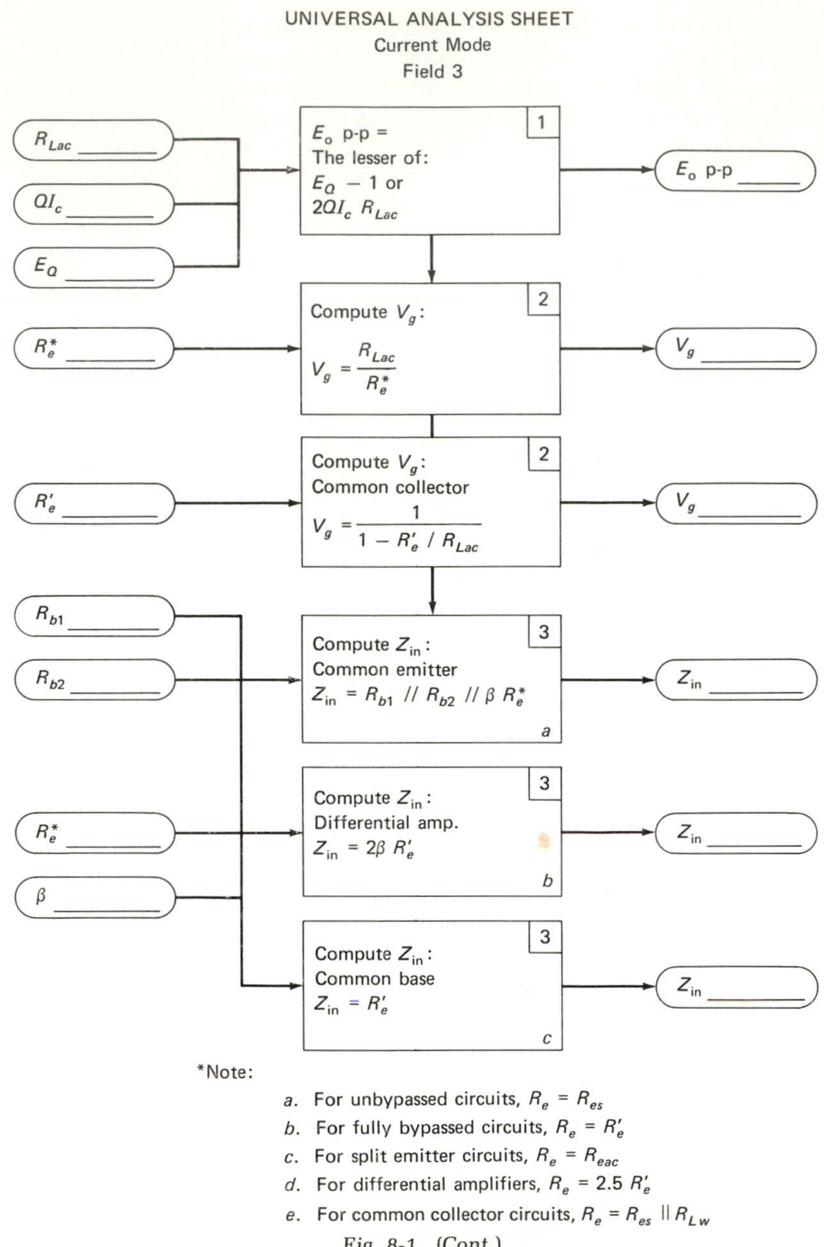

*Note:

a. For unbypassed circuits, $R_e = R_{es}$
b. For fully bypassed circuits, $R_e = R_e'$
c. For split emitter circuits, $R_e = R_{eac}$
d. For differential amplifiers, $R_e = 2.5\ R_e'$
e. For common collector circuits, $R_e = R_{es} \parallel R_{Lw}$

Fig. 8-1 (Cont.)

In order to give you some intuitive feel for the differences between the three approximations, I will work an example: Example 8-1*a*, without including β; 8-1*b*, with a β of 30; and 8-1*c*, with a β of 60. I will also have the circuit built and provide you with a table comparing the measured values with the values computed in all three cases.

Example 8-1a The First Example and Explanation

1. Given the schematic diagram in Fig. 8-2, compute the following:

   E_{Rb1}, E_{Rb2}, E_{Res}, E_{RL}, E_Q, R_e', and R_{Lac}

2. Compute the voltage gain, the maximum output voltage, and Z_{in}.

Because this is a base-biased circuit, we will start with Field 1. For emitter-biased circuits we would skip Field 1 altogether and begin with the emitter bias branch (Frame 1) in Field 2. See Fig. 8-3.

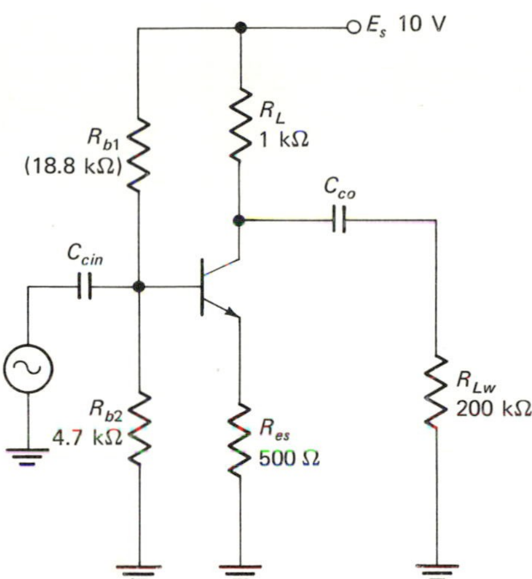

Fig. 8-2 The schematic for Example 8-1a

There is only one reasonable starting place. Knowing the supply voltage and that R_{b1} and R_{b2} form a voltage divider and that a voltage divider can be analyzed without knowing the current, we can solve for the voltage drop across either R_{b1} or R_{b2}. We will begin with the voltage across R_{b1}. Figure 8-4 shows a voltage divider, which in our case would be the divider composed of R_{b1} and R_{b2}, and $\beta \times R_{es}$.

In this instance we assume that we do not know the β of the transistor, and so we cannot determine the partial input resistance R_{int}. This is the first approximation case. We can solve the proportion

$$\frac{R_{b1}}{R_{b1} + R_{b2}} = \frac{E_{Rb1}}{E_s}$$

or we can write it in the more convenient form generally used for unloaded voltage dividers:

$$E_{Rb1} = \frac{R_{b1}}{R_{b1} + R_{b2}} \times E_s$$

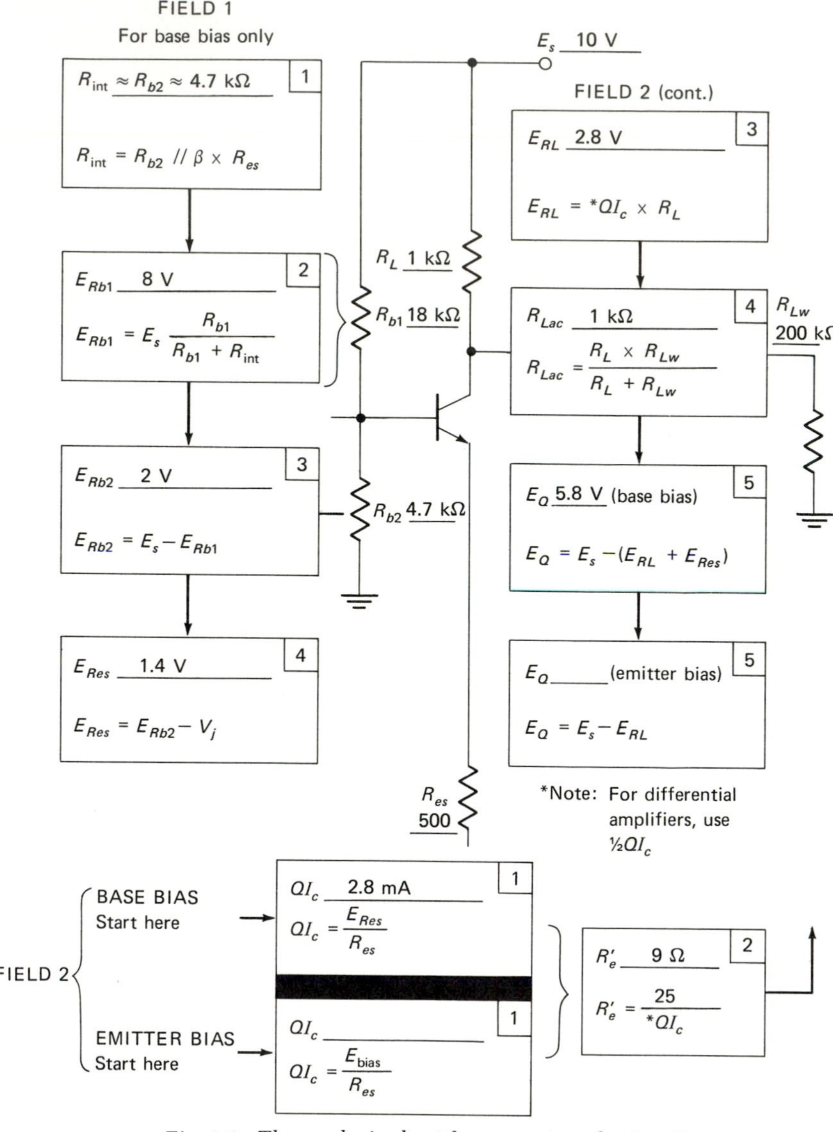

Fig. 8-3 *The analysis sheet for current mode circuits*

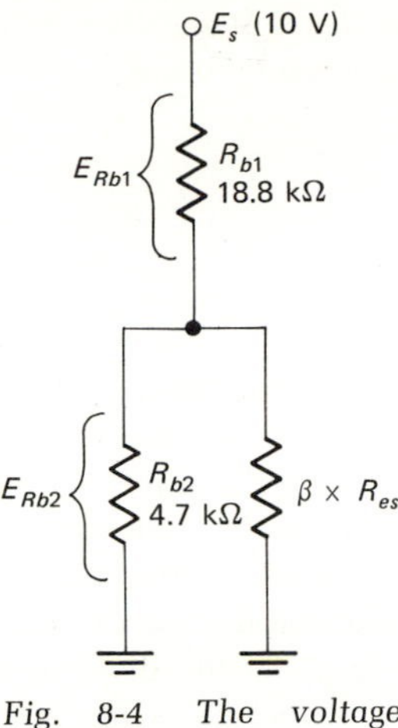

Fig. 8-4 The voltage divider R_{b1}-R_{b2}

Now, let us look at Frame 1 in Fig. 8-3 (the analysis sheet) and begin the example in earnest.

FIELD 1 BASE SIDE COMPUTATIONS (BASE BIAS CIRCUITS ONLY)

FRAME 1

Compute the value of the partial input resistance R_{int}.

$R_{int} = R_{b2} \| \beta R_{es}$

In this case, we are assuming that we do not know β. If β is sufficiently high, as it often is, it will be reasonably accurate to assume that $R_{int} \approx R_{b2}$. We will make that assumption in this case in order to demonstrate that fairly large errors in our knowledge of β result in only small errors in our analyses of practical circuits. We shall always assume that $R_{int} \approx R_{b2}$, when we use the *first* analysis. $R_{int} \approx 4.7$ kΩ.

FRAME 2

Solve for the voltage drop across R_{b1}.

$$E_{Rb1} = E_s \frac{R_{b1}}{R_{b1} + R_{int}}$$

$$E_{Rb1} = 10 \frac{18.8 \text{ k}\Omega}{18.8 \text{ k}\Omega + 4.7 \text{ k}\Omega}$$

$$E_{Rb1} = 8.0 \text{ V}$$

FRAME 3

Compute E_{Rb2}, the voltage drop across R_{b2}.

At this stage, we know the supply voltage and the voltage drop across R_{b1}; so it is a simple matter of subtracting the voltage drop across R_{b1} from the supply voltage, which leaves us with the voltage drop across R_{b2}.

$$E_{Rb2} = E_s - E_{Rb1}$$
$$E_{Rb2} = 10 - 8 = 2\ \text{V}$$
$$E_{Rb2} = 2\ \text{V}$$

FRAME 4

Find E_{Res}, the voltage drop across the emitter resistor.

The voltage drop across the emitter resistor will be either 0.6 or 0.2 V lower than the voltage drop across R_{b2}. It will be 0.6 V lower for silicon and 0.2 V for germanium devices. If it should happen that you do not know whether the transistor is germanium or silicon, you can measure the base-to-ground voltage, measure the emitter-to-ground voltage, and take the difference between the two. This voltage will almost always be close enough to make the determination, even in a badly malfunctioning circuit. $E_{Res} = E_{Rb2} - V_j$, where V_j is the junction voltage drop. We will assume that this unit is a silicon device.

$$E_{Res} = 2\ \text{V} - 0.6\ \text{V}$$
$$E_{Res} = 1.4\ \text{V}$$

FIELD 2 COLLECTOR SIDE COMPUTATIONS

FRAME 1

Compute the value of the quiescent collector current QI_c.

Because we are dealing with a base-biased circuit, we will take the base bias branch. We ignore the emitter bias branch altogether; it does not apply.

Here we must make the assumption that the collector current is the same as the emitter current, which is a very close approximation for most practical circuits. If we know the voltage drop across the emitter resistor and its resistance, Ohm's law will give us the emitter current; and since this is about the same as the collector current, we will know that as well.

$$QI_c \approx \frac{E_{Res}}{R_{es}} \qquad \text{(Ohm's law)}$$

$$QI_c \approx \frac{1.4\ \text{V}}{500\ \Omega}$$

$$QI_c \approx 2.8\ \text{mA}$$

FRAME 2

Compute R_e', the transistor's internal junction resistance.

In some instances we will not need this value, but in most cases we will need it when we compute input impedance and voltage gain. The calculation is a simple one and worth the effort in the interest of forming a consistent order of analysis steps. This example is a case in point, but we will do it anyway.

$$R'_e = \frac{25}{QI_c}$$

$$R'_e = \frac{25}{2.8}$$

$$R'_e \approx 9\ \Omega$$

FRAME 3

Compute E_{RL}, the voltage drop across the collector load resistor.

We now know the collector current and the load resistance, and again Ohm's law will tell us the voltage drop.

$$E_{RL} = QI_c \times R_L \qquad \text{(Ohm's law)}$$
$$E_{RL} = 2.8\ \text{mA} \times 1\ \text{k}\Omega$$
$$E_{RL} = 2.8\ \text{V}$$

FRAME 4

Compute the value of the ac collector load R_{Lac}.

$$R_{Lac} = R_L \| R_{Lw}$$

$$R_{Lac} = \frac{R_L \times R_{Lw}}{R_L + R_{Lw}}$$

In this case, R_{Lw} is so high that R_{Lac} is effectively equal to R_L alone.

$$R_{Lac} = 1\ \text{k}\Omega$$

FRAME 5

Compute the transistor's collector-emitter voltage drop E_Q.

We have to get at this value indirectly, because we do not know the actual collector-emitter resistance. But we do know that when we have three resistances in series, and know the voltage drops across two of them, the balance of the supply voltage *must* appear across the remaining resistance.

$$E_Q = E_s - (E_{Res} + E_{RL})$$
$$E_Q = 10 - (2.8 + 1.4)$$
$$E_Q = 5.8\ \text{V}$$

FIELD 3 SIGNAL PARAMETERS (SEE FIG. 8-1)

FRAME 1

Compute the maximum output voltage E_o.

E_o is the maximum peak-to-peak output voltage without distortion. The maximum output voltage is *a* $(E_Q - 1)$ or *b* $(2QI_cR_{Lac})$, whichever is less.

a. $E_Q - 1 = 5.8 - 1 = 4.8$ V p-p
b. $2QI_cR_{Lac} = 2(2.8 \times 10^{-3} \times 1 \times 10^3) = 5.6$ V

The lesser of the two is 4.8 V p-p.

$E_o \approx 4.8$ V p-p

FRAME 2

Compute the voltage gain V_g.

$$V_g = \frac{R_{Lac}}{R_e}$$

In this case the emitter resistor is unbypassed, and so $R_e = R_{es}$.

$$V_g = \frac{1\ \text{k}\Omega}{500} = 2$$

FRAME 3

Compute the input impedance Z_{in}.

$Z_{in} = R_{b1} \| R_{b2} \| \beta R_{es}$
$Z_{in} = 18.8\ \text{k}\Omega \| 4.7\ \text{k}\Omega \| ??$ (β is unknown)

At this point we come to another branch, one for the differential amplifier, a special case, which we will discuss in a later chapter. We will ignore that branch and use the one that applies to all other current-mode circuits.

Here again we do not know β, and so we will consider only $R_{b1} \| R_{b2}$.

$Z_{in} \approx R_{b1} \| R_{b2}$
$Z_{in} \approx 3.76\ \text{k}\Omega$

That is about everything we might need to analyze, or know, to troubleshoot a stage. The completed analysis sheets are shown in Figs. 8-3 and 8-5.

Now, let us work the same example including the one significant transistor parameter, β. We will assume a β of 30. (Follow Figs. 8-6 and 8-7.)

Example 8-1*b*

Given the schematic in Fig. 8-2, find the following:

E_{Rb1}, E_{Rb2}, E_{Res}, E_{RL}, E_Q, QI_c, V_g, and E_o.

FIELD 1 BASE SIDE COMPUTATIONS (SEE FIG. 8-6)

FRAME 1

Compute the value of the partial input impedance R_{int}.

This partial input impedance includes R_{b2} and $\beta \times R_{es}$, but does not include R_{b1}.

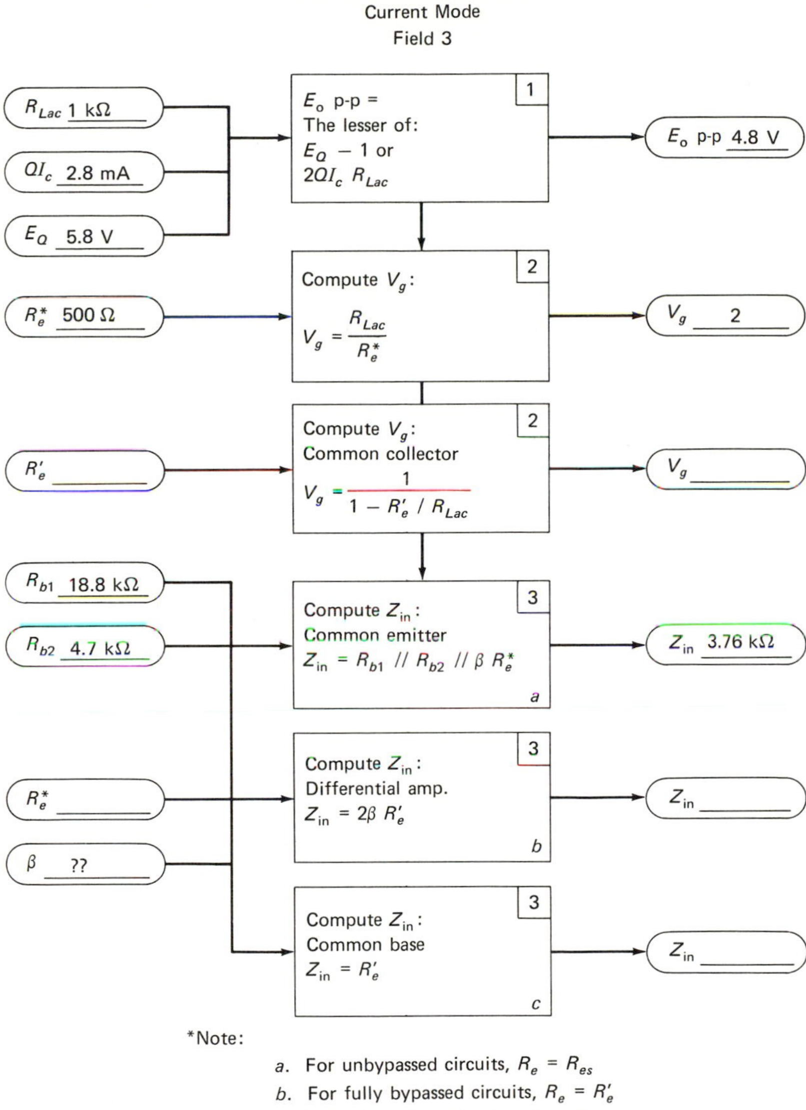

*Note:

a. For unbypassed circuits, $R_e = R_{es}$
b. For fully bypassed circuits, $R_e = R_e'$
c. For split emitter circuits, $R_e = R_{eac}$
d. For differential amplifiers, $R_e = 2.5\ R_e'$
e. For common collector circuits, $R_e = R_{es} \parallel R_{Lw}$

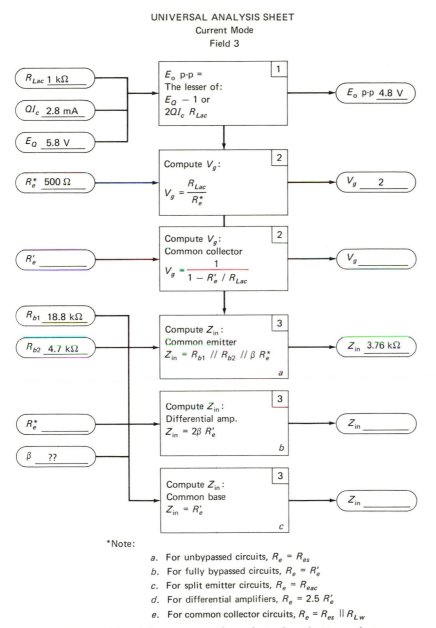

Fig. 8-5 *Field 3 of the current-mode analysis sheet for Example 8-1a*

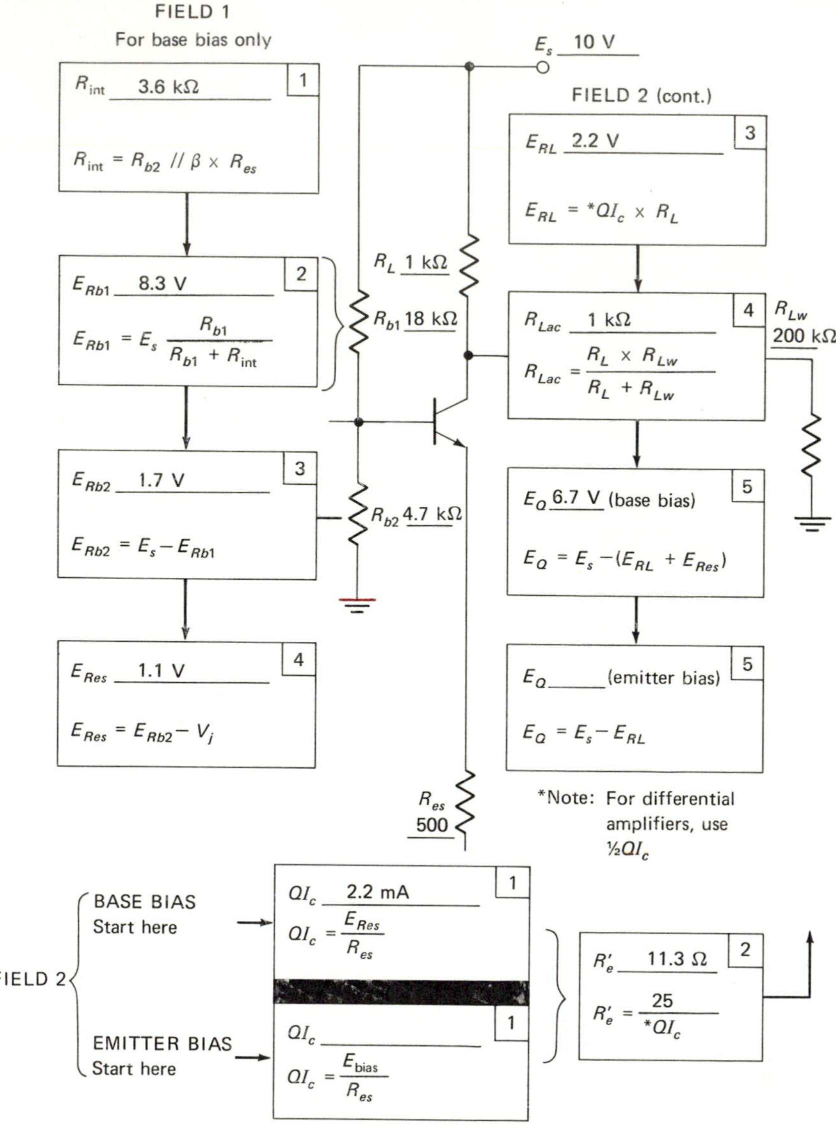

Fig. 8-6 *Analysis sheet for Example 8-1b*

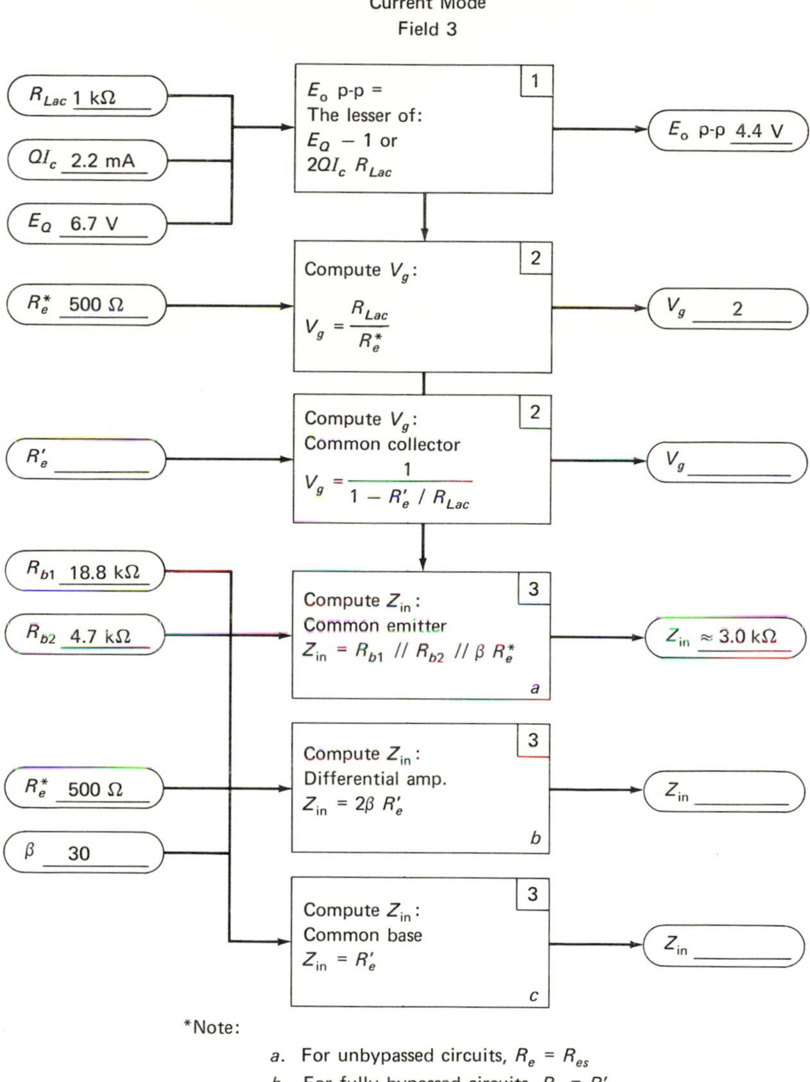

*Note:

a. For unbypassed circuits, $R_e = R_{es}$

b. For fully bypassed circuits, $R_e = R_e'$

c. For split emitter circuits, $R_e = R_{eac}$

d. For differential amplifiers, $R_e = 2.5\ R_e'$

e. For common collector circuits, $R_e = R_{es} \parallel R_{Lw}$

Fig. 8-7 *Field 3 for Example 8-1b*

$R_{int} = R_{b2} \| \beta R_{es}$

$$R_{int} = \frac{R_{b2} \times (\beta R_{es})}{R_{b2} + (\beta R_{es})}$$

$$R_{int} = \frac{4.7\ \text{k}\Omega\ (30 \times 500)}{4.7\ \text{k}\Omega + (30 \times 500)}$$

$R_{int} \approx 3.6\ \text{k}\Omega$

FRAME 2

Compute the voltage drop across R_{b1}.

$$E_{Rb1} = \frac{R_{b1}}{R_{b1} + R_{int}} \times E_s$$

$$E_{Rb1} = \frac{18.8\ \text{k}\Omega}{18.8\ \text{k}\Omega + 3.6\ \text{k}\Omega} \times 10$$

$E_{Rb1} = 8.3\ \text{V}$

FRAME 3

Compute the voltage drop from base to ground.

$E_{Rb2} = E_s - E_{Rb1}$
$E_{Rb2} = 10 - 8.3$
$E_{Rb2} = 1.7\ \text{V}$

FRAME 4

Find the voltage drop across the emitter resistor.

$E_{Res} = E_{Rb2} - V_j$
$E_{Res} = 1.7\ \text{V} - 0.6\ \text{V}$
$E_{Res} = 1.1\ \text{V}$

FIELD 2 COLLECTOR SIDE COMPUTATIONS

FRAME 1

Compute the quiescent collector current QI_c.

$QI_c = E_{Res}/R_{es}$ (Ohm's law)

(Here again we ignore the emitter bias branch.)

$$QI_c = \frac{1.1\ \text{V}}{500} = 2.2\ \text{mA}$$

$QI_c = 2.2\ \text{mA}$

FRAME 2

Compute R_e', the base-emitter junction resistance.

$$R'_e = \frac{25}{QI_c}$$

$$R'_e = \frac{25}{2.2}$$

$$R'_e \approx 11.3\ \Omega$$

FRAME 3

Compute the voltage drop across the collector load resistor.

Knowing the current and resistance, we can find the voltage drop by Ohm's law.

$$E_{RL} = R_L \times QI_c$$
$$E_{RL} = 1\ \text{k}\Omega \times 2.2\ \text{mA}$$
$$E_{RL} = 2.2\ \text{V}$$

FRAME 4

Compute the value of the ac load R_{Lac}.

$$R_{Lac} = \frac{R_L \times R_{Lw}}{R_L + R_{Lw}}$$

$$R_{Lac} = \frac{1\ \text{k}\Omega \times 200\ \text{k}\Omega}{1\ \text{k}\Omega + 200\ \text{k}\Omega} \approx 1\ \text{k}\Omega$$

FRAME 5

Find E_Q, the voltage drop across the transistor (from collector to emitter).

$$E_Q = E_s - (E_{RL} + E_{Res})$$
$$E_Q = 10 - (2.2 + 1.1)\ \text{V}$$
$$E_Q = 6.7\ \text{V}$$

FIELD 3 SIGNAL PARAMETERS (SEE FIG. 8-7)

FRAME 1

Compute the maximum output voltage without distortion.

The maximum output voltage is *a* $(E_Q - 1)$ or *b* $(2QI_cR_{Lac})$, whichever is less.

a. $E_Q - 1 = 6.7 - 1 = 5.7$ V p-p
b. $2QI_cR_{Lac} = 2(2.2 \times 10^{-3} \times 1 \times 10^3) = 4.4$ V

Therefore $E_o \approx 4.4$ V p-p

FRAME 2

Compute the voltage gain V_g.

$$V_g = \frac{R_{Lac}}{R_{es}} \quad \text{(for the unbypassed emitter circuit } R_e = R_{es}\text{)}$$

$$V_g = \frac{1{,}000}{500}$$

$$V_g = 2$$

FRAME 3

Compute the input impedance Z_{in}.

$$Z_{in} = R_{b1} \| R_{b2} \| \beta R_e \qquad \text{where } R_e = R_{es}$$

$$\frac{1}{Z_{in}} = \frac{1}{18.8\ \text{k}\Omega} + \frac{1}{4.7\ \text{k}\Omega} + \frac{1}{15\ \text{k}\Omega} \qquad \text{(resistors in parallel)}$$

$$Z_{in} \approx 3.0\ \text{k}\Omega$$

Now we shall do the same problem, using a β of 60. This time we shall skip the step-by-step procedure and show the work on the analysis sheets in much the same way you would if you were doing it yourself, using the analysis sheets.

Example 8-1*c*

Given the schematic in Fig. 8-2, find the following:

E_{Rb1}, E_{Rb2}, E_{Res}, E_{RL}, E_Q, QI_c, V_g, and E_0.

Assume a β of 60.

Table 8-1 summarizes the results of our computations in Example 8-1*a*, *b*, and *c* and compares them with measured results for the real circuit.

TABLE 8-1. SUMMARY OF COMPUTED AND MEASURED RESULTS FOR EXAMPLE 8-1*a*, *b*, *c*

Parameter	Without β	β of 30	β of 60	Measured
E_{Rb1}	8.0 V	8.3 V	8.2 V	7.92
E_{Rb2}	2 V	1.7 V	1.8 V	2.09 V
E_{Res}	1.4 V	1.1 V	1.2 V	1.5 V
QI_c	2.8 mA	2.2 mA	2.4 mA	2.6 mA
E_{RL}	2.8 V	2.2 V	2.4 V	2.6 V
E_Q	5.8 V	6.7 V	6.4 V	5.9 V
V_g	2	2	2	1.83
Z_{in}	3.76 kΩ	3.0 kΩ	3.3 kΩ	3.4 kΩ

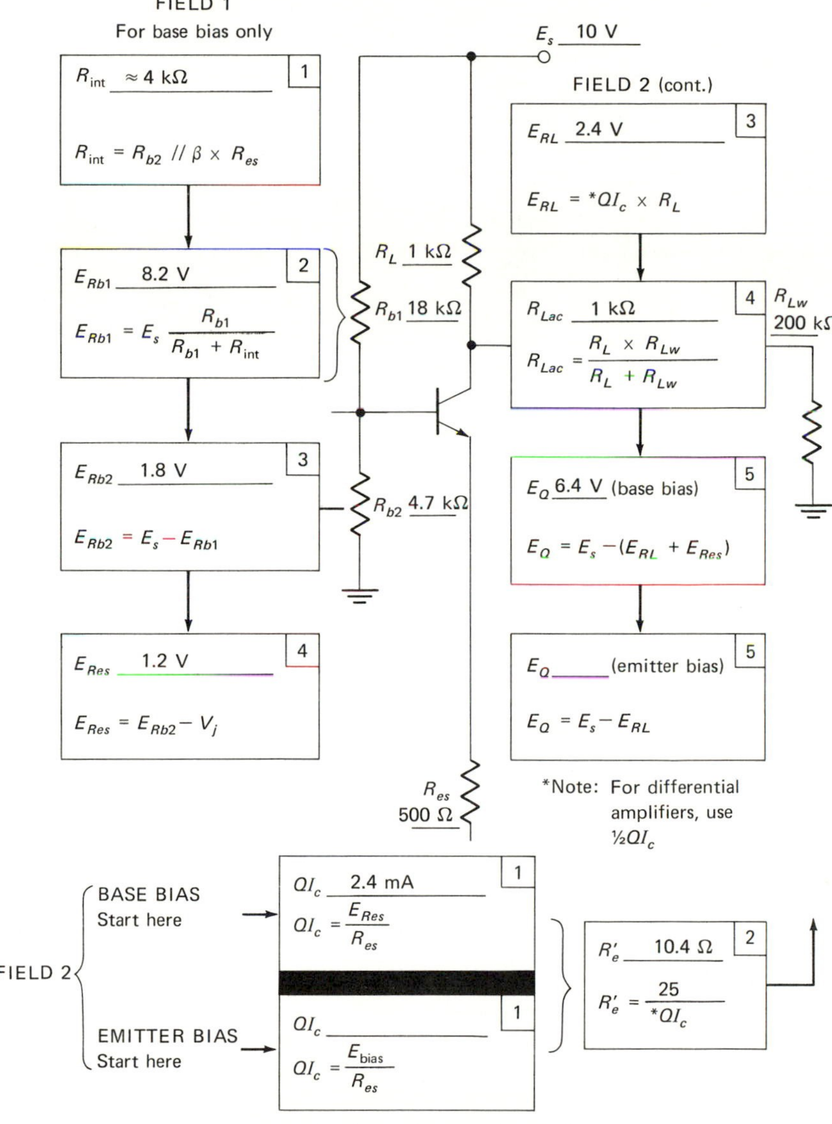

Fig. 8-8 Analysis sheet for Example 8-1c

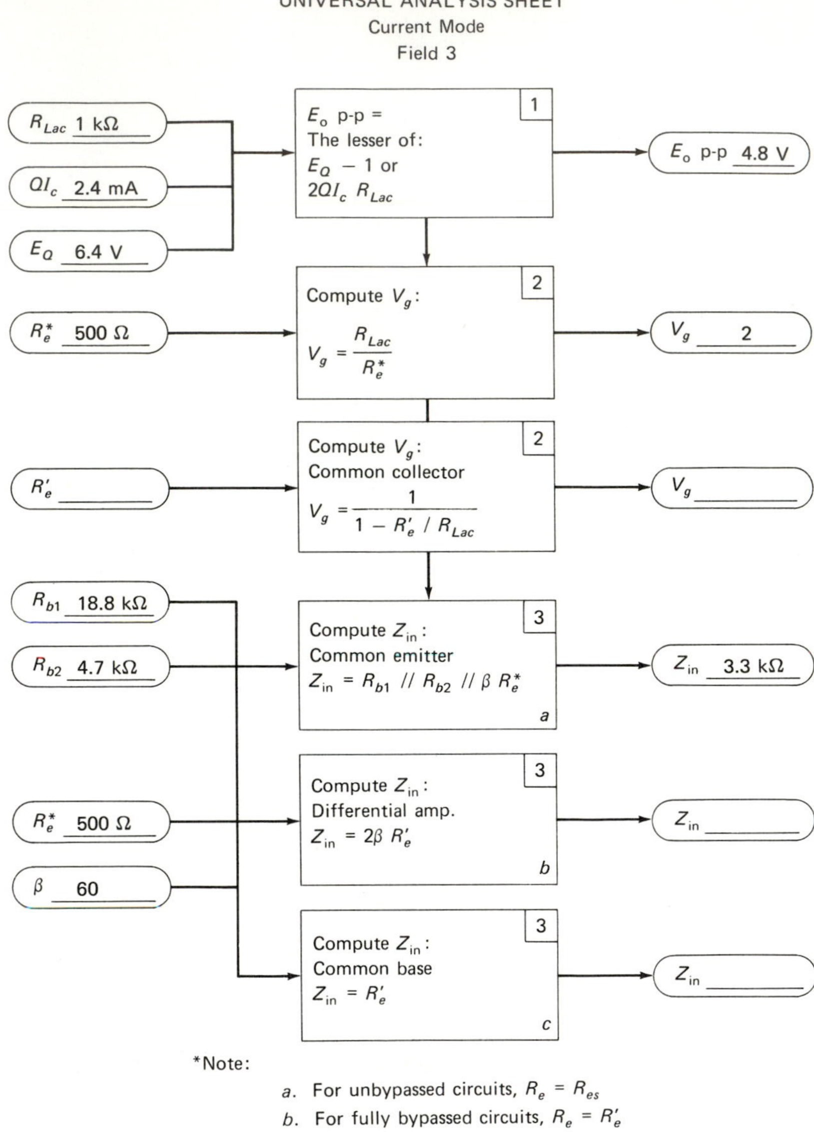

*Note:

a. For unbypassed circuits, $R_e = R_{es}$
b. For fully bypassed circuits, $R_e = R_e'$
c. For split emitter circuits, $R_e = R_{eac}$
d. For differential amplifiers, $R_e = 2.5\ R_e'$
e. For common collector circuits, $R_e = R_{es} \parallel R_{Lw}$

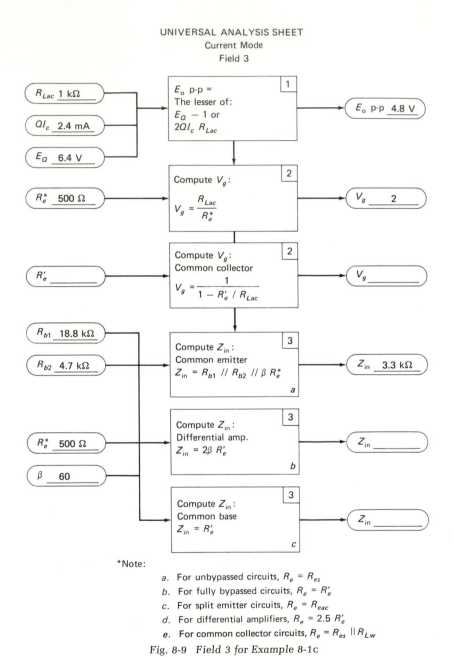

Fig. 8-9 *Field 3 for Example 8-1c*

On the face of it, a comparison of the calculated values and the measured values in Table 8-1 would seem to indicate that the omission of β might be a good standard practice. That is not really the case, except where β is known to be quite high. However, the table should convince you that we can perform a pretty accurate circuit analysis without knowing *anything* at all about the transistor involved.

With a reasonable estimate of β, we can do even better, in spite of the results of this particular set of measurements. I would have been happier, of course, if these measurements confirmed that greater accuracy resulted from a more accurate knowledge of β, but the student who made the measurements is a careful technician, and I must accept his results as they stand.

8-2 THE SECOND EXAMPLE AND EXPLANATION

Now, I would like to go through another example with you. In this example we will take a basic, unbypassed circuit and make two variations of it. One of the variations will involve bypassing the entire emitter resistor, and the other will involve splitting the emitter resistor in two and bypassing a portion of it. The Field 1 and Field 2 parameters for all three variations will be the same, and the use of an emitter bypass capacitor will affect *only some* of the signal parameters. Those signal parameters will be vastly different for the three variations. Figure 8-10 shows the schematics of the three variations of the common-emitter current mode (base bias circuit).

Up to this point, we have paid no attention to the fact that, in reality, we are dealing with three different circuits. And so we are, but they are different only in their individual voltage gains and input impedances. We can, of course, expect to find different values of coupling capacitors, and the bypass capacitors for the fully bypassed and split emitter circuits will also be different. The capacitor sizes are not our responsibility; they merely reflect the variations in input impedance and the size of the emitter resistance being bypassed. Frames 3 and 4 (Field 3) are concerned with voltage gain and input impedance, and here we must diverge into three separate sets of calculations.

The analysis sheet is set up to accommodate the voltage gain and input impedance for each of the three configurations on a single sheet. We can compare them easily. The *Note entries designated *a* will pertain to the bypassed configuration. Entries designated *b* will pertain to the fully bypassed configuration. Entries designated *c* will concern the split emitter configuration. We will follow the analysis sheet and explain the differences in the three configurations. There is also a branch shown on the analysis sheet for the differential amplifier. The differential amplifier is such an important circuit that we will devote an entire chapter to it. For now we will ignore that branch on the analysis sheet.

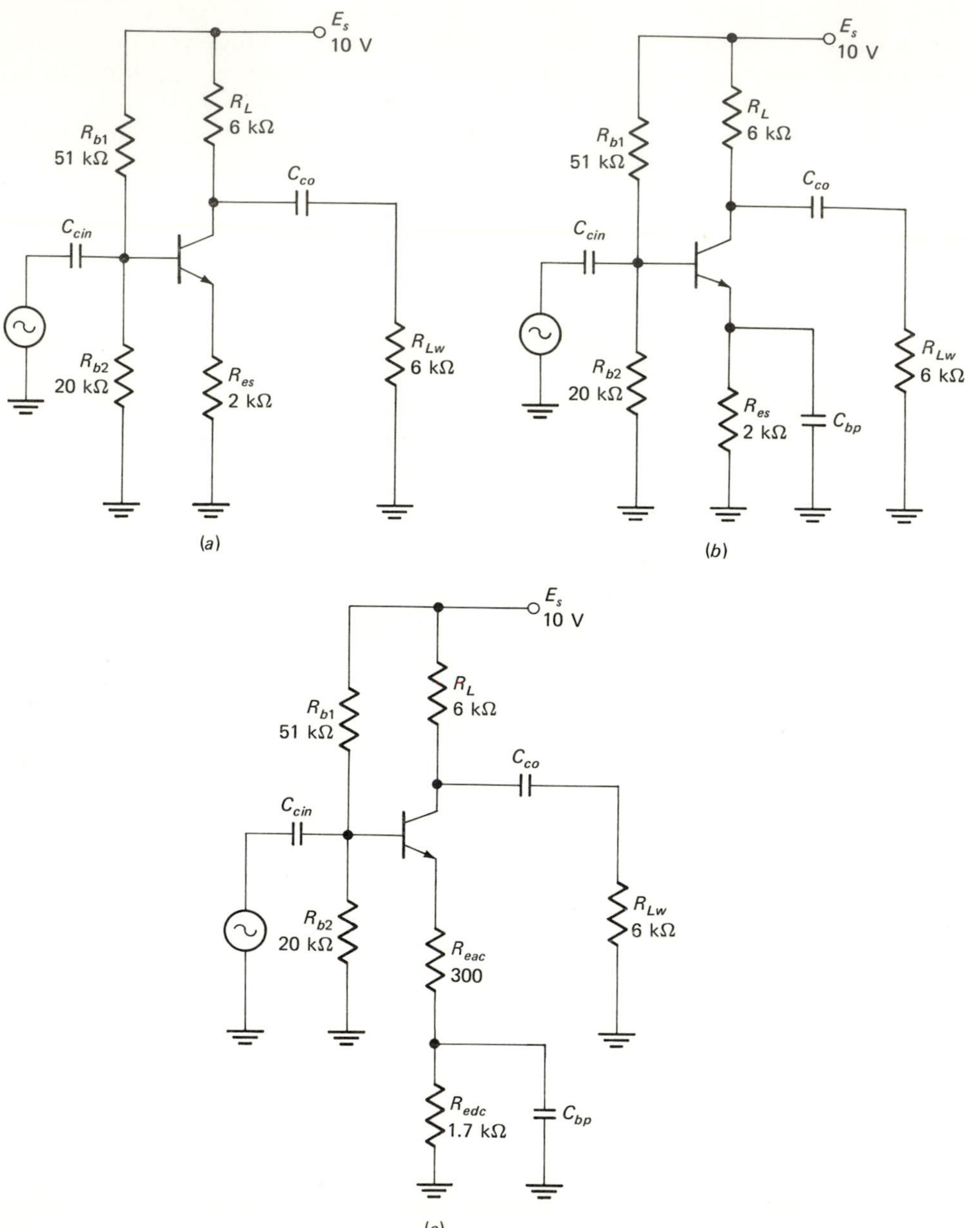

Fig. 8-10 The three variations of the base-biased common-emitter circuit for Example 8-2. (a) *The unbypassed circuit;* (b) *the fully bypassed circuit;* (c) *the split emitter circuit*

Example 8-2

FIELD 3 (SEE FIG. 8-10*a*, *b*, AND *c*)

FRAME 2

Compute the voltage gain:

a. The unbypassed configuration. In this case $R_e = R_{es} = 2\ \text{k}\Omega$.

$$V_g = \frac{R_{Lac}}{R_{es}}$$

$$V_g = \frac{3\ \text{k}\Omega}{2\ \text{k}\Omega}$$

$$V_g = 1.5$$

b. The fully bypassed configuration. In this case R_{es} is shorted at signal frequencies by the bypass capacitor; R_e is, in this case, only the transistor's internal emitter resistor R_e'.

$$R_e' \approx \frac{25}{QI_c} \qquad QI_c \text{ in mA; from Field 2, Frame 2, Fig. 8-1}$$

$$QI_c = 1\ \text{mA}$$

$$R_e' \approx \frac{25}{1} \approx 25\ \Omega \qquad V_g = \frac{3\ \text{k}\Omega}{25} = 120$$

c. The split emitter configuration. The unbypassed part (R_{eac}) of the emitter resistor is 300 Ω. We could add the 25 Ω of R_e' to this figure, but because the load is quite heavy, we will have to make a gain correction with the Young-Albertson curve; and the curve already includes the correction for the amount of R_e'.

$$V_g = \frac{R_{Lac}}{R_{eac}} = \frac{3\ \text{k}\Omega}{300} = 10$$

The uncorrected $V_g = 10$. Looking at the Young-Albertson curve (Fig. 6-5), we see that the computed gain must be reduced by 30 percent.

$$V_g = 7$$

FRAME 3

Compute the input impedance Z_{in}.

$$Z_{in} = R_{b1} \| R_{b2} \| \beta R_e$$

Since $R_{b1} \| R_{b2}$ is common to all three configurations, we can save ourselves a little work by solving that part first.

TABLE 8-2. SUMMARY OF CALCULATIONS FOR EXAMPLE 8-2

Parameter	Unbypassed Configuration	Fully Bypassed Configuration	Split Emitter Configuration
E_{Rb1}	7.4 V	7.4 V	7.4 V
E_{Rb2}	2.6 V	2.6 V	2.6 V
E_{Res}	2.0 V	2.0 V	2.0 V
QI_c	1 mA	1 mA	1 mA
E_{RL}	6 V	6 V	6 V
E_Q	2 V	2 V	2 V
E_o	3 V p-p	3 V p-p	3 V p-p
R_{Lac}	3 kΩ	3 kΩ	3 kΩ
V_g	1.5	120	7*
Z_{in}	14 kΩ	2.1 kΩ	9.8 kΩ

* Value corrected by Young-Albertson curve.

$R_{b1} = 51\ \text{k}\Omega$ and $R_{b2} = 20\ \text{k}\Omega$

$$R_{b1} \| R_{b2} = \frac{51 \times 20}{51 + 20} = 14.3\ \text{k}\Omega$$

a. The unbypassed configuration.
$Z_{in} = (R_{b1} \| R_{be}) \| \beta \times R_{es}$
$Z_{in} = 14.3\ \text{k}\Omega \| 100 \times 2\ \text{k}\Omega$
$Z_{in} = 14.3\ \text{k}\Omega \| 200\ \text{k}\Omega$
$Z_{in} \approx 14\ \text{k}\Omega$

b. The fully bypassed circuit.
$Z_{in} = (R_{b1} \| R_{b2}) \| \beta \times R'_e$
$Z_{in} = 14.3\ \text{k}\Omega \| 100 \times 25$
$Z_{in} \approx 2.1\ \text{k}\Omega$

c. The split emitter circuit.
$Z_{in} = (R_{b1} \| R_{be}) \| \beta \times R_{eac}$
$Z_{in} = 14.3\ \text{k}\Omega \| 100 \times 300$
$Z_{in} \approx 9.8\ \text{k}\Omega$

Table 8-2 summarizes the results of the computations for the three configurations shown in Fig. 8-10.

PROBLEMS

8-1 Analyze the base-biased common-emitter circuit:
Given the circuit in Fig. 8-11, compute the following parameters:

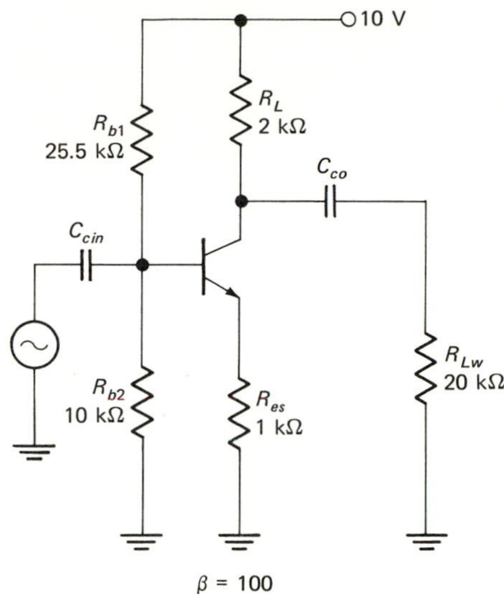

Fig. 8-11 Schematic for analysis Problem 8-1

a. E_{Rb1}
b. E_{Res}
c. E_{RL}
d. E_Q
e. QI_c
f. Z_{in}
g. V_g
h. E_o p-p
i. Voltage from collector to ground
j. Voltage from base to ground

Add a bypass capacitor to Fig. 8-11 and compute the following parameters:

a. Voltage gain
b. E_o p-p
c. Input impedance

Add an 82-Ω resistor to R_{es} and bypass the 1-kΩ resistor (R_{es}) and compute the following:

a. E_o p-p
b. Voltage gain
c. Input impedance

8-3 ANALYZING THE EMITTER-BIASED CIRCUIT (AN EXAMPLE)

The emitter-biased common-emitter circuit normally appears in two forms: the fully bypassed and the differential amplifier circuit. The unbypassed variation is not generally practical because of its inherent low voltage gain, of

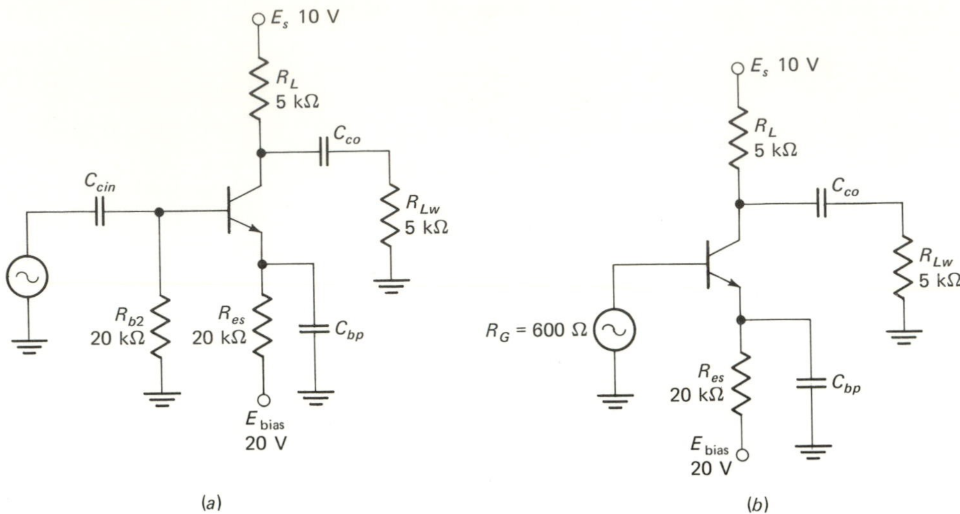

Fig. 8-12 The emitter-biased common-emitter circuit for Example 8-3. (a) *With capacitor-coupled generator;* (b) *the direct-coupled generator*

unity or less. The differential amplifier will be treated in a chapter devoted to it alone. The split emitter version is rare but fairly practical. I will not give you an example of the split emitter circuit analysis, but if you should ever run across one to analyze, you can follow Field 2 on the current-mode analysis sheet exactly, and use the split emitter option for Field 3 on the analysis sheet. Now let us try our hand at analyzing an emitter-biased fully bypassed circuit. See Fig. 8-12.

Field 1 is for base bias circuits *only*. For emitter bias circuits we simply skip all of Field 1.

Example 8-3

FIELD 2 COLLECTOR SIDE CALCULATIONS

FRAME 1

Compute QI_c.

$$QI_c = \frac{E_{bias}}{R_{es}}$$

This equation is a good approximation for all adequate stability factors, where bias voltages of 10 V or more are used. If a bias voltage of less than 10 V is used, the following equation may be used as a closer approximation:

$$QI_c = \frac{E_{bias} - (V_j + I_b R_{b2})}{R_{es}}$$

This equation allows for both the junction voltage drop and the voltage drop across R_{b2}.

From a measurement standpoint the voltage from emitter to ground will be approximately 1 V, the base-emitter junction voltage of 0.6 V (for silicon) plus the voltage drop across R_{b2}, where $E_{Rb2} = I_b R_{b2}$. For nearly all cases the simple form $QI_c = E_{\text{bias}}/R_{es}$ is satisfactory.

$$QI_c = \frac{E_{\text{bias}}}{R_{es}}$$

$$QI_c = \frac{20\ \text{V}}{20\ \text{k}\Omega}$$

$$QI_c = 1\ \text{mA}$$

FRAME 2

Compute R_e', the junction resistance.

This is our old friend the Shockley relationship.

$$R_e = \frac{25}{QI_c}$$

$$R_e = \frac{25}{1\ \text{mA}} = 25\ \Omega$$

FRAME 3

Compute the voltage drop across the collector load resistor.

This is simply Ohm's law.

$$E_{RL} = QI_c \times R_L$$
$$E_{RL} = 1\ \text{mA} \times 5\ \text{k}\Omega = 5\ \text{V}$$

FRAME 4

Compute the value of the ac load R_{Lac}.

$$R_{Lac} = \frac{R_L \times R_{LW}}{R_L + R_{LW}}$$

$$R_{Lac} = \frac{5\ \text{k}\Omega \times 5\ \text{k}\Omega}{5\ \text{k}\Omega + 5\ \text{k}\Omega}$$

$$R_{Lac} = 2.5\ \text{k}\Omega$$

FRAME 5

Compute E_Q.

In the emitter-biased circuits the supply voltage E_s does not contribute to the voltage drop across R_{es}. E_s is distributed between E_{RL} and E_Q. E_Q is simply the difference between E_s and E_{RL}.

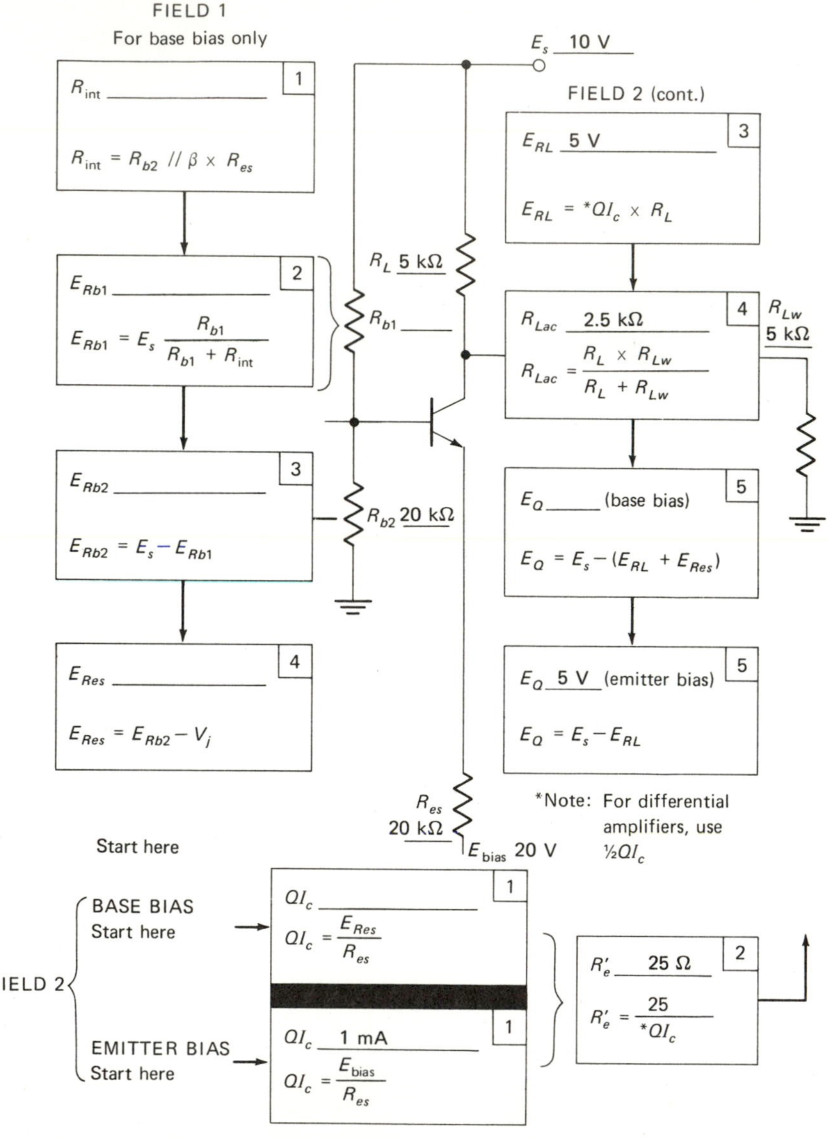

Fig. 8-13 *Field 2 analysis sheet for Example 8-3*

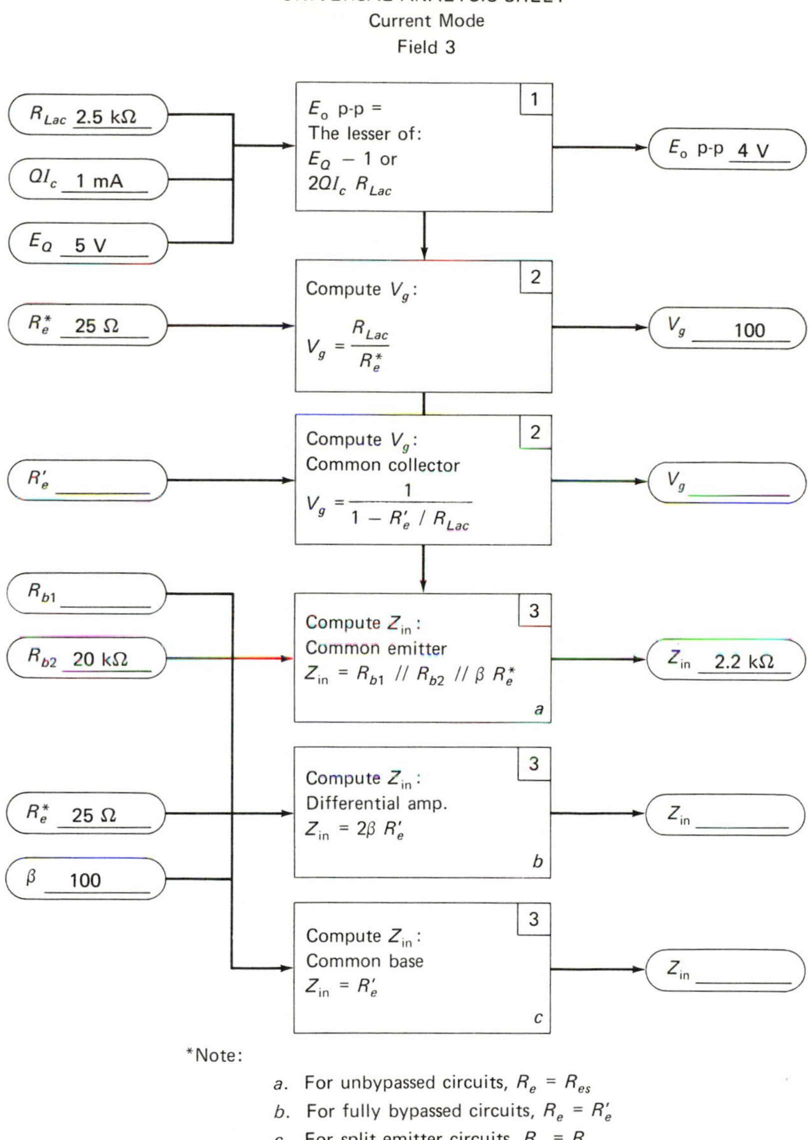

*Note:

a. For unbypassed circuits, $R_e = R_{es}$
b. For fully bypassed circuits, $R_e = R_e'$
c. For split emitter circuits, $R_e = R_{eac}$
d. For differential amplifiers, $R_e = 2.5\ R_e'$
e. For common collector circuits, $R_e = R_{es} \parallel R_{Lw}$

Fig. 8-14 *Field 3 analysis sheet for Example 8-3*

$E_Q = E_s - E_{RL}$
$E_Q = 10\text{ V} - 5\text{ V} = 5\text{ V}$

Figure 8-13 shows the completed analysis sheet for Field 2.

FIELD 3 SIGNAL PARAMETERS

FRAME 1

Compute the maximum signal output voltage.
E_o p-p $= 2QI_cR_{Lac}$ or $E_Q - 1$ whichever is less

$2QI_cR_{Lac} = 5\text{ V}$
$E_Q - 1 = 4\text{ V}$
E_o p-p $= 4\text{ V}$

FRAME 2

Compute the voltage gain.
Since this circuit is fully bypassed, we will use R_e'.

$$V_g = \frac{R_{Lac}}{R_e'}$$

$$V_g = \frac{2.5\text{ k}\Omega}{25} = 100$$

FRAME 3

Compute the input impedance Z_{in}.

$Z_{in} = R_{b2} \| \beta \times R_e'$

The circuit has no R_{b1}.

$Z_{in} = 20\text{ k}\Omega \| 100 \times 25$
$Z_{in} \approx 2.2\text{ k}\Omega$

Figure 8-14 shows the completed analysis sheet (Field 3) for Example 8-3.

PROBLEMS

8-2 Analyze the circuit in Fig. 8-15 for the following parameters:

a. E_{Rb2} b. E_{Res}
c. E_{RL} d. E_Q
e. QI_c f. Z_{in}
g. V_g h. E_o p-p

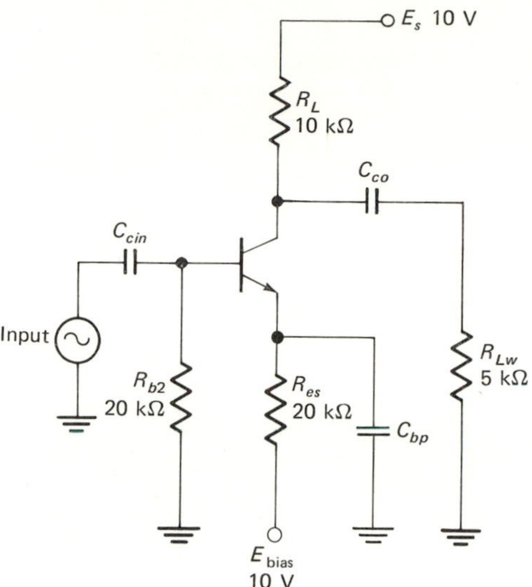

Fig. 8-15 The circuit for Problem 8-2

8-4 ANALYZING THE VOLTAGE-MODE COMMON-EMITTER CIRCUIT (AN EXAMPLE)

Figure 8-16 shows a voltage-mode analysis sheet.

Example 8-4

Given the circuit in Fig. 8-17, compute the following circuit parameters:

QI_c, E_{RL}, E_Q, E_o p-p, V_g, and Z_{in}.

Figure 8-18 shows the completed voltage-mode analysis sheet.

FIELD 1

FRAME 1

Compute I_{Rb2}, the current through R_{b2}.

$$I_{Rb2} = \frac{V_j}{R_{b2}}$$

This is simply Ohm's law. We know the resistance of R_{b2}, and we know that, because it is in parallel with the base-emitter junction, the voltage across it is the junction voltage.

Since the device is silicon, $V_j = 0.6$ V.

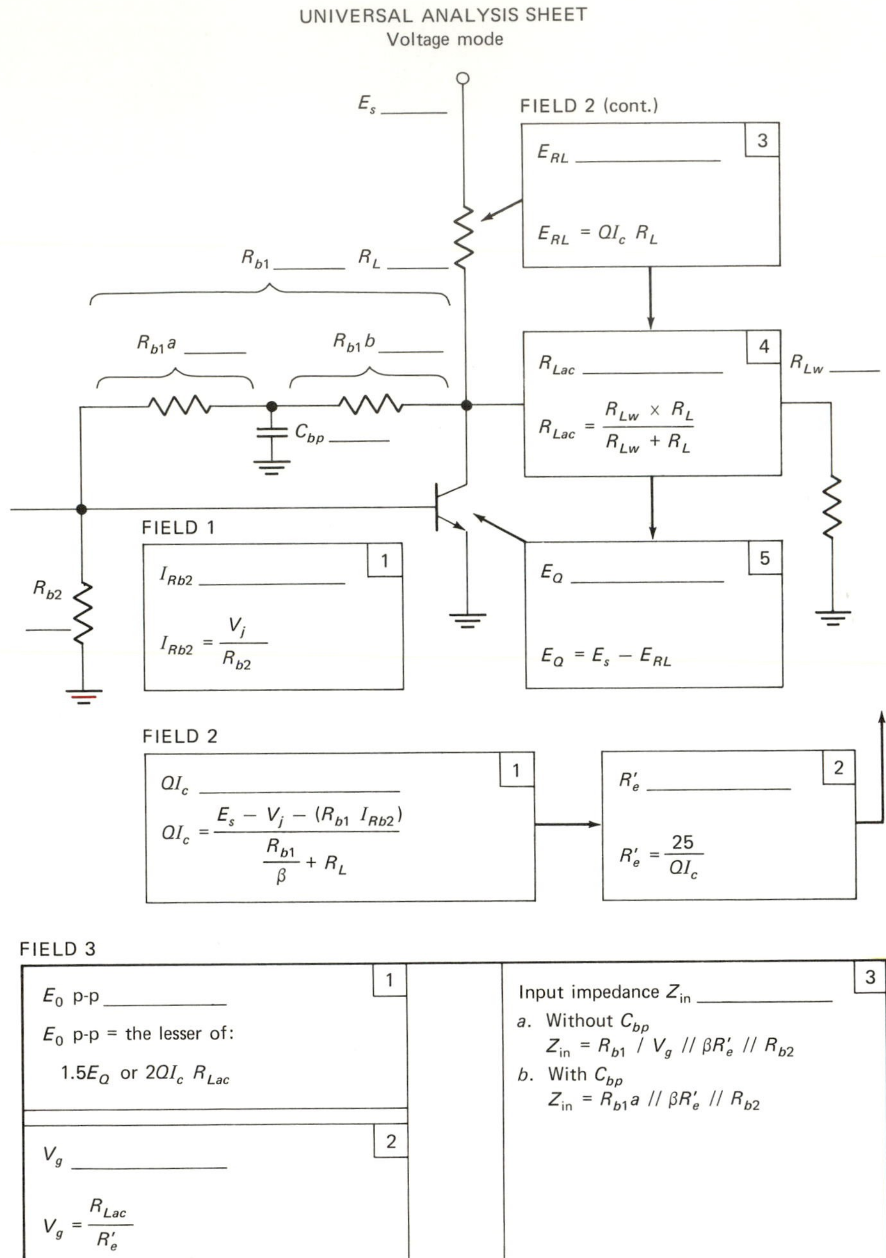

Fig. 8-16 The voltage-mode universal analysis sheet

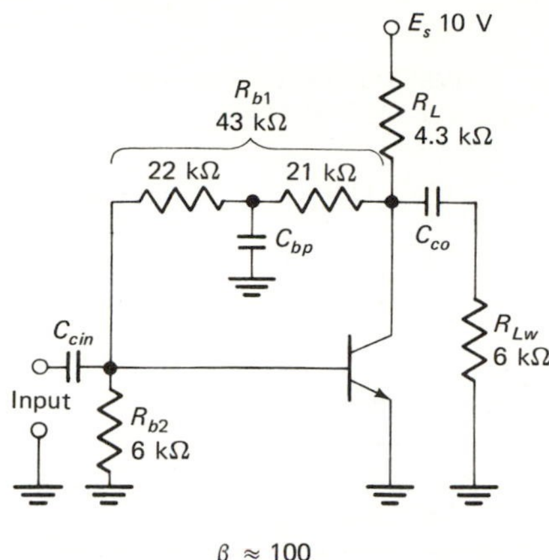

Fig. 8-17 Analyzing the voltage-mode feedback circuit (Example 8-4)

$$I_{Rb2} = \frac{0.6\ \text{V}}{6\ \text{k}\Omega} = \frac{6 \times 10^{-1}}{6 \times 10^{3}} = 1 \times 10^{-4}\ \text{A}.$$

NOTE: There is only one frame in this field.

FIELD 2

FRAME 1

Compute the value of the quiescent collector current QI_c.

$$QI_c = \frac{E_s - V_j - (R_{b1}I_{Rb2})}{R_{b1}/\beta + R_L}$$

This monster needs a bit of explanation. It is actually nothing more than the loop equation through R_L, R_{b1}, R_{b2}, etc., solved for QI_c. I will set up the loop equation for you so that you can see where it came from.

The Kirchhoff loop analysis looks like this:

$$E_s - QI_cR_L - I_bR_{b1} - (I_{Rb2}R_{b1}) - V_j = O$$

The term in parentheses represents the voltage drop across R_{b1} due to the current through R_{b2}. For circuits which have no R_{b2}, the loop equation would be the same except that this term (in parentheses) would be left out.

If we rearrange the loop equation, we get

$$I_bR_{b1} + (I_{Rb2}R_{b1}) + V_j = E_s - QI_cR_L$$

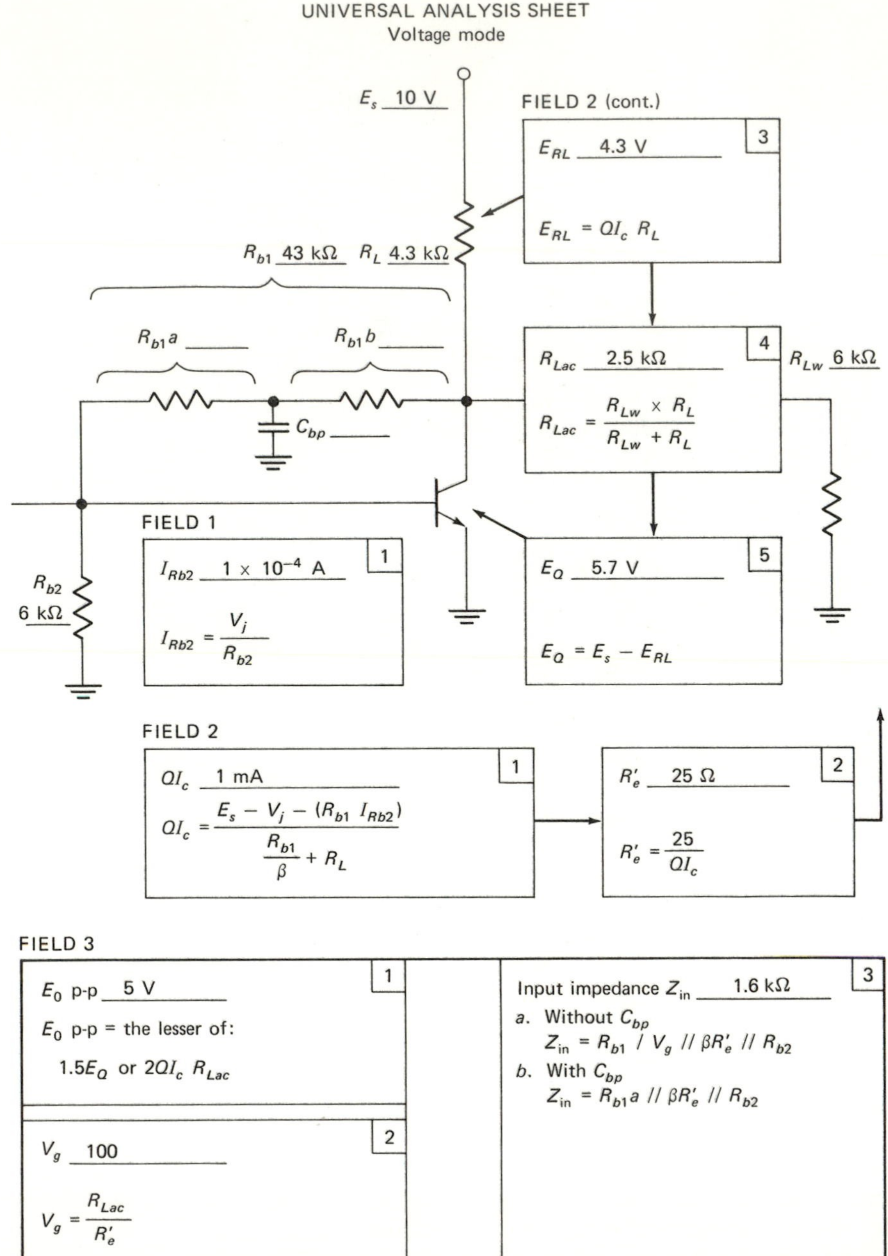

Fig. 8-18 The completed voltage-mode analysis sheet for Example 8-4

Because $I_b = QI_c/\beta$, let us substitute QI_c/β for I_b:

$$\frac{QI_c}{\beta} R_{b1} + (I_{Rb2}R_{b1}) = V_j = E_s - QI_cR_L$$

Solving for QI_c, we get

$$QI_c = \frac{E_s - V_j - (R_{b1}I_{Rb2})}{R_{b1}/\beta + R_L}$$

Notice that the term $(R_{b1}I_{Rb2})$ appears in the final equation. If we are analyzing a case 1 circuit, which has no R_{b2}, I_{Rb2} will equal zero and this term drops out. Now back to the example:

$$QI_c = \frac{E_s - V_j - (R_{b1}I_{Rb2})}{R_{b1}/\beta + R_L}$$

$$QI_c = \frac{10 - 0.6 - (4.3 \times 10^4 \times 1 \times 10^{-4})}{43\ \text{k}\Omega/100 + 4.3\ \text{k}\Omega}$$

$$QI_c = \frac{5.1}{4.73 \times 10^3}$$

$$QI_c \approx 1\ \text{mA}$$

FRAME 2

Compute R_e.

$$R_e = \frac{25}{QI_c} \qquad \text{(the Shockley relationship)}$$

$$R_e = \frac{25}{1} = 25\ \Omega$$

FRAME 3

Compute E_{RL}, the voltage drop across the collector load resistor.

$$E_{RL} = QI_cR_L \qquad \text{(Ohm's law)}$$
$$E_{RL} = 1\ \text{mA} \times 4.3\ \text{k}\Omega = 4.3\ \text{V}$$

FRAME 4

Compute the value of the ac load R_{Lac}.

$$R_{Lac} = R_L \| R_{LW}$$

$$R_{Lac} = \frac{R_{LW} \times R_L}{R_{LW} + R_L} \qquad \text{(two resistors in parallel)}$$

$$R_{Lac} = \frac{6\ \text{k}\Omega \times 4.3\ \text{k}\Omega}{6\ \text{k}\Omega + 4.3\ \text{k}\Omega}$$

$$R_{Lac} \approx 2.5\ \text{k}\Omega$$

FRAME 5

Compute the transistor collector-to-emitter voltage drop E_Q.

$E_Q = E_s - E_{RL}$ simple subtraction
$E_Q = 10 - 4.3$ V
$E_Q = 5.7$ V

FIELD 3

FRAME 1

Compute the maximum peak-to-peak output voltage E_o p-p

E_o p-p $= 1.5E_Q$ or $2QI_cR_{Lac}$, whichever is less
$1.5\ E_Q = 1.5 \times 5.7 = 8.55$ V
$2QI_cR_{Lac} = 2 \times 1 \times 10^{-3} \times 2.5 \times 10^3 = 5$ V
E_o p-p $= 5$ V

FRAME 2

Compute the voltage gain V_g.

$$V_g = \frac{R_{Lac}}{R_e}$$

$$V_g = \frac{2.5\ \text{k}\Omega}{25} = 100$$

FRAME 3

Compute the input impedance Z_{in}.

Because R_{b1} is split and bypassed, the Miller effect does not enter into the equation and we use the following equation:

$Z_{in} = R_{b1}a \| \beta R_e' \| R_{b2}$
$Z_{in} = 22\ \text{k}\Omega \| 2.5\ \text{k}\Omega \| 6\ \text{k}\Omega$
$Z_{in} \approx 1.6\ \text{k}\Omega$

8-5 ANALYZING THE PARAPHASE AMPLIFIER (SEE FIG. 8-19)

I won't give you an example of a paraphase analysis because it is exactly the same procedure as that used for ordinary common-emitter circuits, with the exception of Field 3, and even here the differences are fairly minor. I shall summarize them.

FIELD 3

FRAME 1

Compute E_o p-p.

This frame is OK as it stands.

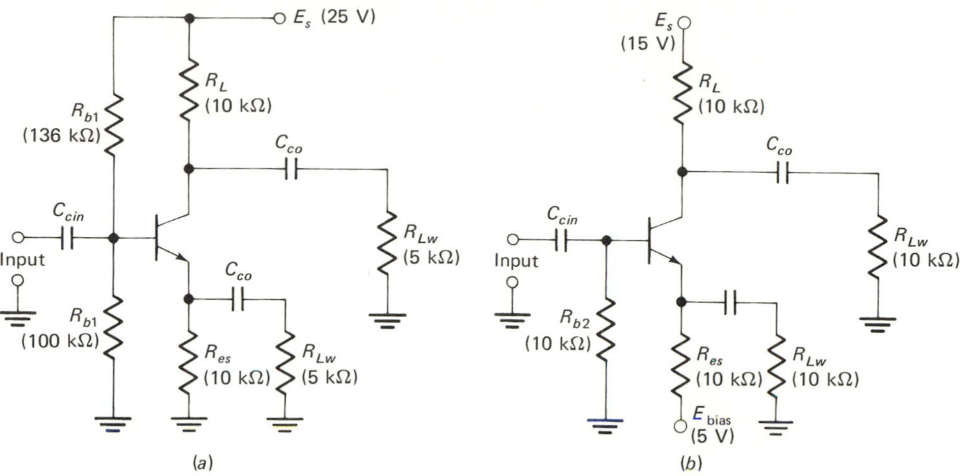

Fig. 8-19 The paraphase amplifier circuit. (a) *The base-biased version;* (b) *the emitter-biased version*

FRAME 2

Compute V_g.

Here

$$V_g \approx \frac{R_L \| R_{Lw}a}{R_{es} \| R_{Lw}b} \approx 1$$

where $R_L = R_{es}$ and $R_{Lw}a = R_{Lw}b$

If the paraphase amplifier is base-biased, follow the procedure for base-biased common-emitter circuits up to Field 3. If it is emitter-biased, follow the common-emitter emitter-biased procedure up to Field 3.

FRAME 3

Compute Z_{in}.

In this case, from a signal standpoint, the emitter resistor (R_{es}) has another resistor (R_{Lw}) in parallel with it. This alters the input impedance.

1. *For base bias circuits:*

$$Z_{in} \approx R_{b1} \| R_{b2} \| \beta(R_{es} \| R_{Lw})$$

2. *For emitter bias circuits with a capacity-coupled input and an* R_{b2}:

$$Z_{in} \approx R_{b2} \| \beta(R_{es} \| R_{Lw})$$

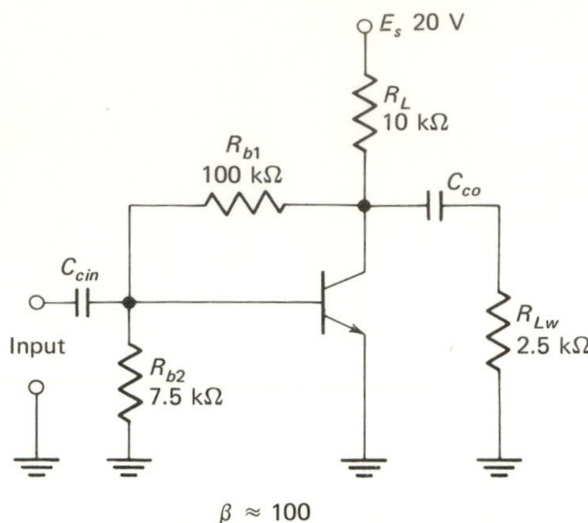

$\beta \approx 100$

Fig. 8-20 The circuit for Problem 8-4

3. *For emitter bias circuits with a direct-coupled generator and no* R_{b2}:

$$Z_{in} \approx \beta(R_{es} \| R_{Lw})$$

Notice that these input impedance equations are identical to the input impedance equations for the emitter-follower (common-collector) circuit.

It is as simple as that.

PROBLEMS

8-3 Analyze the circuit shown in Fig. 8-19*a*. Values are shown in the parentheses on the drawing.

8-4 Analyzing the voltage-mode feedback circuit, compute the following parameters:

a. E_{Rb1} — *b.* E_{RL}
c. E_Q — *d.* QI_c
e. Z_{in} — *f.* V_g
g. If R_{b1} is split into two equal parts and the junction returned to ground through a large capacitor, compute the new input impedance. See Fig. 8-20.

8-6 ANALYZING THE COMMON-COLLECTOR CIRCUIT WITH BASE BIAS

The common-collector circuit can be analyzed as though it were a common-emitter circuit with some minor differences. The main difference is that there is no collector load resistor. The voltage drop across the collector load resistor is 0 V because there is *no* collector load resistor.

The voltage gain equation that we used for the common-emitter circuit

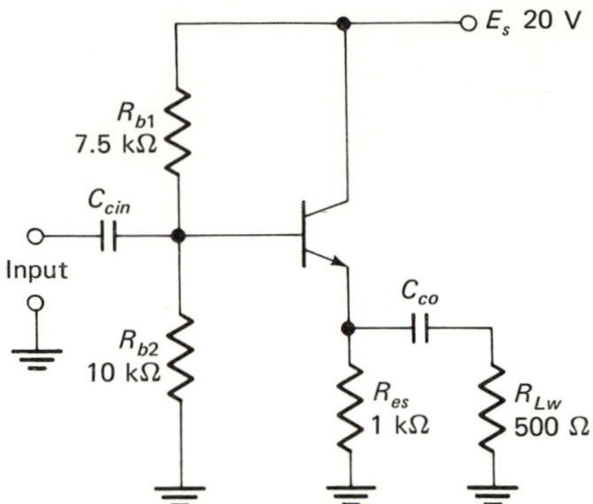

Fig. 8-21 The base-biased common-collector circuit for Example 8-5

is not valid for the common-collector circuit, and I will give you a new voltage gain equation for it. The voltage gain of a common-collector circuit is approximately unity, generally a little less, and is dependent upon R_{Lac} and R'_e.

The input impedance equation must also be modified to account for the fact that, for ac, R_{Lw} is in parallel with R_{es}.

These changes apply to both base-biased and emitter-biased emitter-follower circuits.

Example 8-5

Given Fig. 8-21, compute the following circuit parameters:

E_{Res}, QI_c, E_Q, E_o p-p, V_g, and Z_{in} (see also Fig. 8-22).

FIELD 1 (UNIVERSAL ANALYSIS SHEET, CURRENT MODE; SEE FIG. 8-22)

FRAME 1

Compute the partial input resistance R_{int}. (Assume a β of 100.)

$$R_{int} = R_{b2} \| \beta R_{es}$$
$$R_{int} = 10 \text{ k}\Omega \| 100 \times 1 \text{ k}\Omega$$
$$R_{int} \approx 9.1 \text{ k}\Omega$$

At this point we are dealing with dc parameters and R_{Lw} is isolated for dc by a coupling capacitor; so it is not included in this calculation.

FRAME 2

Compute the voltage drop across R_{b1}, E_{Rb1}.

$$E_{Rb1} = E_s\left(\frac{R_{b1}}{R_{b1} + R_{int}}\right) \qquad \text{(the voltage divider equation)}$$

$$E_{Rb1} = 20\left(\frac{7.5\ \text{k}\Omega}{7.5\ \text{k}\Omega + 9.1\ \text{k}\Omega}\right)$$

$$E_{Rb1} = 9\ \text{V}$$

FRAME 3

Find E_{Rb2}, the voltage drop across R_{b2}.

$$E_{Rb2} = E_s - E_{Rb1}$$
$$E_{Rb2} = 20 - 9 = 11\ \text{V}$$

FRAME 4

Compute E_{Res}, the voltage drop across R_{es}.

$$E_{Res} = E_{Rb2} - V_j \qquad \text{simple subtraction}$$
$$E_{Res} = 11 - 0.6 = 10.4\ \text{V}$$

FIELD 2

FRAME 1

Compute the quiescent collector current QI_c.

$$QI_c = \frac{E_{Res}}{R_{es}} = \frac{10.4\ \text{V}}{1\ \text{k}\Omega}$$

$$QI_c = 10.4\ \text{mA}$$

FRAME 2

Compute the base-emitter junction resistance R_e'.

$$R_e' = \frac{25}{QI_c}$$

$$R_e' = \frac{25}{10.4} = 2.4\ \Omega$$

FRAME 3

Compute the voltage drop across the collector load resistor.

There is no collector load resistor, and so the voltage drop is 0 V.

$$E_{RL} = 0\ \text{V}$$

FRAME 4

Calculate the effective ac load resistance R_{Lac}.

$$R_{Lac} = \frac{R_L \times R_{Lw}}{R_L + R_{Lw}} \qquad \text{or} \qquad \frac{R_{es} \times R_{Lw}}{R_{es} + R_{Lw}}$$

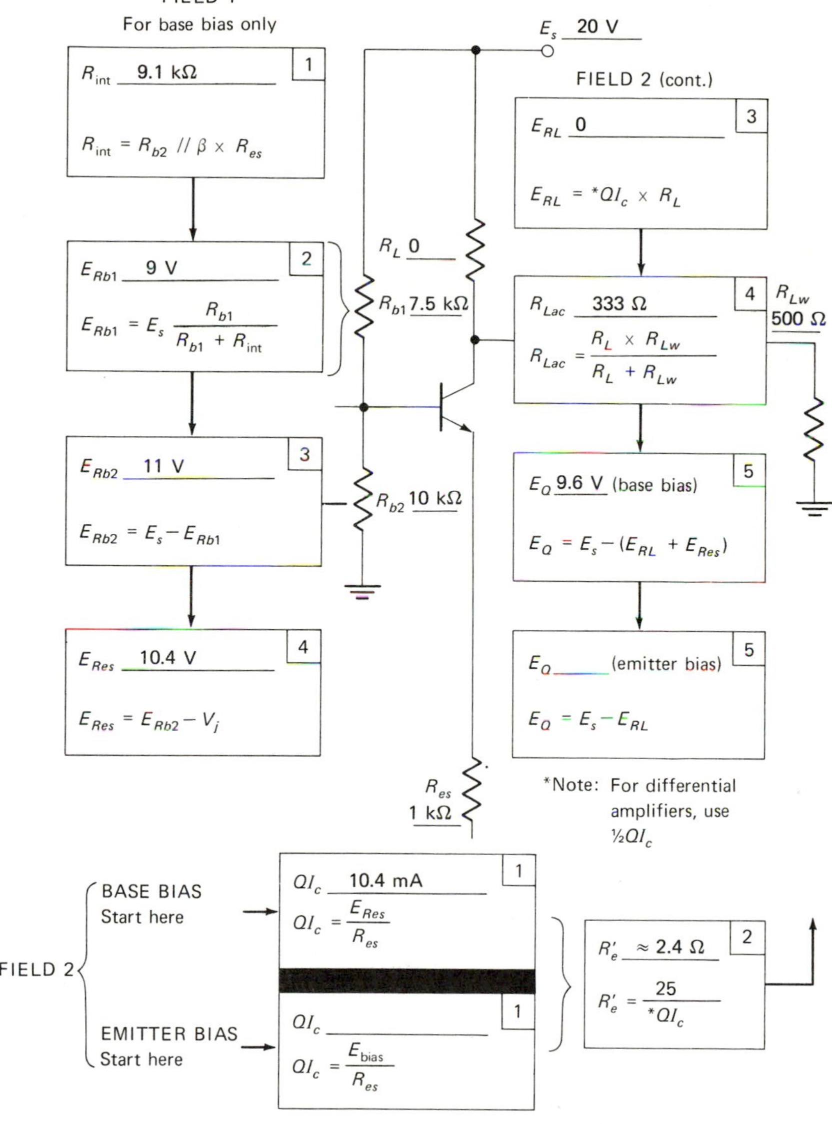

Fig. 8-22 The completed analysis sheet for Example 8-5 (Fields 1 and 2)

The load is now in the emitter and is composed of R_{es} and R_{Lw} in parallel. For the sake of continuity we can call R_{es} by two names: R_{es} and R_L. This should cause no confusion since there is no *collector* load resistor.

$$R_{Lac} = \frac{1\ \text{k}\Omega \times 500}{1\ \text{k}\Omega + 500}$$

$$R_{Lac} = 333\ \Omega$$

FRAME 5

Compute the transistor collector-emitter voltage drop E_Q.

$$E_Q = E_s - (0 + 10.4)$$
$$E_Q = 20 - (0 + 10.4)$$
$$E_Q = 9.6\ \text{V}$$

Figure 8-22 shows the completed analysis sheet for Fields 1 and 2.

FIELD 3 SIGNAL PARAMETERS

FRAME 1

Compute the maximum output voltage E_o p-p.

E_o p-p $= E_Q - 1$ or $2QI_cR_{Lac}$ whichever is less

$$E_Q - 1 = 9.6 - 1 = 8.6\ \text{V}$$
$$2QI_cR_{Lac} = 2 \times 10.4 \times 10^{-3} \times 3.33 \times 10^2 = 6.9\ \text{V}$$
$$E_o\ \text{p-p} = 6.9\ \text{V}$$

FRAME 2

Compute the voltage gain V_g.

$$V_g = \frac{1}{1 - (R_e'/R_{Lac})} = \frac{1}{1 - (2.4/2{,}500\ \text{k}\Omega)}$$

$$V_g \approx 1$$

V_g for this configuration will be very nearly equal to 1 except when R_{Lac} approaches R_e. Anytime R_{Lac} is 10 times (or more) as large as R_e, it is safe to assume that $V_g = 1$.

FRAME 3

Compute the input impedance Z_{in}.

$Z_{in} = R_{b1} \| R_{b2} \| \beta R_e$ where $R_e \approx R_{Lac}$

$$Z_{in} = 7.5\ \text{k}\Omega \| 10\ \text{k}\Omega \| 100 \times 333$$
$$Z_{in} = 7.5\ \text{k}\Omega \| 10\ \text{k}\Omega \| 33\ \text{k}\Omega$$
$$Z_{in} \approx 3.8\ \text{k}\Omega$$

Figure 8-23 shows the completed analysis sheet for Field 3.

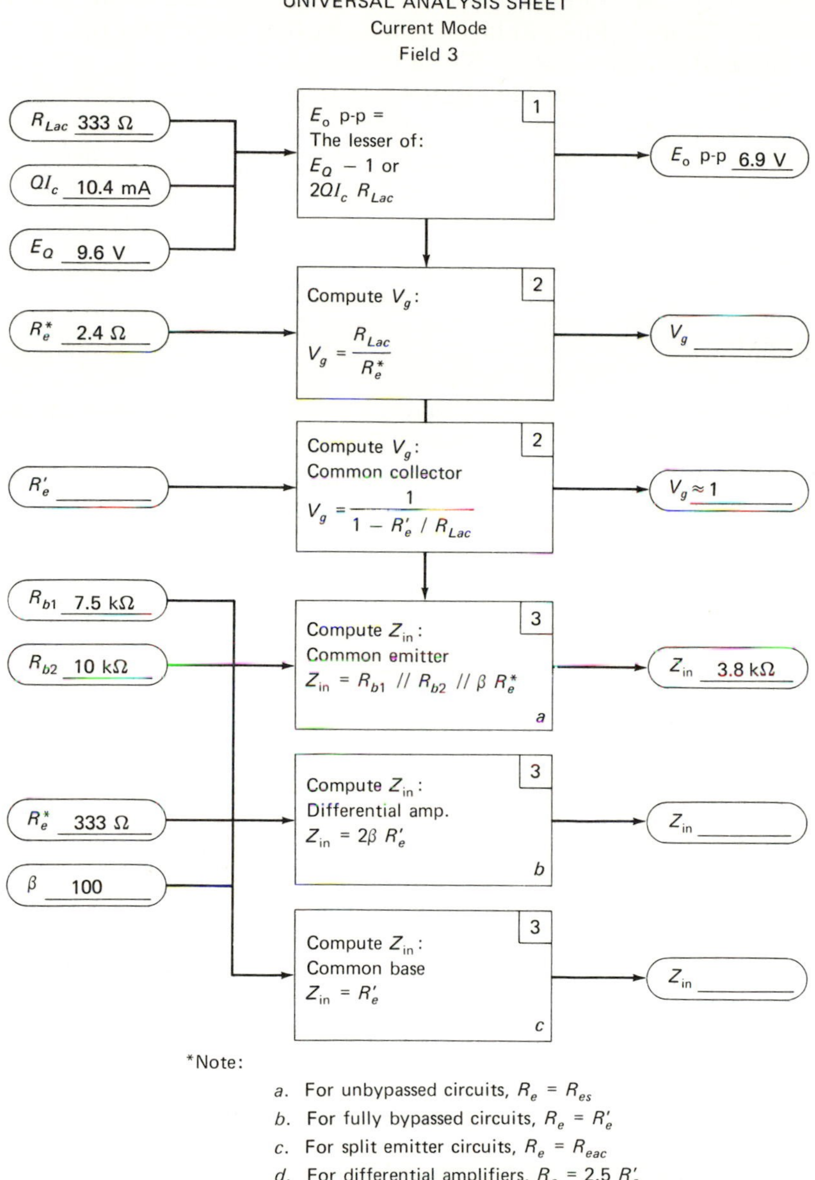

*Note:

a. For unbypassed circuits, $R_e = R_{es}$

b. For fully bypassed circuits, $R_e = R'_e$

c. For split emitter circuits, $R_e = R_{eac}$

d. For differential amplifiers, $R_e = 2.5\ R'_e$

e. For common collector circuits, $R_e = R_{es} \parallel R_{Lw}$

Fig. 8-23 The Field 3 analysis sheet for Example 8-5

What to Do about Emitter-biased Common-collector Circuits

To analyze emitter-biased circuits, skip Field 1, take the emitter bias branch to solve for QI_c, and proceed to Frame 5 just as we did in the previous example. At Frame 5 (Field 2) take the emitter bias branch for E_Q.

In Field 3 proceed as we did in the previous example up to Frame 3. If the circuit is capacitor-coupled to the generator, there will be no R_{b1}, and so R_{b1} can be dropped from the equation. Use the first (common-emitter) branch, but omit R_{b1} from the equation. If the generator is direct-coupled, both R_{b1} and R_{b2} will not exist, and both will be dropped from the input impedance equation. The *generator* will see an impedance of $\beta(R_{es}\|R_{Lw})$. See Fig. 8-25.

PROBLEMS

8-5 Given the circuit in Fig. 8-24, analyze the circuit for the following parameters.

a. E_{Rb1}
b. E_{Res}
c. E_{RL}
d. E_Q
e. QI_c
f. Z_{in}

8-6 Analyze the circuit in Fig. 8-25. The values are shown in parentheses on the drawing.

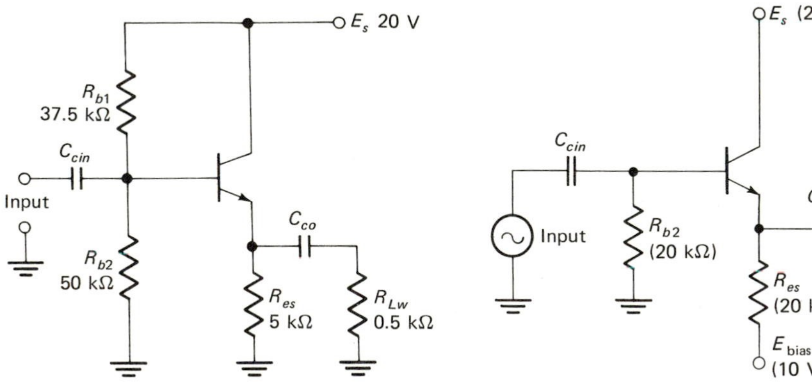

Fig. 8-24 The circuit for Problem 8-5

Fig. 8-25 The emitter-biased common-collector circuit

8-7 ANALYZING THE COMMON-BASE CIRCUIT

The common-base circuit generally takes one of two practical circuit forms, both of which utilize current-mode feedback for bias stabilization. The circuit shown in Fig. 8-26 is actually a base-biased common-emitter circuit for dc.

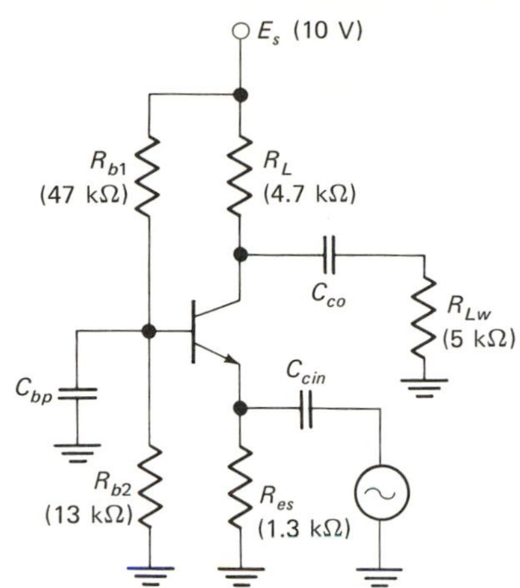

Fig. 8-26 The base-biased common-base circuit

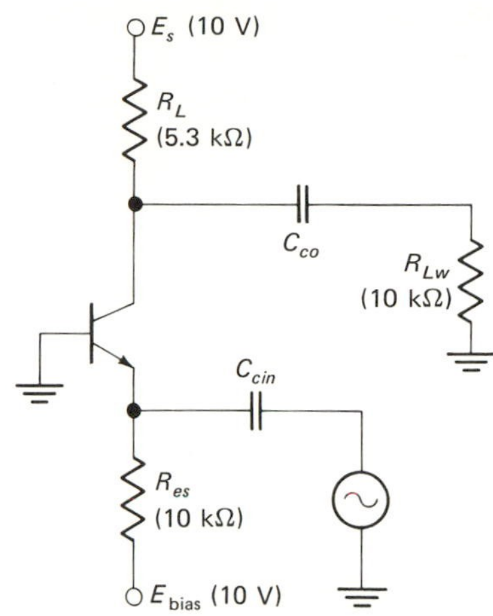

Fig. 8-27 The emitter-biased common-base circuit

The capacitor C_{bp} places the base at ground potential for the ac signal, and the input signal is applied between emitter and ground, making the circuit common-base for signal frequencies.

The analysis of this configuration is exactly the same as the analysis procedure used for the common-emitter base-biased current-mode circuit, with a single exception: the input impedance. In nearly all practical versions of this configuration $Z_{in} \approx R'_e$.

The second practical common-base circuit is emitter-biased and is common-base for both signal and bias. The circuit is shown in Fig. 8-27. The analysis of this circuit is exactly the same as the analysis of an emitter-biased common-emitter circuit, again with the sole exception of the input impedance, which is $Z_{in} \approx R'_e$.

The current gain is not included in the analysis process because, for all configurations, I_g, the current gain, is approximately equal to **S**, the stability factor. However, in the hybrid arrangement, where the bias configuration is not common-base, a small complication arises (see Fig. 8-26).

The impedance from base to ground for dc is approximately the value of R_{b2}, but for ac the impedance to ground, because of the bypass capacitor, is near 0 Ω. The result is two different stability factors and two different current gains, one for dc and one for signal. We have defined **S** as the ratio R_{b2}/R_{es}, which is still OK for the dc value, but for ac it becomes

$$\mathbf{S} \approx \frac{\text{impedance, base-ground}}{R_{es}}$$

However, if we use this equation, we are going to find ourselves in an arithmetical bind that yields an **S** of zero and a corresponding current gain of zero, which is just not true. In fact, the current gain of the common-base circuit is always approximately equal to unity. You will remember that back in Chap. 4 I said that **S** ranged roughly from R_b/R_{es} to $1 + R_b/R_{es}$ along that part of the curve in which we were interested. We selected R_b/R_{es} to use for convenience. Here, however, to keep us out of trouble, we have to use $1 + R_b/R_{es}$. Remember I_g common-base ≈ 1.

PROBLEMS

8-7 Analyze the circuit in Fig. 8-26. Values are shown in parentheses on the drawing. Assume a β of 100.

8-8 Analyze the circuit in Fig. 8-27. Values are shown in parentheses on the drawing. Assume a β of 100.

8-8 HOW TO ANALYZE A CIRCUIT WHERE BOTH VOLTAGE-MODE AND CURRENT-MODE FEEDBACK ARE USED

Occasionally a base-biased circuit, common-emitter, common-collector, or common-base, will have a resistor in the emitter circuit and return R_{b1} to the collector. Both current-mode and voltage-mode feedback are used at the same time. See Fig. 8-28.

To analyze this hybrid, treat it as a *voltage-mode* circuit. However, the input impedance equation will have to be modified and R_{es} will have to be included in the QI_c equation.

The original equation for QI_c on the analysis sheet (Fig. 8-16) (Field 2, Frame 1) is

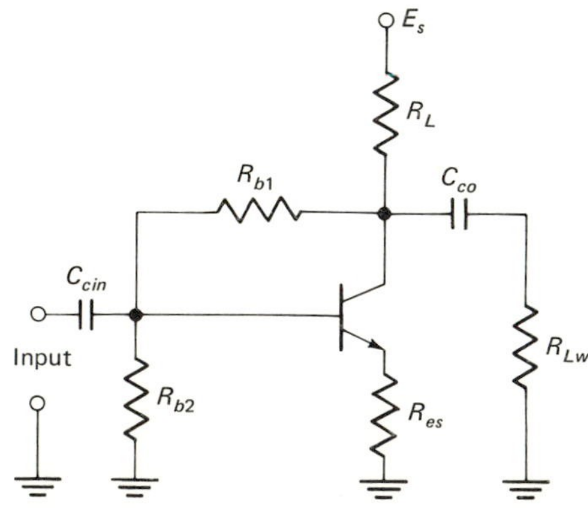

Fig. 8-28 The current-mode and voltage-mode hybrid

$$QI_c = \frac{E_s - V_j - (R_{b1}I_{Rb2})}{R_{b1}/\beta + R_L}$$

We can alter this equation to include R_{es} by simply adding R_{es} to the denominator:

$$QI_c = \frac{E_s - V_j - (R_{b1}I_{Rb2})}{R_{b1}/\beta + R_L + R_{es}}$$

You can verify this if you write the loop equation, including R_{es}, and solve for QI_c.

The input impedance equation modification is also a simple one. There are two input impedance equations on the voltage-mode design sheet Field 3, Frame 3). Both contain the term $\beta R_e'$. For the hybrid circuit, R_{es} must be included (if it is unbypassed) and the term becomes $\beta(R_e' + R_{es})$.

If R_{es} is large enough with respect to R_e', R_e' can be dropped from the equation. However, you can't depend upon that always being the case in this hybrid circuit. If the emitter resistor is bypassed, the input impedance equations are correct as they stand on the analysis sheet.

PROBLEMS

8-9 Make a table titled "A Summary of Differences in Current-mode Analysis Procedures."

9

ac-coupled multistage circuits

OBJECTIVES:

(Things you should be able to do upon completion of this chapter)
NOTE: Specific design problems are included in the chapter.

1. Design a multistage iterative *R-C*-coupled amplifier.
2. Design a multistage split emitter amplifier circuit.
3. Design a two-stage *R-C*-coupled circuit with two-stage current-mode feedback.
4. Design a multistage (two or more stages) audio-frequency (untuned) transformer-coupled amplifier.
5. Design a tuned RF transformer-coupled multistage (two or more stages) amplifier.

EVALUATION:

1. Self-evaluation will be based on your ability to design the circuits in the *problem* sections of the chapter.
2. The final evaluation will be based upon your ability to design *functional* amplifiers according to the specifications and criteria in the laboratory manual.

INTRODUCTION

One of the most popular methods of coupling transistor amplifier stages is capacitor coupling. One of the chief advantages of this kind of coupling is that the driven stage loads the driving stage only for ac and does not affect the quiescent collector current of the driving stage. Capacitor coupling also isolates the bias resistors in the driven stage from the collector voltages and resistances in the driving stage. The net result is that each stage, driving and driven, can be designed, insofar as all dc parameters are concerned, independently of the other stages. The reason for the importance of this independence will become more clear later, when we examine *direct*-coupled amplifiers. In effect, capacitor coupling eliminates the need for a number of often not-too-happy compromises between dc and signal parameters. Capacitor coupling has the disadvantage that it is not usable much below 10 Hz. This frequency limitation is not very often a problem. For radio frequencies, transformer coupling is sometimes a distinct advantage, especially when tuned circuits are involved, but often capacitor coupling *can* be used at radio frequencies. It is important to understand that in capacitor-coupled circuits there is no effort made to match impedances, and that impedance matching is of little importance unless maximum power transfer is essential, and this is seldom the case. What is important in capacitor-coupled stages is that the input impedance of the driven stage becomes the R_{Lw} (ac load) for the driving stage. This fact leads to the iterative concept of cascading stages, where the bulk of the voltage amplification in an amplifier is provided by a string of *identical* stages. This not only makes the designer's work easier, but it also is the only practical way to go for fully bypassed stages, and is often a good choice for split emitter stages.

For inexpensive audio equipment the fully bypassed circuit is generally in order, and it is also in order for much RF circuitry, other than some high-quality broadcast equipment. For true high-fidelity audio equipment the split emitter circuit is often a good choice. The unbypassed circuit should generally be reserved for instrumentation circuitry where its very low distortion and good bandwidth are *worth* the sacrifice in gain.

The fully bypassed circuit almost *must* be designed as an iterative circuit. If it is designed a stage at a time, the differences from one stage to the next will be quite superficial, and you will have done a lot of extra work for very little profit. The fully bypassed circuit has the largest voltage gain, the highest distortion, and the poorest predictability of the three configurations (gain and input impedance). The fully bypassed circuit is cheap and dirty and is quite naturally, in our mass-production-oriented society, the most common of the three circuits.

The split emitter circuit may be designed a stage at a time or it may be iterated. In this chapter we will examine the iterative design problems because you already know how to design the split emitter circuit a stage at a time. In order for a one-stage-at-a-time design to be practical, each stage (working from the last to the first) must have a higher input impedance than the stage following it. Because we can control the value of the input impedance by varying R_{eac}, we can have it either way, ascending input impedances or the same input impedance for all stages in the string. I will give you a special formula for finding an appropriate value for R_{eac} for iterative design purposes.

The unbypassed circuit does not lend itself well to iterative design procedures, and because of the kind of applications for which it is most suitable, it is not a good practice anyway. We will design it one stage at a time.

In summary, the fully bypassed circuit (capacitor-coupled) almost has to be designed on an iterative basis. The split emitter circuit can be designed either one stage at a time or on an iterative basis. The unbypassed circuit does not lend itself well to iterative methods and is almost always designed on a one-stage-at-a-time basis.

The total voltage gain of a cascade of coupled stages is the *product* of the individual voltage gains. For iterative stages the total gain is $V_g{}^{N-1} \times V_g$ (last stage) where N is the number of iterative stages. The last stage is treated separately because, even though it is identical to the rest of the stages generally, it may well have a working load value which is different from that of the rest of the stages. We will see shortly why that happens.

The input impedance of the entire circuit is simply the input impedance of the first (input stage).

Transformer coupling can be used to achieve higher voltage gains per stage than can be had with capacitor coupling. Transformer coupling is quite important in radio-frequency circuits, because transformers at these frequencies are inexpensive and provide essential bandwidth control. The design of transformer-coupled stages is no more difficult than the design of capacitor-coupled circuits except for problems and limitations imposed by commercially available transformers.

We will examine the design of transformer-coupled circuits in Secs. 9-5 and 9-6.

9-1 THE FULLY BYPASSED *R-C*-COUPLED AMPLIFIER

When fully bypassed stages are coupled by a capacitor, the input Z of each driven stage is the R_{Lw} for the driving stage. Because, in the fully bypassed circuit, the value of R_e' controls the input impedance, and the quiescent collector current controls R_e', we find the options on the table of possibilities for the fully bypassed circuit are limited to a single row in Table 6-2. The input impedance of a fully bypassed circuit with a 1-mA QI_c is approximately 2 kΩ, including the bias resistors R_{b1} and R_{b2}. At 0.5 mA the input impedance is approximately 4 kΩ, for 2 mA the input impedance is about 500 Ω, and so on. If in Table 6-2 we examine the row R_{Lw} at 1 mA equals 2 kΩ, we see that the highest listed voltage gain is 60, which yields an E_{RL} of 6 V. At a 1-mA QI_c this would yield an R_L value of 6 kΩ. Most iterative fully bypassed circuits which are biased at a 1-mA QI_c have an R_L of 10 kΩ. 10 kΩ for R_L has become so common as to be almost standard, and yet the table of possibilities does not show a 10 V E_{RL} which would yield a 10 kΩ R_L at 1 mA. I have duplicated that part of the table of possibilities (Table 6-2) which applies to iterative fully bypassed circuits and have added an entry, E_{RL} at $V_g = 66$, which yields a 10-V E_{RL} and a 10-kΩ R_L at 1 mA. A voltage gain of 66 per stage is the *limit* for iterative fully bypassed circuits. The modified table of possibilities is Table 9-1.

TABLE 9-1. THE EXPANDED TABLE OF POSSIBILITIES FOR ITERATIVE FULLY BYPASSED CIRCUITS

Working Loads at Various Values of QI_c							**Voltage Drop Across R_L for Various Voltage Gains**		
R_{Lw} at 0.5 mA	**R_{Lw} at 1.0 mA**	**R_{Lw} at 2.0 mA**	**R_{Lw} at 4.0 mA**	**R_{Lw} at 10 mA**	**R_{Lw} at 20 mA**	**R_{Lw} at 40 mA**	**E_{RL} at $V_g = 40$**	**E_{RL} at $V_g = 60$**	**E_{RL} at $V_g = 66$**
4K	2K	1K	500	200	100	50	2.0	6.0	10

There is not a great deal of difference between $V_g = 60$ and $V_g = 66$, especially in view of the variability of R_e', but now the table conforms to what has become common practice in the industry. If you will take another look at Table 9-1, you will observe that $V_g = 66$ is the magic number no matter what QI_c is selected, because when we increase the quiescent collector current, we reduce R_e' (and consequently the stage input impedance) proportionally, putting us exactly back where we started, insofar as voltage gain is concerned.

The only problem with iterative stages is that the string must eventually end, and the load for the very last stage is likely to be different from the input impedance of the iterative stages. There are three basic solutions to this problem:

1. Abandon the iterative idea altogether. This is really not (in this case) a solution, for two reasons: First, each stage in the chain would have to be designed individually, and second, it would take *many* stages in cascade to reduce QI_c by a factor of 2, and even then the input stage could have an unsatisfactorily low input impedance.
2. Design the entire iterative chain to operate at a QI_c commensurate with the *final* load R_{Lw}. This is a satisfactory solution when the final R_{Lw} is between 4 and 1 kΩ. The problem here for heavy loads is that amplifier input impedance is likely to end up so low that an input emitter follower will be required. An *input* emitter follower should be avoided if possible because the iterative stages will consume larger currents. If an emitter follower must be used, it is best to put it as near to the last stage as possible.
3. Use an emitter follower to elevate the final load (R_{Lw}) resistance to about the same impedance as the input impedance of the iterative stages, 2 kΩ for this circuit. The last two approaches are generally the most satisfactory. Chapter 7 details the method for designing emitter followers with more or less specific input impedances. There is no great need for precision in the input impedance of the impedance elevating emitter follower because the individual voltage gains are not really very dependable anyway; and because the total voltage gain is the product of the individual voltage gains, these errors are also multiplied. The best policy is to make the input impedance of the emitter follower buffer somewhat

higher—as much as 100 percent is OK—than the input impedance of the iterative stages. It is not usually hard to live with a bit too much voltage gain, but too little gain can be a problem. It may seem a little wasteful to add an emitter follower, which provides no voltage gain at all, but it just doesn't work out that way in most cases. Generally, adding an emitter follower is economically sound if the final load is lower than 1 kΩ or so.

Suppose we try an example with the assumption that the final load was either intrinsically about 2 kΩ or has been elevated to that value by a buffer emitter follower. We will follow the usual current-mode design sheet.

Example 9-1 The Iterative Fully Bypassed Current-mode Circuit with Base Bias

Design an iterative-amplifier fully bypassed circuit, using the following specifications.

Transistor Parameters

1. β	100
2. I_c (max)	100 mA
3. E_c (max)	25 V

Circuit Parameters

1. Voltage gain per stage	66
2. E_o p-p	4 V
3. R_{Lw} (final)	2 kΩ
4. Stability factor	5
5. $\%\Delta QI_c$ to be allowed	10%
6. Maximum operating temperature	100°C
7. f_o	20 Hz

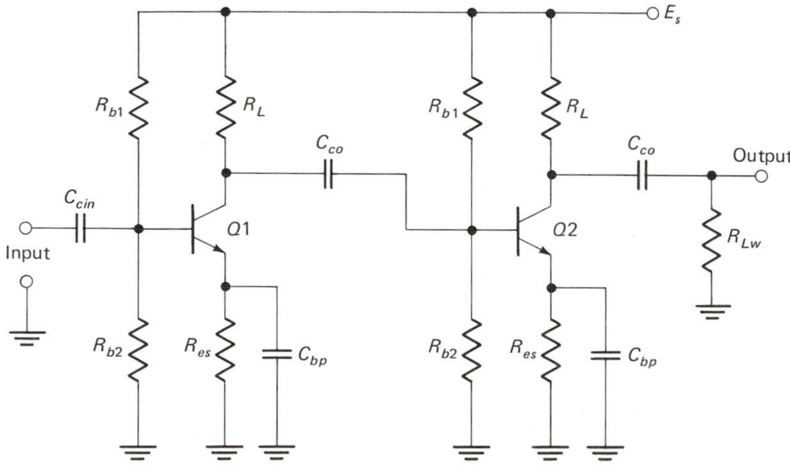

Fig. 9-1 The R-C-coupled iterative fully bypassed common-emitter cascaded amplifier

UNIVERSAL DESIGN SHEET

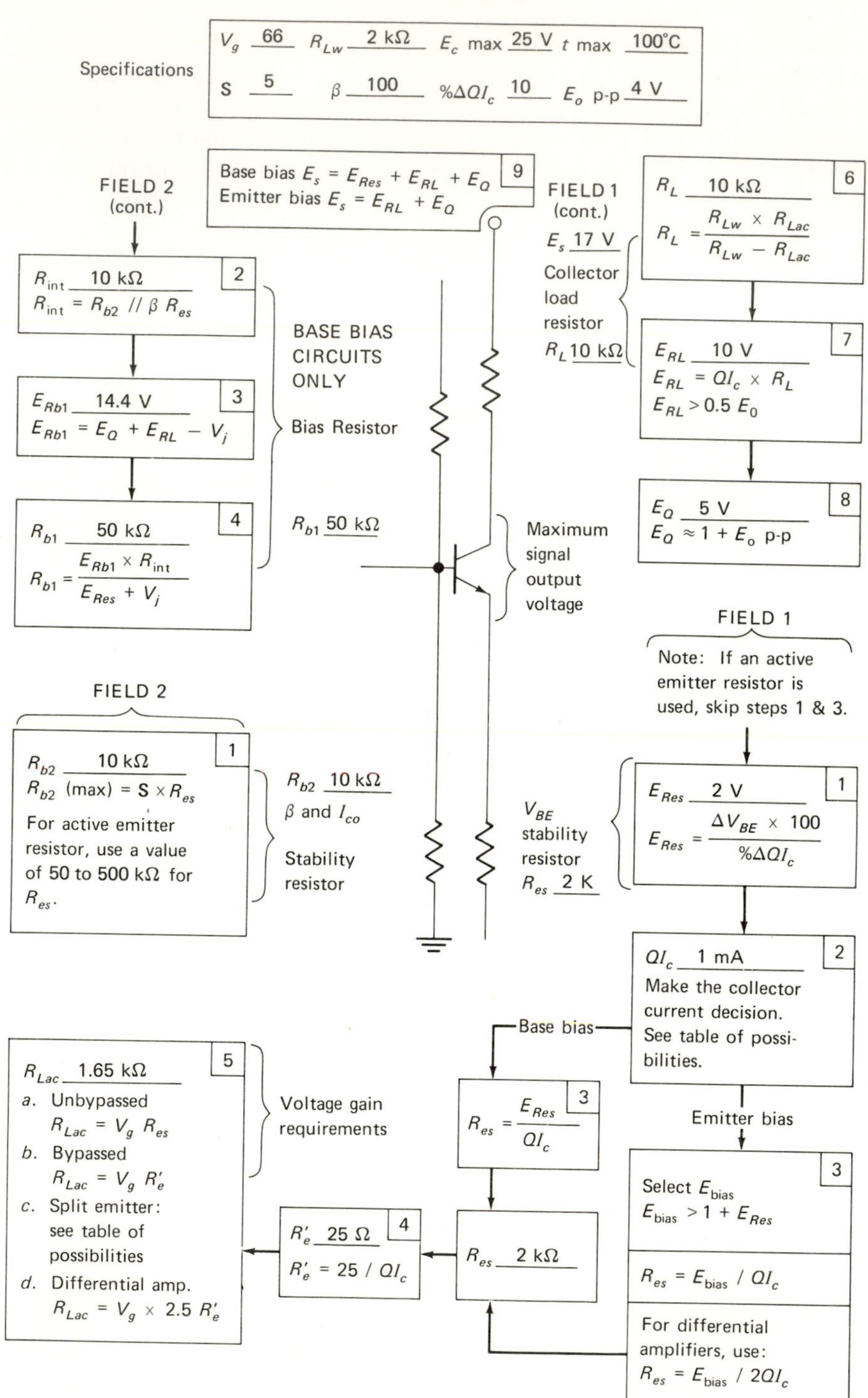
Specifications
V_g 66 R_{Lw} 2 kΩ E_c max 25 V t max 100°C
S 5 β 100 %ΔQI_c 10 E_o p-p 4 V
Base bias $E_s = E_{Res} + E_{RL} + E_Q$
Emitter bias $E_s = E_{RL} + E_Q$
9
FIELD 2 (cont.)
FIELD 1 (cont.)
E_s 17 V
Collector load resistor R_L 10 kΩ
6
R_L 10 kΩ
$R_L = \frac{R_{Lw} \times R_{Lac}}{R_{Lw} - R_{Lac}}$
2
R_{int} 10 kΩ
$R_{int} = R_{b2} \,//\, \beta R_{es}$
BASE BIAS CIRCUITS ONLY
Bias Resistor
7
E_{RL} 10 V
$E_{RL} = QI_c \times R_L$
$E_{RL} > 0.5\ E_0$
3
E_{Rb1} 14.4 V
$E_{Rb1} = E_Q + E_{RL} - V_j$
8
E_Q 5 V
$E_Q \approx 1 + E_o$ p-p
4
R_{b1} 50 kΩ
$R_{b1} = \frac{E_{Rb1} \times R_{int}}{E_{Res} + V_j}$
R_{b1} 50 kΩ
Maximum signal output voltage
FIELD 1
Note: If an active emitter resistor is used, skip steps 1 & 3.
FIELD 2
1
R_{b2} 10 kΩ
R_{b2} (max) = S × R_{es}
For active emitter resistor, use a value of 50 to 500 kΩ for R_{es}.
R_{b2} 10 kΩ
β and I_{co}
Stability resistor
V_{BE} stability resistor
R_{es} 2 K
1
E_{Res} 2 V
$E_{Res} = \frac{\Delta V_{BE} \times 100}{\%\Delta QI_c}$
2
QI_c 1 mA
Make the collector current decision. See table of possibilities.
Base bias
Emitter bias
5
R_{Lac} 1.65 kΩ
a. Unbypassed
$R_{Lac} = V_g\ R_{es}$
b. Bypassed
$R_{Lac} = V_g\ R'_e$
c. Split emitter: see table of possibilities
d. Differential amp.
$R_{Lac} = V_g \times 2.5\ R'_e$
Voltage gain requirements
3
$R_{es} = \frac{E_{Res}}{QI_c}$
4
R'_e 25 Ω
$R'_e = 25 / QI_c$
R_{es} 2 kΩ
3
Select E_{bias}
$E_{bias} > 1 + E_{Res}$
$R_{es} = E_{bias} / QI_c$
For differential amplifiers, use:
$R_{es} = E_{bias} / 2QI_c$

Fig. 9-2 The completed design sheet for Example 9-1

The circuit for Example 9-1 is shown in Fig. 9-1. The completed design sheet is shown in Fig. 9-2.

FIELD 1 COLLECTOR SIDE CALCULATIONS

FRAME 1

Compute E_{Res}, the emitter stability voltage drop.

$$E_{Res} = \frac{\Delta V_{BE} \times 100}{\%\Delta QI_c}$$

Figuring 2.5 MV/C°, we have a total Δt of $100° - 20° = 80°C$ and a total ΔV_{BE} of $80° \times 2.5$ mV or 200 mV.

$$E_{Res} = \frac{200 \times 10^{-3} \times 10^2}{10}$$

$$E_{Res} = 2 \text{ V}$$

It is worthwhile to note here that a 10 percent variation in QI_c due to ΔV_{BE} is quite satisfactory for this kind of circuit provided that E_Q is greater than 3 V.

FRAME 2

Make the collector current (QI_c) decision. (See Table 9-1.)

$$QI_c = 1 \text{ mA}$$

FRAME 3

Compute R_{es}, the value of the emitter stability resistor.

$$R_{es} = \frac{E_{Res}}{QI_c} = \frac{2 \text{ V}}{1 \text{ mA}} \qquad \text{(Ohm's law)}$$

$$R_{es} = 2 \text{ k}\Omega$$

FRAME 4

Compute R_e'.

$$R_e' = \frac{25}{QI_c} = \frac{25}{1} = 25\ \Omega \qquad \text{(the Shockley relationship)}$$

FRAME 5

Compute R_{Lac}. (Use Branch *b*.)

$$R_{Lac} = V_g R_e' = 66 \times 25 = 1.65 \text{ k}\Omega$$

FRAME 6

Compute R_L.

$$R_L = \frac{R_{Lw} \times R_{Lac}}{R_{Lw} - R_{Lac}}$$

$$R_L = \frac{2\ \text{k}\Omega \times 1.65\ \text{k}\Omega}{2\ \text{k}\Omega - 1.65\ \text{k}\Omega}$$

$$R_L \approx 10\ \text{k}\Omega$$

This, of course, does not come out to exactly 10 kΩ, but in view of the probable inaccuracy in R_e', it is a respectable, and quite useful, figure. If you want to calculate it a little closer, go ahead; you will still end up building the circuit with a 10-kΩ resistor for R_L.

FRAME 7

Compute E_{RL}.

$$E_{RL} = QI_c \times R_L = 1\ \text{mA} \times 10\ \text{k}\Omega = 10\ \text{V}$$

$$E_{RL} = 10\ \text{V}$$

FRAME 8

Compute E_Q.

$$E_Q = 1 + E_o\ \text{p-p}$$
$$E_Q = 1 + 4 = 5\ \text{V (min)}$$

FRAME 9

Compute E_s (base bias).

$$E_s = E_{Res} + E_{RL} + E_Q$$
$$E_s = 2 + 10 + 5 = 17\ \text{V}$$

FIELD 2 BASE SIDE COMPUTATIONS

FRAME 1

Compute R_{b2}.

$$R_{b2} = \mathbf{S} \times R_{es} = 5 \times 2\ \text{k}\Omega = 10\ \text{k}\Omega$$

FRAME 2

Compute R_{int}. (Base bias, so we go on.)

$$R_{int} = R_{b2} \| \beta \times R_{es}$$
$$R_{int} = 10\ \text{k}\Omega \| 100 \times 2\text{k}\Omega$$
$$R_{int} = 10\ \text{k}\Omega \| 200\ \text{k}\Omega$$
$$R_{int} \approx 10\ \text{k}\Omega$$

FRAME 3

Compute E_{Rb1}.

$$E_{Rb1} = E_Q + E_{RL} - V_j$$
$$E_{Rb1} = 5 + 10 - 0.6$$
$$E_{Rb1} = 14.4 \text{ V}$$

FRAME 4

Compute R_{b1}.

$$R_{b1} = \frac{E_{Rb1} \times R_{int}}{E_{Res} + V_j}$$

$$R_{b1} = \frac{14.4 \times 10 \text{ k}\Omega}{2.6}$$

$$R_{b1} \approx 50 \text{ k}\Omega$$

FIELD 3 INPUT IMPEDANCE AND CAPACITOR CALCULATIONS

FRAME 1

Compute Z_{in}, the input impedance.

$$Z_{in} \approx R_{b1} \| R_{b2} \| \beta R'_e$$

The input impedance of the entire string is simply the input impedance of the first stage.

$$Z_{in} \approx 50 \text{ k}\Omega \| 10 \text{ k}\Omega \| 100 \times 25$$
$$Z_{in} \approx 2 \text{ k}\Omega$$

If you will remember, we started this design assuming an input impedance of about 2 kΩ.

FRAME 2

Compute C_{cin}.

$$X_c \text{ of } C_{cin} \approx 0.1 Z_{in}$$
$$X_c \text{ of } C_{cin} \approx 200 \ \Omega \text{ at } f_o$$

Consulting Table 6-3,

$$C \approx 50 \ \mu\text{F}$$

FRAME 3

Compute C_{co}.

In this case C_{co} is an interstage coupling capacitor, and it works into the input impedance of the driven stage which is R_{Lw} for the driving stage and is the same as the input impedance of the first stage. Therefore it will have the same value as C_{cin}.

$C_{cin} = 50\ \mu F$

FRAME 4

Compute C_{bp}.

$X_c \approx R'_e$ at 20 Hz $= 25\ \Omega$

Refer to the discussion of the bypass capacitor calculations in Chap. 6 if you have forgotten.

X_c at 20 Hz $= 25$

Consulting Table 6-3, we find that 25 Ω of X_c at 20 Hz requires a capacitor of 500 μF.

$C_{bp} = 500\ \mu F$

Figure 9-3 shows the complete circuit (two stages).

Table 9-2 is a rapid-design table for fully bypassed iterative stages. The table also includes some extra columns for the design of single-stage unbypassed circuits. The table is constructed around a QI_c of 1 mA, but it can be scaled up or down easily for other values. R_{Lw} is the same as Z_{in}. The table is actually fairly limited but can be very handy at times.

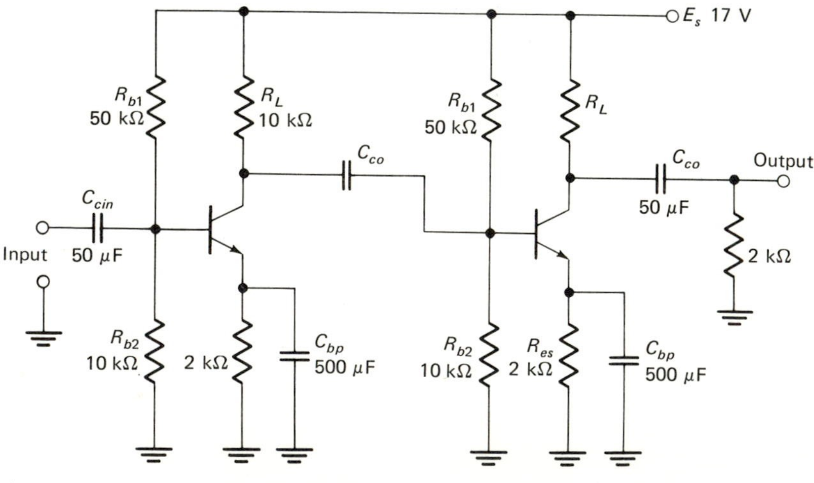

Fig. 9-3 *The complete circuit with values for Example 9-1*

TABLE 9-2. RAPID-DESIGN TABLE

E_s	E_{RL}	E_Q	E_{Res}	R_{B2}	R_{b1}	R_L	**Bypassed** Z_{in}	**Unbypassed** Z_{in}	**Bypassed** V_g	**Unbypassed** V_g	R_{es}
6	2	3	1	10K	27K	2K	1.8K	6.8K	38	1.5	1K
9	5	3	1	10K	43K	5K	2K	7.5K	56	3	1K
9	4	3	2	20K	43K	4K	2K	16K	53	3.2	2K
12	8	3	1	10K	62K	8K	2K	9K	64	4	1K
12	7	4	1	10K	62K	7K	2K	9K	60	4	1K
12	7	3	2	20K	68K	7K	2K	16K	60	4	2K
18	13	4	1	10K	91K	13K	2K	9K	66	5.3	1K
18	12	4	2	20K	110K	12K	2K	16K	66	4	2K
21	16	4	1	10K	110K	16K	2K	9K	66	5.6	1K
21	15	4	2	20K	130K	15K	2K	16K	66	4	2K
24	19	4	1	10K	130K	19K	2K	9K	66	6	1K
24	18	4	2	20K	150K	18K	2K	16K	66	4	2K
28	23	4	1	10K	150K	23K	2K	9K	66	6	1K
28	22	4	2	20K	180K	22K	2K	16K	66	4	2K
32	27	4	1	10K	160K	27K	2K	9K	66	6	1K
32	26	4	2	20K	200K	26K	2K	16K	66	4	2K
48	42	4	2	20K	300K	42K	2K	16K	66	5	2K

PROBLEMS

Problem 9-1

TITLE: **Designing the Two-stage Iterative Fully Bypassed Circuit**
PROCEDURE: **Design the circuit, using the following specifications. See Fig. 9-1.**

Transistor Parameters

1. β (min)	100
2. I_c (max)	100 mA
3. E_c (max)	30 V
4. Transistor type	*npn* silicon

Circuit Parameters

1. Voltage gain	40 per stage
2. E_o p-p	5 V
3. R_{Lw}	1 kΩ
4. Stability factor	10
5. $\%\Delta QI_c$ to be allowed	5%
6. Maximum operating temperature	65°C
7. f_o	25 Hz

NOTE: Use the expanded table of possibilities, Table 9-1.

9-2 CASCADING SPLIT EMITTER STAGES

There are two ways in which to cascade split emitter stages. In the case of the fully bypassed circuit there is no real choice but to use iterative stages, because even if they are designed one at a time, they will come out *almost* identical anyway. So there is no point to it.

The split emitter circuit can be designed one at a time, or it can be designed in an iterative fashion with an emitter follower at the end of the chain if necessary. In Chap. 6 we examined the process for designing individual (one-at-a-time) split emitter stages, and so in this chapter we will examine only the iterative approach. The big problem here is the final working load value, R_{Lw}. In the case of the fully bypassed circuit we knew in advance that the final load had to be raised to about 2 kΩ, for a QI_c of 1 mA in the iterative stages, so that the gain of each stage, including the last one, would be the same, making the problem of *designing in* the desired *total* voltage gain an easy one. Here we cannot, so easily, predict the input impedance in advance, and so we cannot know to what value to raise the final working load to make the last stage have the same voltage gain as the rest of the iterative stages. There are two solutions to the problem, depending upon the circumstances. For this problem we will use the split emitter table (Table 6-4) and proceed with the design just as we would with a single stage, with the exception that we have a special problem with R_{eac}. The iterative string is unlike a single-stage circuit in that we do not know R_{Lw}, except for the last stage, until after we have computed R_{b1} and R_{b2}, because the input impedance of the driven stage is R_{Lw} for the driving stage.

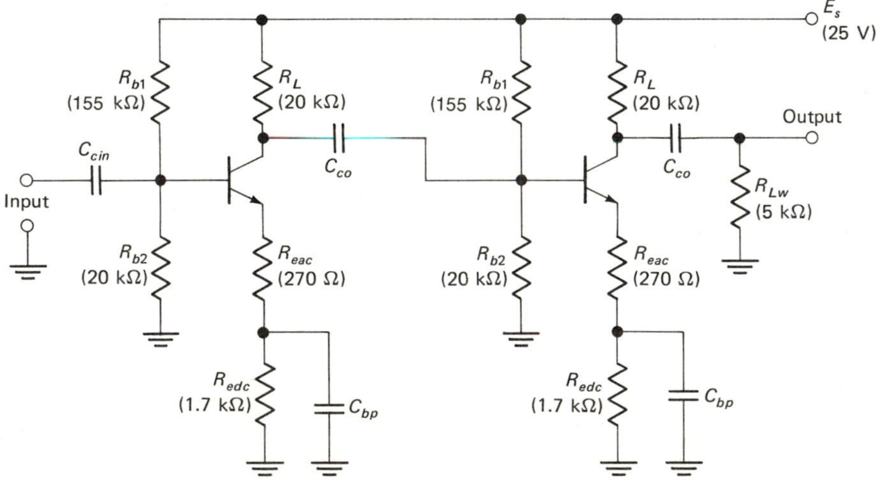

Fig. 9-4 The split emitter cascade (figures in parentheses are values for the completed Example 9-2)

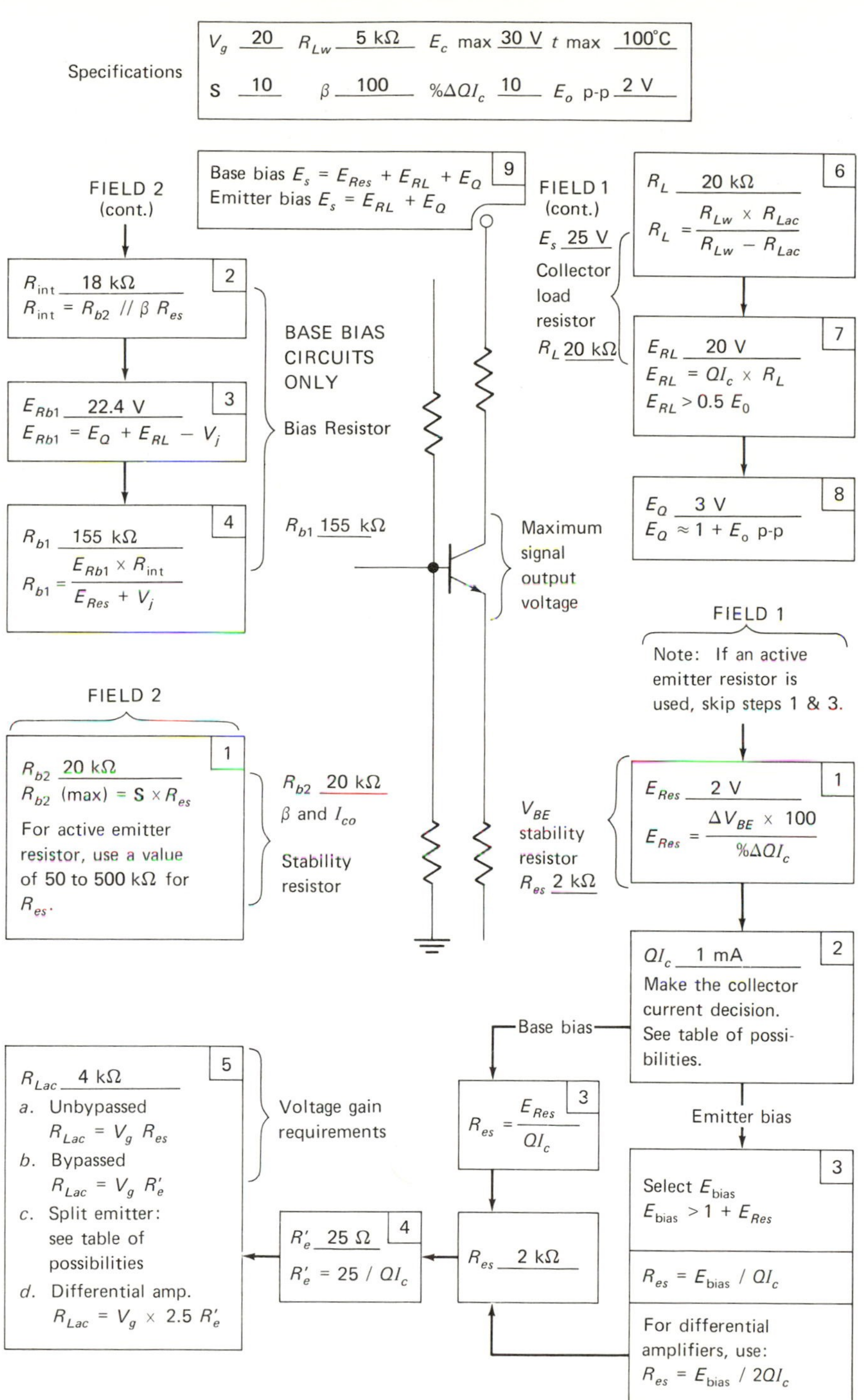

Fig. 9-5 The completed design sheet for Example 9-2

The input impedance is

$$Z_{in} = R_{Lw} = R_{b1} \| R_{b2} \| \beta R_{eac}$$

Even after we have values for R_{b1} and R_{b2}, we are far from being out of the woods because R_{eac} is a part of both Z_{in} and R_{Lac}.

Remember

$$R_{Lac} = R_L \| R_{Lw} \qquad R_{Lw} = Z_{in}$$

and

$$R_{eac} = \frac{R_{Lac}}{V_g}$$

With a little substitution we get

$$R_{eac} = \frac{R_L \| R_{b1} \| R_{b2} \| \beta R_{eac}}{V_g} \tag{9-1}$$

Now isn't that a fine mess? R_{eac} appears on both sides of the equation. Obviously we must manipulate this thing so that R_{eac} no longer appears on both sides. Now that you see the problem, you should be prepared for the fact that the revised equation for finding R_{eac} will be a bit hairy. Some of my students throw up their hands in despair the first time they see it, but by the time you actually have to use the equation, there is no more to it than substituting in all the values and doing a little arithmetic. It is not really bad at all. Here it is, the great split emitter formula:

$$R_{eac} = \frac{1 - V_g/\beta}{V_g/R_L + V_g/R_{b1} + V_g/R_{b2}} \tag{9-2}$$

V_g is the gain of *each* stage.

If you have enough time, patience, and curiosity, you can practice your algebra and manipulate Eq. (9-1) into Eq. (9-2).

Now let us try a design example. The schematic is shown in Fig. 9-4. Figure 9-5 shows the completed design sheet.

Example 9-2 Designing a Two-stage Iterative Split Emitter Circuit

Design the circuit, using the following specifications.

Transistor Parameters

1. β (min)	100
2. I_c (max)	50 mA
3. E_c (max)	30 V
4. Type	*npn* silicon

Circuit Parameters

1. Voltage gain	20 per stage (400 total)
2. E_o p-p	2 V
3. R_{Lw}	5 kΩ
4. Stability factor	10
5. $\%\Delta QI_c$ to be allowed	10%
6. Maximum operating temperature	100°C
7. f_o	20 Hz

FIELD 1 COLLECTOR SIDE CALCULATIONS

FRAME 1

Compute E_{Res}, the voltage drop across the emitter stability resistor.

$$E_{Res} = \frac{\Delta V_{BE} \times 100}{\%\Delta Q_{Ic}} \qquad (\text{total } \Delta V_{BE} = 80° \times 2.5 \text{ MV})$$

$$E_{Res} = \frac{200 \times 10^{-3} \times 10^2}{10}$$

$$E_{Res} = 2 \text{ V}$$

FRAME 2

Make the quiescent collector current decision.

Consulting the table of possibilities (Table 6-4), we find we can use 1 mA with 20 V for E_{RL}.

$$QI_c = 1.0 \text{ mA}$$

FRAME 3

Compute R_{es}. (Use the base bias branch.)

$$R_{es} = \frac{E_{Res}}{QI_c} = \frac{2 \text{ V}}{1 \text{ mA}} = 2 \text{ k}\Omega$$

FRAME 4

Compute R_e' (the Shockley relationship).

$$R_e' = \frac{25}{QI_c} = \frac{25}{1} = 25\ \Omega$$

FRAME 5

Compute R_{Lac}, the total ac load resistance.

Here, we go back to the table of possibilities (Table 6-4) and look in the table's superstructure to find an R_{Lac} value of 4 kΩ. We must wait until later to compute the values for R_{eac} and R_{edc}.

$$R_{Lac} = 4 \text{ k}\Omega$$

FRAME 6

Compute the value of the dc collector load R_L.

$$R_L = \frac{R_{Lw} \times R_{Lac}}{R_{Lw} - R_{Lac}} = \frac{5\ \text{k}\Omega \times 4\ \text{k}\Omega}{5\ \text{k}\Omega - 4\ \text{k}\Omega}$$

$$R_L = 20\ \text{k}\Omega$$

FRAME 7

Compute the voltage drop across R_L.

$$E_{RL} = QI_c \times R_L = 1\ \text{mA} \times 20\ \text{k}\Omega$$
$$E_{RL} = 20\ \text{V}$$

FRAME 8

Compute the transistor's collector-emitter voltage drop E_Q.

$$E_Q = 1 + E_o\ \text{p-p} = 1 + 2$$
$$E_Q = 3\ \text{V}$$

FRAME 9

Compute E_s, the power supply voltage.

$$E_s = E_{Res} + E_{RL} + E_Q \qquad \text{(base bias)}$$
$$E_s = 2 + 20 + 3 = 25\ \text{V}$$

FIELD 2 BASE SIDE CALCULATIONS

FRAME 1

Compute R_{b2}.

$$R_{b2} = \mathbf{S} \times R_{es} = 10 \times 2\ \text{k}\Omega$$
$$R_{b2} = 20\ \text{k}\Omega$$

Because this is a base bias circuit, we will go on to Frame 2.

FRAME 2

Compute R_{int}.

$$R_{int} = R_{b2} \| \beta \times R_{es} = 20\ \text{k}\Omega \| 100 \times 2\ \text{k}\Omega$$
$$R_{int} \approx 18\ \text{k}\Omega$$

FRAME 3

Compute the voltage drop across R_{b1}.

$$E_{Rb1} = E_Q + E_{RL} - V_j$$
$$E_{Rb1} = 3 + 20 - 0.6$$
$$E_{Rb1} = 22.4\ \text{V}$$

FRAME 4

Compute R_{b1}.

$$R_{b1} = \frac{E_{Rb1} \times R_{int}}{E_{Res} + V_j} = \frac{22.4 \times 18 \text{ k}\Omega}{2 + 0.6}$$

$$R_{b1} = 155 \text{ k}\Omega$$

Now that we have all the values we need for the great split emitter formula, we can solve for R_{eac} and R_{edc}.

$$R_{eac} = \frac{1 - V_g/\beta}{V_g/R_L + V_g/R_{b1} + V_g/R_{b2}}$$

$$R_{eac} = \frac{1 - (20/100)}{20/20{,}000 + 20/155{,}000 + 20/20{,}000}$$

$$= \frac{8 \times 10^{-1}}{(1 \times 10^{-3}) + (0.13 \times 10^{-3}) + (1 \times 10^{-3})}$$

$$R_{eac} \approx \frac{8 \times 10^{-1}}{2.00 \times 10^{-3}} \approx 4 \times 10^2$$

$$R_{eac} = 400 \ \Omega$$

FIELD 3 INPUT IMPEDANCE AND CAPACITORS

FRAME 1

Compute the input impedance.

$$Z_{in} = R_{b1} \| R_{b2} \| \beta R_{eac}$$
$$Z_{in} = 155 \text{ k}\Omega \| 20 \text{ k}\Omega \| 100 \times 400$$
$$Z_{in} \approx 13 \text{ k}\Omega$$

We need this figure to compute the input capacitor value, and we need it also to determine whether a Young-Albertson loading correction is necessary. We can also use it to verify R_{eac} because $Z_{in} = R_{Lw}$. First let us check the value of R_{eac}.

$$V_g = \frac{R_{Lac}}{R_{eac}}$$

R_{Lac} for the iterative stages is $R_L \| R_{Lw}$, where $R_{Lw} = Z_{in} = 13 \text{ k}\Omega$.

$$R_{Lac} = R_L \| Z_{in} = 20 \text{ k}\Omega \| 13 \text{ k}\Omega$$
$$R_{Lac} \approx 8 \text{ k}\Omega$$
$$R_{eac} = 400 \ \Omega$$

$$V_g = \frac{8 \text{ k}\Omega}{400 \ \Omega} = 20$$

Now let us see about the Young-Albertson correction.

$R_L = 20\ \text{k}\Omega \qquad R_{Lw} = 13\ \text{k}\Omega$

$$\frac{R_L}{R_{Lw}} = \frac{20\ \text{k}\Omega}{13\ \text{k}\Omega} \approx 1.5$$

If we look at the Young-Albertson curve (Fig. 6-5) at the correction for $R_L = 1.5R_{Lw}$, we see that we need a correction of about 33 percent. This means a reduction in R_{eac} of 33 percent.

$400 \times 0.33 = 132\ \Omega$

Reducing R_{eac} by 132 Ω, we get a new R_{eac} of

$400 - 132 = 268\ \Omega$
$R_{eac} \approx 270\ \Omega$

Now for R_{edc}.

$R_{edc} = R_{es} - R_{eac}$
$R_{edc} = 2\ \text{k}\Omega - 270\ \Omega$
$R_{edc} = 1{,}730\ \Omega$

The nearest standard value is 1.8 kΩ.

FRAME 2
Compute C_{cin}.

X_c of $C_{cin} \approx 0.1Z_{in}$

X_c of $C_{cin} \approx 1.3\ \text{k}\Omega$ at 20 Hz

Consulting Table 6-3, we find we will need a capacitor of 10 μF.

$C_{cin} = 10\ \mu\text{F}$

If you haven't already asked yourself a couple of questions, it is time to ask them together. First, where did the 5 kΩ R_{Lw} for the last stage come from? The truth is, I made it up, almost, but not quite, off the top of my head. If you will examine the split emitter table of possibilities (Table 6-4), you can readily see that there is nothing whatever to be done at 1 mA for loads less than about 2 kΩ. At 5 kΩ we have three columns to work in with a fair variety of possible voltage gains and voltage drops across R_L. For these reasons I picked 5 kΩ out of an educated hat. I suppose we are back to my prejudice in favor of a 1-mA QI_c. I like 1 mA because it makes the arithmetic easy, but there is no way I can tell you what QI_c to use. All I can say is that if you use the lowest practical current, you will end up with a higher input impedance and a lower total current drain. Sometimes these two factors are not particularly important and you may select any current the table of possibilities will allow. After you have had a bit of experience, you need no longer go along with my prejudices, but it has been my own *experience* that has formed my 1-mA prejudice, and it is not entirely an arbitrary one. I might mention here that it is generally poor practice to get the highest possible voltage gain in any particular column under the heading "V_g range." For example, if we wanted a V_g of 15, I would not try to get it out of the first-column V_g range, 2.5 to 15. I would move over to the second-column V_g range, 3 to 20, and take it out of the middle, as it were. The second question is, how do we get around the fact that the gain in the final stage of the iterative chain is going to be different from the gain of all the rest of the stages in the chain? And more important, how do we cope with it? In the case of our present example the voltage gain of the last stage will be

$$V_g = \frac{R_{Lac}}{R_{eac}} = \frac{5\ \text{k}\Omega \| 20\ \text{k}\Omega}{400} = \frac{4\ \text{k}\Omega}{400} = 10$$

R_{Lac} for the rest of the stages will be equal to $Z_{in} \| R_L$, which is about 8 kΩ, for a gain of 20 for each of the iterative stages. We can consider that 5 kΩ final load as an arbitrary figure as long as 5 kΩ is no more than about fifty times as large as the *real* final load, and use a common-collector (emitter-follower) circuit to elevate the *real* final load to the same value as R_{Lac} for the iterative stages, in this case about 8 kΩ.

PROBLEMS

Problem 9-2

TITLE: **Designing the Split Emitter Iterative Circuit**
PROCEDURE: **Design the circuit, using the following specifications. See Fig. 9-4.**

Transistor Parameters

1. β (min)	100
2. I_c (max)	100 mA
3. E_c (max)	85 V
4. Transistor type	*npn* silicon

Circuit Parameters

1. Voltage gain	30 per stage
2. E_o p-p	6 V
3. R_{Lw}	6 kΩ
4. Stability factor	10
5. $\% \Delta QI_c$ to be allowed	5%
6. Maximum operating temperature	70°C
7. f_o	30 Hz

9-3 *R-C*-COUPLED CIRCUIT WITH TWO-STAGE FEEDBACK

Negative feedback of a current-mode type is sometimes used with two *R-C*-coupled stages. Negative feedback is seldom used around more than two stages because even though the feedback is negative at mid-frequencies, phase shifts due to reactive elements in the system may be such that some frequency will be fed back as positive feedback, causing the amplifier to oscillate at that frequency. The circuit is capable of very predictable and very stable voltage gains. The cost is a loss of at least 75 percent of the voltage gain possible without R_f, the feedback resistor. The circuit is shown in Fig. 9-6.

The principal disadvantage of this circuit is that it is too easy to fall in love with. One manual of preferred circuits contains the circuit in Fig. 9-6 with component values such that the total voltage gain of the entire circuit is only 50. As you can see, there is a lot of hardware here for an average voltage gain of less than 8 per stage. Sufficiently stable and predictable gains of around 50 could just as well be had with two stages of the unbypassed common-emitter configuration, at lower cost and probably with better low-frequency response. The two-stage feedback arrangement does have one special advantage over circuits with single-stage feedback, and that is that the gain is easily altered by changing *only* the feedback resistor R_f. No additional circuit re-design is normally required. The resistor R_f can even be installed externally so that it can be experimentally adjusted while the circuit is in operation.

Basically the circuit consists of a split emitter circuit driving a fully

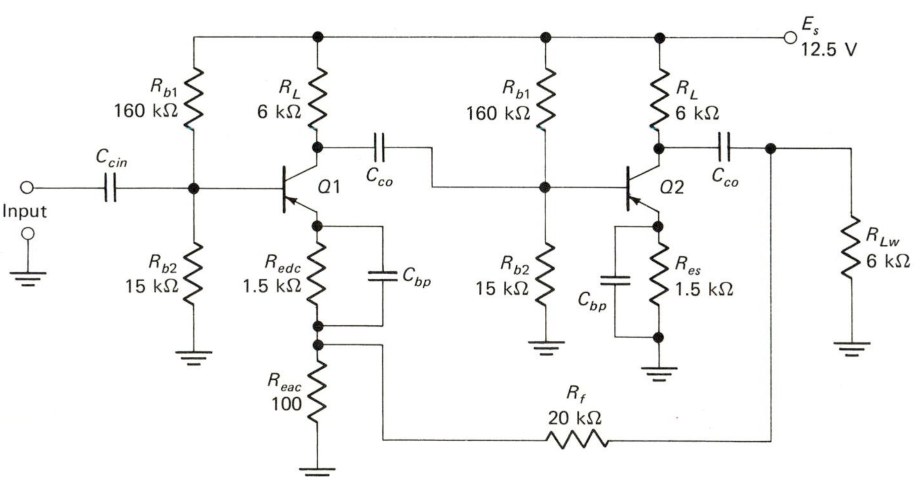

Fig. 9-6 *The two-stage R-C-coupled circuit with two-stage feedback*

bypassed circuit with an *added* feedback resistor. To design the circuit, you simply design (first) a fully bypassed circuit with a V_g as high as practical—at least a V_g of 100—and then design a split emitter circuit with a V_g as high as practical. The voltage gain of the split emitter should be *about* 15 with the input impedance of the fully bypassed stage, which makes up R_{Lw}, being considered as a constant 2 kΩ (at $QI_c = 1$ mA).

The total gain V_g (Q1) × V_g (Q2) must be at least 1,000 and preferably more, if this circuit is to be worthwhile. This means that this circuit is not able to drive a final load (R_{Lw} in Fig. 9-6) of 2 kΩ or less, *profitably*. The input impedance of a fully bypassed circuit (at 1 mA) is about 2 kΩ, and so it it not economical to try to drive a fully bypassed circuit with this feedback *pair*. Of course, economics is not the sole design criterion, but often the economics of the situation can keep us from other foolishness. For example, it would be foolish as well as poor economics to drive a fully bypassed circuit with this feedback pair. Why give up 75 percent or more of the brute voltage gain for stable and predictable voltage gain, only to use this highly precise voltage gain to drive a circuit which is known for its V_g unpredictability? The result is the *product* of a known voltage gain and an unknown voltage gain. However, it is not foolish to use this two-stage feedback circuit to drive a fully bypassed circuit *if* R_f can be adjusted with the circuit in operation for the purpose of obtaining the proper overall gain in a system where the gain is not *inherently* very predictable. Rather than take up your time going through design procedure for the two stages, I will just design them for you and concentrate on the feedback loop and its effect on circuit parameters. Before I show you the designed circuit, let me take a moment to provide you with some guidelines to follow when you have to do the design work yourself.

1. Remember that R_{Lw} for the split emitter stage is going to be about 2 kΩ when the fully bypassed stage is operated at 1 mA. According to the split emitter table of possibilities (Table 6-4), the maximum (V_g) gain possible is 15 when the split emitter circuit is operated at 1 mA. We will design for this maximum gain. I must mention here that a fairly common version of this circuit is found in several manuals of preferred circuits where the split emitter stage uses an R_L value of 10 kΩ instead of the 6 kΩ that the table of possibilities dictates. The difference between the circuit I have designed for you and the one with a 10 kΩ R_L is that my circuit, using the table of possibilities, has a voltage gain of 15, whereas the circuit from the manual has a voltage gain of 16. Whatever quiescent current you decide upon here, design for the maximum figure in the V_g range column you are working in. This will generally ensure that R_{eac} will be small compared to R_f.
2. The fully bypassed stage should be designed for the highest *practical* voltage gain. The overall final voltage gain will be determined by the

feedback resistor and *not* by the voltage gain of the individual stages. However, the total gain of the two stages without the feedback resistor should be 1,000 or more, if possible, and it must *always* be at least four times the desired total voltage gain with the feedback resistor installed.

Example 9-3

Design a two-stage R-C -coupled feedback amplifier with the following specifications.
Stage 1
1. V_g 15
2. Maximum operating temperature 50°C
3. Allowable quiescent current drift at 50° 5%
4. Stability factor 10
5. Working load (input impedance of the fully bypassed stage) 2 kΩ
6. f_o 30 Hz
7. E_o p-p 0.4 V or greater

Stage 2
1. V_g 100 or greater
2. Maximum operating temperature 50°C
3. Allowable quiescent current drift at 50°C 5%
4. Stability factor 10
5. Working load (R_{Lw}) 6 kΩ
6. f_o 30 Hz
7. E_o p-p 4 V or greater

Desired total voltage gain *with feedback* is 200.

Figure 9-6 shows the values for the two stages. The voltage gain of the split emitter stage is 15, and the voltage gain for the fully bypassed stage is 120. The total open-loop voltage gain (without R_f) is 15 × 120 = 1,800. The figure 1,800 is considerably larger than four times the required closed-loop gain of 200, and so our specifications are realistic.

Computing the Value of R_f

In Chap. 4 we developed the general equation for the voltage gain with current-mode feedback as

$$V_g' = \frac{1}{F_f} \tag{9-3}$$

In this case

$$F_f = \frac{R_{eac}}{R_f + R_{eac}} \tag{9-4}$$

Because we are going to make it a policy to make R_{eac} small compared to R_f, we can drop it from Eq. (9-4) and rewrite Eq. (9-4) as

$$F_f = \frac{R_{eac}}{R_f} \tag{9-5}$$

Equation (9-4) is simply our old friend the voltage divider equation. Now we can write Eq. (9-3) in a form which is relevant to our problem here as

$$V_g' = \frac{1}{R_{eac}/R_f} \tag{9-6}$$

$$V_g' = \frac{R_f}{R_{eac}} \tag{9-7}$$

Equation (9-7) is fine if we are computing the voltage gain of a circuit designed by someone else. In our case we know V_g' because it is a design specification, and we know R_{eac} because it was part of the design of the split emitter stage. What we need is a value for R_f. So, let us rearrange Eq. (9-7) to solve for R_f.

$$R_f = V_g' \times R_{eac} \tag{9-8}$$

Let's plug our values into Eq. (9-8).

$$R_f = 200 \times 100$$
$$R_f = 20\ \text{k}\Omega$$

The Input Impedance

The addition of the negative feedback loop alters the input impedance to

$$Z_{in} = R_{b1} \| R_{b2} \| \frac{V_g}{V_g'} \beta R_e'$$
$$Z_{in} = 160\ \text{k}\Omega \| 15\ \text{k}\Omega \| (1{,}800/200)(100 \times 25)$$
$$Z_{in} = 160\ \text{k}\Omega \| 15\ \text{k}\Omega \| 22.5\ \text{k}\Omega$$
$$Z_{in} \approx 8\ \text{k}\Omega$$

PROBLEMS

Problem 9-3

TITLE: **Designing the Two-stage Capacitor-coupled Amplifier with Overall Current-mode Feedback**

PROCEDURE: **Design the circuit, using the following specifications. See Fig. 9-6.**

Transistor Parameters	Q1 and Q2
1. β (min)	100
2. I_c (max)	100 mA
3. E_c (max)	25 V
4. Transistor type	*npn* silicon

Circuit Parameters	Q1	Q2
1. Voltage gain	10	200
2. E_o p-p	2 V	5 V
3. R_{Lw}	2 kΩ	12 kΩ
4. Stability factor	10	10
5. $\%\Delta QI_c$ to be allowed	10%	10%
6. Maximum operating temperature	100°C	
7. f_o	30 Hz	
8. Total voltage gain with feedback	400	

9-4 INDUCTIVE COUPLING

Sections 9-4, 9-5, and 9-6 are not intended to be a completely comprehensive study of transformer coupling, particularly for higher radio frequencies, but I think you will find them satisfactory for most audio and much radio frequency work.

Transformer design is a highly specialized area, and we shall assume that we are using properly designed commercial inductors and transformers. We shall even assume that the transformers are *ideal* transformers. The errors incurred by such an assumption are generally not very significant except at higher frequencies, where additional considerations arise.

The procedures presented here have been used with good results up to 2 MHz and with no additional problems arising except for the selection of a transistor with high-frequency capabilities. The transistor manual can guide you in this selection. Frequencies higher than 1 or 2 MHz give rise to a number of problems which are peculiar to RF circuits, and which are also beyond the scope of this book. The techniques in this section are applicable to higher frequencies, but those RF peculiarities must also be dealt with. If you try some RF designs, even at frequencies of 0.5 MHz or so, special care is required in breadboarding the circuit in the lab. Leads must be kept short, and careful layout practices must be followed.

Figure 9-7 shows a capacitor-coupled circuit which uses an inductor as a collector load. This circuit is essentially a single-frequency circuit (actually a high-pass circuit), because the voltage gain is a function of the inductive reactance of the coil, and X_L is frequency-dependent.

The design procedure is basically the same as that for the ordinary fully bypassed current-mode circuit with the following differences:

1. The X_L of the inductor must be used as R_L for all signal calculations.
2. The dc resistance of the inductor must be used for R_L for all bias (quiescent) calculations.
3. The dc resistance of the choke will probably be only a few ohms, so that, at quiescence, nearly all of E_s appears across E_Q and R_{es}. The collector-to-ground voltage will be approximately E_s. When large inductors are

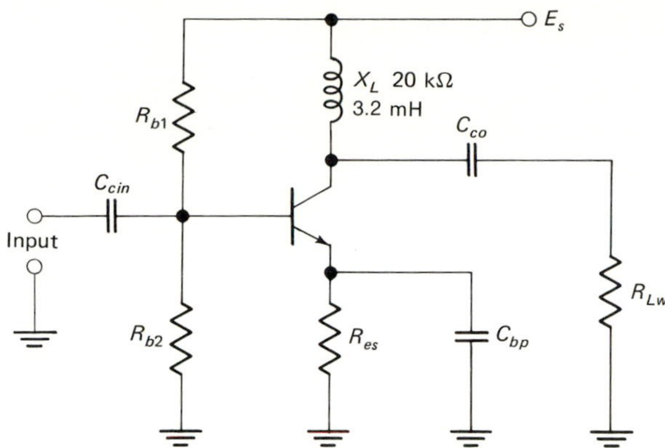

Fig. 9-7 The inductive-coupled circuit

used, as they might be in a 60-Hz servo amplifier, the dc resistance of the choke (inductor) might be large enough to cause some appreciable quiescent voltage drop. When this is the case, that voltage drop must be included in computing R_{b1} in Field 2, Frame 3 of the current mode design sheet.

Calculating the Value of the Inductor

Suppose Frame 6, Field 1, on the universal current-mode design sheet computes out to an R_L

$$R_L = \frac{R_{Lw} \times R_{Lac}}{R_{Lw} - R_{Lac}}$$

of 20 kΩ. Assuming that we are to operate the circuit at 1 mHz, find the value for L.

$$X_L = 2\pi fL$$

Solving for L,

$$L = \frac{X_L}{2\pi f}$$

$$L = \frac{20\text{ k}\Omega}{6.28 \times 1 \times 10^6} = 3.2\text{ mH}$$

At this point we would normally have to stop and find a suitable inductor in the catalog so that we will know the dc resistance and can calculate the quiescent voltage drop across it.

9-5 UNTUNED TRANSFORMER COUPLING

Transformer coupling has some considerable advantages in RF circuits because transformers are cheap and much higher gains per stage can be obtained.

The advantage of transformer coupling in audio-frequency circuits is greater gain per stage, but AF transformers are not ideal because they invariably have a ferromagnetic core of some sort, and they are subject to saturation problems and their attendant distortion. Poor frequency response (limited bandwidth) is also a problem. Truly high-quality broadband audio-frequency transformers are almost always prohibitively expensive. Japan produces cheap, but poor-quality, audio transformers, for use in pocket radios and low-fidelity amplifiers.

Maximum Power Transfer

There is a great deal of fallacious reasoning about the importance of impedance matching for the purpose of *maximum power transfer*. The literature often implies that maximum power transfer is always important and the sole reason for transformer coupling. In reality, we are only *occasionally* concerned with maximum power transfer in electronics, and even where we are concerned with it, fairly large impedance mismatches result in only *relatively* small power losses.

Table 9-3 shows the power loss for various degrees of mismatch. As you can see, a 50 percent mismatch results in a power loss of only 14 percent, or less than 3 dB. A 100 percent mismatch results in 56 percent, or a little more than a 6-dB power loss. The table shows fairly large mismatches, and you can see that small deviations from the *ideal* in a practical impedance-matching transformer will yield only small losses.

However, transformer coupling can also provide a significant advantage in terms of voltage gain in stages where voltage gain is far more important than maximum power transfer as such.

Let us use fully bypassed iterative stages, as an example, and compare the practical gain per stage with capacitor coupling and the practical gain per

TABLE 9-3. IMPEDANCE MISMATCH VS. POWER LOSS IN THE LOAD

Ratio of Load to Source Z	Percent Mismatch	Percent of Power Loss to Load
10:1	1,000%	83%
4:1	400%	68%
2:1	100%	56%
1.5:1	50%	14%
1:1	0%	0%

stage with transformer coupling. Remember that capacitor-coupled fully bypassed iterative stages provide a maximum voltage gain per stage of about 66. Suppose we compare that figure with what is possible with transformer coupling.

Just to review the situation: the input impedance of a fully bypassed stage is about 2 kΩ at a QI_c of 1 mA. If we have a stage driving that fully bypassed stage (also with a QI_c of 1 mA), the voltage gain of the driving stage is severely limited by the low input impedance of the driving stage. See Fig. 9-8.

If we make R_L 10 kΩ, the total value of R_{Lac} is 10 kΩ‖2 kΩ, or 1.66 kΩ. The voltage gain is

$$V_g = \frac{1.66 \text{ k}\Omega}{25} \approx 66$$

If we use a transformer with a 10:1 ratio as shown in Fig. 9-8*b*, we can make the collector see a much larger *reflected* impedance which increases the voltage gain. Let's look at some figures.

Elementary transformer theory tells us that the impedance reflected from the primary back into the secondary is

$$R_{Lac} \text{ (reflected)} \approx n^2 R_{Lw}$$

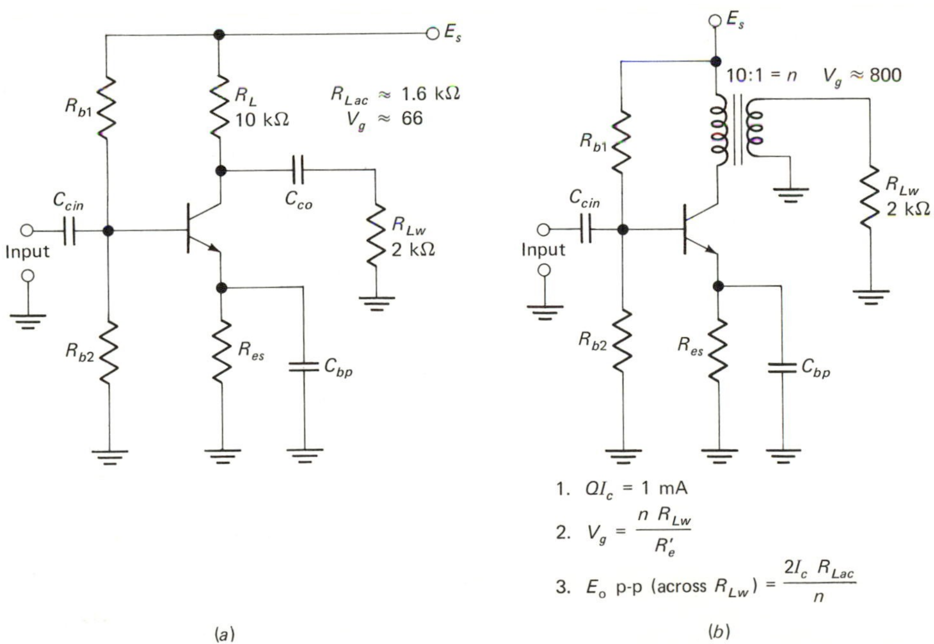

Fig. 9-8 *Transformer coupling.* (a) *The loading problem;* (b) *the transformer-coupled circuit*

where R_{Lac} is the *effective* load impedance at the collector and n is the turns ratio of the transformer, in this case 10:1. The effective collector load impedance with the transformer in Fig. 9-8*b* is

$$R_{Lac} \approx 10^2 \times 2\ \text{k}\Omega = 200\ \text{k}\Omega$$

The potential gain *at the collector* is

$$V_g \approx \frac{R_{Lac}\ (\text{effective})}{R_e'} = \frac{200\ \text{k}\Omega}{25} = 8{,}000$$

Please don't get too excited yet. Remember that we have a 10:1 voltage step-down transformer between the collector and the load, so that the actual voltage gain, as delivered to the load, is $^{8{,}000}/_{10}$, or 800. Still, compared to a voltage gain of 66 for a capacitor-coupled circuit, a voltage gain of about 800 is not bad. As good as this looks, we must weigh the cost of the transformer against the available gain, and we must realize that truly high-fidelity audio-frequency transformers are very expensive, and they take up a lot of space.

Suppose we write a practical voltage gain equation which takes into account both the impedance step-up and the voltage step-down:

$$V_g = \frac{n^2 R_{Lw}}{nR_e'} = \frac{nR_{Lw}}{R_e'}$$

In the case of radio-frequency circuits transformer coupling is closer to an ideal solution because RF transformers are less susceptible to saturation and bandwidth problems. However, most RF circuits use tuned transformers for selectivity; we shall look at tuned transformers in the next section.

Selecting the Quiescent Collector for the General Case of Any Value R_{Lw} (See Fig. 9-8*b*)

In this case the dc resistance of the transformer will normally be quite low, so that the quiescent voltage drop across the transformer primary is not a limiting factor.

Because the entire load in this case is an ac load, the table of possibilities cannot be used. The voltage gain is simply the reflected collector load impedance divided by R_e'.

We can find the required QI_c by selecting it to provide the appropriate value for R_e' to yield the desired voltage gain:

1. Compute R_e' required for the specified voltage gain.

$$R_e' = \frac{R_{Lac}}{V_g}$$

2. Compute the value of QI_c required for the calculated value of R_e'.

$$QI_c = \frac{25}{R_e'} \qquad QI_c \text{ in mA}$$

Another Consideration

In addition to selecting a QI_c to meet the voltage gain requirements, we must also have sufficient available current to develop the desired E_o p-p voltage across the actual R_{Lw} on the secondary side of the transformer. With resistive loads (R_L) our quest for voltage gain and the required dc voltage drop across the load almost automatically takes care of this problem. Things are a bit more complex with transformer coupling, and so we must check to see whether the QI_c we have selected is large enough to produce the desired voltage swing across R_{Lw}. If you will remember as far back as Chap. 8, we used an equation containing QI_c to compute the available E_o p-p:

$$E_o \text{ p-p} = 2QI_cR_{Lac}$$

In a transformer-coupled circuit we must modify this a bit because the E_o p-p specified here is at the collector. We are interested in the E_o p-p delivered to the final load R_{Lw}, which is stepped down by the turns ratio n of the transformer. We must also rearrange the equation to solve for QI_c. The modified equation looks like this:

$$QI_c \text{ (min)} = \frac{nE_o \text{ p-p}}{2R_{Lac}}$$

In the case of iterative stages it is not generally necessary to go through all this if we stick to QI_c values of 0.5 to 1 mA. We are pretty well restricted by available interstage transformers, and they are generally designed to operate in this current range. A higher collector current will saturate the transformer, and then the best we can hope for is a *badly* distorted signal. A much lower current will increase R_e', thus lowering the voltage gain, and can possibly get us into other trouble.

Selecting a Transformer

I have just opened my catalog to select a transformer for the next example. It is not uncommon for the catalog to list only primary and secondary impedances without comment on the turns ratio n. There are perhaps a dozen transformers listed here, and any one of them would do, depending upon the desired voltage gain. We must select one *prior* to starting the design. The actual impedances are only slightly important, and a secondary impedance of anywhere from 1 to 4 kΩ (±100 percent) will work fine into a 2-kΩ load (the input impedance of a fully bypassed stage). The primary impedance is of no importance within the range of impedances listed for this interstage service, 1 to 100 kΩ. Even up to 1 MΩ would serve, although you are not likely to find one with a primary impedance that high. I might add here that if we were really interested in a perfect impedance match, we would need a transformer with a 2-kΩ secondary and a 0.5 to 2-MΩ primary. Apparently nobody is really interested in matched impedances in this kind of circuit, for I have never seen such a beast, and I suspect the technical problems involved in the design of such a transformer would be outrageous.

Converting Impedance Ratios to Turns Ratios

Because most catalogs list only the impedances and we are interested in the turns ratio n to compute voltage gains, we must go back to basic transformer theory for the following relationship:

$$n = \sqrt{\frac{Z_{\text{Pri}}}{Z_{\text{Sec}}}}$$

Suppose I select a couple of transformers from this catalog and see what voltage gains are possible, working into a 2-kΩ load (R_{Lw}).

Transformer 1
 Primary impedance 20 kΩ
 Secondary impedance 2 kΩ

$$n = \sqrt{\frac{20\ \text{k}\Omega}{2\ \text{k}\Omega}} = \sqrt{10} \approx 3$$

Substituting this value of n in the voltage gain equation (assuming an R_e' of 25 Ω), we get

$$V_g = \frac{R_{Lw}}{R_e'}$$

$$V_g = \frac{3 \times 2\ \text{k}\Omega}{25\ \Omega} \approx 240$$

We will use this particular transformer for the next example.

Transformer 2
 Primary impedance 100 kΩ
 Secondary impedance 1 kΩ

$$n = \sqrt{\frac{100\ \text{k}\Omega}{1\ \text{k}\Omega}} = \sqrt{100} = 10$$

Substituting this value of n in the voltage gain equation (again assuming an R'_e of 25 Ω), we get

$$V_g = \frac{R_{Lw}}{R'_e}$$

$$V_g = \frac{10 \times 2\ \text{k}\Omega}{25\ \Omega} = \frac{20\ \text{k}\Omega}{25\ \Omega} = 800$$

Both transformers listed a maximum dc current of 2 mA, and so a 1-mA operating current should be fairly safe. Now for an example.

Example 9-4 The Iterative Fully Bypassed Transformer-coupled Current-mode Circuit with Base Bias

Design an iterative transformer-coupled fully bypassed circuit, using the following specifications.

Transistor Parameters

1. β	100
2. I_c (max)	100 mA
3. E_c (max)	25 V

Circuit Parameters

1. Voltage gain	240
2. E_o p-p	4 V
3. R_{Lw} (final)	2 kΩ
4. Stability factor	10
5. $\%\Delta QI_c$ to be allowed	10%
6. Maximum operating temperature	100°C
7. f_o	20 Hz

The circuit for Example 9-4 is shown in Fig. 9-9. The completed design sheet is shown in Fig. 9-10.

FIELD 1 COLLECTOR SIDE CALCULATIONS

FRAME 1

Compute E_{Res}, the emitter stability voltage drop.

$$E_{Res} = \frac{\Delta V_{BE} \times 100}{\%\Delta QI_c}$$

Figuring 2.5 MV/C°, we have a total Δt of $100° - 20° = 80°\text{C}$ and a total ΔV_{BE} of $80° \times 2.5$ mV, or 200 mV.

$$E_{Res} = \frac{200 \times 10^{-3} \times 10^2}{10}$$

$$E_{Res} = 2\ \text{V}$$

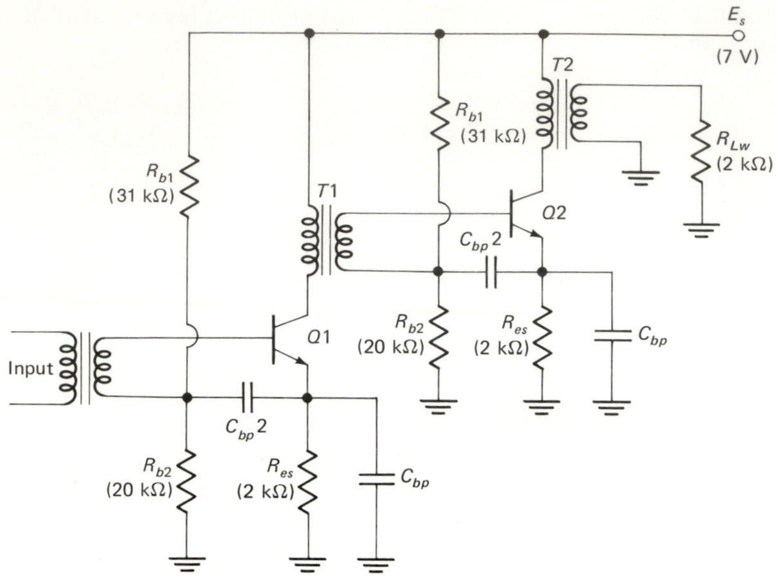

Fig. 9-9 The base-biased transformer-coupled common-emitter iterative circuit

It is worthwhile to note here that a 10 percent variation in QI_c due to ΔV_{BE} is quite satisfactory for this kind of circuit, provided that E_Q is greater than 3 V.

FRAME 2

Make the collector current decision (dictated mostly by the transformer and common practice).

$QI_c = 1$ mA

FRAME 3

Compute R_{es}.

$$R_{es} = \frac{E_{Res}}{QI_c} = \frac{2\text{ V}}{1\text{ mA}} = 2\text{ k}\Omega$$

FRAME 4

Compute R'_e.

$$R'_e = \frac{25}{QI_c} = \frac{25}{1} = 25\ \Omega$$

FRAME 5

Compute R_{Lac}.

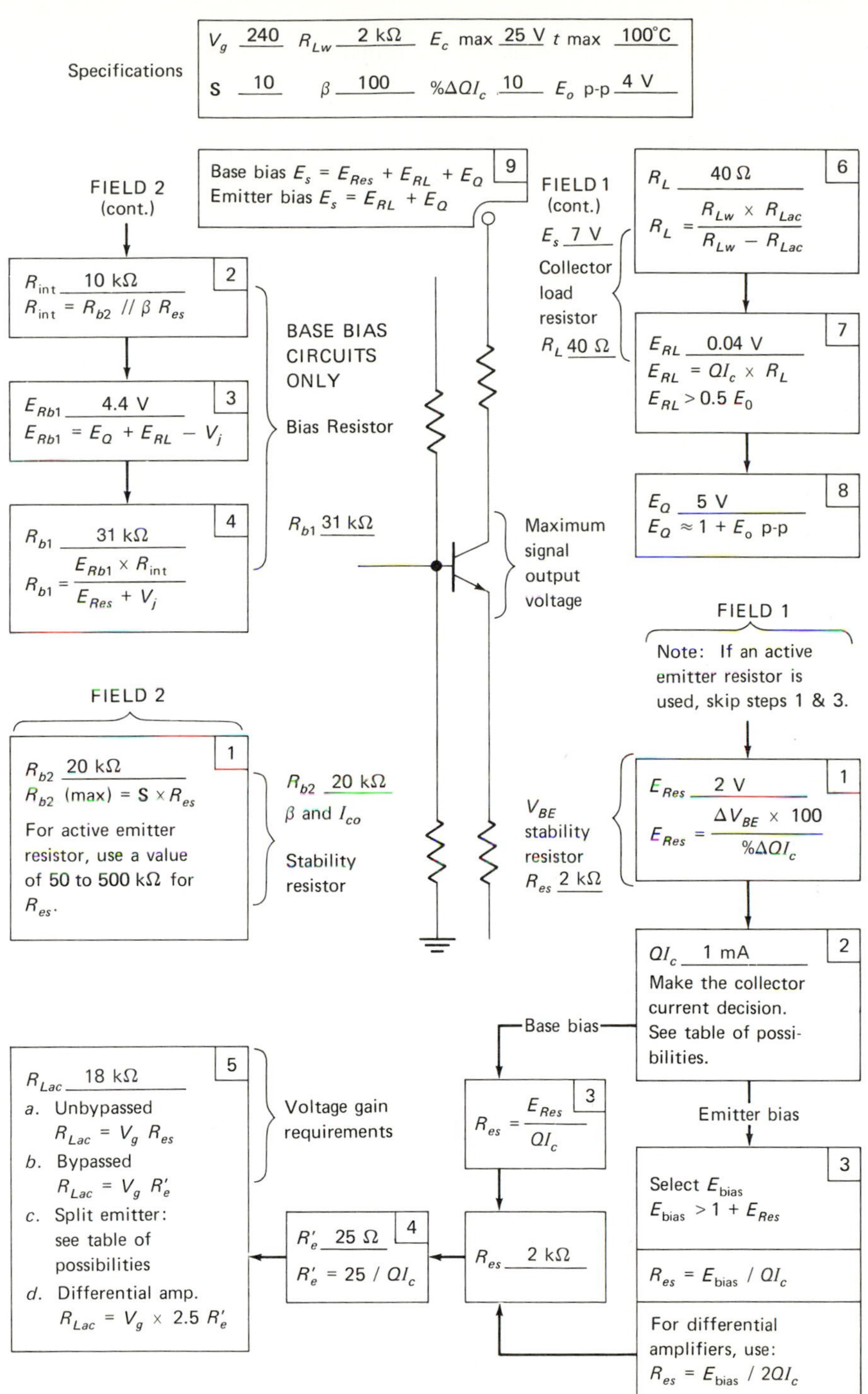

Fig. 9-10 *The completed design sheet for the transformer-coupled circuit in Example 9-4*

This has already been determined by the transformer we selected, but we will compute it anyway.

$R_{Lac} = n^2 R_{Lw} = 3^2 \times 2\text{ k}\Omega = 9 \times 2\text{ k}\Omega = 18\text{ k}\Omega$

FRAME 6

Compute R_L.

In this case R_L is simply the dc resistance of the transformer, which in this case was listed in the catalog as 40 Ω.

FRAME 7

Compute E_{RL}.

$E_{RL} = QI_c \times R_L = 1 \times 10^{-3} \times 4 \times 10^1 = 4 \times 10^{-2}$ V
$E_{RL} = 0.04$ V

This voltage drop is so small we will call it zero in future calculations.

FRAME 8

Compute E_Q.

$E_Q\ (\text{min}) = 1 + E_o$ p-p
$E_Q\ (\text{min}) = 1 + 4 = 5$ V

FRAME 9

Compute E_s.

$E_s = E_{Res} + E_{RL} + E_Q$
$E_s = 2 + 0 + 5 = 7$ V

FIELD 2 BASE SIDE COMPUTATIONS

FRAME 1

Compute R_{b2}.

$R_{b2} = \mathbf{S} \times R_{es} = 10 \times 2\text{ k}\Omega = 20\text{ k}\Omega$

FRAME 2

Compute R_{int}. (Base bias, so we go on.)

$R_{int} = R_{b2} \| \beta \times R_{es}$
$R_{int} = 10\text{ k}\Omega \| 100 \times 2\text{ k}\Omega$
$R_{int} = 10\text{ k}\Omega \| 200\text{ k}\Omega$
$R_{int} \approx 10\text{ k}\Omega$

FRAME 3

Compute E_{Rb1}.

$E_{Rb1} = E_Q + E_{RL} - V_j$
$E_{Rb1} = 5 + 0 - 0.6$
$E_{Rb1} = 4.4 \text{ V}$

FRAME 4

Compute R_{b1}.

$$R_{b1} = \frac{E_{Rb1} \times R_{int}}{E_{Res} + V_j}$$

$$R_{b1} = \frac{4.4 \times 10 \text{ k}\Omega}{1.4}$$

$$R_{b1} \approx 31 \text{ k}\Omega$$

You will notice that we ignored the fact that the secondary winding resistance of the transformer is between the junction of the bias network and the base. This resistance is almost always negligible.

We have an additional capacitor to compute, $C_{bp}2$; its reactance at the lowest frequency of interest should be about $0.1R_{b2}$. This capacitor is often returned to ground instead of to the emitter; electrically it doesn't make any difference.

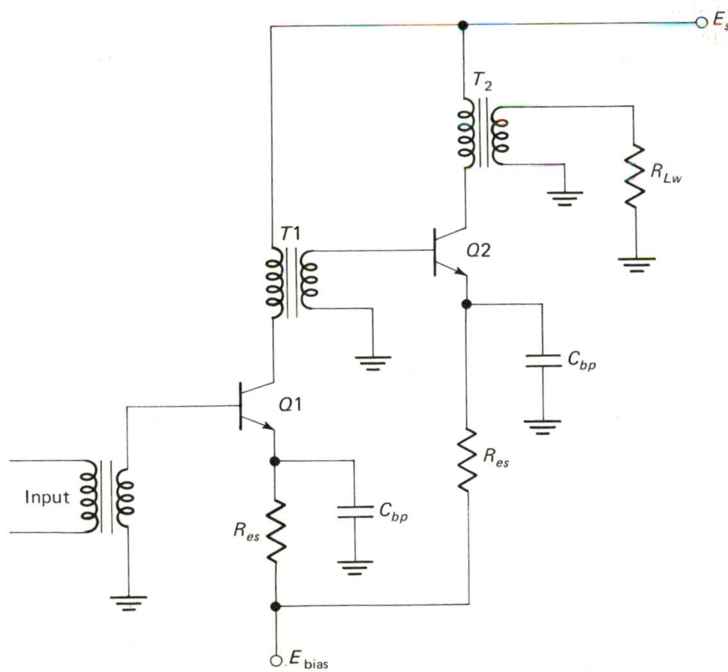

Fig. 9-11 The emitter-biased transformer-coupled circuit

The Emitter-biased Version

The emitter-biased version is handled in the same way as the base-biased circuit except that the emitter bias branch is used and no calculations are required for R_{b1}. The transformer winding resistance will nearly always be low enough to satisfy the stability requirement. See Fig. 9-11.

9-6 TUNED TRANSFORMER COUPLING

Most radio-frequency circuits are designed to operate within a relatively narrow band of frequencies, while sharply rejecting frequencies outside of that band. The tuned capacitor-coupled circuit shown in Fig. 9-12 shows one approach.

The resonant circuits in the collectors serve the function of frequency selection. If you remember your basic theory, the equation for the resonant frequency of the tuned circuit in Fig. 9-12 is

$$f_o \approx \frac{1}{2\pi\sqrt{LC}}$$

At resonance the impedance of the tuned circuit will be very high, and so will the voltage gain of the stage. At frequencies off resonance the impedance of the tuned circuit falls rapidly, and so does the voltage gain. The bandwidth limits are measured at the frequencies on either side of the resonant frequency at which the gain is down by about 25 percent, or 3 dB.

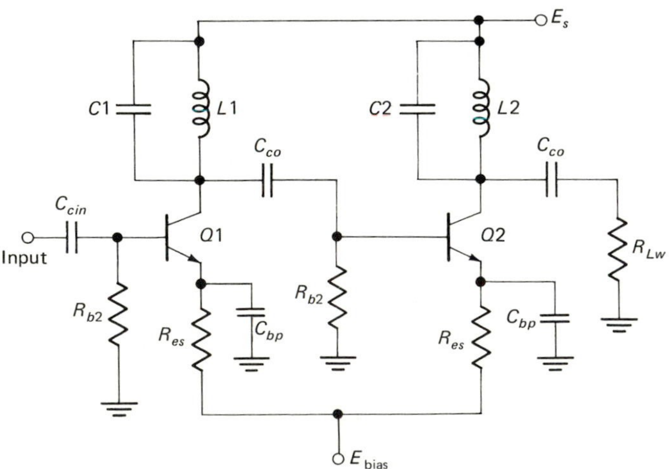

Fig. 9-12 The tuned capacitor-coupled circuit

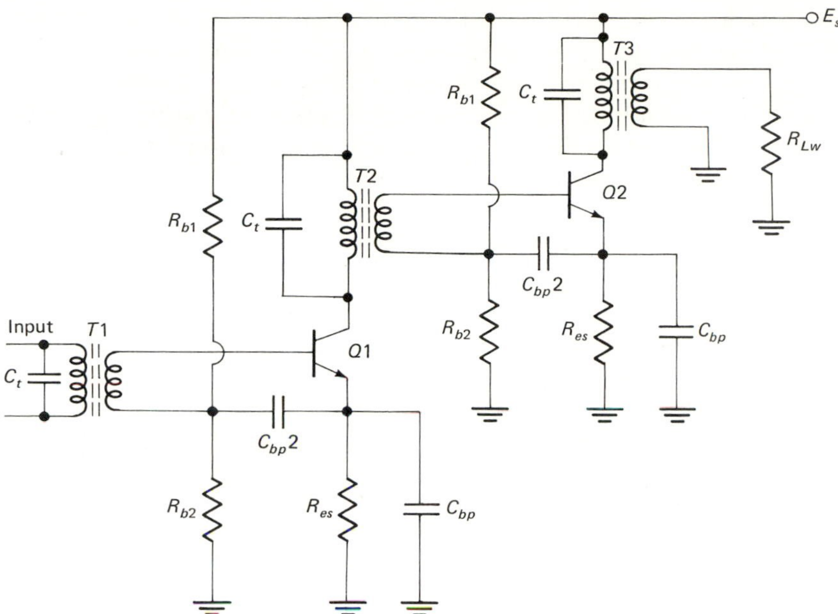

Fig. 9-13 *Tuned transformer-coupled circuit with base bias*

The bandwidth equation is

$$(3\ \text{dB})\ Bw = \frac{f_o}{Q}$$

where f_o is the resonant frequency and $Q = R_{Lac}/X_L$ (note $X_L = X_c$ at resonance). The value of R_{Lac} in this configuration is likely to be fairly low, especially in base-biased versions, and this leads to low values for Q and, consequently, broad bandwidths. The bandwidth is likely to be not only broad, but also not too well defined because of the uncertainty in the value of R'_e. As a result this circuit is not very often used.

A better approach is to use transformer coupling where the value of R_{Lac} reflected into the tuned primary (see Fig. 9-13) can be made high enough to allow high values for Q and narrow bandwidth. Even if a broad-band circuit *is* required, it is best to shunt the primary with a fixed resistor of any desired value to get a very predictable bandwidth, rather than depend upon the uncertain value of R'_e to arrive at a broader bandwidth.

Designing a Tuned Transformer-coupled Circuit

Basically the procedure demonstrated in Sec. 9-4, the last transformer-coupled (AF) circuit, is entirely applicable to the design of radio-frequency circuits.

However, there are things about RF circuits that you must know in addition to what I have told you. Many of these special problems are subtle and require special study beyond the scope of this book. However, if you are familiar with those additional problems, you can readily apply the techniques presented in Sec. 9-4 to RF circuits. I will suggest some topics to study if you are interested in them.

1. Neutralization techniques and parasitics
2. Gain control methods in RF circuits
3. The effect of the Miller effect upon tuned circuits
4. Radio-frequency transformer characteristics and problems
5. Methods of bandwidth control
6. Physical problems in RF circuit layout

An additional problem in RF circuits is the frequent lack of data available about commercial RF and IF (intermediate-frequency) transformers. Sometimes we have no choice but to use circuit designs supplied by the transformer manufacturer.

Still, with all these admonitions, my students have had good success in designing 455 kHz IF strips with the procedures in Sec. 9-4 coupled only with what they had learned in basic ac circuits and some additional informa-

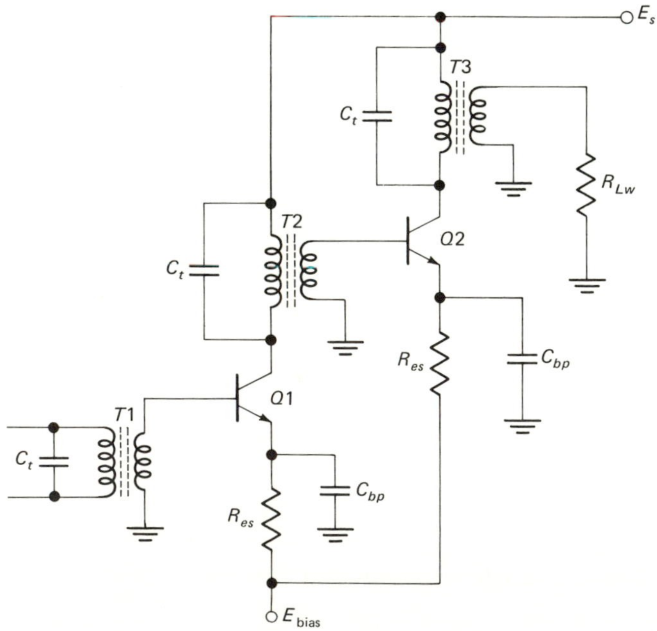

Fig. 9-14 Tuned transformer-coupled circuit with emitter bias

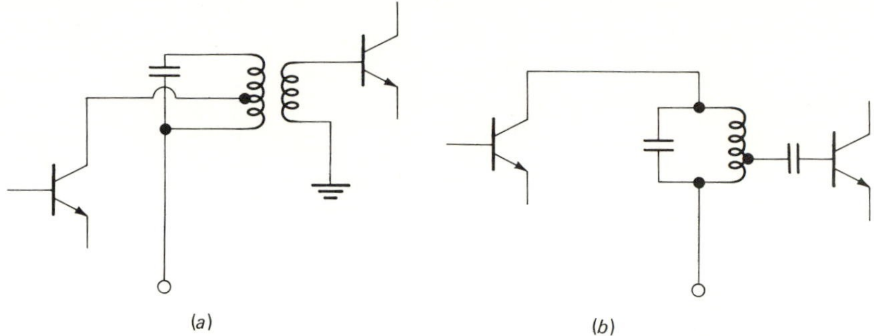

Fig. 9-15 Two common variations of RF transformer configurations. (a) *Tapped primary;* (b) *the autotransformer*

tion about proper RF breadboarding, and some data about the IF transformers they were using.

Figure 9-14 shows an emitter-biased tuned transformer-coupled circuit. Often RF transformers are autotransformers or other tapped configurations. Figure 9-15 shows two of the most common variations.

PROBLEMS

Problem 9-4

TITLE: **Designing the Base-biased Fully Bypassed AF Transformer-coupled Iterative Common-emitter Amplifier**

PROCEDURE: **Design the circuit, using the following specifications. See Fig. 9-9.**

Transistor Parameters	Q1 and Q2
1. β (min)	100
2. I_c (max)	100 mA
3. E_c (max)	30 V
4. Transistor type	*npn* silicon

Circuit Parameters	
1. Voltage gain	$\approx$800 per stage
2. E_o p-p	5 V
3. R_{Lw}	2 kΩ
4. Stability factor	10
5. $\%\Delta QI_c$ to be allowed	5%
6. Maximum operating temperature	60°C
7. f_o (dictated by transformer)	50 Hz
8. Transformer primary impedance	20 kΩ
9. Transformer secondary impedance	2 kΩ
10. Transformer dc resistance, primary	97 Ω

10

direct-coupled circuits

OBJECTIVES:

(Things you should be able to do upon the completion of this chapter)

1. Design a two-stage direct-coupled (collector-to-base) amplifier.
2. Design a two-stage diode-coupled amplifier.
3. Design a two-stage direct-coupled (collector-to-base) amplifier with two-stage voltage-mode feedback.
4. Design a Darlington pair emitter-follower (common-collector) circuit.

EVALUATION:

1. There are problems in the chapter.
2. As usual the real test, the final evaluation, is your ability to design circuits that work in the laboratory, according to the criteria and specifications in the laboratory manual.

10-1 THE COLLECTOR-TO-BASE DIRECT-COUPLED CIRCUIT

The first circuit we will study is similar to the *R-C*-coupled common-emitter circuits that we discussed in the last chapter, except that there are no coupling capacitors. The *collector* of the first stage is coupled directly to the *base* of the second stage. See Fig. 10-1. Figure 10-1 shows all the voltage drops and component values for the first example in this chapter and is typical of circuits of this type. There are several possible variations of this basic circuit when it is used with an ac signal, and where bypassing of the emitter resistors is allowable. Table 10-1 (p. 367) summarizes the available voltage gains for each variation for the example we will work shortly. The values are fairly typical.

The particular advantage of this direct-coupled circuit is the absence of R_{b1} and R_{b2} in the second stage. If the second stage is unbypassed, the input

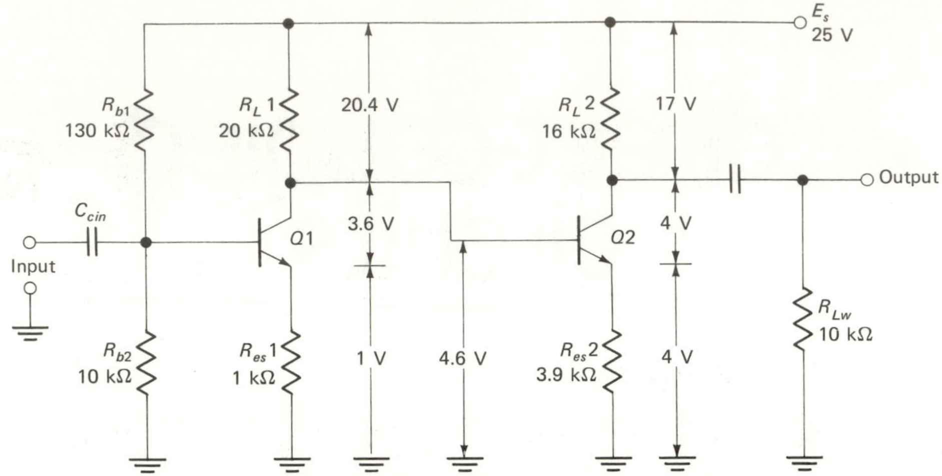

Note: All resistance values are standard values and may differ slightly from the calculated values.

Fig. 10-1 Collector-to-base direct-coupled circuit

impedance for both ac and dc for stage 2 is very high (βR_{es}); this leaves the first stage with a very light load, often in excess of 200 kΩ. This means that the quiescent operating conditions of the first stage are almost completely independent of second-stage loading effects. There are cases where the second stage *must* be unbypassed with a very restricted voltage gain, or even an emitter follower with a voltage gain of unity. In these cases we must get all the voltage gain (or nearly all of it) in the first stage. Direct coupling can provide a factor of 10 (or so) increase in the voltage gain of stage 1 simply because R_{b1} and R_{b2} are eliminated. Table 10-1 indicates that bypassing both stages yields the highest total voltage gain, and here we can get total voltage gains of about the same order as those obtained by two capacitor-coupled stages, but with fewer components.

As you can see by the table, the unbypassed version has very little to offer in the way of available voltage gain, but there are times when response down to dc is required and capacitors just aren't allowed. Later, I will show you a diode-coupled version which can provide a bit higher voltage gain when capacitors are not allowed. Probably the most important version of this circuit incorporates two-stage voltage-mode feedback. That variation is used extensively in regulators and power amplifiers, because of the excellent stability possible with the addition of the feedback loop. This circuit as it stands in Fig. 10-1 is not overwhelmingly stable, partly because any drift in the first stage is amplified by the current gain ($I_g \approx \mathbf{S}$) of the second stage, and partly because of some necessary compromises which are imposed by the direct coupling.

In general V_{BE} stability in the second stage is excellent because E_{Res} is necessarily high, while it tends to be only fair in the first stage. The stability factor for the pair of stages is the *product* of the stability factors of the individual stages. For that reason this kind of direct coupling should *not* be used for more than two stages. The direct-coupled *pair* is often isolated by input and output capacitors when the signal is ac.

Before we try a sample design problem, let us take time to look at some limitations imposed by coupling directly from the collector of the first transistor to the base of the second. If you will look at Fig. 10-1, you will see that I have noted some typical voltage drops and I will explain why they must be so. (The resistance values will be explained in Example 10-1.)

1. The voltage drop from the collector of Q1 to ground *must* be the same as the voltage drop between the base of Q2 and ground because they are connected *directly* together.
2. Because of the danger of saturation of Q1 as the temperature drifts, we must have an E_Q for Q1 of at *least* 2 V. Three volts (or more) would be safer. There is a small flexibility in how to divide the 4.6 V between E_Q (Q1) and E_{Res} (Q1), but we must use at *least* two of the 4.6 V for E_Q and preferably a bit more. (Please look again at Fig. 10-1.)
3. There is always a 0.6-V drop across the junction of Q2.
4. If we have 4.6 V from the base of Q2 to ground, and a 0.6-V drop across the base-emitter junction of Q2, we must have 4 V for E_{Res} of Q2. (Please study Fig. 10-1 until all this makes sense.)

One of the problems in designing this circuit is that it is almost impossible to meet *all of a set of arbitrary specifications*. Because of the direct coupling there are some natural conflicts between the resistance values required for necessary dc voltage drops and those demanded by desired signal parameters. The design process *can* be a modified version of the current-mode design sheet process with required quiescent voltage drops placing some limits on the possible signal parameters. Unfortunately this approach will not allow you to meet all initial specifications in the very predictable fashion you have come to expect. In addition, it is too much work to take that route if it won't yield the kind of results we are used to. It would be possible to develop a design procedure for this circuit which would take all these factors into account, define all the limits, and allow you to meet all specifications (as long as they fall within specified limits), but it would be a complex procedure, and I doubt that the circuit as presented here is important enough to merit your learning it and my developing it. What we will do here instead is to use a simplified voltage-drop-oriented design procedure for this circuit which will yield the maximum voltage gain for a given load and supply voltage, and then use negative feedback around both stages to get closer to the desired gain figures if it is necessary, or to provide better stability if that happens to be the requirement.

Example 10-1

Design a two-stage direct-coupled amplifier with the following specifications. See Fig. 10-1.

Transistor Parameters	Q1	Q2
1. β (min)	100	100
2. I_C (max)	100 mA	100 mA
3. E_C (max)	30 V	30 V

Circuit Parameters	
1. Voltage gain (total)	See Table 10-1
2. E_o p-p (stage 2)	3 V
3. R_{Lw} (stage 2)	10 kΩ
4. Stability factor (total)	50 max
5. $\%\Delta QI_c$ (each stage)	10%
6. Maximum operating temperature	55°C
7. f_o	20 Hz
8. E_s (power supply voltage)	25 V

PHASE 1

Because it is the *quiescent voltages* which set *all* the limits on voltage gain, we will begin by computing all the quiescent voltages in such a way that we will end up with the *maximum* voltage gain possible and still meet the stability requirements.

FRAME 1

Compute E_{Res} of Q1.

$$E_{Res} = \frac{\Delta V_{BE} \times 100}{\%\Delta QI_c}$$

$$E_{Res} = \frac{105 \times 10^{-3} \times 10^2}{10\ (\%)} = 1.05 \text{ V}$$

Let's call it 1.0 V even, since five-hundredths of a volt is negligible and it will make the arithmetic easy.

$$E_{Res} = 1 \text{ V}$$

If this figure turns out to be much more than a volt, the specification for V_{BE} variations is not realistic for this circuit if the second stage is unbypassed, because this will force us up to a higher value for R_{es} in stage 2, reducing its voltage gain.

FRAME 2

Assign a value for E_Q in Q1.

This is more or less an arbitrary decision, but the range here is limited to from 2 V to about 3.6 V. Even though the decision is arbitrary, you can hardly go far wrong with so limited a range. The p-p output voltage is still $E_Q - 1$, as it is in all current-mode circuits. This is normally not a problem, but if it is, we really can't do a great deal about it in this configuration. If the second stage is bypassed, we can elevate E_Q of Q1 to allow for a greater E_o p-p. This will seldom be necessary and should not be done unless it *is necessary;* we shall assume here that it is not. Let us use 3.6 V.

E_Q (Q1) $= 3.6$ V

FRAME 3

Select a power supply voltage.

Here we will get higher gains for higher supply voltages, because we can make R_L in each stage larger. The power supply voltage cannot, however, exceed the maximum collector voltage rating of the transistor. In this example E_c (max) is 30 V. Suppose we use a 25-V power supply.

$E_s = 25$ V

FRAME 4

Compute E_{RL}, the voltage drop across the collector resistor (Q1).

$$E_{RL} = E_s - (E_Q + E_{Res})$$
$$E_{RL} = 25 - (3.6 + 1)$$
$$E_{RL} = 20.4 \text{ V}$$

FRAME 5

Compute E_{Res} of Q2.

$$E_{Res} = E_Q + E_{Res}\ (\text{Q1}) - V_j$$
$$E_{Res} = (3.6 + 1)\ \text{V} - 0.6\ \text{V}$$
$$E_{Res} = 4\ \text{V}$$

FRAME 6

Compute E_Q for Q2.

$$E_Q = 1 + E_o\ \text{p-p}$$
$$E_Q = 1 + 3 = 4\ \text{V}$$
$$E_Q = 4\ \text{V}$$

FRAME 7

Compute E_{RL} of Q2.

$$E_{RL} = E_s - (E_Q + E_{Res})$$
$$E_{RL} = 25 - (4\ \text{V} + 4\ \text{V})$$
$$E_{RL} = 17\ \text{V}$$

FRAME 8

Select a quiescent collector current for both transistors.

Consult the table of possibilities, using the final R_{Lw} and the configuration of the *last* stage, unbypassed or fully bypassed (split emitter is also sometimes used). If we use the same QI_c for both stages, Q1 will always be reasonably lightly loaded, for dc, but the final R_{Lw} is almost invariably a much lower impedance than the input resistance of Q2, which is R_{Lw} for Q1 (when Q2 is unbypassed). If we are to use the same quiescent current for both transistors, we generally cater to the heaviest load (lowest resistance) of the two R_{Lw}'s. Let us examine both cases, where R_{es} in the last stage is unbypassed, and where it is bypassed (according to the specifications the final R_{Lw} is 10 kΩ).

Unbypassed. In the unbypassed case we should try for a gain of 2 or so; this means a QI_c of about 2 mA. If we went up to 10 mA (see the unbypassed table of possibilities, Table 5-5), we could get a V_g in the last stage of a little over 3. At 1 mA we get a gain of about 1.5; so to double the gain, we need (here) a 10-fold increase in QI_c. In the majority of cases it is not worth it, and unless I were really pressed, I would stick to 1 or 2 mA. Let's use 1 mA for this example just to keep the arithmetic simple.

Fully bypassed. If we examine the fully bypassed Table 6-2, we find that at 1 mA we can get a voltage gain between 240 and 320 with an E_{RL} of 17.6 V for Q2. So let us design at 1 mA of QI_c for both stages.

$QI_c = 1$ mA

PHASE 2

Compute all resistance values (all simple Ohm's law)

FRAME 1

Compute R_{es} of Q2.

NOTE: The order here is not important.

$$R_{es}\ (Q2) = \frac{E_{Res}}{QI_c} = \frac{4\ \text{V}}{1\ \text{mA}} = 4\ \text{k}\Omega$$

FRAME 2

Compute R_L (Q2).

$$R_L = \frac{E_{RL}}{QI_c} = \frac{17.0\ \text{V}}{1\ \text{mA}} = 17.0\ \text{k}\Omega$$

FRAME 3

Compute R_{es} (Q1).

$$R_{es}\ (Q2) = \frac{E_{Res}}{QI_c} = \frac{1\ \text{V}}{1\ \text{mA}} = 1\ \text{k}\Omega$$

FRAME 4

Compute R_L (Q1).

$$R_L\ (Q1) = \frac{E_{RL}}{QI_c} = \frac{20.4}{1\ \text{mA}} = 20.4\ \text{k}\Omega$$

FRAME 5

Compute R_{b2} (Q1).

$R_{b2} = \mathbf{S} \times R_{es} = 10 \times 1\ \text{k}\Omega = 10\ \text{k}\Omega$

FRAME 6

Compute R_{b1} (Q1).

Here we use the same three-step procedure we used with the current-mode design sheet.

1. Compute R_{int}.
   $R_{int} = R_{b2} \| \beta R_{es}$
   $R_{int} = 10\ \text{k}\Omega \| 100 \times 1\ \text{k}\Omega$
   $R_{int} \approx 9\ \text{k}\Omega$

2. Compute E_{Rb1}.
   $E_{Rb1} = E_Q + E_{RL} - V_j$
   $E_{Rb1} = 3.6 + 20.4 - 0.6$
   $E_{Rb1} = 23.4\ \text{V}$

3. Compute R_{b1}.

$$R_{b1} = \frac{E_{Rb1} \times R_{int}}{E_{Res} + V_j}$$

$$R_{b1} = \frac{23.4 \times 9 \times 10^3}{1 + 0.6}$$

$$R_{b1} \approx 130\ \text{k}\Omega$$

Figure 10-1 shows the circuit with all component values and voltage drops. Input impedance, coupling capacitors, and bypass capacitors are computed in the same way as for R-C-coupled circuits.

The stability factor for stage 1 (Q1) is 10. The stability factor for stage 2 is

$$\mathbf{S} = \frac{R_L\ (\text{Q1})}{R_{es}\ (\text{Q2})} \approx \frac{20\ \text{k}\Omega}{4\ \text{k}\Omega}$$

TABLE 10-1. VOLTAGE GAIN POSSIBILITIES FOR EXAMPLE 10-1

Condition	V_g Stage 1	V_g Stage 2	Total
Both stages unbypassed	20	1.5	30
Stage 1 *only* bypassed	800	1.5	1,200
Stage 2 *only* bypassed	2.2	204	449
Both stages bypassed	88	204	17,952

NOTE: 1. R_{Lac} stage 1 $\approx$ 2.2 kΩ when stage 2 is bypassed.
2. R_{Lac} stage 1 $\approx$ 400 kΩ when stage 2 is unbypassed.
3. R_{Lac} stage 2 $\approx$ 6 kΩ

$S = 5$ (well within the specifications). R_L (Q1) is the total base-to-ground return path for Q2. The power supply impedance is considered negligible compared to 20 kΩ.

10-2 USING DIODES AS COUPLING ELEMENTS

Ordinary silicon diodes can be used as coupling elements between the stages of a collector-to-base direct-coupled circuit, for improving the voltage gain of the second stage. If we could reduce the E_{Res} voltage drop in Q2 from 4 V (see Fig. 10-1) to 1 V, we could increase the voltage gain of the second stage from 1.5 to a little over 6, without bypassing $R_{es}2$. We can do this by using coupling diodes as shown in Fig. 10-2 or by using a zener diode as shown in Fig. 10-3.

In Figs. 10-2 and 10-3 the coupling diodes function as a sort of electronic sponge, soaking up that 3 V that we can best do without as a drop across R_{es} of Q2. The ordinary forward-biased diodes (or the zener diode) develop a constant 3-V drop, so that any *change* in collector (output) voltage of Q1 *must* appear across the input (base-to-ground) of Q2. Because the coupling diodes soak up 3 V of the required 4.6 V, and the base-emitter junction of Q2 soaks up another 0.6 V, we can get by with only 1 V across the emitter resistor of Q2. At 1 mA QI_c, for Q2, this means a 1-kΩ resistor for R_{es} (Q2) and this means an increased voltage gain for Q2; it also means that we have 3 V left over in the collector-to-ground circuit of Q2. We *can* increase E_Q by 3 V to use up the surplus voltage, but we have already met the output

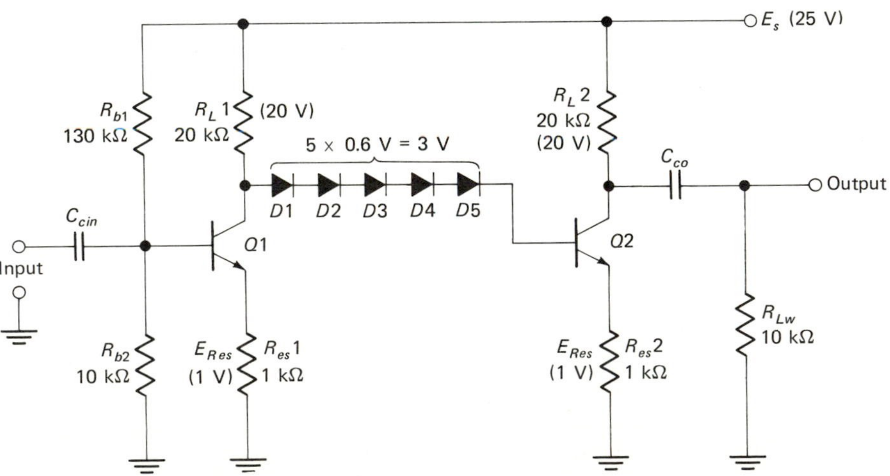

Fig. 10-2 Coupling with diodes for improved voltage gain in dc amplifiers

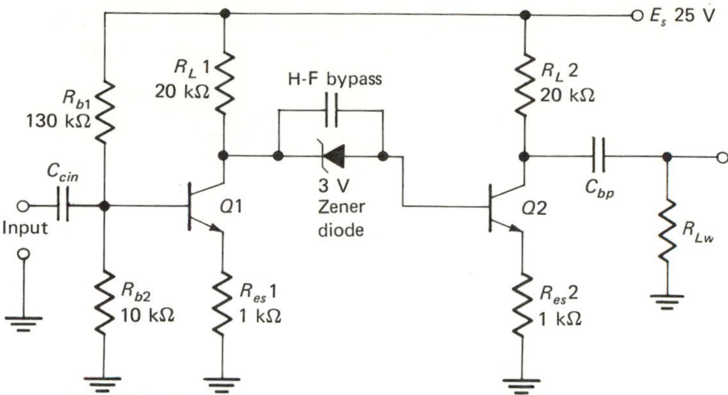

Fig. 10-3 Using zener diode coupling for increased voltage gain in dc amplifiers

swing requirements as specified in Example 10-1; so let us use the extra voltage as part of the voltage drop across R_L and pick up a wee bit more voltage gain because of it. If we do that, it raises the voltage drop across R_L (Q2), in Figs. 10-2 and 10-3, from 17 V to 20 V, and at 1 mA it raises R_L from 17 kΩ to 20 kΩ. The new value for R_{Lac} is 20 kΩ in parallel with 10 kΩ or 6.6 kΩ, which yields a voltage gain, in stage 2, of

$$V_g = \frac{R_{Lac}}{R_{es}} = \frac{6.6 \text{ k}\Omega}{1 \text{ k}\Omega} = 6.6$$

The overall gain of the two-stage circuit with the diodes added is 20 × 6.6, or 132, as compared to a voltage gain of 30 without the diodes. Even if we had kept the 17 kΩ R_L so that the only difference between the two circuits was the change in value of R_{es} (Q2) and the addition of the diodes, these two changes would have increased the overall voltage gain from 30 to 120.

The zener diode can serve the same function as *forward-biased* silicon diodes, but zener diodes operate in an avalanche mode and as a result they are inherently noisy. The 1-μf capacitor across the zener in Fig. 10-3 helps to reduce the zener-generated noise, but it may well be inadequate for low-level or otherwise critical applications. Occasionally the literature refers to these diodes as temperature-compensating diodes, but in fact they are not, and the use of several of them can add to the overall stability problem. The diodes can just as well be placed in the emitter leg, as shown in Fig. 10-4. In that case the most accurate terminology is voltage-elevating diodes. If the diodes are used between the collector of one stage and the base of the next, the term *coupling diodes* is most commonly used. See Figs. 10-2 and 10-3.

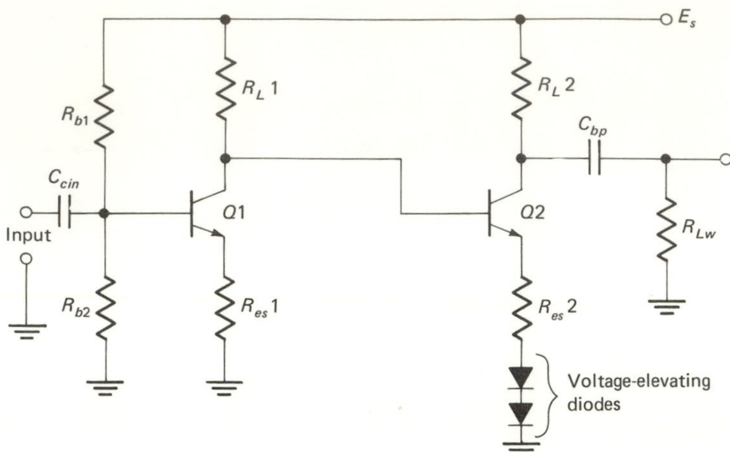

Fig. 10-4 Collector-to-base direct-coupled circuit with voltage-elevating diodes

PROBLEMS

Problem 10-1

TITLE: **Designing the Collector-to-Base Direct-coupled Circuit (Unbypassed)**
PROCEDURE: **Design the circuit, using the following specifications. See Fig. 10-1 or 10-2, your choice.**

Transistor Parameters	Q1 and Q2
1. β (min)	100
2. I_c (max)	100 mA
3. E_c (max)	30 V
4. Transistor type	*npn silicon*

Circuit Parameters	Q1	Q2
1. Voltage gain	maximum possible	
2. E_o p-p	—	4 V
3. R_{Lw}	—	10 kΩ
4. Stability factor	10	10
5. $\%\Delta QI_c$ to be allowed	10%	10%
6. Maximum operating temperature	50°C	50°C
7. f_o	0 Hz	—

Problem 10-2

Compute the total (two-stage) voltage gain and input Z for the circuit in Problem 10-1 for the following conditions:

1. Q1 and Q2 unbypassed
2. Q1 (only) bypassed
3. Q2 (only) bypassed
4. Both Q1 and Q2 bypassed

10-3 COLLECTOR-TO-BASE DIRECT-COUPLED CIRCUIT WITH TWO-STAGE VOLTAGE-MODE FEEDBACK

As I pointed out a little earlier, the collector-to-base circuit is not likely to win any trophies for overall Q-point stability as it was presented in Secs. 10-1 and 10-2, but the addition of two-stage voltage-mode feedback can make the circuit *quite* stable. The circuit is particularly useful when one stage is difficult to stabilize, and this is often the case when the second stage (Q2) is a power transistor. We will examine that particular application and its associated problems in Chap. 12. The circuit is shown in Fig. 10-5.

This circuit has much in common with the more versatile case (case 2) of single-stage voltage-mode feedback, but it yields greater stability, larger voltage gains, and greater flexibility. The circuit makes a particularly effective driver (Q1) and power amplifier (Q2) combination. The second stage (Q2) can be either a common-emitter or a common-collector configuration. When Q2 is a power stage, it is generally an emitter follower.

The circuit suffers from the basic limitations imposed by both voltage-mode feedback and direct coupling.

Voltage-mode feedback tends to produce low input impedances, and direct coupling makes it difficult to design to a given set of arbitrary specifications. An additional problem is generator loading of the feedback network (R_f). In spite of these limitations the circuit has some applications where it can serve better than any other circuit.

If you will examine Fig. 10-5, you will see that E_{Res} serves both stages

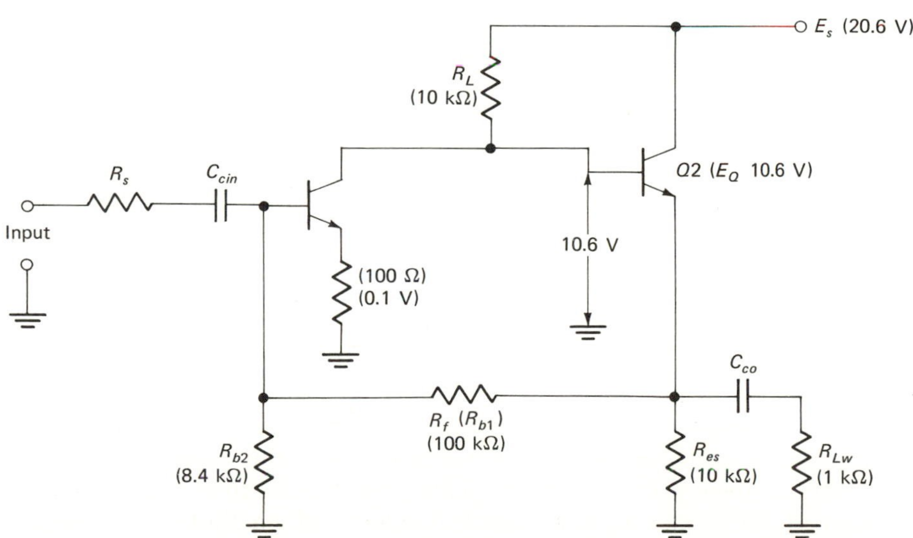

Fig. 10-5 The common-emitter-to-common-collector circuit with two-stage voltage-mode feedback

in the stabilization of V_{BE}. It is a practical, although not entirely accurate, assumption that the V_{BE} stability will be the same for both stages. The usual equation

$$E_{Res} = \frac{\Delta V_{BE} \times 100}{\% \Delta QI_c}$$

will be used and will apply to each stage.

The stability factor for this circuit is $\mathbf{S} = R_f/R_L$ (Q1), and in most cases R_{b2} is required to prevent the overbiasing of Q1 when R_f is made small enough to satisfy the stability requirement. This is essentially the same problem we had in the single-stage voltage-mode feedback circuit, and we will use the same equations in the biasing of Q1 that we used in the more versatile case of voltage-mode feedback in Chap. 7.

The feedback voltage is actually derived from R_L of Q1. The emitter-follower output simply *follows* the collector voltage of Q1, providing a feedback voltage which is in phase with and equal in amplitude to the collector voltage of Q1. Because the circuit is direct-coupled, the quiescent collector voltage variations are also reflected through the emitter follower to R_f. Any independent change in the quiescent output voltage of the emitter follower is also reflected through R_f to the input of Q1, where it is amplified, inverted, and delivered as a correction voltage to the base of the emitter follower. Including the emitter follower in the feedback loop makes the entire circuit function as though it were a single-stage voltage-mode feedback amplifier, providing β stability for the circuit as a whole. The voltage-mode stability factor in this case applies to the entire circuit. The emitter follower also has a stability factor for the current-mode feedback of $\mathbf{S} \approx R_L$ (Q1)/R_{es}. The total stability factor for the circuit including the voltage-mode stability factor $\mathbf{S}_v$ and the current-mode stability factor $\mathbf{S}_c$ is

$$\mathbf{S}\,(Q2) = \frac{\mathbf{S}_v \times \mathbf{S}_c}{\mathbf{S}_v + \mathbf{S}_c}$$

Here, as in nearly all noniterative circuits, we will start our design of the last stage and work toward the first stage. In some cases not all the specifications for the first stage (Q1) can be decided upon until we have designed the second stage. This may not seem to be a desirable situation, but direct coupling does impose such limits, and we must learn to live with them. However, a little experience will allow you to make some pretty expert guesses about what limits are likely to prevail. This circuit is most often used with an emitter-follower second stage, but it can be used with a common-emitter second stage. When a common-collector second stage is used, the total voltage gain is simply the voltage gain of stage 1. This gain can be quite high because the collector of Q1 is generally so lightly loaded. When the second stage is a common-emitter circuit, it is generally most profitable to make Q1 a fully bypassed stage. Let's try an example using a common-collector output stage.

Example 10-2

Design a two-stage common-emitter-to-common-collector direct-coupled amplifier with overall voltage-mode feedback, using the following specifications.

Transistor Parameters	Stage 1	Stage 2
1. β	100	100
2. I_c (max)	100 mA	100 mA
3. E_c (max)	30 V	30 V
4. Type	*npn* silicon	*npn* silicon
Circuit Parameters		
1. Voltage gain	80	≈ 1
2. E_o p-p	5 V	5 V
3. R_{Lw}	To be computed	1 kΩ
4. Stability factor	10 (loop)	1
5. $\% \Delta QI_c$ to be allowed	10%	5%
6. Maximum operating temperature	60°C	60°C
7. f_o	25 Hz	25 Hz

Designing the Emitter Follower (See Sec. 7-4)

FIELD 1

FRAME 1

Compute E_{Res} (min).

$$E_{Res}\ (\text{min}) = \frac{\Delta V_{BE} \times 100}{\% \Delta QI_c}$$

$$E_{Res}\ (\text{min}) = \frac{2.5\ \text{mV} \times 40 \times 100}{5}$$

$$E_{Res}\ (\text{min}) = 2\ \text{V}$$

FRAME 2

Make the quiescent collector current decision.

Let us use a K of 10, making R_{es} $10 \times R_{Lw}$ or 10 kΩ. Consulting the common-collector table of possibilities (Table 7-3), we find that a QI_c of 1.0 mA will do nicely.

$QI_c = 1.0$ mA

FRAME 3

Compute R_{es}.

Here we have decided to make R_{es} $10 \times R_{Lw}$, so that $R_{es} = 10$ kΩ. At 1 mA, a 10 kΩ resistor provides a voltage drop of 10 V, which is well in excess of the minimum E_{Res} of 2 V required for V_{BE} stability.

FRAME 4

Compute R'_e.

$$R_e \approx \frac{25}{QI_c} \approx 25\ \Omega$$

FRAME 5

Compute R_{Lac}.

Skip this frame.

FRAME 6

Compute R_L.

There is no R_L; so skip this frame also.

FRAME 7

Compute E_{RL}.

Skip this frame.

FRAME 8

Compute E_Q.

It is standard operating procedure to make E_Q equal to or greater than E_{Res}, in common-collector circuits, when it is convenient. At this point E_Q can, as usual, be increased to accommodate an available power supply voltage. In this case we will use a voltage of 20.6 V for E_s. The reason for this selection is simply that the regulator in the power supply we have been using to test these circuits went out and the power supply now provides only one voltage: 20.6 V.

This 20.6 V turns out to be a happy accident because it makes the arithmetic involved in the Q1 base side calculations most convenient.

$$E_Q = 10.6\ \text{V}$$

SPECIAL FIELD COLLECTOR SIDE CALCULATIONS FOR Q1

FRAME 1

Compute the value of R_L for Q1.

The value of R_L is equal to or less than $\mathbf{S} \times R_{es}$ (Q2).

$$R_L\ (\text{max}) = 1 \times 10\ \text{k}\Omega = 10\ \text{k}\Omega$$

FRAME 2

Compute QI_c for Q1.

1. The voltage drop from collector of Q2 to ground is 10.6 V.
2. This leaves 10 V across R_L for Q1. E_{RL} (Q1) is equal to E_s minus the collector-to-ground voltage of Q1.

$$E_{RL} = 20.6 - 10.6 = 10\ \text{V}$$

3. Compute QI_c.

$$QI_c = \frac{E_{RL}}{R_L} = \frac{10\text{ V}}{10\text{ k}\Omega} = 1\text{ mA}$$

$$QI_c = 1\text{ mA}$$

FRAME 3

Compute R_{es} for Q1.

$$V_g = \frac{R_{Lac}}{R_{es} + R_e'}$$

If we rearrange this familiar equation to solve for R_{es}, we get

$$R_{es} = \frac{R_{Lac}}{V_g} - R_e'$$

We include R_e' in this case because it is not *generally* going to be small enough in comparison to R_{es} to ignore. Remember that, in this circuit, we are not depending upon R_{es} (Q1) to provide V_{BE} stability for Q1. V_{BE} stability for both Q1 and Q2 is provided by E_{Res} of Q_2 through the feedback resistor R_{b1} (R_f): R_{es} of Q1 is used only to provide added predictability and stability for the voltage gain of Q1. Of course, it does add to the V_{BE} stability, but we can consider the additional stability as simply a bonus and not be too concerned with it.

$$R_{Lac}\ (Q1) \approx R_L\ (Q1) \| \beta R_{Lac}\ (Q2)$$
$$R_{Lac}(Q1) \approx 10\text{ k}\Omega \| \beta\ (10\text{ k}\Omega) \| 1\text{ k}\Omega$$
$$R_{Lac}\ (Q1) \approx 10\text{ k}\Omega \| 100\text{ k}\Omega$$
$$R_{Lac} \approx 10\text{ k}\Omega \approx R_L\ (Q1)$$

Here the input resistance of the second stage was so high that we could ignore it with only a 10 percent error, but there are times when the error can be more significant, and at times we might even want to work around this small an error; in this case we won't take the trouble.

$$R_{es} = \frac{R_{Lac}}{V_g} - R_e'$$

$$R_{es} = \frac{10\text{ k}\Omega}{80} - 25 \qquad E_{Res} = QI_cR_{es} = 1\text{ mA} \times 100 = 0.1\text{ V}$$

$$R_{es} = 100\ \Omega \qquad E_{Res} = 0.1\text{ V}$$

The contribution of R_{es}, in this case, is almost insignificant, but many designers do include it anyway. Personally I generally leave it out.

Base Side Calculations for Q1

FIELD 2

These calculations are essentially the same as the base side calculations for the more versatile case (case 2) of voltage-mode feedback. The only difference is that the bias (feedback) voltage is derived from E_{Res} of Q2 rather than from the collector of Q1. The design sheet (Fig 10-6) includes this case, and you may wish to use it. I will use it here for illustrative purposes. Perhaps, by now, you are too sophisticated to need it.

FRAME 1

Compute R_{b1}.

$R_{b1} = \mathbf{S} \times R_L$ (Q1)
$R_{b1} = 10 \times 10\ \text{k}\Omega \qquad R_{b1} = 100\ \text{k}\Omega$

FRAME 2

Compute E_{Rb1}.

$E_{Rb1} = E_{Res}$ (Q2) $- [E_{Res}$ (Q1) $+ V_j]$
$E_{Rb1} = 10 - (0.1 + 0.6)$
$E_{Rb1} = 10 - 0.7 = 9.3$ V

FRAME 3

Compute I_{Rb1}.

$$I_{Rb1} = \frac{E_{Rb1}}{R_{b1}} = \frac{9.3\ \text{V}}{1 \times 10^5} = 9.3 \times 10^{-5}\ \text{A}$$

FRAME 4

Compute I_b.

$$I_b = \frac{QI_c}{\beta} = \frac{1 \times 10^{-3}}{10^2} = 1 \times 10^{-5}\ \text{A}$$

FRAME 5

Compute I_{Rb2}.

$I_{Rb2} = I_{Rb1} - I_b$
$I_{Rb2} = (9.3 \times 10^{-5}) - (1 \times 10^{-5})$
$I_{Rb2} = 8.3 \times 10^{-5}$

FRAME 6

Compute R_{b2}.

VOLTAGE MODE DESIGN SHEET (case two)

SPECIFICATIONS

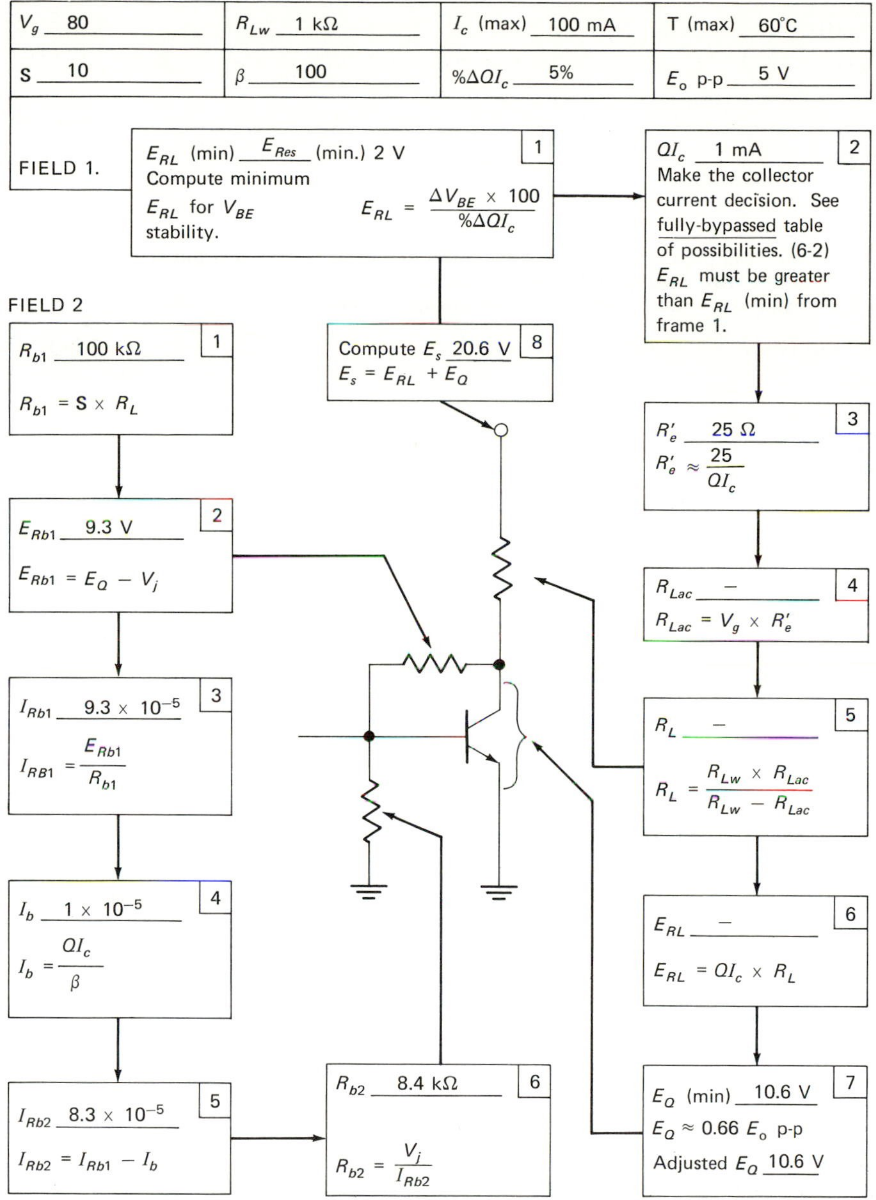

Fig. 10-6 Completed case 2 voltage-mode design sheet for Example 10-2

$$R_{b2} = \frac{E_{Res}\,(Q1) + V_j}{I_{Rb2}}$$

$$R_{b2} = \frac{0.1 + 0.6}{8.3 \times 10^{-5}}$$

$$R_{b2} = 8.4\ \text{k}\Omega$$

PROBLEMS

Problem 10-3

TITLE: **Designing the Common-emitter-to-common-collector Direct-coupled Circuit with Two-stage Voltage-mode Feedback**

PROCEDURE: **Design the circuit, using the following specifications. See Fig. 10-5.**

Transistor Parameters	Q1 and Q2
1. $\beta =$ (min)	100
2. I_c (max)	100 mA
3. E_c (max)	25 V
4. Transistor type	*npn* silicon

Circuit Parameters	Q1	Q2
1. Voltage gain	100	1
2. E_o p-p	—	5 V
3. R_{Lw}	—	600 Ω
4. Stability factor	1	
5. $\%\Delta QI_c$ to be allowed	5%	
6. Maximum operating temperature	60°C	
7. f_o	25 Hz	

NOTE: Use 1 kΩ for R_s.

10-4 THE DARLINGTON PAIR COMPOUND

One of the simplest and most popular direct-coupled circuits is called the Darlington pair. The circuit is shown in Fig. 10-7.

The *pair* shown in the dotted box can be treated as a single transistor with an extremely high β. The β for the compound transistor is the product of the individual β's. For example, if each transistor had a β of 100, the β of the compound would be $\beta_1 \times \beta_2 = 100 \times 100 = 10{,}000$.

What can we do with a transistor which has a β of 10,000? Well, we can make a quite spectacular emitter follower with such a device. Look again at Fig. 10-7*a*. The input impedance, as seen by the generator, is $\beta_1 \times \beta_2 \times (R_{es} \| R_{LW})$. If we assume $\beta_1 = 100$, $\beta_2 = 100$, $R_{es} = 10$ kΩ, $R_{LW} = 1$ kΩ, the generator will see an input impedance of about 10,000 × 1 kΩ, or a whopping 10 MΩ. In order to actually realize this high an input impedance, we would have to bootstrap the collector junction as we did with the emitter follower in Chap. 7.

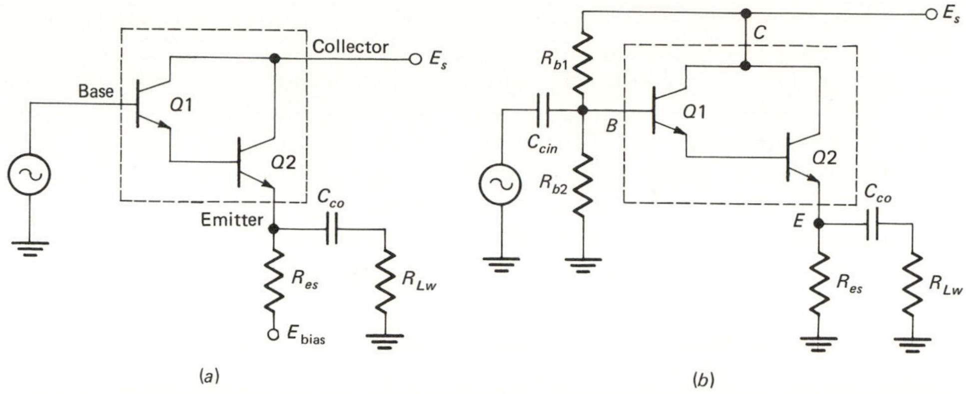

Fig. 10-7 The Darlington pair. (a) *The emitter-biased version;* (b) *The base-biased version*

The most remarkable thing about this simple circuit is that the input base-to-ground resistance can be several hundred times R_{es} without raising the stability factor of the pair above 10 or so. When we examined single-stage emitter followers, we found the practical input resistance limited by the necessary base-to-ground resistance to meet the stability factor requirement. If we needed higher input impedances, we had to resort to bootstrapping. Some designers obtain higher input impedances by simply violating the stability factor limits, but this can be successful only when very large Q-point drifts can be tolerated, and it is an approach best left to the very very experienced designer.

The ideal combination of discrete transistors for Darlington pairs consists of a power transistor driven by a low- or medium-power transistor. An examination of the circuit shows that the *base* current for Q2 *is* the *collector* current of Q1. When a low- or medium-power transistor drives a power transistor, the Q-point current of both transistors can generally be 1 to 25 percent of QI_c max (Q1), so the β of each transistor can be fairly high. When a pair of identical transistors are used, Q1 must operate quite low on its collector current curve, and in most cases this results in a relatively low β for Q1. This can result in a fairly low $\beta_1 \times \beta_2$ product. If you will remember the β vs. QI_c curve, you will recall that β falls significantly at very low collector currents. Commercial Darlington pairs (in a single package) are available in which Q1 has been specifically designed to operate at very small collector currents with high β values. With these commercial units it is generally a fairly safe assumption that β of Q1 is approximately equal to β of Q2.

The principal advantage of such a high β compound is the potentially very high input impedance of $\beta_1 \times \beta_2 \times R_{es}$. However, the stability factor required for stable operation, at first, seemed to negate this advantage because of the low value required for R_b. In early experiments in my laboratory we found that "apparent" stability factors of 200 or so actually yielded stability behaviors of an **S** of 10 or so.

It soon became obvious that we could treat a Darlington pair as a single *high* β transistor in every respect, with the exception of the *real* stability factor. My students insisted that I try to find some theoretical justification for this discrepancy, and I think I have done it, although I can find no authority that even mentions the *problem*. The analysis that I present here fits very well with reality, and is theoretically logical; although I can't, in all honesty, claim that it is the complete answer, it does work consistently in the lab.

The Stability Analysis for the Darlington Pair

The following discussion is based on Fig. 10-8.

The *stability factor for Q1 is*

$$\mathbf{S} \approx \frac{R_b \text{ (actual)}}{R_{es} \text{ (effective)}}$$

This is the usual stability factor equation with the exception that the effective R_{es} for Q1 is β (Q2) $\times$ R_{es}.

$$\mathbf{S}_1 \approx \frac{R_b}{\beta \text{ (Q2) } R_{es}}$$

The stability factor for Q2 is

$$\mathbf{S}_2 = \frac{R_b \text{ (effective)}}{R_{es} \text{ (actual)}}$$

This, again, is the usual stability factor equation except that the effective value of R_b is lowered by the factor β (of Q1). Q1 is essentially an emitter

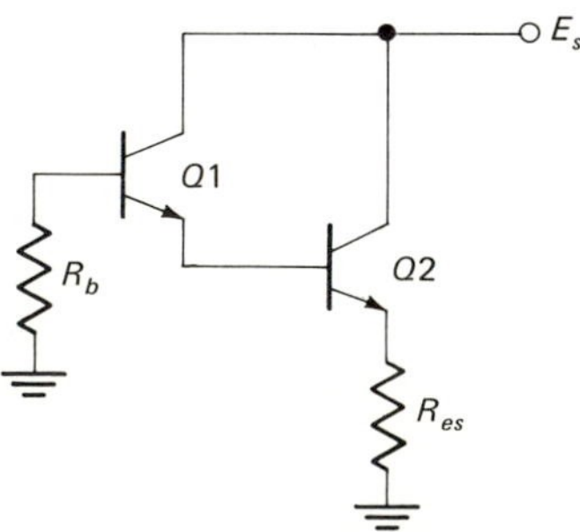

Fig. 10-8 The Darlington pair stability factor analysis

follower with a very high value for R_{es} of Q1. R_{es} is not actually infinite as it might seem to be in the drawing, where Q1 seems to have *no* individual emitter resistor of its own. The collector-base junction impedance is still there and provides an effective R_{es} for Q1 of 1 MΩ or so (in addition to the input resistance of Q2). For practical purposes we can ignore this and assume that the output resistance of Q1 is simply

$$\frac{R_b \text{ actual}}{\beta \text{ (Q1)}}$$

This yields an **S** equation for Q2 of

$$\mathbf{S}\text{ (Q2)} \approx \frac{R_b/\beta \text{ (Q1)}}{R_{es}}$$

THE STABILITY FACTORS

$$(\mathbf{S}_{Q1}) \quad \mathbf{S}_1 \approx \frac{R_b}{\beta_2 R_{es}} \qquad \mathbf{S}\text{ total} = \mathbf{S}_1 \times \mathbf{S}_2$$

$$(\mathbf{S}_{Q2}) \quad \mathbf{S}_2 \approx \frac{R_b/\beta_1}{R_{es}}$$

VALUES FOR R_b FOR A GIVEN S_1 OR S_2. In design work we will need to solve for an appropriate value for R_b to meet *both* of the stability requirements (Q1 and Q2).

$$(\mathbf{S}_1) \quad R_b = \mathbf{S}_1 \beta_2 R_{es}$$
$$(\mathbf{S}_2) \quad R_b = \mathbf{S}_2 \beta_1 R_{es}$$

Example 1

Let $\mathbf{S}_1 = 3$, $\mathbf{S}_2 = 3$, $\beta_1 = 100$, $\beta_2 = 100$, $R_{es} = 10$ kΩ.

$$(\mathbf{S}_1) \quad R_b = 3 \times 100 \times 10 \text{ k}\Omega$$
$$R_b = 3 \text{ M}\Omega$$
$$(\mathbf{S}_2) \quad R_b = 3 \times 100 \times 10 \text{ k}\Omega$$
$$R_b = 3 \text{ M}\Omega$$

So, 3 MΩ for R_b will satisfy the stability requirements for both transistors and yield a *total* stability factor $(\mathbf{S}_1 \times \mathbf{S}_2) = 3 \times 3 = 9$.

Example 2

Let $\mathbf{S}_1 = 3$, $\mathbf{S}_2 = 3$, $\beta_1 = 50$, $\beta_2 = 100$, $R_{es} = 10$ kΩ.

$$(\mathbf{S}_1) \quad R_b = 3 \times 100 \times 10 \text{ k}\Omega = 3.0 \text{ m}\Omega$$
$$(\mathbf{S}_2) \quad R_b = 3 \times 50 \times 10 \text{ k}\Omega = 1.5 \text{ m}\Omega$$

Here we must select the smaller of the two values for R_b, 1.5 MΩ. Again, we have a total stability factor for the compound of 9, but R_b must be 1.5 MΩ to get it.

Remember that there are two base-emitter junctions in series in the input circuit. This means R'_e (effective) $= R'_e$ (Q1) $+ R'_e$ (Q2), and a total V_{BE} of 0.6 + 0.6, or 1.2 V. Remember that ΔV_{BE} will also be ΔV_{BE} (Q1) $+ \Delta V_{BE}$ (Q2).

Although the Darlington pair can be used for any circuit, there are many cases where it offers no advantages and may even be a liability. For example, in a fully bypassed voltage amplifier circuit we would normally have no need for the high current gain.

The Darlington compound can be used in any circuit where very high current gain or a higher input impedance (or both) is required. All the design and analysis procedures already presented for single transistors can be used with the Darlington pair if the following considerations are taken into account.

1. Use the β for the entire pair $\beta_1 \times \beta_2$, or the β given for a commercial pair, wherever β appears in an equation.
2. Use the equations presented in this section to find an appropriate value for R_b (or R_{b2}; see Fig. 10-7).

The Darlington Pair with a Fixed Emitter Resistor for Q1

Figure 10-9 shows a Darlington pair common-collector circuit with an emitter resistor for Q1 added. Technically this is no longer called a Darlington pair, but rather a compound emitter follower. In a couple of cases the addition of the extra resistor is worthwhile:

1. The resistor can be added to form a current divider which will allow a higher collector current for Q1, a condition more favorable to a higher β.
2. Where the final load (R_{es}) varies over a wide range, or where the value of R_{es} (Q2) is uncertain.

The addition of R_{es} for Q1 lowers the stability factor for Q2, and increases it for Q1, for a given value of R_b, ($R_b = R_{b2} \| R_{b1}$). The stability equations for this case are

1. $\mathbf{S}\ (\text{Q1}) \approx \dfrac{R_b}{\beta_2 R_{es}1 \| R_{es}2}$

2. $\mathbf{S}\ (\text{Q2}) \approx \dfrac{(R_b/\beta_1) \| R_{es}1}{R_{es}2}$

The value of $R_{es}1$ is far from critical and can vary from 1 to 100 times the final load value, $R_{es}2$. For low-level stages, QI_c of Q2 less than 250 mA, a

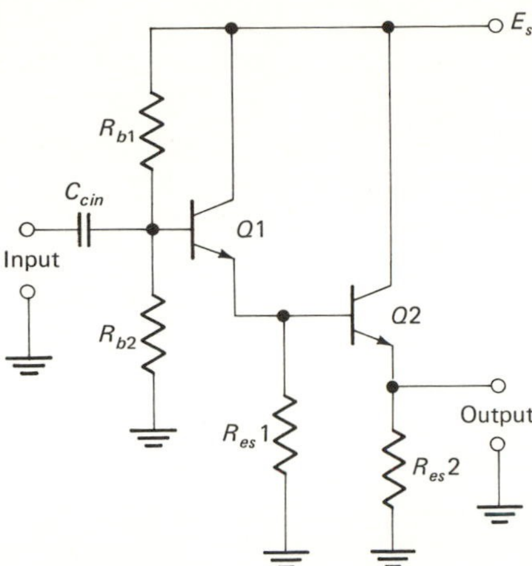

Fig. 10-9 The addition of an emitter resistor for Q1

good rule of thumb is to make $R_{es}1 \approx 10 \times R_{es}2$. For higher-level stages where the QI_c of Q2 is greater than 250 mA, make $R_{es}1 \approx 50 \times R_{es}2$. This rule of thumb allows a rather large variation in the value of $R_{es}2$ with very little change in circuit stability. I built a partial table to show the effect of $R_{es}1$ on stability as it varied over a wide range. That table indicated that the actual value of $R_{es}1$ was not very critical, but that a high β for Q1 was far more important. Since this particular configuration is largely used as a power amplifier pair, we will hold the example until we get to the chapter on power amplifiers.

PROBLEMS

Problem 10-4

TITLE: **Designing the Darlington Pair Emitter Follower (Base-biased)**
PROCEDURE: **Design the circuit, using the following specifications. See Fig. 10-7b.**

Transistor Parameters

1. β (min)	100
2. I_c (max)	100 mA
3. E_c (max)	25 V
4. Transistor type	*npn* silicon

Circuit Parameters

1. Voltage gain	≈ 1
2. E_o p-p	5 V
3. R_{Lw}	10 Ω
4. Stability factor	9
5. $\% QI_c$ to be allowed	10%
6. Maximum operating temperature	65°C

11

the amazing differential amplifier

OBJECTIVES:

(Things you should be able to do upon completion of this chapter)
NOTE: Specific problems are provided in this chapter.

1. Design the basic differential amplifier.
2. Design an *active emitter* resistor for any differential amplifier.
3. Analyze a differential amplifier circuit with an active emitter resistor.

EVALUATION

1. There are design problems in the chapter.
2. The *final* evaluation will be based upon your ability to design and analyze differential amplifiers according to the criteria and specifications in the laboratory manual.

INTRODUCTION

In the world of compromise, it is seldom possible to find a single "best compromise" which will serve for all occasions.

Since you have progressed this far in the study of transistors, you are no stranger to compromise, and you appreciate its importance and meaning.

There is a transistor circuit which, for most occasions, is the best of all circuits in one. It is called by several names, the difference amplifier, the differential amplifier, the long-tailed pair, and a couple of less common names.

Whatever it is called, it is a remarkable performer, and perhaps the best circuit, on all counts, available. The circuit is shown in Fig. 11-1. So that you do not immediately discard all other circuits in favor of this little gem, I must tell you that there is one small catch. The very outstanding perform-

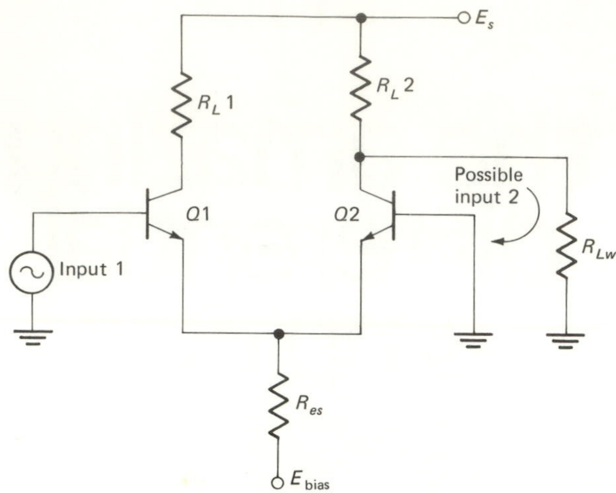

Fig. 11-1 The basic differential amplifier

ance of this circuit depends upon *pairs of closely matched transistors.* The closer the match, the better the performance will be. Unfortunately, this is no small catch, because matched pairs of anything cost a good deal extra, and the specter of cost rears its ugly head. It would be nice to be able to use the differential amplifier for almost everything, but we (except in integrated circuits) just can't afford to do it. The circuit has one other minor problem associated with it; that is the tendency toward fussiness that is so often the hallmark of high-performance designs in all areas of technology. There are some areas where high-performance technology is just not practical. I think it is about time to define "high performance" in terms of this particular circuit.

The circuit is basically a dc amplifier, and so its frequency response is flat all the way down to 0 Hz. Furthermore, it can achieve voltage gains on the order of the gains obtained with the fully bypassed common-emitter circuit, all the way down to 0 Hz.

In addition, the circuit is extraordinarily stable. Q-point drift due to V_{BE} variations can be held down to extremely low values because the emitter resistor voltage drop is limited only by the available bias supply voltage.

Very low stability factors are possible, without uncomfortably low input impedances. Current gains approaching β are also possible.

In summary:

1. Voltage gains of 25 to 400
2. Frequency response from dc to the transistor's upper frequency limit
3. Unusually good stability

4. Current gains approaching β
5. Relatively high input impedances—and all at the same time!

But that is not all. In addition to these performance characteristics, the circuit has some built-in flexibilities that no other common circuit is capable of.

You will notice that the amplifier has two inputs. Either one or both may be used. One input provides an output which, like the common-emitter circuit, is 180° out of phase with the input signal; the other input produces an in-phase output signal. If both inputs are used, the amplifier will amplify the algebraic difference between the two input signals. In multistage amplifiers one input can be used for the input signal and one for a negative feedback input or for some other special purpose. There is output flexibility too, in that the output may be taken between one of the collectors (either one) and ground, or it may be taken *between* the two collectors as a balanced, floating output, or one signal can be taken from each collector.

Because the biasing is done via the emitter, there is a great deal of flexibility in base circuitry. There are no biasing resistors in the base circuit to complicate things.

Although our concern here is not necessarily integrated circuits, it is not out of place to mention that this circuit is the undisputed king of linear monolithic ICs. The nature of the monolithic manufacturing process practically guarantees closely matched transistors, on a given chip. The price of matched pairs of transistors similarly fabricated, on a single chip, should decrease as time goes on. The secret of this circuit is two matched transistors, which share a very large emitter resistor.

The differential amplifier comes in an amazing number of variations, and even though an entire chapter is devoted to this *one* circuit, many of those variations will get left out. The reason I have chosen to use an entire chapter is to try to provide you with a thorough enough understanding of the fundamentals of differential amplifier lore so that you will find yourself in a position to deal with any variation that comes along.

11-1 SOME BASIC THEORY

The differential amplifier is basically a common-base stage driven by a common-emitter or common-collector stage, with both stages sharing a single emitter resistor. Figure 11-1 shows the basic single-ended version of the differential amplifier. The differential amplifier may be base-biased, but it rarely is. It is generally emitter-biased for several very good reasons.

The circuit in Fig. 11-1 shows the input at the base of Q1 and the output from the collector of Q2. Because the collector load resistor of Q1 is not being used to deliver signal to a load, we can consider Q1 to be a common-collector circuit driving a common-base circuit (Q2).

The Action with a Signal

Suppose the base of Q1 goes more positive:

1. The base emitter current of Q1 increases.
2. The collector current of Q1 increases.
3. The emitter of Q1 goes more positive (emitter follower action).
4. The emitter of Q2 goes more positive with respect to the base of Q2.
5. This reduces the forward bias base-emitter voltage of Q2, which reduces the forward bias base-emitter current of Q2.
6. The collector current of Q2 decreases proportionately.

If the two transistors were *absolutely* identical (which of course will never happen), the increase in the collector current of Q1 would result in *exactly* the same decrease in the collector current of Q2. In fact, any *change* in the collector current of Q1 would result in *exactly* an equal and opposite change in the collector current of Q2. Since any increase in the Q1 current would be met by an equal decrease in the Q2 current (and vice versa), there would be *no* net change in the current through the shared emitter resistor.

If the current through a resistor does not vary, the voltage drop across it does not vary. If the voltage drop across R_{es} never varies, no signal voltage can be developed across it. If no signal voltage can be developed across it, it *cannot* function as a negative feedback (current-mode) resistor for the signal. The resistor R_{es}, insofar as the signal is concerned, appears as an open circuit. It just doesn't seem to be there at all.

The input signal applied at the base of Q1 sees two junctions in series so that the theoretical voltage gain is

$$V_g \approx \frac{R_{Lac}}{2R_e'}$$

In practice the empirical equation

$$V_g \approx \frac{R_{Lac}}{2.5R_e'}$$

yields more realistic values, and that is what we will use. The differential amplifier yields about half the voltage gain of a similar fully bypassed circuit, but it does not require a bypass capacitor, and the frequency response goes down to 0 Hz.

An additional fringe benefit which is common to all emitter-biased circuits, but more often used with the differential amplifier, is that the voltage gain can be varied by simply changing the bias voltage or the emitter resistance. This happens because a change in bias voltage causes a change in the quiescent collector current which changes R_e'. Some packaged differential amplifiers have several taps on the emitter resistor to provide voltage gain flexibility.

The Action with Temperature

One of the requirements for stable operation is a good thermal connection between the two transistors so that they both operate at the same temperature. The ideal is to have them both on a single chip. This ideal situation exists in integrated circuits and in some commercially available pairs. There are also heat sinks designed to provide good thermal coupling of the two transistors.

Suppose the ambient temperature increases. V_{BE} in both transistors decreases, and the collector current of both transistors increases. This results in an increased voltage drop across R_{es}. The increased voltage drop across R_{es} reduces the forward bias potential of the base-emitter junctions of *both* transistors, reducing the collector current of both transistors. This is negative feedback, and the larger we make R_{es}, the smaller the change in QI_c due to temperature-induced variations in V_{BE}. The beautiful part of this arrangement is that we can increase R_{es} to any desired value, for any degree of V_{BE} stability, without reducing the voltage gain a bit.

The emitter resistor provides degenerative (negative) feedback when both collector currents try to change at the same time (and same direction), but it does *not* provide degenerative feedback when only one transistor changes collector current.

The circuit is essentially symmetrical so that the generator could just as well have been connected to the base of Q2 with the base of Q1 grounded. The output can be taken from the collector of either Q1 or Q2 and ground, or it may be taken between the two collectors (floating). When the output is taken between the two collectors, the voltage swing is twice that of the single-ended connection.

Both bases may be used for input signals, and the circuit will then amplify the *difference between the two signals*. This mode of operation provides the name *difference* or *differential* amplifier. If the two signals are identical in phase and amplitude, the emitter resistor will provide so much negative feedback that the output signal will approach 0 V, just as a temperature-induced change in both transistors causes no change in collector current. With two signals the amplifier amplifies only the *difference* voltage between the two signals. Because there is no such thing as a perfect differential amplifier, the application of two identical signals may result in some fractional (much less than unity) gain. This is called the *common-mode* gain and is used as a figure of merit for differential amplifiers. The smaller the *common-mode gain*, the better the amplifier. A low common-mode figure also signifies good V_{BE} stability because temperature-induced QI_c changes are in the common mode.

The Common-mode Rejection Ratio

Another often-used figure of merit is called the common-mode rejection ratio; it is simply the ratio of the differential gain to the common-mode gain:

$$\text{CMRR} = \frac{V_g}{V_g\text{CM}}$$

where V_g is the desired gain with a signal on one input and with the other input grounded. The common-mode gain cannot be conveniently computed in advance. It is something measured after the circuit is designed and built. The reasons for this are explained below in "Comments on Stability."

WHAT CAN WE DO AS DESIGNERS TO IMPROVE THE COMMON-MODE REJECTION?

1. We can use transistors which are as closely matched as possible; actual parameters are not very important so long as they are closely matched in the two transistors.

2. We can make certain that there is a good thermal connection between the two transistors.

3. We can make E_{Res} as high as practical. We can use the V_{BE} stability equation that we have been using all along, and the same formula for computing E_{Res}. However, I will provide you with a new table (Table 11-1) with

TABLE 11-1. TABLE OF BASE-EMITTER VOLTAGE VARIATIONS WITH TEMPERATURE

Δt	R_{EF}**20°C** Temp., C°	**3 MV/C° at a QI_c of 0–0.1 mA** ΔV_{BE} **MV**	**2.75 MV/C° at a QI_c of 0.1–1.0 mA** ΔV_{BE} **MV**	**2.5 MV/C° at a QI_c of 1.0–10 mA** ΔV_{BE} **MV**	**2 MV/C° at a QI_c of 10–100 mA** ΔV_{BE} **MV**
5	25	15	13.75	12.5	10
10	30	30	27.50	25.0	20
15	35	45	41.25	37.5	30
20	40	60	55.00	50.0	40
25	45	75	68.75	62.5	50
30	50	90	82.50	75.0	60
35	55	105	96.25	87.5	70
40	60	120	110.00	100.0	80
45	65	135	123.75	112.5	90
50	70	150	137.50	125.0	100
55	75	165	151.25	137.5	110
60	80	180	165.00	150.00	120
65	85	195	178.75	162.5	130
70	90	210	192.50	175.0	140
75	95	225	206.25	187.5	150
80	100	240	220.00	200.0	160

total ΔV_{BE} precalculated for added convenience and to aid in the design discussion to follow.

4. We can have stability factors of unity or so. The usual stability factor equation applies:

$$R_{b2} = \mathbf{S} \times R_{es}$$

Comments on Stability

Because we use the same design formulas for the stability of V_{BE} and β, you may be inclined to assume that the differential amplifier is no more stable than a similar single stage, but that is not actually the case.

When the output is taken between one collector and ground, we can expect 10 to 100 times the stability of a single stage, and when the output is taken between the two collectors, we can expect up to 1,000 times the stability of a single stage.

How much more stable a given differential amplifier will be than a similar single-stage amplifier depends upon such intangibles as transistor match, how closely the transistors maintain that match at different temperatures and collector currents (this is called tracking), and the thermal resistance between the junctions of the two transistors. It is difficult, if not very nearly impossible, to measure all these things and include them in the design equations. In practical situations we simply overdesign and measure the behavior of the circuit. Unless we need a superstable circuit, we will usually come out fine. Figure 11-2 shows the possible input and output combinations for the dif-

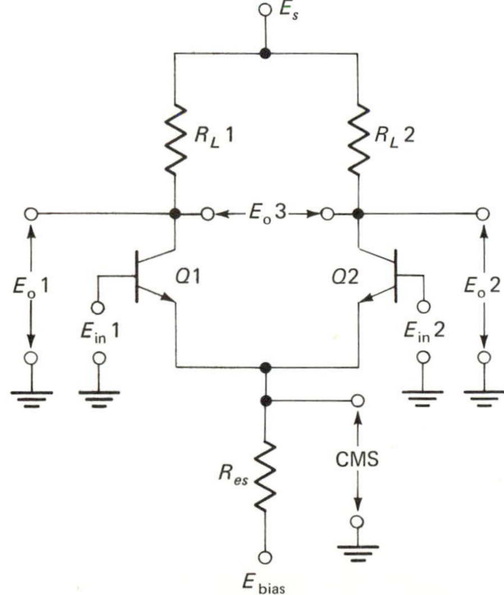

Fig. 11-2 Possible input and output configurations for the differential amplifier

ferential amplifier. It shows a common-mode signal (CMS) from emitter to ground. This is an excellent place to put a scope when a differential amplifier is newly built or malfunctioning. With one input grounded and a signal applied to the other, any significant CMS signal voltage indicates poorly matched transistors, and a large CMS signal voltage usually indicates a defective transistor. It is a good quick trouble check.

Suppose we try a design example.

11-2 A DESIGN EXAMPLE: THE BASIC DIFFERENTIAL AMPLIFIER

Example 11-1 The Differential Amplifier

Design a differential amplifier with the parameters of the circuit, using the following specifications. See Fig. 11-1.

Transistor Parameters

1. β (min)	100-100 (matched pair)
2. I_c (max)	100 mA
3. E_c (max)	30 V
4. Type	*npn* silicon

Circuit Parameters

1. Voltage gain	64
2. E_o p-p	10 V
3. R_{Lw}	16 kΩ
4. Stability factor	1
5. $\%\Delta QI_c$ to be allowed	1%
6. Maximum operating temperature	90°C
7. f_o	20 Hz

We will use the universal current-mode feedback design sheet. So far we have bypassed those interesting boxes in the design sheet that give special instructions for the differential amplifier; now we shall use them. The completed design sheet for the example is shown in Fig. 11-3.

FIELD 1

FRAME 1

Compute the value of E_{Res}.

$$E_{Res} = \frac{\Delta V_{BE} \times 100}{\%\Delta QI_c}$$

The new table I introduced will save you a bit of arithmetic and provide a better insight into very stable circuits. (See Table 11-1.)

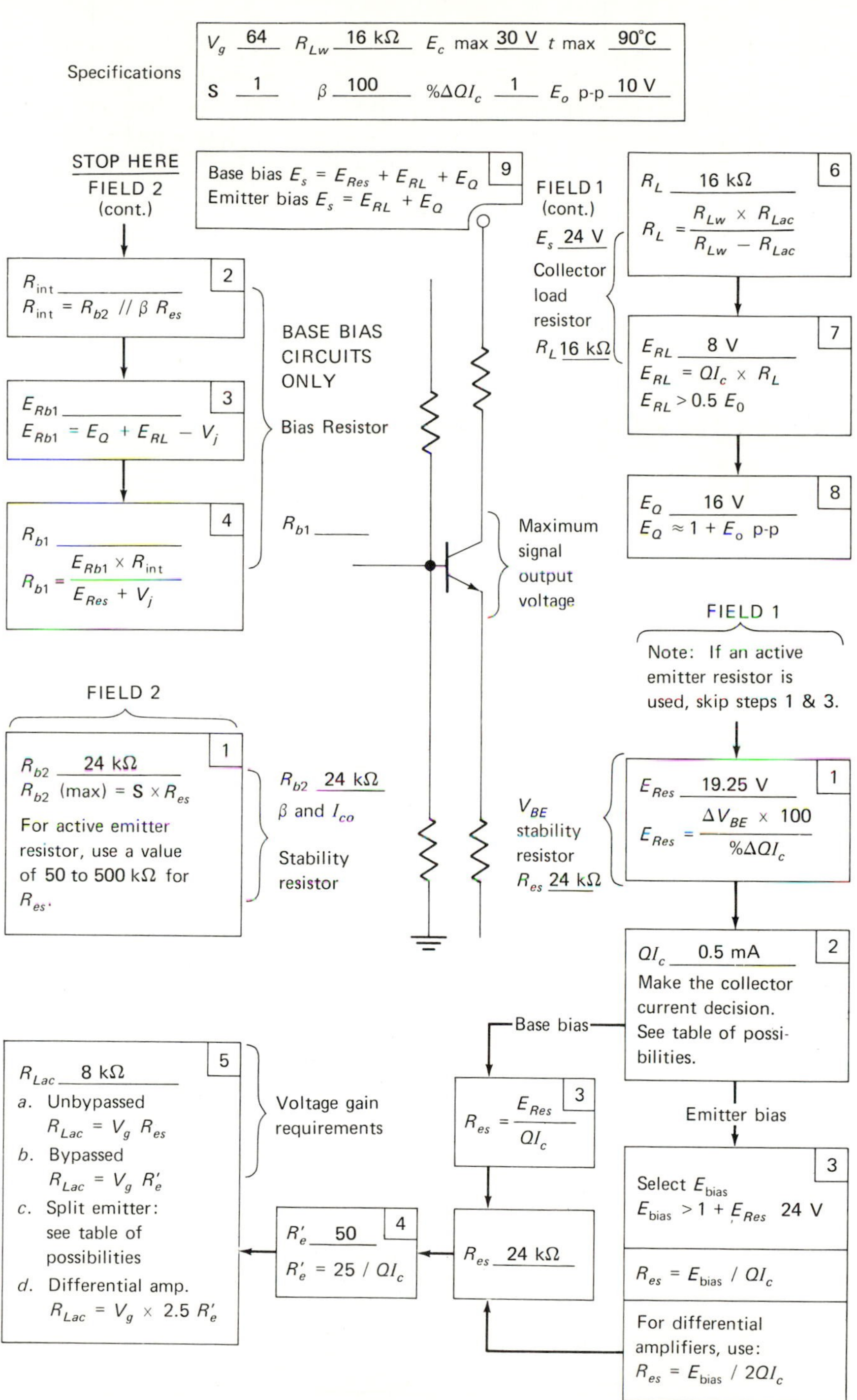

Fig. 11-3 The completed design sheet for Example 11-1

TABLE 11-2. TABLE OF POSSIBILITIES FOR THE DIFFERENTIAL AMPLIFIER CIRCUIT

Working Loads at Various Values of QI_c							Voltage Drop Across the Collector Load Resistor (R_L) for Various Voltage Gains								
R_{Lw} at 0.5 mA	R_{Lw} at 1.0 mA	R_{Lw} at 2.0 mA	R_{Lw} at 4.0 mA	R_{Lw} at 10 mA	R_{Lw} at 20 mA	R_{Lw} at 40 mA	E_{RL} at $V_g =$ **16**	E_{RL} at $V_g =$ **24**	E_{RL} at $V_g =$ **32**	E_{RL} at $V_g =$ **40**	E_{RL} at $V_g =$ **56**	E_{RL} at $V_g =$ **64**	E_{RL} at $V_g =$ **96**	E_{RL} at $V_g =$ **128**	E_{RL} at $V_g =$ **160**
2K	1K	500	250	100	50	25	*	*	*	*	*	*	*	*	*
4K	2K	1K	500	200	100	50	2.0	6	*	*	*	*	*	*	*
6K	3K	1.5K	750	300	150	75	1.5	3	6	15	*	*	*	*	*
8K	4K	2K	1K	400	200	100	1.3	2.4	4	6.6	28	*	*	*	*
10K	5K	2.5K	1.25K	500	250	125	1.25	2.1	3.3	5.0	11.6	20	*	*	*
12K	6K	3K	1.5K	600	300	150	1.2	2.0	3	4.3	8.4	12	*	*	*
14K	7K	3.5K	1.75K	700	350	175	1.16	1.9	2.8	3.9	7.0	9.3	42	*	*
16K	8K	4K	2K	800	400	200	1.1	1.8	2.7	3.6	6.2	8	24	*	*
18K	9K	4.5K	2.25K	900	450	225	1.1	1.8	2.6	3.5	5.7	7.2	18	72	*
20K	10K	5K	2.5K	1K	500	250	1.1	1.7	2.5	3.3	5.4	6.7	15	40	*
24K	12K	6K	3K	1.2K	600	300	1.1	1.7	2.4	3.1	5.0	6.0	12	24	60
30K	15K	7.5K	3.75K	1.5K	700	350	1.1	1.6	2.3	3.0	4.5	5.5	10	17	30
40K	20K	10K	5K	2K	1K	500	1.1	1.6	2.2	2.8	4.2	5.0	8.5	13	20
50K	25K	12.5K	6.25K	2.5K	1.2K	625	1.1	1.6	2.2	2.7	4.0	4.7	8.5	11	17
100K	50K	25K	12.5K	5K	2.5K	1.25K	1.0	1.6	2	2.6	3.7	4.3	6.8	9.5	12.5
200K	100K	50K	25K	10K	5K	2.5K	1.0	1.6	2	2.5	3.6	4	6.4	8.7	11

* An asterisk signifies an impossible condition.

Consulting Table 11-1, we find that the total ΔV_{BE} is 192.5 mV.

$$E_{Res} = \frac{192.5 \times 10^{-3} \times 10^{2}}{1}$$

$$E_{Res} = 19.25 \text{ V}$$

FRAME 2

Make the collector current decision.

Here we will use the special table of possibilities for the differential amplifier, Table 11-2. This table is identical to the fully bypassed table except that the voltage gain figures have all been reduced by a factor of 2.5.

An examination of the table indicates that 0.5 mA is the lowest practical collector current; we will use that figure.

$$QI_c = 0.5 \text{ mA}$$

FRAME 3

Select a bias voltage and compute a value for R_{es}.

1. Select a bias voltage.

   $$E_{\text{bias}} > 1 + E_{Res}$$
   $$E_{\text{bias}} > 19.25 + 1 = 20.25$$

   Let us select 24 V for E_{bias}.

2. Compute R_{es}.
   For differential amplifiers use

   $$R_{es} = E_{\text{bias}}/2QI_c$$

   Each transistor has a quiescent collector (emitter) current of 0.5 mA. Both collector currents flow through R_{es}, and so we must multiply QI_c by 2 to get the actual current through R_{es}.

   $$R_{es} = \frac{24}{2 \times 0.5 \text{ mA}} = 24 \text{ k}\Omega$$

FRAME 4

Compute R_e'.

$$R_e' \approx \frac{25}{QI_c} = \frac{25}{0.5} = 50$$

FRAME 5

Compute R_{Lac}.

Use Branch *d* for the differential amplifier.

$$R_{Lac} = V_g \times 2.5R_e'$$
$$R_{Lac} = 64 \times 125 = 8 \text{ k}\Omega$$

FRAME 6

Compute R_L.

$$R_L = \frac{R_{Lw} \times R_{Lac}}{R_{Lw} - R_{Lac}} \qquad R_L = \frac{16 \times 8}{16 - 8}$$

$R_L = 16\ \text{k}\Omega$

FRAME 7

Compute E_{RL}.

$E_{RL} = QI_c \times R_L$
$E_{RL} = 0.5\ \text{mA} \times 16\ \text{k}\Omega = 8\ \text{V}$

FRAME 8

Compute E_Q.

$E_Q\ (\text{min}) \approx 1 + E_o\ \text{p-p} = 1 + 10 = 11$

Suppose we raise this to 16 V, to make $E_s = 24$ V.

$E_Q = 16$ V

FRAME 9

Compute E_s.

$E_s = E_{RL} + E_Q = 8 + 16 = 24$ V

Figure 11-4 shows the designed circuit for Example 11-1 with values.

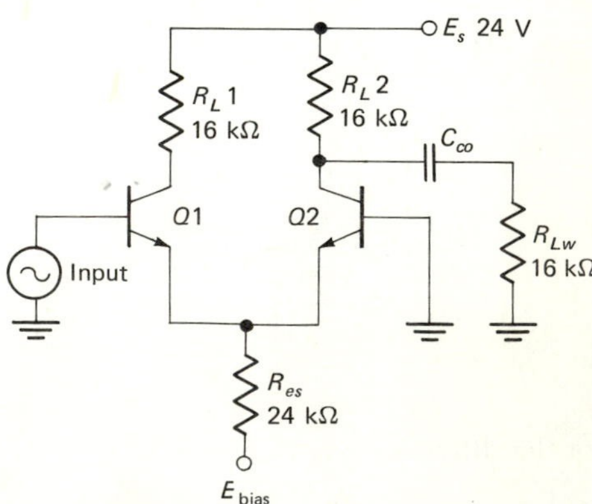

Fig. 11-4 The designed circuit for Example 11-1

Computing the Input Impedance

The generator sees an input impedance of

$$Z_{in} \approx 2\beta R_e'$$

Because the signal currents through R_{es} cancel each other, there is no effective signal current through R_{es}. By Ohm's law $R = E/I$; if $R = E/O$, $R \approx \infty$. Since the value of R_{es} approaches infinity, as far as the signal is concerned, the generator simply sees $\beta(R_e'\ (Q1) + R_e'\ (Q2))$. The two collector currents are equal, and that makes the two R_e''s equal, giving us $Z_{in} \approx \beta \times 2R_e'$. By the custom of algebra we put the constant first, and we get $Z_{in} \approx 2\beta R_e'$. For Example 11-1, $Z_{in} \approx 2 \times 100 \times 50$. $Z_{in} \approx 10\ \text{k}\Omega$.

PROBLEMS

Problem 11-1

TITLE: **The Basic Differential Amplifier**
PROCEDURE: **Design the basic differential amplifier circuit, using the following specifications. See Fig. 11-1.**

Transistor Parameters	Q1	Q2
1. β (min)	100	100
2. I_c (max)	100 mA	100 mA
3. E_c (max)	25 V	25 V

Circuit Parameters	
1. Voltage gain	96
2. E_o p-p	5 V
3. R_{Lw}	24 kΩ
4. Stability factor	1
5. $\%\Delta QI_c$ to be allowed	1%
6. Maximum operating temperature	70°C
7. f_o	20 Hz

Analyzing the Differential Amplifier

Let us analyze the circuit we have just designed, using the current-mode universal analysis sheet (Fig. 11-5). Here are some comments.

1. Because the circuit is *emitter-biased*, we will skip Field 1 and begin with Field 2.
2. In Frames 2 and 3 we will refer to the note (see asterisk) because the current we calculated in Frame 1 is actually divided equally between two transistors.
3. The voltage gain is

$$V_g \approx \frac{R_{Lac}}{2.5R_e'} = \frac{8\ \text{k}\Omega}{2.5 \times 50} = 64$$

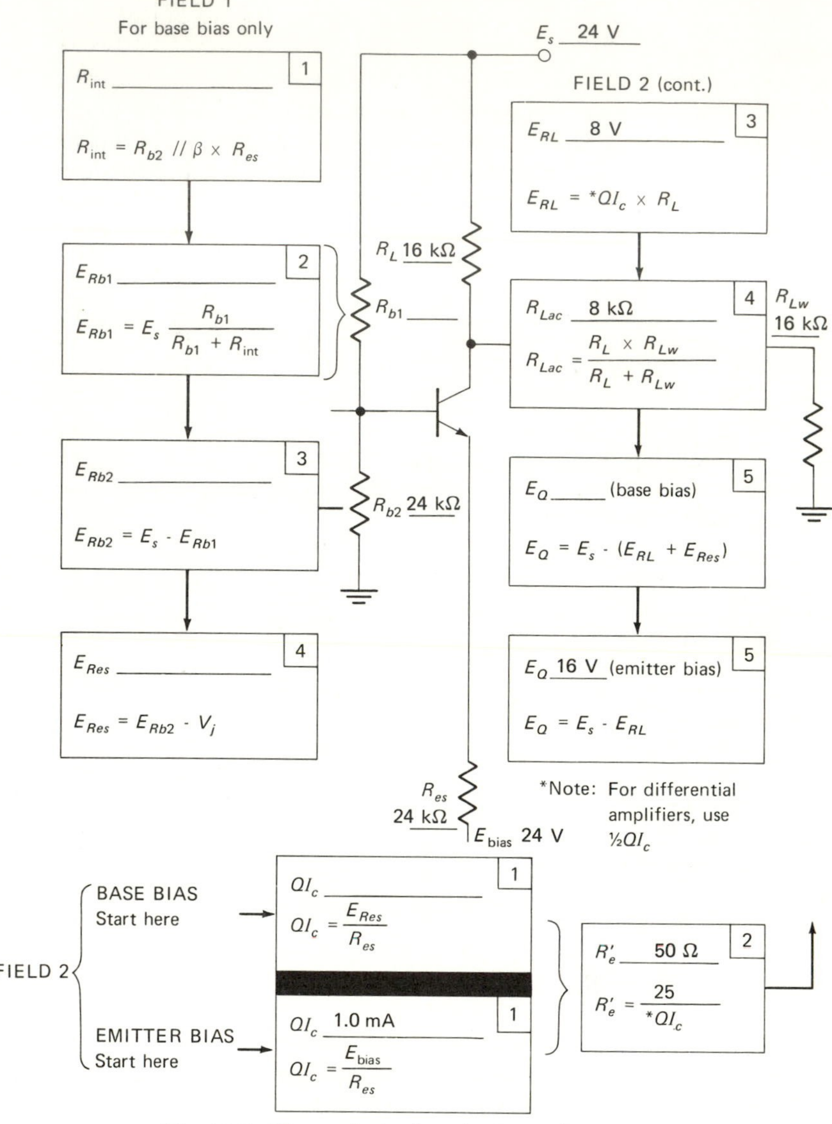

Fig. 11-5 *The analysis sheet for Example 11-1*

4. The input impedance is

$$Z_{in} \approx 2\beta R'_e = 2 \times 100 \times 50 = 10\ \text{k}\Omega$$

Figure 11-5 shows the completed analysis sheet for Example 11-1.

11-3 THE ACTIVE EMITTER RESISTOR (ALSO CALLED CURRENT SINK AND CONSTANT CURRENT SOURCE)

It is desirable to have as large an E_{Res} as possible to control V_{BE} drift. This is usually a convenient way of saying that we want R_{es} to be as large as possible. As long as we use an ordinary resistor, E_{Res} and R_{es} are proportional and we cannot have a high value of R_{es} without having a large voltage drop across it. In the last example we managed a very respectable V_{BE} stability, but it took a large value of emitter resistor and a high bias voltage (24 V) to get it. Large bias voltages are not always practical, particularly where batteries are used. Large bias voltages and large resistor values are especially difficult in integrated circuitry. However, we can use a transistor, biased into its linear region, as a substitute for an ordinary emitter resistor, and get very high effective resistance with a relatively low voltage drop across it. When we use a transistor as an emitter resistor, we can take advantage of the high dynamic collector-base junction resistance of the transistor, and still have a relatively small quiescent collector-to-emitter voltage drop. In effect we can have a 0.5- to 2-MΩ emitter resistor with only a few volts' (two or more) drop across it. With a current of 1 mA and 1 MΩ of ordinary resistance we would need $E = IR = 10^{-3} \times 10^{6} = 10^{3}$ V. Now, a 1,000-V bias supply is a bit much, but with an active emitter resistor we can get the same 1-MΩ effective resistance with a meager 2-V drop across the active emitter resistor, and a bias voltage of only 3 V or so. This scheme is almost universally used in integrated circuitry and is very common in discrete circuitry.

There are some difficulties, of course, but most of these simply involve our scientific desire to have exact values for everything. The actual collector-base junction resistance is another of those intangibles that we must work around, and because of that uncertainty we don't really know the value of the effective emitter resistance when we use an active emitter resistor. Still, we can expect an improvement of one or two orders of magnitude over an ordinary resistor without an excessive voltage drop.

At this point we introduce a new concern, that of the Q-point stability of the transistor we are using as an active emitter resistor.

Because there is no input signal to the base of the transistor, we can have a base-to-ground resistor as small as necessary for β and I_{co} stability, but we cannot have a very large voltage drop across the V_{BE} stability resistor, for the transistor we are using as an active emitter resistor, without defeating the entire purpose of the *active* emitter resistor. The best solution seems to be a combination of current-mode feedback and diode stabilization. See Fig. 11-6.

β Stabilization (and I_{co} Stabilization)

The technique we have used before to stabilize β, using current-mode feedback, is applicable here: keep the *circuit* current gain down by making the

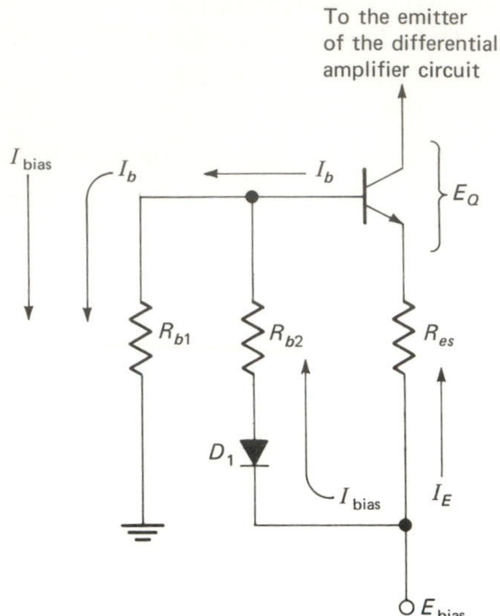

Fig. 11-6 The active emitter resistor circuit

base-to-ground resistance of R_b between 1 and 10 times R_{es}. The *design sheet* (Fig. 11-7) specifies R_{b1}, R_{b2} voltage divider currents of 1/3 to 1 times the emitter current to ensure suitable values of R_b to satisfy the stability requirement.

V_{BE} Stabilization

V_{BE} stabilization is achieved by a combination of emitter resistor swamping voltage and diode compensation. The "bias" current through $D1$ must operate the diode at a current level somewhere near the current through the base-emitter diode of the transistor, QI_c.

Because junction diodes are nonlinear, it is important to operate the compensating diode ($D1$) at a current near that of the emitter current of the transistor being stabilized. Fortunately, operating the diode at from 0.3 to 1 times the transistor emitter current automatically results in a low stability factor for the *active emitter resistor* transistor.

Example 11-2 Designing an Active Emitter Circuit

Suppose we design an active emitter circuit to connect to the simple differential amplifier from the previous example.

We don't need a long list of specifications here because this transistor *does not* process a signal. All that needs to be done is to select a silicon transistor with a reasonably good β which has an E_c (max) greater than the E_Q value we intend to use, and which can handle the collector current required; bias it to the desired emitter (collector) current and hold it there.

The diode $D1$ is often the base-emitter junction of a transistor of the same type number as that used for the active emitter resistor, with base and collector tied together. This technique not only provides the optimum tracking with temperature but also eliminates the problem of *selecting* an appropriate diode.

FIELD 1

(There is only one field in this procedure.)

FRAME 1

Record the total emitter current required by the amplifier.

For this example let us suppose we are designing this active emitter resistor to replace the emitter resistor in Example 11-1. The total emitter current is 1 mA, 0.5 mA per transistor. This emitter current will be the QI_c for the active emitter resistor, but we will write the equations in terms of quiescent emitter current.

$QI_E = 1$ mA

FRAME 2

Select the base voltage E_b.

$E_b \geqslant 2$ V
$E_b = 3$ V

FRAME 3

Select E_Q.

$E_Q \geqslant 2$ V
$E_Q = 3$ V

FRAME 4

Compute E_{bias}.

$E_{\text{bias}} = 1 + E_b + E_Q$
$E_{\text{bias}} = 7$ V

Here E_Q can be increased to yield a more convenient value for E_{bias} if it is desirable. The 1-V constant in the equation accounts for the base-emitter junction drops of the active emitter transistor *and* the differential amplifier it feeds. It should be 1.2 V, but 1 V is more convenient and produces a negligible error.

FRAME 5

Compute R_{es}.

$$R_{es} = \frac{E_{\text{bias}} - (E_b + V_j)}{I_E} = \frac{7 - (3 + 0.6)}{1 \text{ mA}}$$

$R_{es} = 3.4$ kΩ

THE ACTIVE EMITTER RESISTOR DESIGN SHEET

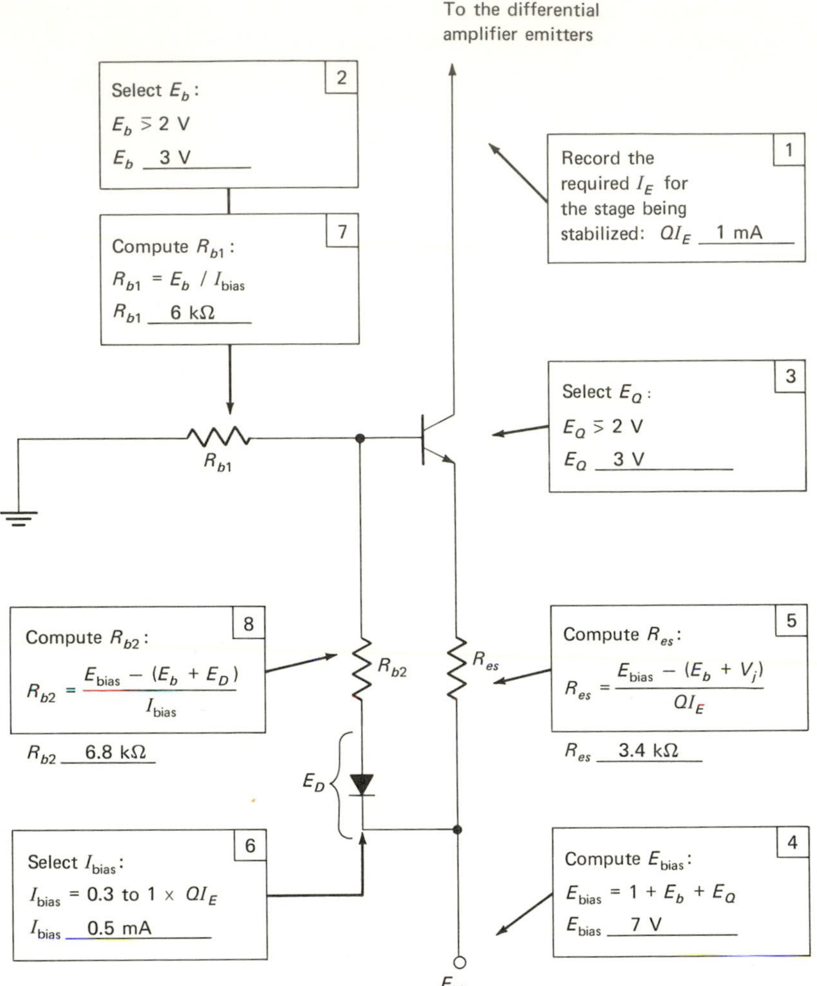

Fig. 11-7 The completed design sheet for Example 11-2

FRAME 6

Select I_{bias}.

I_{bias} = from 0.3 to 1 times QI_E
$I_{bias} = 0.5 \times QI_E = 0.5$ mA

Operation in this range is necessary to provide the desired stability factor. **S** is approximately equal to the reciprocal of the factor by which I_E is multiplied. $K = 1/3$ to 1.

$I_{bias} = KI_E$

$$\mathbf{S} \approx \frac{1}{K}$$

Let $I_{bias} = 0.5I_E = 0.5$ mA.

$\mathbf{S} \approx 1/0.5 \approx 2$

FRAME 7

Compute R_{b1}.

$$R_{b1} = \frac{E_b}{I_{bias}}$$

$$R_{b1} = \frac{3\text{ V}}{0.5\text{ mA}} = 6\text{ k}\Omega$$

FRAME 8

Compute R_{b2}.

$$R_{b2} = \frac{E_{bias} - (E_b + E_D)}{I_{bias}}$$

$$R_{b2} = \frac{7 - (3 + 0.6)}{0.5\text{ mA}}$$

$R_{b2} = 6.8\text{ k}\Omega$

Figure 11-7 shows the completed design sheet for Example 11-2.

Analyzing the Active Emitter Resistor Circuit

Suppose we analyze the active emitter circuit we have just designed. There is nothing more involved than simple Ohm's law and simple addition and subtraction of voltages.

We begin by analyzing the voltage divider consisting of R_{b1}, R_{b2}, and $D1$. We treat it as a simple series circuit and find the total current (I_{bias}). It is then a simple matter to find the divider voltage drops. The only assumption here is that the base current is small compared to the current (I_{bias}) through R_{b1}, R_{b2}, and $D1$. This is a valid assumption because the resistances of R_{b1} and R_{b2} must be small compared to the input resistance of the transistor if the circuit is to be stable and the diode operated at an appropriate current. See Fig. 11-8.

1. Record E_{bias}.

   $E_{bias} = 7$ V

2. Compute I_{bias}.

$$I_{bias} = \frac{E_{bias} - E_D}{R_{b1} + R_{b2}}$$

$$I_{bias} = \frac{7 - 0.6}{6\ k\Omega + 6.8\ k\Omega}$$

$$I_{bias} = 0.5\ mA$$

3. Compute E_b.

$$E_b = I_{bias} \times R_{b1}$$
$$E_b = 0.5\ mA \times 6\ k\Omega$$
$$E_b = 3\ V$$

4. Compute E_{Res}.

The voltage drop across R_{es} is approximately the same as the voltage drop across R_{b2}. If we were to write the loop equation, we would find E_D and V_j canceling each other, assuming the voltage drops across both junctions are identical. For bias purposes our selection of a bias current through the diode of about one-half the base-emitter junction current is a good choice because it makes the diode junction voltage drop just very slightly higher than the transistor junction voltage drop, which helps ensure forward bias.

Let's assume for analysis purposes that $V_j = E_D$ and write the E_{Res} equation as

$$E_{Res} = I_{bias} \times R_{b2} = 0.5\ mA \times 6.8\ k\Omega$$
$$E_{Res} = 3.4\ V$$

5. Compute E_Q.

$$E_Q = E_{bias} - (1 + E_b)$$
$$E_Q = 7 - (1 + 3)$$
$$E_Q = 3\ V$$

6. Compute I_E (QI_c of the active emitter transistor).

$$I_E = \frac{E_{Res}}{R_{es}} = \frac{3.4\ V}{3.4\ k\Omega}$$

$$I_E = 1\ mA$$

This is the emitter current delivered to the differential amplifier to which the circuit is connected. Each transistor in the differential circuit will have a QI_c of 0.5 mA. Figure 11-8 shows the completed analysis sheet for this example.

Figure 11-9 shows the circuit from Example 11-1 combined with the active emitter resistor we designed in Example 11-2.

THE ACTIVE EMITTER RESISTOR ANALYSIS SHEET

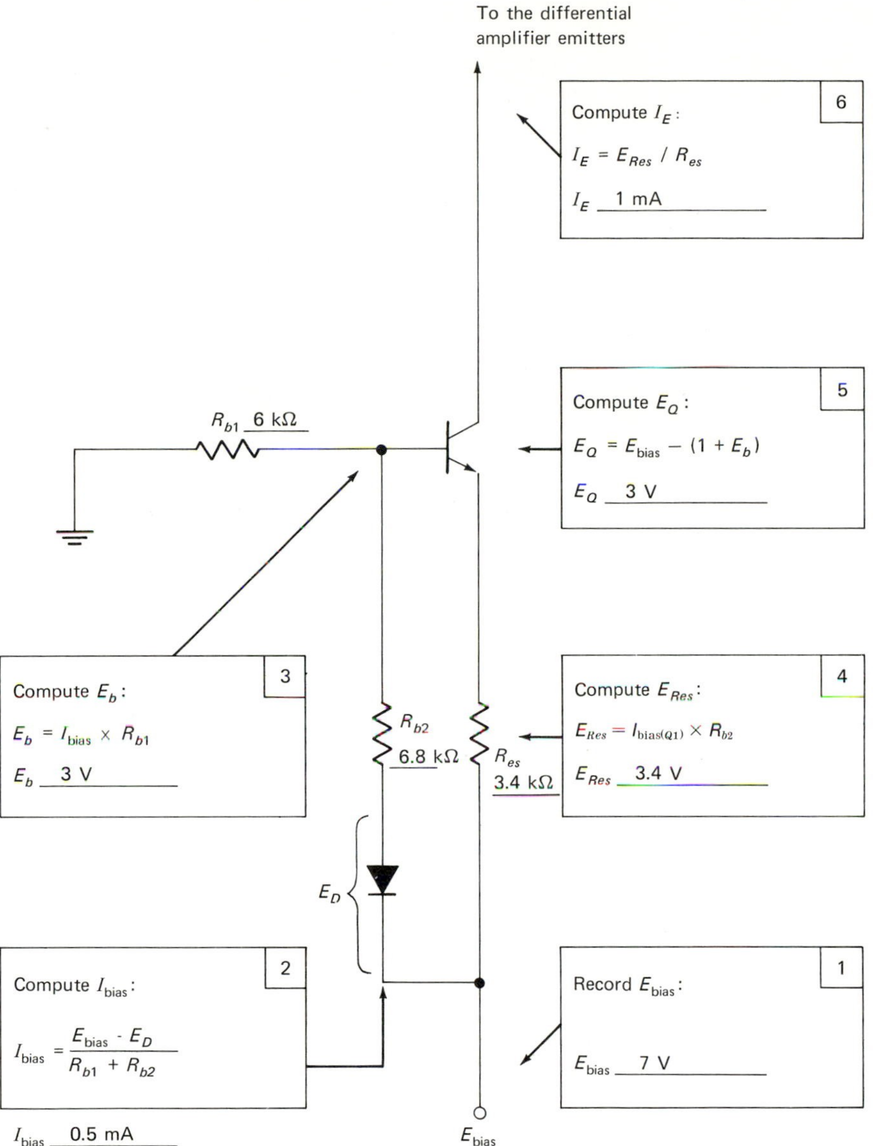

Fig. 11-8 The completed analysis sheet for Example 11-2

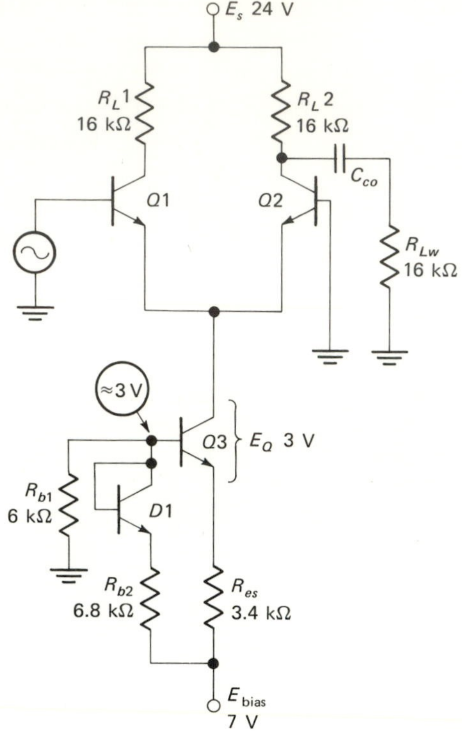

Fig. 11-9 The complete differential amplifier with an active emitter resistor

PROBLEMS

Problem 11-2

TITLE: **The Active Emitter Resistor**

PROCEDURE: **Design an active emitter circuit to be used with the circuit you have designed in Prob. 11-1. You will have to take circuit parameters from your design of Prob. 11-1.**

Transistor Parameters

1. β (min)	100
2. I_c (max)	100 mA
3. E_c (max)	40 V

Problem 11-3

Analyze the circuit you have just designed in Prob. 11-2. Use the analysis sheet in Fig. 11-8 as a guide.

11-4 PRACTICAL GAIN CONTROL AND BALANCE CONTROL

The gain of a differential amplifier can be controlled by varying the bias voltage or the emitter resistance R_{es}. The bias voltage can be an AGC voltage derived from some other part of the system. Many packaged units use a tapped emitter resistor to allow for various selected R_{es} values to provide a practical range of voltage gains with a given bias voltage.

The voltage gain of the basic circuit of Fig. 11-1, and all the variations we have studied so far, is dependent upon the value of R'_e, which, as you are aware, is something less than accurately predictable. Practical gain control often demands more predictable voltage gains than are possible when R'_e dominates the system. In addition, we often desire more linearity, and consequently less distortion, than the basic transistor transfer curve can provide. We can have more predictable voltage gains and less distortion by introducing *some* negative feedback into the circuit. Theoretically the *common* emitter resistance introduces no negative feedback at all, but we can add some negative feedback into the circuit by adding some resistance in the emitter leg of each transistor that is *not* common to both transistors. This arrangement is shown in Fig. 11-10; it is analogous to the split emitter configuration for single-stage current-mode feedback circuits.

To design this circuit, we use the current-mode design sheet and the split emitter table of possibilities (Table 6-4). The voltage gain is

$$V_g = \frac{R_{Lac}}{R_{eac}}$$

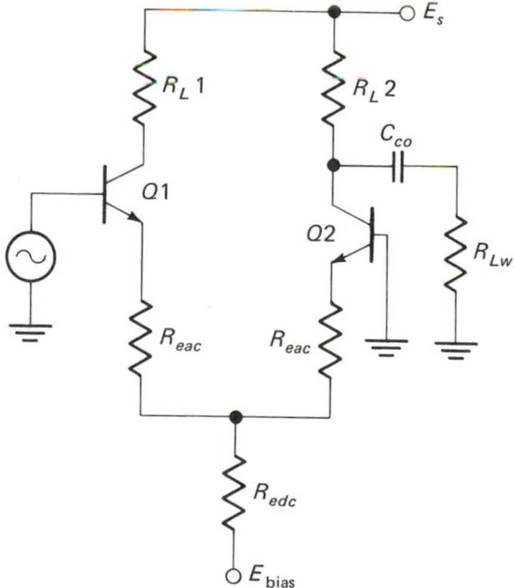

Fig. 11-10 The split emitter variation of the differential amplifier

In nearly all practical cases we can simply tack R_{eac} on top of the value we computed for R_{es} without any significant change in the quiescent current. Remember that $2QI_c$ flows through R_{edc} while only $1 \times QI_c$ flows through R_{eac}.

The voltage gain can be controlled by making both R_{eac}'s variable.

Because there are two R_{eac}'s, a dual potentiometer would be required. The dual potentiometer can be replaced by a single potentiometer by making a wye-to-delta transformation involving both R_{eac}'s and R_{edc}. The procedure is to design the circuit as shown in Fig. 11-11*a* and then to transform the emitter circuit resistances from the wye configuration (Fig. 11-11*a*) into the delta configuration shown in Fig. 11-11*b*.

To transform the circuit of Fig. 11-11*a* into the circuit of Fig. 11-11*b*, we can use the following equations taken from classic literature on wye-to-delta transformations. The equations refer to Fig. 11-11*a* and *b*.

To transform a wye circuit into a delta circuit, use the following relationships:

$$R_1 = \frac{(R_a \times R_b) + (R_b \times R_c) + (R_a \times R_c)}{R_c}$$

$$R_2 = \frac{(R_a \times R_b) + (R_b \times R_c) + (R_a \times R_c)}{R_a}$$

$$R_3 = \frac{(R_a \times R_b) + (R_b \times R_c) + (R_a \times R_c)}{R_b}$$

At first glance this may look like a lot of work, but if you will notice, the numerator is the same in all three equations, and the denominator couldn't be simpler. It is certainly worth the effort to get away from the need

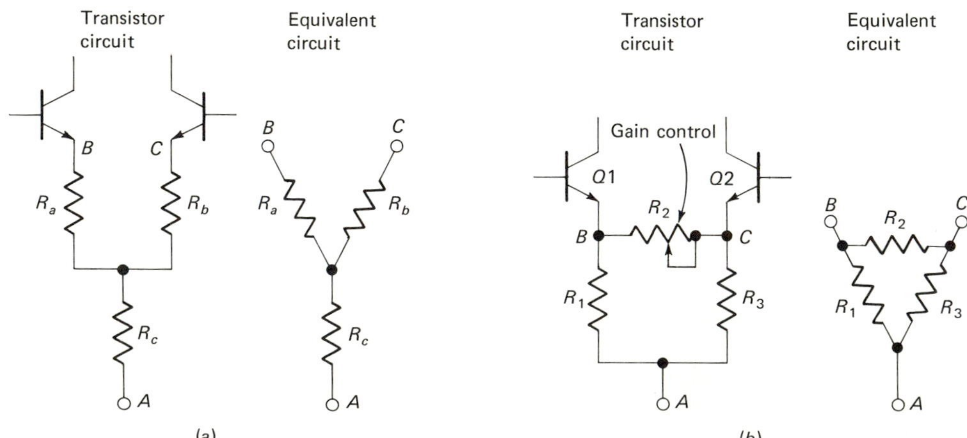

Fig. 11-11 *Wye-to-delta transformations. (a) The wye configuration; (b) the delta configuration*

for a dual potentiometer. Not only is a dual more expensive, but it would also probably be very unsatisfactory, because the two potentiometers would not track perfectly.

Finding the Voltage Gain with the Delta Emitter Circuit

When the emitter circuit is in the wye form, we can use the split emitter equation for V_g:

$$V_g = \frac{R_{Lac}}{R_{eac}}$$

to analyze a circuit designed by someone else. R_{eac} is shown in Fig. 11-11*a* as R_b and R_c.

However, when the emitter circuit is in the delta form, we have a problem: R_{eac} is not clearly visible. The problem here is simply to find the wye equivalent value of R_{eac} (called R_b and R_c in Fig. 11-11*a*). We can do this by using one of the equations for transforming a delta circuit into a wye circuit. Suppose we use the one for finding R_b in Fig. 11-11*a* when we have the circuit in Fig. 11-11*b*.

$$R_{eac} = R_b = \frac{R_1 \times R_2}{R_1 + R_2 + R_3}$$

(R_b is from Fig. 11-11*a*; it has nothing to do with base resistance, etc.) For bias purposes we will need to know R_c:

$$R_c = \frac{R_1 \times R_3}{R_1 + R_2 + R_3}$$

Balancing the Circuit

It is sometimes necessary to provide an adjustment to compensate for minor imbalance between the two sides of the differential amplifier; this is particularly true when the floating, balanced (collector-to-collector) output is used. Minor imbalances can be caused by differences in resistor values, loading imbalances, and aging, as well as initial differences between the transistors and slight differences in junction temperatures. Occasionally some controlled imbalance is desirable; for example, in an electronic voltmeter the meter pointer zero may not be a true electrical zero. Also, it has been determined experimentally that equal V_{BE} in the two transistors is more important to stability than equal values for QI_c for the two transistors, and in some cases a small difference in QI_c can provide improved stability. Figure 11-12 shows the two most common methods of providing a balance control. In both parts

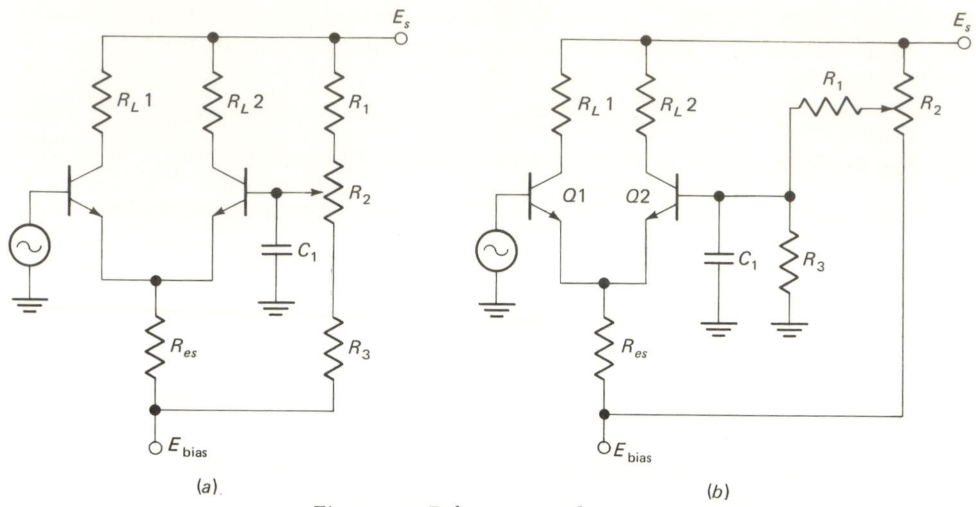

Fig. 11-12 *Balance control circuits*

of the figure, R_1 and R_3 serve to limit the range of control to a few percent of the total possible. Without these limits the balance control would be far too touchy. Figure 11-12*b* is the more popular configuration.

In Fig. 11-12*a*, $R_1 + \frac{1}{2}R_2$ and $R_3 + \frac{1}{2}R_2$ must both be less than $\mathbf{S} \times R_{es}$ to maintain the stability factor.

In Fig. 11-12*b*, it is customary to make R_3 equal to $\mathbf{S} \times R_{es}$ to satisfy the stability requirement.

A capacitor (C_1) is often added from base to ground to filter out such things as power-supply ripple, potentiometer noise, and stray ac pickup. Electrolytic capacitors should generally be avoided here because of inherent instability and variable leakage currents.

11-5 INCREASING THE INPUT IMPEDANCE OF THE DIFFERENTIAL AMPLIFIER

The input impedance of the simple differential amplifier is $Z_{in} \approx 2\beta R_e^*$, where $R_e^* = R_e'$ for the circuit in Fig. 11-1 and $R_e^* = R_{eac}$ in the split emitter version in Fig. 11-10. For very high input impedances we can use the split emitter version plus a super-β transistor. The super-β transistor is, of course, our old friend the Darlington pair, where the composite β is $\beta_1 \times \beta_2$. The circuit is shown in Fig. 11-13.

We now have four junctions in series so that the input impedance equation is

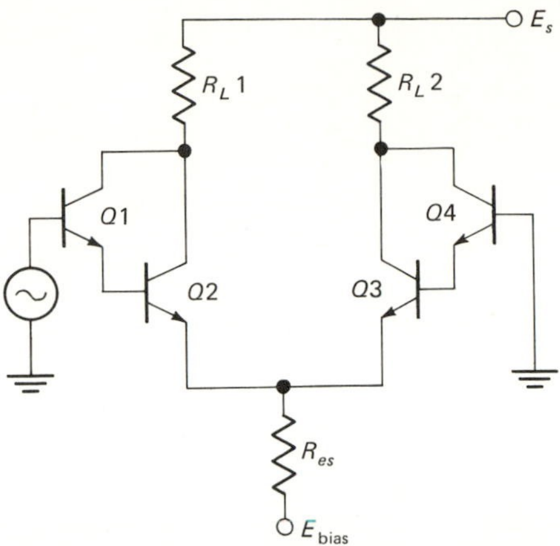

Fig. 11-13 The Darlington pair differential amplifier

$$Z_{in} \approx \beta'[R'_e\ (Q1) + R'_e\ (Q2) + R'_e\ (Q3) + R'_e\ (Q4)]$$

where β' is the total β of the Darlington compound.

We can simplify this equation a little because QI_c of Q2 of Q3 are the same and the QI_c of Q1 and Q4 are equal. We can write the equation as

$$Z_{in} \approx \beta'[2R'_e\ (Q1) + 2R'_e\ (Q2)]$$

The collector current of Q1 and Q4 is probably only 1 percent (or so) of the collector current of Q2 and Q3.

PROBLEMS

Problem 11-4

Analyze the circuit in Fig. 11-14. Provide the following information:

1. *DC Voltages Computed*
   - E_{Res}: Q1 ______ Q2 ______
   - E_{RL}: Q1 ______ Q2 ______
   - E_Q: Q1 ______ Q2 ______
2. *Signal Parameters Computed*
   - V_g: Q1 ______ Q2 ______
   - E_o p-p: Q1 ______ Q2 ______
   - Z_{in}: Q1 ______ Q2 ______
3. *Quiescent Current Computed*
   - Q1 ______ Q2 ______

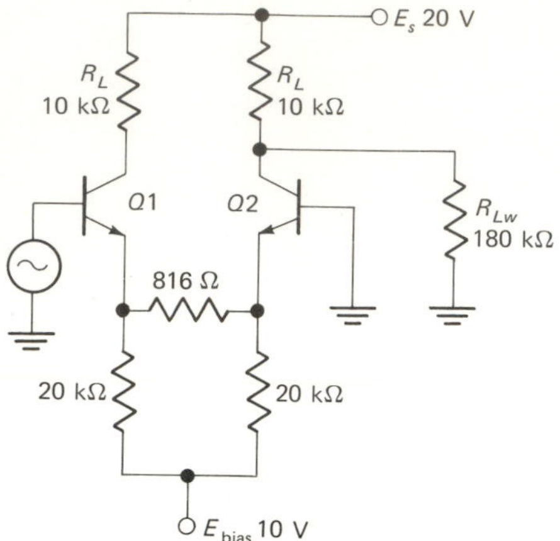

Fig. 11-14 *The circuit for Problem 11-4*

11-6 CASCADING STAGES OF THE DIFFERENTIAL AMPLIFIER

The differential amplifier is nearly always direct-coupled, which means that the working load resistance R_{Lw} should be 10 or more times R_L. It is standard practice to ensure that the loading is relatively light. When the loading is not light, it is common practice to put an emitter follower between R_{Lw} and the final output. The circuit we are going to design is shown in Fig. 11-16.

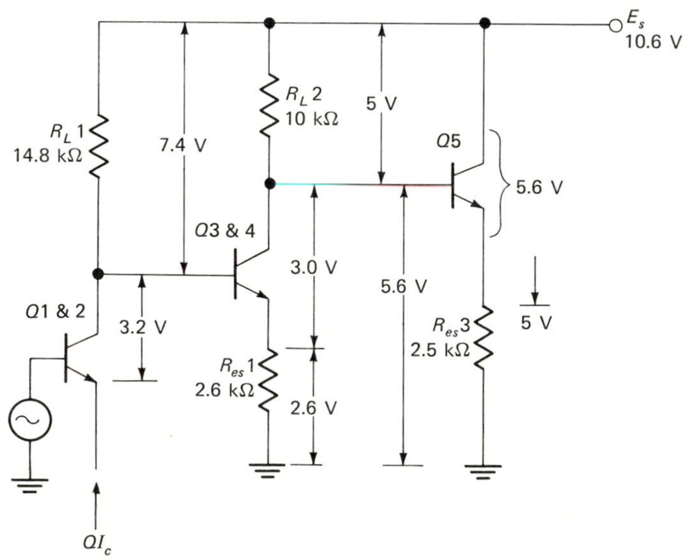

Fig. 11-15 *The simplified partial schematic for Example 11-3*

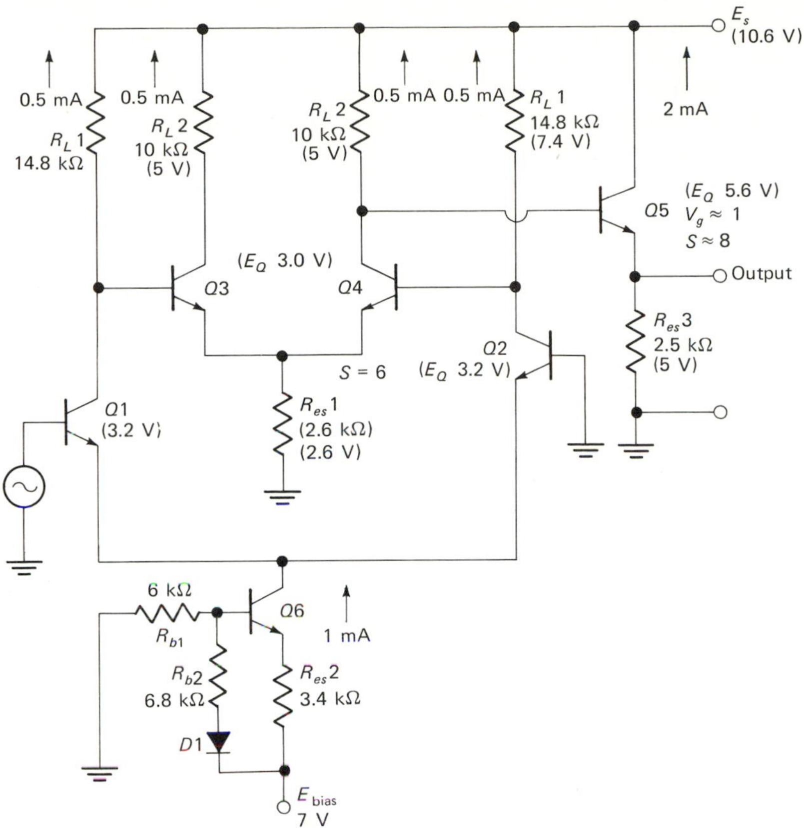

Fig. 11-16 The complete two-stage differential amplifier for Example 11-3

The circuit in Fig. 11-16 looks very formidable, and it would be except that we can break the circuit down into several simple parts. Because of the basic symmetry of the circuit, we can redraw it, leaving out one transistor in each differential pair as long as we keep in mind that $2QI_c$ will be flowing in $R_{es}1$ when we put the parts back together. The active emitter resistor can be designed after we have decided on QI_c for Q1 and Q2.

The simplified circuit is shown in Fig. 11-15.

There are a number of interacting factors involved in the design, and it is nearly impossible to meet all of an arbitrary list of specifications. Perhaps the best, and most flexible, approach is to design for the *highest* possible voltage gain for each stage with a given supply voltage. Here, we will select at least an approximate value for E_s as a specification and try for the highest possible voltage gain per stage with that supply voltage. After the amplifier has been designed, we can adjust the voltage gain to any lower value we want by adding a single negative feedback resistor from the output back to the input of the complete amplifier. The circuit (Fig. 11-16) is as nearly typical a configuration as I could find in the literature, but there are a great many variations.

Two common variations consist of taking the emitter follower off the collector of Q3 instead of Q4 as we will do in redrawing the circuit, and eliminating the collector load resistor from the collector of the transistor (of the differential pair) from which no signal is taken. In this example the second stage is much less stable than the first, and this is standard procedure. The levels of stability in the example are also pretty typical values.

Example 11-3 will be much easier to understand if you copy Figs. 11-15 and 11-16 onto a couple of pieces of paper and actually go through each step with me, writing all important figures on your hand-drawn copy. Also check your work frequently against the figures in my drawings.

Example 11-3 The Multistage Differential Amplifier

Design the circuit in Fig. 11-15, using the following specifications.

Transistor Parameters	Q1–Q2	Q3–Q4	Q5
1. β (min)	100	100	100
2. I_c (max)	100 mA	100 mA	100 mA
3. E_c (max)	30 V	30 V	30 V
Circuit Parameters			
1. Voltage gain	Max pos.	Max pos.	1
2. E_o p-p	—	3 V	3 V
3. R_{Lw}	?	$\approx\infty$	1.25 kΩ
4. Stability factor	1	5.7	8
5. $\%\Delta QI_c$ to be allowed	Active emitter "R"	10%	10%
6. Maximum operating temperature	100°C	100°C	100°C
7. f_o	0 Hz	0 Hz	0 Hz
8. $E_s \approx$	10 V	10 V	10 V

Collector Side Calculations for Q3 and Q4

We will start with Q3 (Q4), which represents the second (last) stage of differential amplification. The collector side voltages and resistances of Q3 control the E_o p-p specification, the stage voltage gain, and the stage V_{BE} stability. The only demand placed on the emitter follower is that it elevate the effective impedance of R_{Lw} to a high enough value to cause negligible loading on the Q4 collector. We will design it later.

1. Compute E_{Res} (Q4). V_{BE} stability.

$$E_{Res} = \frac{\Delta V_{BE} \times 100}{\%\Delta QI_c}$$

$$E_{Res} = \frac{2.5 \times 80 \times 10^{-3} \times 10^2}{10}$$

$$E_{Res} = 2 \text{ V min}$$

Let's call it 2.6 V just to make some numbers come out more even, later.

2. Compute E_Q for Q3. To provide for E_o p-p,

$$E_Q = 1 + E_o \text{ p-p} \qquad E_Q = 1 + 2 = 3 \text{ V}$$

3. Compute R_L for Q3.
Find the value for R_L which satisfies the stability requirement for the emitter follower Q5.

$$R_L \text{ (max)} = \mathbf{S} \text{ (Q5)} \times R_{es} \text{ (Q5)}$$
$$R_L \text{ (max)} = 4 \times 2.5 \text{ k}\Omega = 10 \text{ k}\Omega$$
$$R_L = 10 \text{ k}\Omega$$

4. Find E_{RL} (R_L2).

$$E_{RL} = E_s - (E_{Res} + E_Q)$$
$$E_{RL} = 10.6 - (2.6 + 3.0)$$
$$E_{RL} = 5 \text{ V}$$

5. Compute the quiescent collector current for Q3 (and Q4).

$$QI_c = \frac{E_{RL}}{R_L} = \frac{5 \text{ V}}{10 \text{ k}\Omega} = 0.5 \text{ mA}$$

The purpose of Q5, the emitter follower, is to provide an almost infinite value of R_{Lw} for Q4.

In most previous circuits our selection of QI_c was dictated by the value of R_{Lw}. In this case R_{Lw} is so high that we are free to select any convenient collector current that is suitable for the transistor used. In this case we computed the collector current simply to give ourselves a predetermined voltage drop (E_{RL}) across a predetermined resistance (R_L).

$$QI_c = 0.5 \text{ mA}$$

6. Compute the resistance of the emitter resistor for Q3 (and Q4), $R_{es}1$. We have a 2.6-V drop across R_{es} and a current of $2QI_c$.

$$R_{es} = \frac{2.6 \text{ V}}{1 \times 10^{-3}} = 2.6 \text{ k}\Omega$$

7. Compute the voltage gain.

$$V_g \approx \frac{R_{Lac}}{2.5\, R'_e}$$

The constant 2.5 is used because the output is single-ended. R_{Lac} is equal to R_L because of the very high input resistance of the emitter follower, which is R_{Lw} for Q3.
$R_{Lw} = Z_{in} \approx \beta\ (Q5) \times R_{es}\ (Q5)$
$Z_{in} \approx 100 \times 2.5\ k\Omega = 250\ k\Omega$
So for all practical purposes R_{Lw} for this stage is infinite, and $R_{Lac} = R_L$.

$$R'_e \approx \frac{25}{QI_c} = \frac{25}{0.5\ mA} = 50$$

$$V_g \approx \frac{10\ k\Omega}{125} = 80 \qquad V_g \approx \frac{R_{Lac}}{2.5\ R'_e}$$

Collector Side Calculations for Q1 and Q2

We will follow the same design sequence we used for stage 2 and point out any differences that come up.

1. Compute E_{Res}.
   We skip this step because we are going to use an active emitter resistor and we cannot design that until we know QI_c for Q1 and Q2. We will save it as a separate part of the problem.
2. Compute E_Q.
   Here we have no choice, and no problem either. E_Q must be 0.6 V greater than the 2.6-V E_{Res} of Q3.

   $E_Q = 2.6 + 0.6 = 3.2$ V

3. Compute R_L to satisfy the stability requirement for Q3 (and Q4).

   $R_L1 = \mathbf{S}\ (Q3) \times R_{es}\ (Q3)$
   $R_L1 = 5.7 \times 2.6\ k\Omega = 14.8\ k\Omega$

The peculiar value for **S** (5.7) needs some explanation. Any stability factor between 4 and 7 would have worked out reasonably well, and 5.7 is between those two limits. But why did I select 5.7 instead of an even 6? Laziness, pure laziness, is the only reason. If I could make the QI_c of Q1 (and Q2) come out to exactly 0.5 mA, I could use the active emitter circuit we designed earlier in the chapter, and avoid the work of designing another one for this example, one which might involve uncomfortable numbers because of an odd QI_c value. So I selected a stability factor which was exactly right to do the job. This takes a little foresight, and it is only a way to make things easier *sometimes*. It is just as satisfactory to follow the procedure exactly and select any arbitrary value for **S**, between 4 and 7, and take the time to design a new active emitter circuit.

You will often find odd values of QI_c in commercial circuits and there is no sin in it at all.

$R_L = 14.8\ k\Omega$

4. Compute E_{RL}.

$$E_{RL} = E_s - E_Q$$
$$E_{RL} = 10.6 - 3.2$$
$$E_{RL} = 7.4 \text{ V}$$

5. Compute QI_c.

$$QI_c = \frac{E_{RL}}{R_L}$$

$$QI_c = \frac{7.4 \text{ V}}{14.8 \text{ k}\Omega} = 0.5 \text{ mA}$$

6. Compute R_{es}.
   This doesn't apply because we are using an *active* emitter resistor.
7. Compute the voltage gain for stage 1 (Q1 and Q2).

$$V_g = \frac{R_{Lac}}{1.25R'_e}$$

The factor 2.5 (in the denominator) does not apply here because the output is taken from between the two collectors, and the input is applied to both bases.

$$R'_e = \frac{25}{0.5} = 50$$

Now we need to find R_{Lw} and R_{Lac}.

Compute the Input Impedance of the Second-stage Differential Amplifier

The input impedance of stage 2 is the working load R_{Lw} for stage 1. Here we have to make a distinction between the static input resistance which is approximately β (Q3) R_{es} (Q3) and the dynamic input impedance where R_{es} is not involved at all. The static input resistance applies *only* to the static QI_c condition. In this case the static input resistance into Q3 (and Q4) is

$$Z_{in} \text{ static} \approx \beta R_{es} = 100 \times 2.6 \text{ k}\Omega = 260 \text{ k}\Omega$$

The implication of this static input resistance is that the quiescent current taken from the collector of Q1 (and Q2) by the input circuit of Q3 (and Q4) is negligible, and we can consider QI_c for Q1 to be the only current (at quiescence) flowing through R_L1. This greatly simplifies the quiescent calculations in the first stage. We have already computed E_{RL} on this basis.

As soon as we begin to deal with signal, we must work with the dynamic impedance of the differential amplifier, which is normally much lower, and will influence the voltage gain of the first stage. The dynamic input impedance of stage 2 (Q3 and Q4) is R_{Lw} for stage 1 and is equal to

$$R_{Lw} = Z_{in} \text{ (Q3 and Q4)} \approx 2\beta R_e'$$

In this case,

$$Z_{in} \approx 2 \times 100 \times 50 = 10 \text{ k}\Omega$$

$$R_{Lac} = \frac{R_L \times R_{Lw}}{R_L + R_{Lw}} = \frac{14.8 \text{ k}\Omega \times 10 \text{ k}\Omega}{14.8 \text{ k}\Omega + 10 \text{ k}\Omega} \approx 6 \text{ k}\Omega$$

$$V_g \approx \frac{6 \text{ k}\Omega}{1.25 \times 50} = 96$$

The total voltage gain of the entire amplifier is

$$V_g \text{ (total)} = V_g1 \times V_g2 \times V_g3$$
$$V_g \text{ (total)} = 96 \times 80 \times 1 = 7{,}680$$

The Active Emitter Resistor

Rather than repeat the design of the active emitter resistor, we will simply use the one we designed earlier in the chapter. See Fig. 11-6 (Example 11-2).

Designing the Emitter Follower

Now we must jump over and take care of the emitter follower. This is necessary because R_L2 is the β stability resistor for the emitter follower, and if we take time out now to take care of the emitter follower, it can save us from possibly using a value for R_L2 which will not meet the stability factor for Q5. Figure 11-16 shows a base-to-ground voltage for Q5 of 5.6 V. This means that the voltage drop across R_{Lw} must be 0.6 V less, or 5 V. The collector current of Q5 is

$$QI_c = \frac{5}{2.5 \text{ k}\Omega} = 2 \text{ mA}$$

Here we must compute the minimum E_{Res} to make certain we have adequate V_{BE} stability for Q5. We are not likely to get into trouble here, but we should check it anyway.

$$E_{Res} \text{ (min)} = \frac{\Delta V_{BE} \times 100}{\% \Delta QI_c} = \frac{80 \times 2.5 \times 10^{-3} \times 10^2}{10}$$

$$E_{Res} \text{ (min)} = 2 \text{ V}$$

The 5-V drop across R_{Lw} is more than adequate.

Figure 11-15 shows the values for the completed circuit design.

PROBLEMS

Problem 11-5

Write a summary of what you have learned in this chapter. This is the only summary of this chapter you will get, so do it carefully. Your instructor may ask you to turn this in.

Problem 11-6

TITLE: **The Multistage Differential Amplifier**
PROCEDURE: **Design a multistage differential amplifier, using the following specifications. See Fig. 11-15.**

Transistor Parameters	Q1–Q2	Q3–Q4	Q5–Q6
1. β (min)	100	100	100
2. I_c (max)	100 mA	100 mA	100 mA
3. E_c (max)	25 V	25 V	30 V

Circuit Parameters	Q1–Q2	Q3–Q4	Q5	Q6
1. Voltage gain	max pos.	max pos.	≈ 1	—
2. E_o p-p	—	—	6 V	—
3. R_{Lw}	—	—	5 kΩ	—
4. Stability factor	1	10	10	≈ 1
5. $\% \Delta QI_c$ to be allowed	active emitter "R"	10	10	≈ 10
6. Maximum operating temperature	. . . .	100°C		
7. f_o	. . . .	0 Hz		
8. E_s	. . . .	approx. 10 to 24 V . .		

Problem 11-7

Analyze the circuit you have just designed in Problem 11-6.

12

power amplifiers and regulators

OBJECTIVES:

(Things you should be able to do upon the completion of this chapter)

1. Design a two-stage basic direct-coupled class A circuit with a Darlington pair emitter follower power output stage, and two-stage voltage-mode feedback.
2. Design a complementary symmetry power amplifier with a Darlington pair (common-collector) output, and voltage-mode feedback from power stage to driver.

EVALUATION:

1. Specific design problems are included in the chapter.
2. The final evaluation will be based upon your ability to design the circuits in this chapter and have them work according to the criteria and specifications in the laboratory manual.

INTRODUCTION

Power amplifier circuits present us with some new problems that dictate approaches a bit different from what we have become used to in the design of small-signal amplifiers. In small-signal amplifiers we used current-mode feedback as our principal method of stabilizing against Q-point drift, and we were not much concerned with power loss across the emitter resistor, large-signal excursions, or internal heat generation. In power transistor circuits we have to face these problems. We will lean heavily on two-stage voltage-mode feedback and diode stabilization, and when it is practical, we may use all three techniques in the same circuit. Modern power transistors offer relatively high β values over most of the useful collector current range, good second breakdown immunity, and good high-frequency response.

Power transistor technology has come a long way in the last few years, and many of the problems mentioned in the literature of a few years ago no longer exist.

Obtaining good stability along with desirable signal-handling properties is still our goal, but because of much lower impedances, higher currents, larger current and voltage swings, and transistor self-heating, these things are harder to achieve. A great many approaches have been tried, but gradually experience has yielded some circuit configurations which have become standard, or nearly so. In this chapter we will examine the most popular configurations and some variations. The emphasis will be on direct-coupled driver power amplifier combinations and push-pull complementary symmetry configurations. We will treat the quasi-complementary symmetry circuit lightly because, although it is much less satisfactory than true complementary symmetry circuits, it is still used in integrated circuit technology because not all the problems of putting matched complementary pairs of transistors on a single monolithic chip have yet been solved. It is probable that these problems will soon be solved and the quasi-complementary will become more or less obsolete. In any event, if you can handle complementary symmetry circuits, quasi-complementary circuits will cause you no problem. We will examine most of the common complementary circuit refinements.

The literature abounds with transformer-coupled circuits, and for that reason and two others we will not cover transformer-coupled power amplifiers here. Transformer-coupled circuits are so dependent upon the design of the transformers that it is generally best to simply use the transformer manufacturer's recommended circuit design. In addition broad-band low-frequency power output transformers (and good quality interstage transformers) are bulky, heavy, expensive, and not very efficient. Transformer coupling is discussed in Chap. 9.

For most audio and servo amplifiers transformers are best not used, especially at higher power levels, if they can be avoided. The principal use of transformer-coupled power amplifiers is in circuits of 1 W or less, which do not pretend to any quality sound reproduction.

12-1 POWER AND EFFICIENCY

In this chapter we will spend most of our time with two classes of amplifier, single-ended direct-coupled class A and modified class B push-pull.

Class A direct-coupled power amplifiers are biased, at quiescence, to about 40 percent of QI_c maximum, with the signal swinging the collector current from near zero collector current to near 90 percent of QI_c maximum. This current swing includes nearly all the transfer curve, and large amounts of negative feedback are necessary to make up for the nonlinearities in the transfer curve. Quite satisfactory audio amplifiers are possible up to a watt or two with class A operation. However, this same direct-coupled class A amplifier is ideal for voltage (or current) regulator service, for quite high power levels, because the entire signal input is the *amplifier output voltage*: up to 100 percent feedback.

At audio power levels above a watt or two we will find that a direct-coupled modified class B push-pull circuit is best.

In the last part of this chapter I will show you how to apply direct-coupled class A amplifiers to power regulator service.

For both kinds of service, and both classes of operation, amplifier and regulator, the common-collector circuit has most of the advantages.

The maximum efficiency for a class A power amplifier is 25 percent, where efficiency is defined as

$$E_{ff} = \frac{P_o\ (\text{max})}{P_s}$$

where P_o (max) is the maximum power output and P_s is the total power delivered by the power supply. For audio power amplifiers up to 2 W this poor efficiency is not too important, but beyond that, such waste becomes expensive.

The basic philosophy behind most regulator circuits is: "Waste (as heat) whatever power you don't need." This puts a burden on the regulator transistor, but not an intolerable one except at very high power levels (50 W or more).

Another difficulty with class A operation is that it demands better filtering for line-operated power supplies than push-pull does (usually class B). Good filtering at currents above ½ A becomes expensive.

The maximum transistor power dissipation in class A operation occurs at quiescence.

$$P_d = QI_cE_Q$$

where E_Q is normally about $\frac{1}{2}E_s$. As in small-signal transistors, QI_c should be kept to 40 percent (or less) of I_c max, although in power transistors QI_c is more often close to or equal to 40 percent of maximum, whereas it is usually only a few percent of maximum in small-signal transistors.

Selecting a transistor for class A single-ended operation is fairly easy. The transistor must be capable of dissipating at least two times the desired load power, in addition to having an adequate I_c max and E_c max rating.

$$P_d = 2P_o$$

where P_d is the transistor dissipation capability in watts and P_o is the desired maximum output power in watts.

Class B Push-Pull Operation

For higher-power audio and servo amplifiers class B operation offers several advantages, the foremost of which is efficiencies up to 78 percent. Higher efficiency means cooler transistors, smaller heat sinks, fewer thermal stability problems, and, most important, lower power supply cost, and bulk.

Push-pull operation does not require as much filtering as single-ended operation because ripple currents tend to cancel in the load.

Even when a continuous sine wave signal is being amplified, the duty cycle of each power transistor is only 50 percent, and when there is no signal, the power transistors don't work at all. In contrast, single-ended class A amplifiers work hardest (dissipate the most heat) when there is no signal. The magnitude of the differences between class A and class B efficiency is important. The maximum power dissipation in the transistors in class B takes place at about 64 percent of maximum power out.

The power dissipation in each transistor in class B is

$$P_d\ (\text{max}) = \frac{2}{\pi^2} P_o\ \text{max}$$

$$P_d\ (\text{max}) \approx 0.2\, P_o\ \text{max}$$

This means that a pair of power transistors in class B can produce five times as much sine wave power to the load as a single class A transistor can produce.

Stated another way, the transistors for a class B amplifier can have a much smaller power dissipation capability for a given power output.

A transistor for class B use must meet the following criteria:

1. Have a high enough I_c (max)
2. Have a high enough E_c (max)
3. Be capable of dissipating at least 0.2 times the desired maximum desired output power P_o

POWER OUTPUT. Let us examine power output in terms of E_o p-p and R_{Lw}.

$$P_o\ (\text{rms}) = \frac{E_o\ \text{p-p}}{2\sqrt{2}} \times \frac{I_o\ \text{p-p}}{2\sqrt{2}}$$

NOTE: $\sqrt{2} \approx 1.4$; rms $\times \sqrt{2} =$ peak.

$$P_o\ (\text{rms}) = \frac{E_o\ \text{p-p}}{8} \times \frac{E_o\ \text{p-p}}{R_{Lw}}$$

because $I = E/R$.

$$P_o\ (\text{rms}) = \frac{(E_o\ \text{p-p})^2}{8R_{Lw}} \tag{12-1}$$

In the design of a power amplifier, the power output is one of the specifications, and we will want to know how much E_o p-p to design in. Most popular circuits use an emitter follower configuration for the output transistors. This means the E_o p-p of the driver must produce the same E_o p-p as the output circuit because of the unity gain of the common-collector circuit.

If we solve Eq. (12-1) for E_o p-p, we get

$$E_o\ \text{p-p} = \sqrt{P_o\ (\text{rms}) \times 8R_{Lw}} \tag{12-2}$$

12-2 THE DIRECT-COUPLED CLASS A EMITTER FOLLOWER POWER AMPLIFIER

The emitter follower configuration is generally preferred to common-base and common-emitter circuits because the load can function as a large current-mode feedback impedance which raises the input impedance and reduces distortion without wasting power in a separate emitter resistor. To illustrate how low the input resistance of a common-emitter circuit can get in a power stage, let us try an example.

Assume a power transistor with a β of 50 and a quiescent collector current of 500 mA with the emitter grounded; the input resistance is $\beta \times R_e'$.

$$R_e' \approx \frac{25}{500} = 0.05$$

and 50 times 0.05 gives $R_{in} \approx 2.5\ \Omega$. This 2.5 Ω would be R_{Lw} for the driver. Such a low value for R_{Lw} would probably dictate an emitter follower for the driver. With a 16-Ω load in the emitter (see Fig. 12-1), the input resistance of the transistor would be $R_{in} \approx 50 \times 16 = 800\ \Omega$. An examination of the fully bypassed table of possibilities (Table 6-2) shows that we could drive this 800-Ω load with a common-emitter circuit and a QI_c of only 10 mA, and with good voltage gain.

Whether we use an emitter follower driven by a common-emitter circuit or a common emitter driven by an emitter follower, we get *voltage* gain in only one stage.

We are not likely to get as much voltage gain in a power common-emitter stage as we can get in a common-emitter driver stage. For one thing, bypass capacitors for any emitter resistance in a power common-emitter stage would be physically very large and very expensive. In addition, the stability problems and distortion problems are greater in the power transistor and the need for feedback is proportionately greater in the power stage. I'm not saying that common-base and common-emitter power stages are never used, but they are not nearly as popular as the common-collector power output stage.

The Importance of Two-stage Voltage-mode Feedback in Power Amplifiers

All the circuits in this chapter will use two-stage voltage-mode feedback, often in conjunction with current-mode feedback, and often with additional signal feedback of one or both types. The driver stage must provide the same E_o p-p as the E_o p-p of the power stage when the power stage is a common-collector circuit. This requirement dictates voltage-mode feedback for the driver because E_o p-p for a voltage-mode driver circuit is approximately E_o p-p $\approx$ 1.5 E_Q whereas the output voltage for a common-emitter driver is E_o p-p $\approx E_Q - 1$. Because in direct-coupled driver-to-power stages the driver voltage distributions are dictated by the voltage distribution in the power amplifier, in both class A single-ended and class B push-pull circuits, the driver ends up with about ½E_s across E_Q and ½E_s across R_L. Since the power output is a function of $(E_o \text{ p-p})^2$, we can get a significantly higher power out-

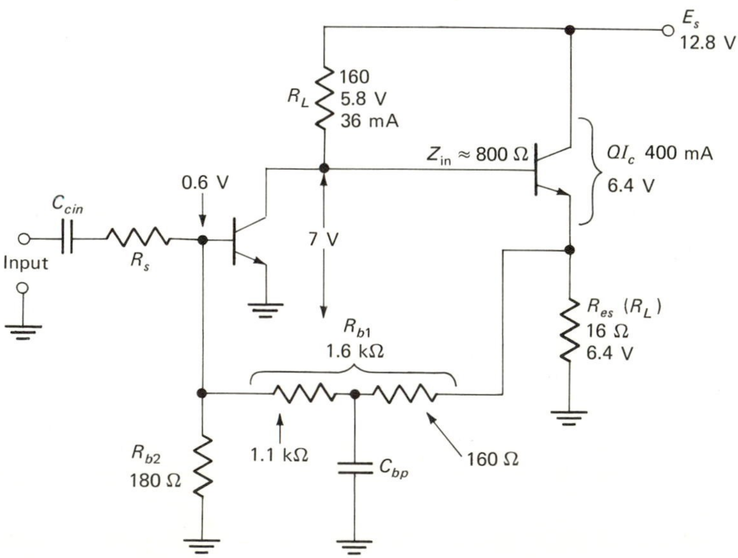

Fig. 12-1 The basic class A power amplifier with two-stage voltage-mode feedback for Example 12-1

put for a given supply voltage using voltage-mode driver feedback than we can with current-mode feedback. Including the power stage in the feedback loop stabilizes it as well. Again there are other arrangements, but experience has shown this one to be one of the best. Figure 12-1 shows the basic class A power amplifier with two-stage voltage-mode feedback.

Example 12-1 Designing the Basic Class A Power Amplifier with Voltage-mode Feedback

Design a class A power amplifier circuit, using the following specifications.

Transistor Parameters	Driver	Power Stage
1. β (min)	100	50
2. I_c (max)	100 mA	1 A
3. E_c (max)	25 V	25 V
4. Power dissipation	—	$2P_0 = 0.78 \times 2 \approx 2$ W min
Circuit Parameters		
1. Voltage gain	Highest possible	1
2. E_o p-p	$\sqrt{P_o\ (\text{rms}) \times 8R_{Lw}}$	$\sqrt{P_o\ (\text{rms}) \times 8R_{Lw}}$
3. R_{Lw}	R_{in} Q2	16 Ω
4. Stability factor	10	10
5. $\%\Delta QI_c$ to be allowed	n.a.	n.a.
6. Maximum operating temperature	n.a.	n.a.
7. f_o	25 Hz	25 Hz
8. Power output	n.a.	780 mW

Designing the Output Stage

1. Compute E_o p-p required for the desired power output, 780 mW.

$$E_o \text{ p-p} = \sqrt{P_o\ (\text{rms}) \times 8R_{Lw}}$$
$$E_o \text{ p-p} = \sqrt{0.78 \times 8 \times 16}$$
$$E_o \text{ p-p} \approx 10 \text{ V}$$

Remember that the driver must be capable of producing this much output swing and so must the power stage. The circuit is shown in Fig. 12-1.

2. Compute E_Q for Q1 necessary for $10 \approx 10$ V E_o p-p.

$$E_o \text{ p-p} \approx 1.5E_Q$$
$$E_Q \approx 0.66E_o \text{ p-p}$$
$$E_Q \approx 0.66 \times 10 = 6.6 \text{ V}$$

3. Make the QI_c decision for Q2.
Actually there are two parts to this decision. The first involves consulting the table of possibilities for the common-collector circuit (Table 7-3) to determine the *minimum* Q-point collector current which will allow the required p-p output volt-

age with the specified value of R_{Lw}. (NOTE: $R_{Lw} = R_{Lac}$.) The value of QI_c obtained from the table of possibilities may not give enough quiescent voltage drop across the load to provide enough E_Q for the driver. Because the driver is direct-coupled, the E_Q is only V_j higher than the voltage drop across the load.

Consulting the table of possibilities (Table 7-3), we find that for 10 V p-p output, we need somewhere between 250 and 380 mA for QI_c.

Now let's see what QI_c we need to get the proper quiescent voltage across the load.

1. Compute the minimum required quiescent voltage drop across the load.

$E_{RLw} = E_Q \text{ (driver)} - V_j$
$E_{RLw} = 6.6 - 0.6$
$E_{RLw} = 6 \text{ V}$

2. Compute the minimum required QI_c to get 6 V drop across R_{Lw}.

$$QI_c = \frac{E_{RL}}{R_L} = \frac{6}{16} = 375 \text{ mA}$$

3. Let's take this on up to 400 mA just to be safe. At 400 mA

$E_{RLw} = 6.4 \text{ V}$

4. Compute E_Q and E_s.
As in most emitter followers $E_Q \approx E_{RLw}$ (E_{Res}). E_{RLw} here equals 6.4 V.

$E_s = 2 \times 6.4 = 12.8 \text{ V}$

Here we could adjust E_Q to accommodate a more convenient power supply voltage. In this case we will leave it at 12.8 V.

Compute E_Q for Q1.

$E_Q \text{ (Q1)} = E_{RLw} + V_j$
$E_Q \text{ (Q1)} = 6.4 + 0.6 = 7 \text{ V}$

5. Compute E_{RL} (Q1).

$E_{RL} \text{ (Q1)} = E_s - E_Q \text{ (Q1)}$
$E_{RL} = 12.8 - 7 = 5.8 \text{ V}$

Actually this is a little in error because the input resistance of Q2 is not quite high enough to assume that it doesn't load R_L at all. We can get by here without taking the loading into account, but just barely.

6. Compute R_L.

$R_L = \mathbf{S} \times R_{Lw}$ (R_{es})
$R_L = 10 \times 16 = 160$

7. Compute QI_c for Q1.
   We know R_L and E_{RL}, so a little Ohm's law gives us

$$I_{RL} = QI_c\ (Q1) = \frac{E_{RL}}{R_L}$$

$$I_{RL} = QI_c\ (Q1) = \frac{5.8\text{ V}}{160} = 36\text{ mA}$$

Designing the Feedback Loop

1. Compute R_{b1} (R_f).

$$R_{b1} = \mathbf{S} \times R_L$$
$$R_{b1} = 10 \times 160 = 1.6\text{ k}\Omega$$

   In splitting R_{b1} for higher input impedance, we split it with the largest part ($R_{b1}a$) toward the input and make $R_{b1}b$ just large enough to avoid loading R_{Lw}. ($R_{b1}b$ = 10 to 100 × R_{Lw}.) For this example we will split it as 160 Ω for $R_{b1}b$ and 1.5 kΩ for $R_{b1}a$, using standard values. If we were going to use $R_{b1}b$ as an adjustable bias control, we would make $R_{b1}b$ as much as 25 percent of R_{b1}.
2. Compute E_{Rb1}.

$$E_{Rb1} = E_{Res} - V_j = 6.4 - 0.6 = 5.8\text{ V}$$

3. Compute I_{Rb1}, the current through R_{b1}.

$$I_{Rb1} = \frac{E_{Rb1}}{R_{b1}} = \frac{5.8}{1.6\text{ k}\Omega} = 3.63 \times 10^{-3}\text{ A}$$

4. Compute I_b (Q1).

$$I_b = \frac{QI_c}{\beta} = \frac{36 \times 10^{-3}}{10^2} = 36 \times 10^{-5}$$

5. Compute I_{Rb2}.

$$I_{Rb2} = I_{Rb1} - I_b = (3.63 \times 10^{-3}) - (0.36 \times 10^{-3})$$
$$I_{Rb2} \approx 3.3 \times 10^{-3}$$

6. Compute R_{b2}.

$$R_{b2} = \frac{V_j}{I_{Rb2}} = \frac{0.6}{3.3 \times 10^{-3}} \approx 180\ \Omega$$

Consult the case 2 voltage-mode design sheet (Fig. 7-6) for the preceding calculations for the feedback-bias loop.

Laboratory Evaluation

The circuit in Fig. 12-1 was constructed and evaluated in the laboratory. The stability was fair, but it had several expected undesirable features.

First, the input impedance of the basic amplifier was very low.

$$Z_{in}\ (\text{amp}) \approx R_{b1}a \| R_{b2} \| \beta R_e'\ (\text{Q1})$$
$$Z_{in}\ (\text{amp}) \approx 1.1\ \text{k}\Omega \| 180 \| 100 \times 0.7\ \Omega$$
$$Z_{in}\ (\text{amp}) \approx 48\ \Omega$$

The measured value was 37.5 Ω. Computed or measured, the input impedance is too low to be very satisfactory for most applications.

Second, the input impedance of the output stage loaded the collector of the driver even more than was expected; this reduced the voltage gain and output swing of the driver more than was expected. And the total undistorted power output fell far short of the mark at 515 mW.

Even the stability was not as good as computed because of the voltage gain reduction due to the loading of the driver by the input impedance of the output stage.

All in all the results were not very satisfactory. The reason for the poor performance is simply the low input impedance of the power stage and the resulting low value of the driver collector load resistor. If the load resistance (R_{Lw}) had been lower, the circuit would have been almost completely unmanageable.

We could have done very slightly better by simply making R_L of the driver larger and depending almost entirely upon the two-stage voltage-mode feedback loop for β stability, that is, by using an unusually large stability factor for the current-mode feedback of Q2 of perhaps 100. This would increase the input impedance and increase the power output by a factor of about 1.5. The stability would probably be reduced slightly. Still, an improvement of only 50 percent or so would hardly make this circuit attractive in most cases, and it seems that the best answer is somehow to increase the input impedance of the power stage. In the next section we will look at the benefits of using a Darlington pair (or cascaded emitter follower) as the output stage.

12-3 IMPROVING THE PERFORMANCE OF THE BASIC DIRECT-COUPLED CLASS A AMPLIFIER

In this section we will examine two simple but highly effective methods of improving the performance of direct-coupled power amplifiers. Although we will apply these techniques in this chapter to class A power amplifiers, the same two techniques will be equally applicable, and perhaps even more important, to the design of class B complementary symmetry (and quasi-complementary symmetry) amplifiers. The first technique involves the use of a Darlington pair (or cascaded common-collector) output stage to elevate the input impedance of the power stage.

The addition of one inexpensive transistor allows a factor of 10 or more increase in voltage gain input impedance and overall stability. In addition, the driver can operate at a significantly lower QI_c, and it becomes possible to drive an 8-Ω load comfortably.

The second modification involves bootstrapping the driver collector load resistor. This is a case of positive feedback which increases the voltage gain by an order of magnitude or so, and at the same time increases the maximum E_o p-p of the driver. The schematic diagram is shown in Fig. 12-2.

We have split the driver collector load resistor and have capacitor-coupled the junction of the split to the amplifier output. The in-phase feedback voltage is applied to the top of R_La, and the voltage at the top of R_La *follows* the voltage changes at the bottom of R_La, maintaining an almost constant voltage drop across R_La. This is the same mechanism as the other bootstrap applications we have examined and the effect is the same; the impedance of R_La appears to be much larger, which increases the voltage gain.

The driver also sees an almost constant collector voltage which helps prevent distortion at low collector voltages. This extends the amount of undistorted E_o p-p and increases the available undistorted output power. The bootstrap capacitor also provides additional filtering of power-supply ripple, and this can be important in some circuits, particularly where the driver is single-ended and the output is push-pull, as is the case in complementary symmetry. The push-pull stage is more tolerant of ripple than the

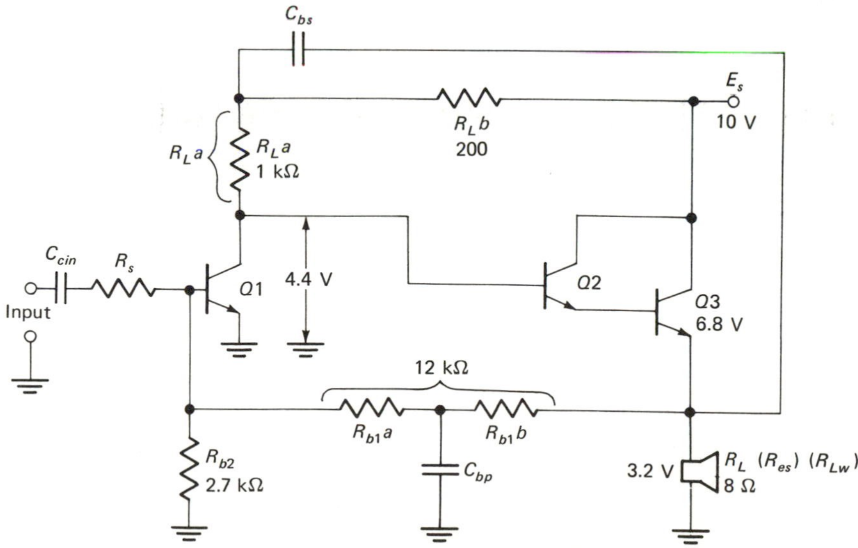

Fig. 12-2 The improved version of the class A direct-coupled power amplifier

single-ended driver, but the addition of the bootstrap capacitor adds additional filtering, allowing the use of a less well filtered basic power supply. The bootstrap capacitor also cushions the annoying thump when the amplifier is first turned on.

The X_c of the bootstrap capacitor should be $\approx 0.1\ R_La$ at the lowest frequency of interest. Now, let's design the improved circuit.

Example 12-2 The Improved Direct-coupled Class A Power Amplifier

Design the circuit in Fig. 12-2, using the following specifications.

Transistor Parameters	Q1	Q2	Q3
1. β (min)	100	100	50
2. I_c (max)	100 mA	100 mA	1 A
3. E_c (max)	25 V	25 V	25 V
Circuit Parameters			
1. Voltage gain	Highest possible	—	1
2. E_o p-p	$\sqrt{P_o \times 8R_{Lw}}$	—	$\sqrt{P_o \times 8R_{Lw}}$
3. R_{Lw}	R_{in} Q2	—	8 Ω
4. Stability factor	10	3	3 Ω
5. $\%\Delta QI_c$ to be allowed	—	—	—
6. Maximum operating temperature	—	—	—
7. f_o	25 Hz	25 Hz	25 Hz
8. Power output (P_o)	—	—	680 mW

Designing the Output Stage

1. Compute E_o p-p required for the desired power output, 680 mW.

$$E_o \text{ p-p} = \sqrt{P_o\,(\text{rms}) \times 8R_{Lw}}$$

$$E_o \text{ p-p} = \sqrt{0.68 \times 8 \times 8}$$

$$E_o \text{ p-p} \approx 6.6 \text{ V}$$

Both power amplifier and driver must be capable of an E_o p-p of at least 6.6 V.

2. Compute E_Q required for the driver for a 6.6-V E_o p-p.

$$E_o \text{ p-p} \approx 1.5\ E_Q\ (\text{Q1})$$
$$E_Q \approx 0.66\ E_o \text{ p-p}$$
$$E_Q \approx 4.4 \text{ V}$$

Theoretically any E_Q larger than this would do, but it is best not to make it much higher than necessary. Because of the direct coupling, the voltage drop across the final load will be only about 1.2 V less than E_Q. The higher we make E_Q, the higher the voltage drop across R_{Lw} and the larger the power stage QI_c.

3. Make the QI_c decision for Q3, the final power stage:
   a. Consult the emitter follower table of possibilities (Table 7-3) for the minimum QI_c for the desired 6.6 V p-p. The table shows we can use any QI_c above 300 mA.
   b. Determine what value of QI_c is required to get the proper quiescent voltage drop across R_{Lw}. In this case the voltage drop across R_{Lw} must be $2V_j$ lower than E_Q of the driver because we are using a Darlington pair and there are two junctions involved.

   $$E_{RLw} = E_Q - 2V_j$$
   $$E_{RLw} = 4.4\text{ V} - 1.2 = 3.2\text{ V}$$

   Now by Ohm's law,

   $$QI_c = \frac{E_{RLw}}{R_{Lw}} = \frac{3.2\text{ V}}{8} = 400\text{ mA}$$

   400 mA is greater than the value specified by the table of possibilities, and so it is acceptable.

   $$QI_c = 400\text{ mA}$$

   Had this figure come out any higher, we would have had to change the maximum collector current specifications for Q3.
4. Compute E_Q and E_s.
   It has been our general custom, in emitter follower circuits, to make E_Q approximately equal to E_{Res} (E_{RLw} in this case). However, it is understood that, should there be some advantage to it, we can make E_Q greater than E_{Res}. In this case there is some advantage to it. We can increase E_Q to allow for a more convenient power supply voltage, and at the same time increase the value of the driver collector load which can provide some increase in voltage gain and input impedance. Actually it doesn't buy us very much—certainly not enough to make the previous circuit (Example 12-1) acceptable—but it can't hurt. Let us use an E_s of 10 V. Therefore,

   $$E_Q\text{ (Q3)} = E_s - E_{RLw} = 10 - 3.2$$
   $$E_Q = 6.8\text{ V}$$

5. Compute E_{RL} for the driver Q1.

   $$E_{RL}\text{ (Q1)} = E_s - E_Q\text{ (Q1)}$$
   $$E_{RL} = 10 - 4.4 = 5.6\text{ V}$$

   Because of the high input impedance of the Darlington pair, we have no worry about the Darlington power stage having such a low input impedance as to upset the quiescent collector voltage of Q1. As you will see shortly, we will be able to consider Q1 to be virtually unloaded for both dc and signal.

6. Compute R_L for the driver.

$$R_L = \mathbf{S} \times R_{Lw}$$

Because of the nature of the Darlington pair, we can get a higher value for R_L than the equation would seem to indicate, and that is another advantage of this arrangement. Let us use an **S** of 3 for both Q2 and Q3 for a total **S** of 9. The Darlington pair stability factor was discussed in Sec. 10-4; if these next few equations don't make complete sense, go back and review that section. For the Darlington pair R_L equals $\mathbf{S}_1\beta_2R_{Lw}$ or $\mathbf{S}_2\beta_1R_{Lw}$, whichever is less.

$$R_L = 3 \times 50 \times 8 = 1.2\ \text{k}\Omega$$
$$R_L = 3 \times 100 \times 8 = 2.4\ \text{k}\Omega$$
$$R_L = 1.2\ \text{k}\Omega$$

Some designers use much higher stability factors here and depend almost entirely upon the two-stage voltage-mode feedback loop to stabilize the entire circuit. Such a procedure is adequate in many cases, but it can cause trouble in hostile environments and does not leave much margin for production errors in heat-sink selection, ventilation, and the inevitable production changes and substitutions. The entire philosophy of this book has been one of conservative design, and until you have had considerable experience and can afford to take some risks, I suggest you follow the more conservative path.

The collector resistor of the driver should be split at about a 5:1 ratio of R_La to R_Lb to form the junction to which C_{bs} is connected. See Fig. 12-2.

7. Compute QI_c for the driver Q1.
We know R_L and E_{Rl}, so a little Ohm's law gives us

$$I_{RL} = QI_c\ (\text{Q1}) = \frac{E_{RL}}{R_L} = \frac{5.6\ \text{V}}{1.2\ \text{k}\Omega}$$

$$QI_c\ (\text{Q1}) = 4.7\ \text{mA}$$

Designing the Feedback Loop

1. Compute R_{b1} (R_f).

$$R_{b1} = \mathbf{S} \times R_L\ (\text{Q1})$$
$$R_{b1} = 10 \times 1.2\ \text{k}\Omega = 12\ \text{k}\Omega$$

Let us split R_{b1} into two parts to increase the input impedance (by filtering out the Miller effect for signal frequencies). Let $R_{b1}b \approx 100\ R_{Lw} \approx 1\ \text{k}\Omega$, $R_{b1}a \approx 11\ \text{k}\Omega$. Again, we could make $R_{b1}b$ adjustable.

2. Compute E_{Rb1}.

$$E_{Rb1} = E_{RLw} - V_j = 3.2 - 0.6 = 2.6\ \text{V}$$

3. Compute I_{Rb1}, the current through R_{b1}.

$$I_{Rb1} = \frac{E_{Rb1}}{R_{b1}} = \frac{2.6 \text{ V}}{1.2 \times 10^4} = 2.16 \times 10^{-4} \text{ A}$$

4. Compute I_b, the base current of Q1.

$$I_b = \frac{QI_c \text{ (Q1)}}{\beta} = \frac{4.7 \times 10^{-3}}{10^2} = 4.7 \times 10^{-5} \text{ A}$$

5. Compute I_{Rb2}, the current through R_{b2}.

$$I_{Rb2} = I_{Rb1} - I_b = (2.66 \times 10^{-4}) - (0.47 \times 10^{-4})$$
$$I_{Rb2} = 2.19 \times 10^{-4} \text{ A}$$

6. Compute R_{b2}.

A little Ohm's law gives us

$$R_{b2} = \frac{V_j}{I_{Rb2}} = \frac{0.6 \text{ V}}{2.19 \times 10^{-4} \text{ A}}$$

$$R_{b2} = 2.7 \text{ k}\Omega$$

Figure 12-2 shows the complete circuit with component values.

12-4 THE COMPLEMENTARY SYMMETRY POWER AMPLIFIER

The most important power amplifier circuit in the transistor industry is the complementary symmetry circuit. This circuit is the basis of nearly all audio power amplifiers, and it is rapidly becoming the *standard* basic power amplifier circuit.

The quest for good high-fidelity power amplifier circuits began almost with the beginning of commercial broadcasting and record making, long before high-fidelity sound caught the public's fancy. And from the very beginning, the transformer used to couple vacuum-tube power amplifiers to relatively low-impedance loudspeakers was the weak link in the system. The fairly high plate currents required and the inductive nature of the transformer caused a lot of problems. Core saturation and the need for a wide band response made such transformers big, heavy, expensive, and inefficient. It was realized quite early in the game that a transformerless power amplifier was an ideal goal, and a remarkable number of schemes were tried, but none of them were ever really successful. There was just no practical way to drive such a low impedance load from the high output impedance of a vacuum tube with any kind of acceptable efficiency.

For a while the major effort was concentrated on improving the characteristics of output transformers, and although some very good transformers were developed, they were still big, heavy, expensive, and inefficient. With this background it is understandable that the coming of the transistor, with its relatively low output impedance, renewed designers' interest in the design of transformerless power amplifiers. The quasi-complementary design was first to come on the scene, and some very good amplifiers of this type were designed, but the circuit has some inherent faults, particularly at high power levels. Even the name *quasi-complementary* underlined the designers' desires to design true complementary symmetry power amplifiers.

The quasi-complementary symmetry circuit was popular in discrete circuitry because *npn-pnp* pairs of well-matched power transistors were not available at a reasonable price. The technological problems involved in manufacturing such complementary power pairs have been solved, and they are now available and inexpensive.

The quasi-complementary circuit is still used in integrated circuitry because of the problem of forming *pnp* and *npn* pairs with closely matched characteristics on an IC chip. However, this problem is also being solved, and true complementary symmetry circuits are being produced in IC form.

The basic circuit with its driver is shown in Fig. 12-3. The circuit is not difficult to analyze once you understand how it operates, but until you do understand it, it can seem quite formidable. The circuit is essentially made up of two complementary emitter followers. One emitter follower is an *npn* circuit which conducts on the more positive-going half cycle, and the other one is a *pnp* circuit which conducts on the negative-going half cycle. Each transistor works for only about ½ cycle in a near class B push-pull mode. The amplifier operates *essentially* as class B, but there must be some small quiescent collector current when there is zero signal. The reason for this small (2 to 20 mA typ.) quiescent collector current is to avoid zero crossing distortion. Without this small quiescent bias, one transistor would begin to turn off at about 0.6 V above the zero reference line (owing to the junction potential), and the other transistor would not turn on until the signal went 0.6 V below the zero line. There would be a gap of no conduction near the zero line. The result would be a waveform similar to that shown in Fig. 12-4, with a sine wave input. This kind of distortion is generally called crossover distortion; it is particularly annoying to human sensibilities.

The diode *D*1 and the resistor R_d in Fig. 12-3 take care of this small quiescent current requirement, and we will examine just how it works shortly. For now we can assume that these two components have virtually no resistance and only a constant 1.2-V drop across them.

Figure 12-5 shows an analogy of the circuit behavior at conditions of *no conduction* and *full conduction* for the driver Q1 (in Fig. 12-3). Assume that the first half cycle turns Q1 off. The voltage at point *A* rises to the supply voltage (E_s) of 30 V. The +30 V on the anode of B_{D1} biases the junction *on* and charges C_{co} as shown in Fig. 12-5*a*. If the charging were allowed to proceed, C_{co} would eventually charge up through R_{Lw} to the supply voltage.

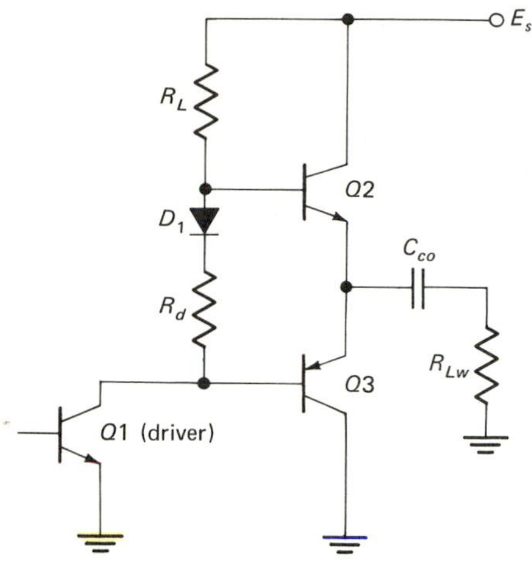

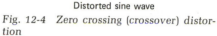

Fig. 12-3 The basic complementary symmetry power amplifier with driver

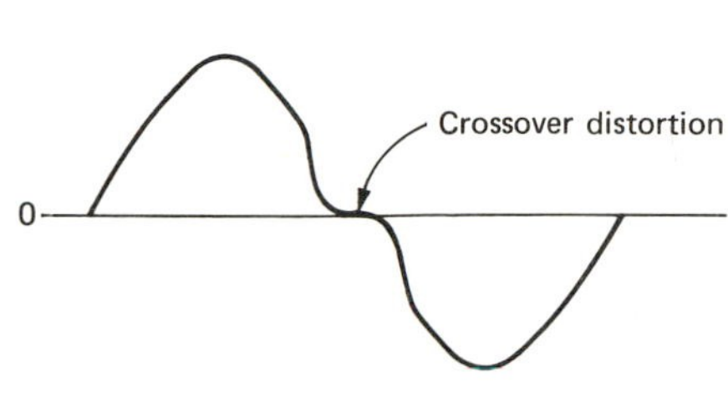

Fig. 12-4 Zero crossing (crossover) distortion

Actually we never let C_{co} take on the full charge because the input signal turns Q1 on before this happens, pulling the voltage at point A down toward zero. B_{D1} is now biased off because the capacitor C_{co} is charged to some voltage in the polarity shown in Fig. 12-5b.

With point A near 0 V and some positive voltage on the cathode (emitter) of B_{D1}, the junction is reverse biased. The discharge half cycle now begins. B_{D2} has nearly 0 V on the cathode (base) and a positive voltage on the anode (emitter) of B_{D2}. B_{D2} is forward-biased, and the capacitor C_{co} begins to discharge through B_{D2} and low resistance of the transistor, Q1. You

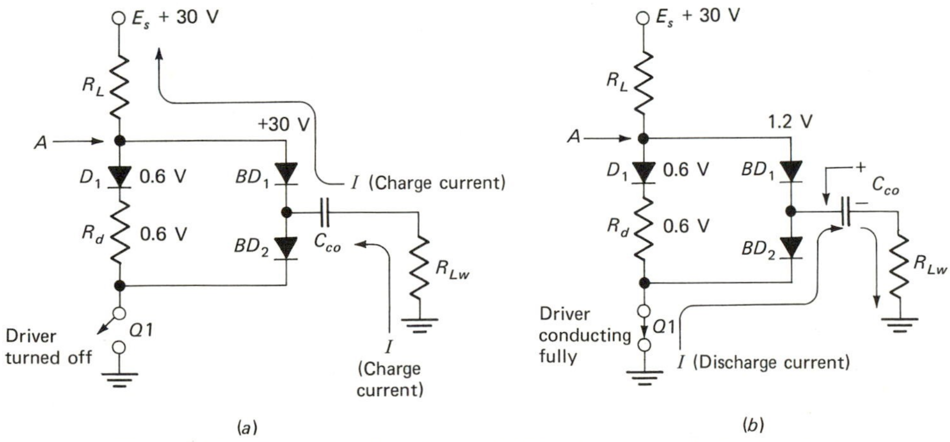

Fig. 12-5 The amplifier input circuit analogy. (a) The charging cycle; (b) the discharge cycle

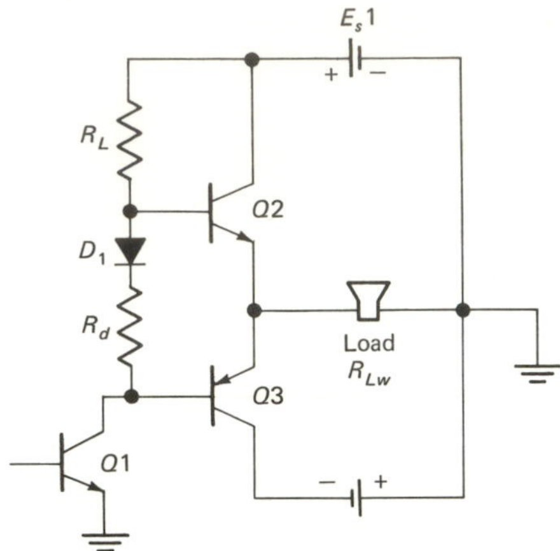

Fig. 12-6 The 0 Hz complementary symmetry amplifier with two power supplies

will notice that current flows first in one direction through R_{Lw} and then in the opposite direction on the other half cycle.

In reality we never actually reach these two extremes of cutoff and saturation of Q1 (unless we happen to like distortion), but the analysis of the two extremes can give us the necessary insight into what happens between these two limits.

What we just went through is a bit of an oversimplification, because we did not take the collector currents of the output transistors into account (see Fig. 12-3). Actually the C_{co} charging current and the C_{co} discharge current consisted of both base and collector currents of Q2 and Q3. The charging current in the first half cycle consists of the collector current plus the base current of Q2. The capacitor, in the second half cycle, must supply both base and collector currents for Q3. The capacitor must supply the total power for the second half cycle; the power supply voltage E_s is not involved in this half cycle. The result is a fairly large value for C_{co}, from 100 μF to several thousand microfarads, depending upon the load Z and desired low-frequency response. The output coupling capacitor is almost universally used in audio equipment, but when frequencies down to 0 Hz are necessary, two power supplies can be used. Figure 12-6 shows the two-power-supply version, which operates in essentially the same fashion as the circuit in Fig. 12-3.

Quiescent-point Stability in the Complementary Symmetry Power Amplifier

As I pointed out earlier, we must have some small quiescent collector current in order to avoid crossover distortion. And as usual we must hold that Q point fairly constant. If we don't hold it fairly constant, we will suffer, and I do mean suffer, from crossover distortion in cold weather, and realize a reduced power capability in hot weather. The crossover distortion is far more serious. The average human ear cannot detect a 20 percent reduction or so in power, but even a music hater can detect, and be irritated by, a few percent of crossover distortion.

Let's Examine the Bias Loop

Power transistors are especially difficult to stabilize because we cannot resort to sufficiently large current-mode feedback resistors without significant power loss and because some self-heating is almost inevitable. So far, the best approach seems to be a combination of diode compensation and driver-power stage (two-stage) voltage-mode feedback. For higher power applications, 25 W rms or greater, an active bias resistor is often added to these two stability controls. In addition, some extra voltage-mode feedback is often incorporated for the purpose of reducing total signal distortion.

Let us draw the equivalent circuit for the bias loop associated with the *npn* transistor in Fig. 12-7*a* and examine it in some detail. See Fig. 12-7*a* and *b*. On the base side we see that *D*1 is in series with R_d. The total voltage drop across these two components is approximately equal to, or slightly

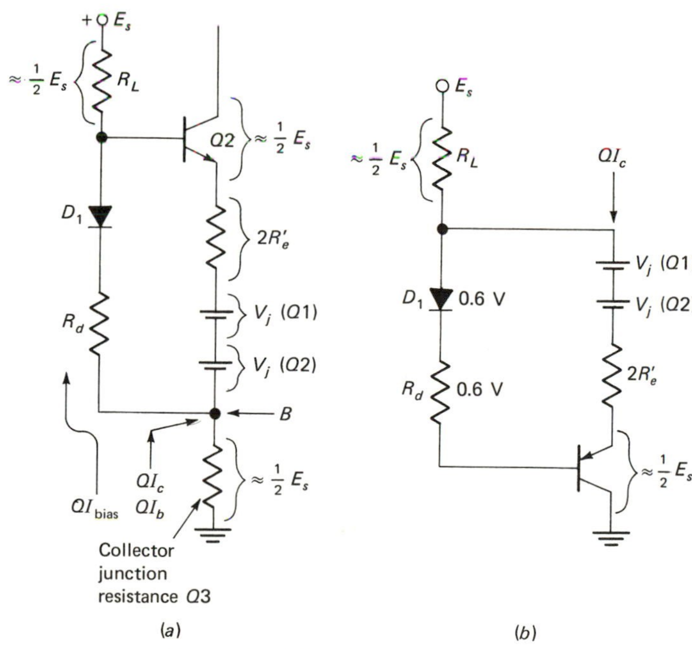

Fig. 12-7 The bias and stability loop. (a) *The* npn *loop;* (b) *the* pnp *loop*

greater than, the two voltage drops across the base-emitter junctions of Q2 and Q3 (Fig. 12-7). On the collector-emitter side of the *npn* transistor (Q2) we include the voltage drops of both junctions (Q2 and Q3), and we include R'_e of both transistors. Only one junction resistance is shown, but the value is multiplied by a factor of 2, because the same current flows through both junctions at quiescence. Point B in Fig. 12-7*a* is a *virtual* ground for Q2 insofar as R_d is concerned. R_b consists of the resistance of $D1$ and R_d. $D1$ is intended to provide compensation for ΔV_{BE}. Because of the particular stability requirements of power transistors, we need all the stability we can get. We will need a low stability factor, but we will supplement this with two-stage voltage-mode feedback. This combination of diode compensation and voltage-mode feedback has become standard practice in the industry.

As in all diode compensation schemes, the current through the compensating diode, or diodes, should be from 1/3 to 1 (often a little more) times the quiescent collector current of the transistor being stabilized. This will ensure a reasonable stability factor, as well as an appropriate operating current for the diodes.

In selecting the diode bias current, we will also be selecting the current through R_L, which will determine the quiescent collector current for the driver transistor Q1 (Fig. 12-3).

The power supply voltage E_s at quiescence is divided equally across the collector-emitter circuits of the two power transistors. E_s is also divided equally between R_L and E_Q of the driver. By the time we have designed the power transistor bias network, we will have taken care of not only the quiescent bias, but also the quiescent stability factor for the power transistors, V_{BE} stability, and the quiescent operating point for the driver. Two-stage voltage-mode feedback from the power transistors back to the base of the driver stabilizes the current gain of the system for the large voltage and current excursions which are essential to large output powers. In addition some small amount of current-mode feedback may be added to the driver to raise the transistor input impedance. Further, an additional *signal* voltage-mode feedback path may be incorporated to reduce total signal distortion. Bootstrapping R_L of the driver is also a very common practice.

Suppose we try a design example.

Example 12-3

Design a 4-W complementary symmetry power amplifier to drive an 8-Ω loudspeaker. The circuit is shown in Fig. 12-8 complete with values.

The driver load is split, and C_{bs} provides the bootstrap feedback to allow a larger undistorted p-p output voltage from the driver Q1 and an increased driver voltage gain.

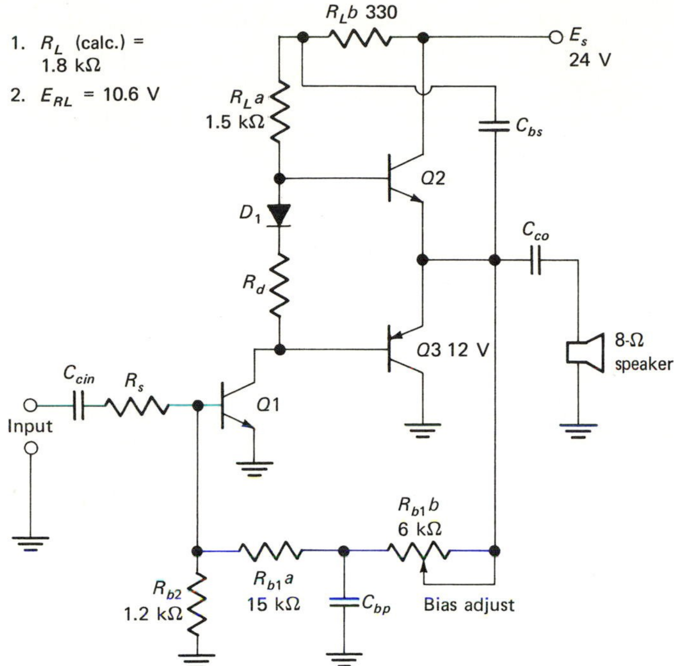

Fig. 12-8 The basic complementary symmetry amplifier for Example 12-3

Design Procedure for the Simple Complementary Symmetry 4-W Power Amplifier

Driver preliminary design considerations.

1. Record the rms power output desired.

   $P_o\ (\text{rms}) = 4\ \text{W}$

2. Record the final load impedance R_{Lw}.

   $R_{Lw} = 8\ \Omega$

3. Compute the required E_o p-p.

   $E_o\ \text{p-p} = \sqrt{P_o\ (\text{rms}) \times 8R_{Lw}}$

   The E_o p-p here is the E_o p-p of the drive, because the power output stage is an emitter follower (actually a complementary pair of them), and the E_o p-p of the emitter follower can be no greater than the p-p input signal. The emitter follower has a unity voltage gain.

E_o p-p $= \sqrt{4 \times 8 \times 8}$
E_o p-p $= 16$ V

4. Compute the minimum E_Q for the driver.

E_Q (min) $= 0.66 \times E_o$ p-p $= 0.66 \times 16$
E_Q (min) $= 10.56$
E_Q (actual) $= 12$ V   See step 5

5. Compute minimum E_s.

E_s (min) $= 2 \times E_Q$
E_s (min) $= (2 \times 10.56) + 2 = 23.12$ V

Let $E_s = 24$ V. The extra 2 V is to allow for V_{BE} drops.

6. Compute E_{RL}.

$$E_{RL} = 0.5\ E_s - (E_d + E_{RD})$$

Here we subtract the voltage drop for the bias stability network from E_{RL}.

$E_{RL} = 0.5 \times 24 - (1.2 \text{ V})$
$E_{RL} = 10.8$ V

We subtract bias loop voltage drops (E_d and E_{RD}) from E_{RL} rather than from E_Q because the possible output swing is equal to $2E_{RL}$ and approximately equal to only 1.5 E_Q, and so it is important not to reduce E_Q.

Designing the Power Transistor Bias and Stability Loop

1. Select the quiescent collector current for the power transistors Q2 and Q3.

This quiescent bias current (for Q2 and Q3) is generally between 2 and 20 mA. Let us select 6 mA.

2. Select I_{bias}, the current through the V_{BE} stability diode $D1$.

In this kind of circuit, where power transistors are involved, good diode compensation requires that the diode bias current be pretty nearly equal to the QI_c of the transistor being stabilized. In circuits where power was not involved, we could get by with from ⅓ to $1QI_c$ for I_{bias} (see Chap. 11), but here it is best to have an I_{bias} which is about equal to QI_c of the transistor(s) being stabilized.

3. Compute R_L for the driver Q1.

$$R_L = \frac{E_{RL}}{I_{bias}}$$

$$R_L = \frac{10.8\text{ V}}{6\text{ mA}} \approx 1.8\text{ k}\Omega$$

Note that the bias network is in series with R_L so that $I_{\text{bias}} \approx QI_c$ of the driver Q1. If the β of the power transistors is too low at 6 mA, we may find power transistor base current loading R_L a bit. This can upset the design-center QI_c voltages but generally not so much that the bias adjustment can't compensate for it. The only really adverse result of such a situation would be a sensitivity to any supply voltage much lower than the computed E_s.

Higher voltages of E_s will normally be tolerated very well even if this problem exists. Unfortunately we may not always be able to find a β figure at these low currents in the manufacturer's literature. Some manufacturers do provide low-current β figures, and these can help considerably.

For higher power levels we will generally go to a Darlington pair configuration for the output transistors and avoid this problem altogether. (Fig. 7-6.)

Designing the Feedback Loop

Here we can use Field 2 of the voltage-mode design sheet.

FRAME 1

Compute R_{b1}.

$R_{b1} = \mathbf{S} \times R_L$ (Q1)
$R_{b1} = 10 \times 1.8\text{ k}\Omega$
$R_{b1} = 18\text{ k}\Omega$

FRAME 2

Compute E_{Rb1}.

$E_{Rb1} = E_Q - V_j$
$E_{Rb1} = 10.6 - 0.6$
$E_{Rb1} = 10.0\text{ V}$

FRAME 3

Compute I_{Rb1}.

$$I_{Rb1} = \frac{E_{Rb1}}{R_{b1}}$$

$$I_{Rb1} = \frac{10\text{ V}}{18\text{ k}\Omega} \approx 5.5 \times 10^{-4}\text{ A}$$

FRAME 4

Compute I_b, the Q1 (driver) base current.

$$I_b = \frac{QI_c}{\beta} = \frac{6 \times 10^{-3}}{1 \times 10^{2}} = 6 \times 10^{-5}\text{ A}$$

FRAME 5

Compute I_{Rb2}.

$$I_{Rb2} = I_{Rb1} - I_b$$
$$I_{Rb2} = (55 \times 10^{-5}) - (6 \times 10^{-5})$$
$$I_{Rb2} = 49 \times 10^{-5}$$

FRAME 6

Compute R_{b2}.

$$R_{b2} = \frac{V_j}{I_{Rb2}} = \frac{6 \times 10^{-1}}{4.9 \times 10^{-4}}$$

$$R_{b2} \approx 1.2\ \text{k}\Omega$$

Practical Considerations

1. Splitting R_{b1}:

Part of R_{b1} is made variable to provide for an initial bias adjustment. This would not normally be a front-panel adjustment, but one to be set initially and, it is hoped, forgotten. One-half of the control resistance should be roughly equal to 25 percent of R_{b1}. $12\ \text{k}\Omega \times 0.25 = 3.0\ \text{k}\Omega$. (The entire resistance of the potentiometer $R_{b1}b$ is 6 kΩ.) Design center is the *center* of the potentiometer, $R_{b1}bR_{b1}a = R_{b1} - R_{b1}b =$ 15 kΩ. The bypass capacitor should be no larger than 0.1 C_{co} to avoid signal loading. If we follow the usual rule of making the reactance of $X_c \approx 0.1R_{b1}a$ at the lowest frequency of interest, we will not get into trouble on this count. I only mention it because, in the lab, there is a tendency to use any capacitor that is handy for bypass purposes as long as it is larger than the calculated value. This is a case where we can go too far in using a generally valid assumption.

2. Heat sinks:

An adequate heat sink is mandatory. Temperature-compensating diodes should be mounted on the power transistor heat sink (or sinks). Values for the completed design are shown in Fig. 12-8.

Computing the Input Impedance

$$Z_{in} = R_s + [R_{b2} \| \beta R'_e\ (Q1) \| R_{b1}a]$$

First let's compute the value of Z_{in} without R_s, that is, the input impedance of the basic amplifier.

$$R'_e = \frac{25}{QI_c} = \frac{25}{6} \approx 4\ \Omega$$

$$Z_{in}\ (\text{amp}) = 1.2\ \text{k}\Omega \| 400 \| 15\ \text{k}\Omega \approx 250\ \Omega$$

R_s is made to be at least 10 times Z_{in} (amp) to prevent loading of the feedback networks. Frequently R_s is made 50 to 100 times Z_{in} (amp) to avoid loading the signal source. All of R_s but 10 times Z_{in} (amp) can be made variable to provide a level control. Suppose we make $R_s = 20Z_{in}$ (amp). $R_s = 20 \times 250 \approx 4.7\ \text{k}\Omega$. The total input impedance as seen by the signal source is $R_s + Z_{in}$ (amp) $= 4.7\ \text{k}\Omega + 250 \approx 4.7\ \text{k}\Omega$.

Computing the Input Signal Required to Drive the Amplifier to Full Power

In order to make this determination, we must know the voltage gain of the driver:

$$V_g = \frac{R_{Lac}}{R'_e}$$

$$R_{Lac} = R_L \| \beta R_{LW} = 1.8\ \text{k}\Omega \| 50 \times 8$$

$$R_{Lac} = 327\ \Omega$$

$$R'_e = \frac{25}{QI_c} = \frac{25}{6} \approx 4\ \Omega$$

$$V_g = \frac{327}{4} \approx 82$$

To get full power out of the amplifier, we need an E_o p-p from the driver of 16 V p-p.

$$E_o = E_{in} \times V_g$$

Solving for E_{in},

$$E_{in} = \frac{E_o}{V_g} = \frac{16}{82} \approx 0.2\ \text{V p-p}$$

However, we have a 20 to 1 (approximately) voltage divider consisting of R_s and Z_{in} (amp) in the input; so the actual signal input from the source to get full power out is approximately 20 times the figure we just calculated.

$$20 \times 0.2 = 4\ \text{V p-p}$$

So 4 V p-p will be required to drive the amplifier to full power.

PROBLEMS

12-1 Compute the input impedance of the basic amplifier in Example 12-2, and select a value for R_s for the example:

$$Z_{in} \approx R_{b1}a \| \beta R'_e\ (Q1) \qquad \text{and} \qquad R_s \approx 10\ Z_{in}(\text{amp})$$

12-2 Compute the voltage gain, assuming a tenfold increase in V_g because of bootstrapping of the driver Q1 in Example 12-2.

12-3 Construct a design sheet, or flow diagram, for the design of class A direct-coupled circuits of the kind discussed in Examples 12-1 and 12-2.

12-4 Give as thorough an explanation as possible why the circuit in Example 12-2 is superior to the circuit in Example 12-1.

Problem 12-5

TITLE: **The Class A Amplifier with a Darlington Pair Output**
PROCEDURE: **Design the circuit in Fig. 12-2, using the following specifications.**

Transistor Parameters	Q1	Q2	Q3
1. β (min)	100	100	50
2. I_c (max)	100 mA	100 mA	1 A
3. E_c (max)	25 V	25 V	30 V
Circuit Parameters			
1. Voltage gain	Highest possible	n.a.	1
2. E_o p-p	$\sqrt{P_o \times 8R_{Lw}}$	—	$\sqrt{P_o \times 8R_{Lw}}$
3. R_{Lw}	R_{in} Q2	—	8 Ω
4. Stability factor	10	3	3
5. $\%\Delta QI_c$ to be allowed	n.a.	n.a.	n.a.
6. Maximum operating temperature	n.a.	n.a.	n.a.
7. f_o	25 Hz	25 Hz	25 Hz
8. Power output (P_o)	n.a.	n.a.	500 mW

"n.a." means not applicable.

12-5 AN IMPROVED VERSION OF THE COMPLEMENTARY SYMMETRY AMPLIFIER, WITH 13 W rms OUTPUT POWER

The circuit we designed in the previous section is a fairly satisfactory circuit up to power levels of 5 to 10 W or so, but the circuit can be drastically improved on almost all counts by making the power stages Darlington pairs or compound emitter followers. The design procedure is quite similar to the circuit in the previous example, with the exception that a small emitter resistor has been added to the driver stage Q1, which allows us to raise the value of R_{b2} and consequently increase the circuit input impedance. See Fig. 12-9. This small emitter resistor is not a critical value, and anything from 100 to 500 Ω is typical. We will use 100 Ω in this example, and I will specify a different value in the student problem involving this design. A comparison of your circuit with mine will give you some insight into the selection of R_{es} for Q1.

Example 12-4

Select the bias current for the power transistor bias diodes, I_{bias}. As in other diode compensation schemes, the diode quiescent current must be close to the same as the quiescent collector (emitter) current of the transistor being stabilized. In power stages it is a good idea to make the diode and power transistor quiescent currents equal. If we make the diode current less, tracking is impaired and the stability factor

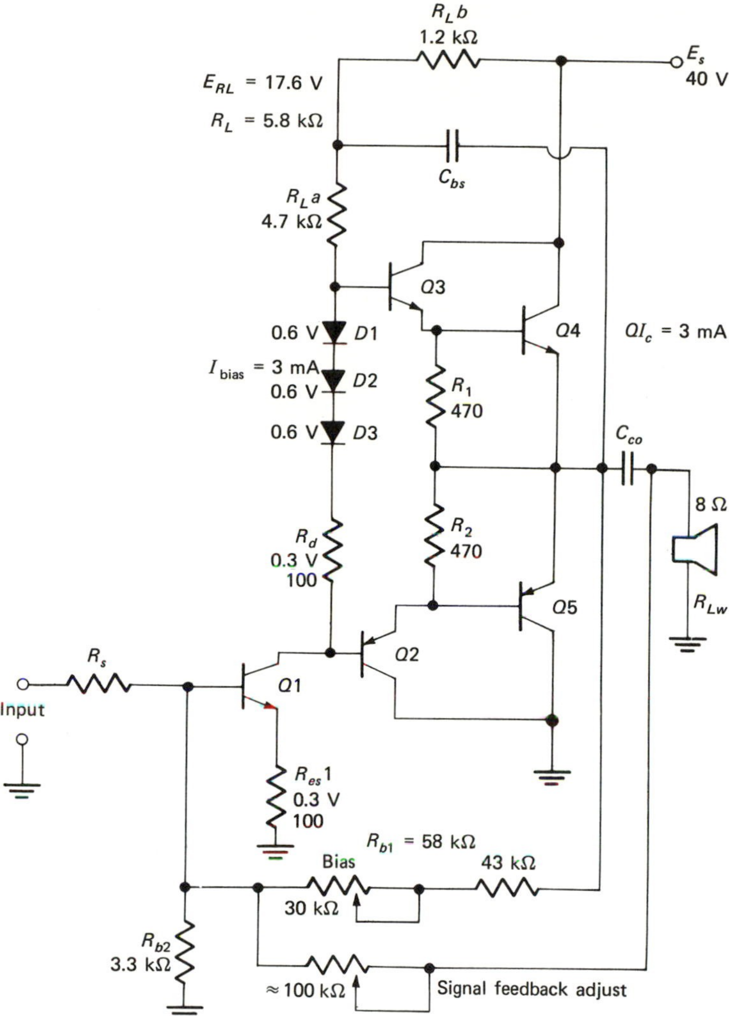

Fig. 12-9 *The improved version of the complementary symmetry amplifier*

is raised. If we make the diode current greater, it reduces the value of the driver collector load resistor, which reduces the voltage gain of the driver. Let $I_{\text{bias}} = 3$ mA.

NOTE: This I_{bias} is also QI_c for the driver.

Driver Preliminary Calculations

1. Record the desired rms power output.

P_o (rms) = 13 W

2. Record the final load impedance R_{Lw}.

   $R_{Lw} = 8\ \Omega$

3. Compute the required E_o p-p of the driver.

   E_o p-p $= \sqrt{P_o \text{ (rms)} \times 8R_{Lw}}$
   E_o p-p $= \sqrt{13 \times 8 \times 8}$
   E_o p-p ≈ 29.6 V ≈ 30 V

4. Compute the minimum E_Q for the driver.

   E_o (p-p) voltage-mode feedback $\approx 1.5\ E_Q$
   E_Q (min) $= 0.66 \times E_o$ p-p $= 0.66 \times 30$
   E_o (min) $= 20$ V

5. Compute E_s (min).

   E_s (min) $= 2 \times E_Q = 40$ V
   $E_s = 40$ V

6. Compute E_{RL}.

   $E_{RL} = 0.5E_s - (E_{D1} + E_{D2} + E_{D3} + E_{Rd} + E_{Res})$

If you will look at Fig. 12-9, you will see three diodes and a resistor in series with R_L (Q1). Each diode will have a 0.6-V drop across it when the circuit is in operation. We will keep the voltage drop across the resistor R_d at less than 0.3 V in this case. A few hundredths of a volt can mean the difference between a lightly forward-biased junction and a very heavily forward-biased junction, and with the very high current gain of the compound emitter follower there is a real danger of overbiasing the power stages beyond the point where the bias adjustment in the feedback loop can compensate for it. When a compound emitter follower is used, E_{Rd} should be no greater than 0.3 V. Some designers prefer from 0.1 to 0.25 V drop for R_d, and some designers use only three diodes and leave R_d out altogether. All these approaches seem to yield good results. Any voltage drop above 0.3 V can be critical; lower voltage drops for R_d are generally not a problem. We will use 0.3 V for E_{Rd}.

In the circuit we examined in the previous example the voltage drop across the resistor R_d was 50 percent of the total bias network voltage drop, and the drop across it could be varied. We had a range, theoretically, of up to 50 percent variation in the bias voltage by varying the collector current through the driver. In this case R_d makes up only 25 percent of the total theoretical bias voltage, and so it will not have as great a range of control over the bias voltage. So we must be very careful not to overbias the power stage initially.

$E_{RL} = 0.5E_s - (0.6 + 0.6 + 0.6 + 0.3 + 0.3)$
$E_{RL} = 17.6$ V

Designing the Power Stage Bias and Stability Loop

1. Select the quiescent collector current for the power transistors Q4 and Q5. This current should be as low as possible, generally 2 to 5 mA. The specification sheet for the transistors we had in mind to test this example with recommended 3 to 10 mA, and so we will go with 3 mA.
2. Select I_{bias}, the current through the V_{BE} stability diodes.

   $I_{\text{bias}} \approx QI_c$ (of the power stage)
   $I_{\text{bias}} = 3\ \text{mA}$

   NOTE: I_{bias} is also QI_c for the driver transistor Q1.
3. Compute R_L for the driver Q1.

$$R_L = \frac{E_{RL}}{I_{\text{bias}}} = \frac{17.6\ \text{V}}{3 \times 10^{-3}} \approx 5.8\ \text{k}\Omega$$

There is a stability factor involving R_L and the dc emitter-to-ground return path of the power transistor pairs, but the dc emitter-to-ground resistance consists of a series of undetermined-value sneak paths in the collector-emitter circuits of the output transistors. Because these resistances are also somewhat variable, we cannot easily compute the stability factor involved. Fortunately we have ensured reasonable bias stability with the diodes in the base loop and the voltage-mode overall feedback. The two emitter resistors in Q2 and Q3 can help a little, but as I mentioned earlier, their contribution is not great and their value is not critical. In the laboratory evaluation of the amplifier we are designing in this example we varied the value of R_1 and R_2 from 100 Ω to 1 kΩ with little or no variation in any performance characteristics. Below 100 Ω we began to upset the bias loop current, and above 1 kΩ we noticed a slight decrease in overall stability. However, even with R_1 and R_2 removed altogether, the overall performance was still good. The quiescent current dropped by about 0.5 mA, and only a slight decrease in stability could be detected. We used 470 Ω for R_1 and R_2, but you will certainly find considerable variation in commercial circuits simply because the value is so uncritical.

Designing the Feedback Loop

Here, again, we can use the procedure in Field 2 of the voltage-mode design sheet (Fig. 7-6).

FRAME 1

Compute R_{b1}.

$R_{b1} = \mathbf{S} \times R_L\ (\text{Q1}) = 10 \times 5.9\ \text{k}\Omega = 59\ \text{k}\Omega$

FRAME 2

Compute E_{Rb1}.

$E_{Rb1} = E_Q - (V_j + E_{Res})$
$E_{Rb1} = 20\ \text{V} - (0.6 + 0.3\ \text{V}) = 19.1\ \text{V}$
$E_{Res} = QI_c\ (Q1)\ R_{es}\ (Q1) = 3\ \text{mA} \times 100\ \Omega = 0.3\ \text{V}$

FRAME 3

Compute I_{Rb1}.

$$I_{Rb1} = \frac{E_{Rb1}}{R_{b1}} = \frac{19.1\ \text{V}}{59\ \text{k}\Omega} = 3.2 \times 10^{-4}\ \text{A}$$

FRAME 4

Compute I_b of Q1, the driver base current.

$$I_b = \frac{QI_c}{\beta} = \frac{3 \times 10^{-3}}{10^2} = 3 \times 10^{-5}\ \text{A}$$

$$I_b = 0.3 \times 10^{-4}\ \text{A}$$

FRAME 5

Compute I_{Rb2}.

$I_{Rb2} = I_{Rb1} - I_B$
$I_{Rb2} = (3.2 \times 10^{-4}) - (0.3 \times 10^{-4})$
$I_{Rb2} = 2.9 \times 10^{-4}\ \text{A}$

FRAME 6

Compute R_{b2}.

$$R_{b2} = \frac{V_j + E_{Res}}{I_{Rb2}} = \frac{E_{Rb2}}{I_{Rb2}}$$

$$R_{b2} = \frac{0.9}{2.9 \times 10^{-4}} \approx 3.3\ \text{k}\Omega$$

The completed circuit with values is shown in Fig. 12-9.

PROBLEMS

12-6 Compute the input impedance of the basic amplifier in Example 12-3 and select a value for R_s.

$$Z_{in} \approx R_{b1}/V_g \| R_{b2} \| \beta(R'_e + R_{es})$$

12-7 Compute the voltage gain of the driver Q1 in Example 12-3. Assume a tenfold increase in voltage gain because of the bootstrapping.

12-8 Give as thorough an explanation as you can of why the circuit in Example 12-3 is superior to the circuit in Example 12-2.

12-9 Design the complementary symmetry power amplifier. PROCEDURE: Design a 20 W (rms) complementary symmetry power amplifier to work into an 8-Ω load. Use a 250-Ω resistor in the emitter of the driver Q1. Use a QI_c value of 2.5 mA for the power stage. See Fig. 12-9 and Example 12-3.

12-6 THE QUASI-COMPLEMENTARY SYMMETRY AMPLIFIER

The direct-coupled version of the quasi-complementary symmetry amplifier is shown in Fig. 12-10.

The basic problem with this circuit is an inherent imbalance because one output transistor is used in a common-collector configuration while the other is used in a common-emitter configuration. This results in different gains and output impedances for each half cycle. Large amounts of negative feedback are generally counted on to compensate for the differences. In the circuit in Fig. 12-10, Q3 is an emitter-follower driver for Q4, and Q2 is a common-emitter driver for Q5. This arrangement provides signals to the power transistor bases that are 180° out of phase with each other.

The design procedure is similar to that for the true complementary symmetry circuit. $R_1 = R_2 \approx 220\ \Omega$ to $680\ \Omega$. The bias loop consisting of $D1$ and R_d is handled in the same way as for the complementary symmetry circuit, except that only the driver transistors Q2 and Q3 are in the bias loop. The final transistors, Q4 and Q4, are biased by the voltage drop across R_1 and R_2. A few manufacturers have used special transformers with two secon-

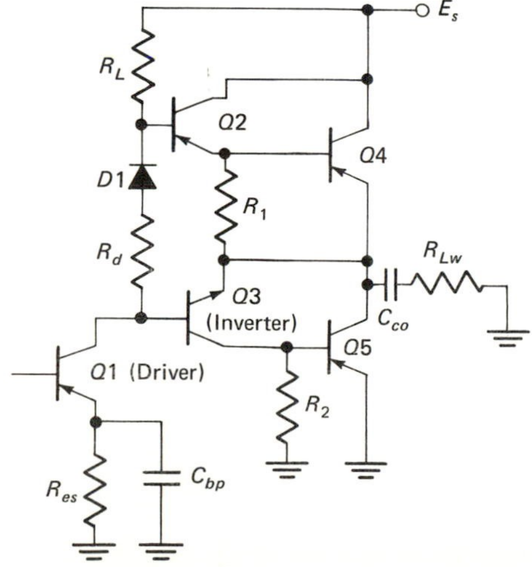

Fig. 12-10 The quasi-complementary circuit

daries to drive the quasi circuit, but it would be a special item, and fairly costly, and would have all the usual disadvantages of audio transformers.

This circuit is rapidly becoming less and less important, and even integrated circuit manufacturers are producing true complementary circuits instead.

12-7 THE CLASS A AMPLIFIER APPLIED TO REGULATOR SERVICE

Figure 12-11*a* shows a class A emitter follower power amplifier adapted for voltage regulator service. Figure 12-11*b* shows the form in which the circuit is conventionally drawn.

The output voltage of the regulator is controlled by the setting of $R_{b1} - R_{b2}$ according to the relationship:

$$E_o \approx V_{\text{zener}}(1 + R_{b1}/R_{b2})$$

The unregulated power supply voltage should be about twice the expected regulated output voltage.

For maximum regulation the voltage gain of the driver should be as high as practical, and the feedback stability factor should be less than unity (even 0.1 or less). Such low stability factors are possible, and practical in this case, because the only input signal is the feedback voltage itself, and the resulting very low input impedance is no problem.

Ripple reduction can be further improved by placing a bypass capacitor between the emitter of Q3 and the wiper of R_{b1}-R_{b2}. This increases the ac

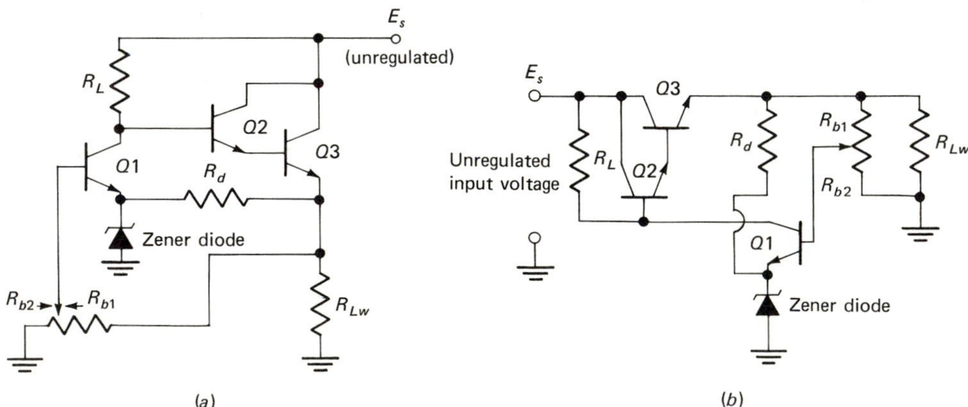

Fig. 12-11 The basic series feedback regulator circuit. (a) *Drawn in familiar form;* (b) *as it is conventionally drawn*

feedback and makes it independent of the setting of the output voltage control, R_{b1}-R_{b2}.

The zener diode resistance (a normally available parameter) should be as low as practical; it should be biased between 20 and 40 percent of its maximum current rating. Bypassing the zener diode can also help to reduce the ripple voltage in cases where the zener diode resistance is near to, or greater than, R_e' of the driver transistor. Bypassing the zener serves the same purpose as bypassing the emitter resistor in a current-mode feedback circuit. It increases the voltage gain.

PROBLEMS

Problem 12-10

Design a voltage regulator circuit for 10 V output at 1 A of current. HINT: $R_{Lw} = 10$ V/1 A (Ohm's law). See Fig. 12-11 and follow the basic design procedure in Sec. 12-3.

appendix

This section contains a summary of nearly all important design and analysis equations and procedures, in the form of standard design and analysis sheets. All frequently needed tables, including all tables of possibilities, are included. This convenient grouping can make the text easier to read if you will refer to it frequently.

PART 1. TABLES OF POSSIBILITIES

TABLE 5-5. TABLE OF POSSIBILITIES FOR THE UNBYPASSED EMITTER CIRCUIT

Voltage Gains

					Emitter resistor voltage drops (E_{Res})	**2.0 V**	1	2	3	4	5	6	7	8	9
						1.5 V	1.3	2.6	4	5.3	6.6	8	9.3	10.6	12
						1.25 V	1.6	3.2	4.8	6.4	8	10.4	11.6	12.8	14.4
						1.0 V	2	4	6	8	10	12	14	16	18
						0.9 V	2.2	3.4	6.6	8.8	11	13	15.5	17.7	20
						0.8 V	2.5	5	7.5	10	12.5	15	17.5	20	23
						0.7 V	2.8	5.7	8.5	11	14	17	20	22	25.7
						0.6 V	3	6.6	10	13	16.6	20	23	25	30
						0.5 V	4	8	12	16	20	24	28	32	36
R_{Lw} at 0.5 mA	R_{Lw} at 1 mA	R_{Lw} at 2 mA	R_{Lw} at 4 mA	R_{Lw} at 10 mA	R_{Lw} at 20 mA	R_{Lw} at 40 mA	E_{RL}	E_{RL}	E_{RL}	E_{RL}	E_{RL}	E_{RL}	E_{RL}	E_{RL}	E_{RL}
2K	1K	500	250	100	50	25	*	*	*	*	*	*	*	*	*
4K	2K	1K	500	200	100	50	*	*	*	*	*	*	*	*	*
6K	3K	1.5K	750	300	150	75	6	*	*	*	*	*	*	*	*
8K	4K	2K	1K	400	200	100	4	*	*	*	*	*	*	*	*
10K	5K	2.5K	1.25K	500	250	125	3.3	20	*	*	*	*	*	*	*
12K	6K	3K	1.5K	600	300	150	3	12	*	*	*	*	*	*	*
14K	7K	3.5K	1.75K	700	350	175	2.8	9.3	42	*	*	*	*	*	*
16K	8K	4K	2K	800	400	200	2.7	8	24	*	*	*	*	*	*
18K	9K	4.5K	2.25K	900	450	225	2.6	7.2	18	72	*	*	*	*	*
20K	10K	5K	2.5K	1K	500	250	2.5	6.7	15	40	*	*	*	*	*
24K	12K	6K	3K	1.25K	600	300	2.4	6.0	12	24	60	*	*	*	*
30K	15K	7.5K	3.75K	1.5K	700	350	2.3	5.5	10	17	30	60	210	*	*
40K	20K	10K	5K	2K	1K	500	2.2	5.0	8.5	13	20	30	47	80	180
50K	25K	12.5K	6.25K	2.5K	1.25K	625	2.2	4.7	7.9	11	17	23	32	44	64
100K	50K	25K	12.5K	5K	2.5K	1.25K	2	4.3	6.8	9.5	12.5	16	19	24	28
200K	100K	50K	25K	10K	5K	2.5K	2	4	6.4	8.7	11	13.5	17	19	22

* An asterisk signifies an impossible condition.

TABLE 6-2. TABLE OF POSSIBILITIES FOR THE FULLY BYPASSED CIRCUIT

Working Loads at Various Values of QI_c							Voltage Drop Across the Collector Load Resistor (R_L) for Various Voltage Gains								
R_{Lw} at 0.5 mA	R_{Lw} at 1.0 mA	R_{Lw} at 2.0 mA	R_{Lw} at 4.0 mA	R_{Lw} at 10 mA	R_{Lw} at 20 mA	R_{Lw} at 40 mA	E_{RL} at $V_g =$ 40	E_{RL} at $V_g =$ 60	E_{RL} at $V_g =$ 80	E_{RL} at $V_g =$ 100	E_{RL} at $V_g =$ 140	E_{RL} at $V_g =$ 160	E_{RL} at $V_g =$ 240	E_{RL} at $V_g =$ 320	E_{RL} at $V_g =$ 400
2K	1K	500	250	100	50	25	*	*	*	*	*	*	*	*	*
4K	2K	1K	500	200	100	50	2.0	6	*	*	*	*	*	*	*
6K	3K	1.5K	750	300	150	75	1.5	3	6	15	*	*	*	*	*
8K	4K	2K	1K	400	200	100	1.3	2.4	4	6.6	28	*	*	*	*
10K	5K	2.5K	1.25K	500	250	125	1.25	2.1	3.3	5.0	11.6	20	*	*	*
12K	6K	3K	1.5K	600	300	150	1.2	2.0	3	4.3	8.4	12	*	*	*
14K	7K	3.5K	1.75K	700	350	175	1.16	1.9	2.8	3.9	7.0	9.3	42	*	*
16K	8K	4K	2K	800	400	200	1.1	1.8	2.7	3.6	6.2	8	24	*	*
18K	9K	4.5K	2.25K	900	450	225	1.1	1.8	2.6	3.5	5.7	7.2	18	72	*
20K	10K	5K	2.5K	1K	500	250	1.1	1.7	2.5	3.3	5.4	6.7	15	40	*
24K	12K	6K	3K	1.2K	600	300	1.1	1.7	2.4	3.1	5.0	6.0	12	24	60
30K	15K	7.5K	3.75K	1.5K	700	350	1.1	1.6	2.3	3.0	4.5	5.5	10	17	30
40K	20K	10K	5K	2K	1K	500	1.1	1.6	2.2	2.8	4.2	5.0	8.5	13	20
50K	25K	12.5K	6.25K	2.5K	1.25K	625	1.1	1.6	2.2	2.7	4.0	4.7	7.9	11	17
100K	50K	25K	12.5K	5K	2.5K	1.25K	1.0	1.6	2	2.6	3.7	4.3	6.8	9.5	12.5
200K	100K	50K	25K	10K	5K	2.5K	1.0	1.6	2	2.5	3.6	4	6.4	8.7	11

* An asterisk signifies an impossible condition.

TABLE 6-4*a*. TABLE OF POSSIBILITIES FOR THE SPLIT EMITTER CIRCUIT

						QI_c	Values of R_{Lac}								
						40 mA	37.5	50	100	150	200	250	300	350	400
						20 mA	75	100	200	300	400	500	600	700	800
						10 mA	150	200	400	600	800	1K	1.2K	1.4K	1.6K
						4 mA	375	500	1K	1.5K	2K	2.5K	3K	3.5K	4K
						2 mA	750	1K	2K	3K	4K	5K	6K	7K	8K
						1 mA	1.5K	2K	4K	6K	8K	10K	12K	14K	16K
						0.5 mA	3K	4K	8K	12K	16K	20K	24K	28K	32K
						$Vg \longrightarrow$ **Range**	2.5–15	3–20	6–40	10–60	13–80	16–100	20–120	23–140	26–160
R_{Lw} at 0.5 mA	R_{Lw} at 1 mA	R_{Lw} at 2 mA	R_{Lw} at 4 mA	R_{Lw} at 10 mA	R_{Lw} at 20 mA	R_{Lw} at 40 mA	E_{RL}	E_{RL}	E_{RL}	E_{RL}	E_{RL}	E_{RL}	E_{RL}	E_{RL}	E_{RL}
2K	1K	500	250	100	50	25	*	*	*	*	*	*	*	*	*
4K	2K	1K	500	200	100	50	6	*	*	*	*	*	*	*	*
6K	3K	1.5K	750	300	150	75	3	6	*	*	*	*	*	*	*
8K	4K	2K	1K	400	200	100	2.4	4	*	*	*	*	*	*	*
10K	5K	2.5K	1.25K	500	250	125	2.1	3.3	20	*	*	*	*	*	*
12K	6K	3K	1.5K	600	300	150	2.0	3	12	*	*	*	*	*	*
14K	7K	3.5K	1.75K	700	350	175	1.9	2.8	9.3	42	*	*	*	*	*
16K	8K	4K	2K	800	400	200	1.8	2.7	8	24	*	*	*	*	*
18K	9K	4.5K	2.25K	900	450	225	1.8	2.6	7.2	18	72	*	*	*	*
20K	10K	5K	2.5K	1K	500	250	1.7	2.5	6.7	15	40	*	*	*	*
24K	12K	6K	3K	1.25K	600	300	1.7	2.4	6.0	12	24	60	*	*	*
30K	15K	7.5K	3.75K	1.5K	700	350	1.6	2.3	5.5	10	17	30	60	210	*
40K	20K	10K	5K	2K	1K	500	1.6	2.2	5.0	8.5	13	20	30	47	80
50K	25K	12.5K	6.25K	2.5K	1.25K	625	1.6	2.2	4.7	7.9	11	17	23	32	44
100K	50K	25K	12.5K	5K	2.5K	1.25K	1.5	2	4.3	6.8	9.5	12.5	16	19	24
200K	100K	50K	25K	10K	5K	2.5K	1.5	2	4	6.4	8.7	11	13.5	17	19

* An asterisk signifies an impossible condition.

TABLE 6-4b. MODIFIED SPLIT EMITTER TABLE OF POSSIBILITIES

						QI_c	← R_{Lac} →																	
						40 mA	37.5		50		100		150		200		250		300		350		400	
						20 mA	75		100		200		300		400		500		600		700		800	
						10 mA	150		200		400		600		800		1K		1.2K		1.4K		1.6K	
						4 mA	375		500		1K		1.5K		2K		2.5K		3K		3.5K		4K	
						2 mA	750		1K		2K		3K		4K		5K		6K		7K		8K	
						1 mA	1.5K		2K		4K		6K		8K		10K		12K		14K		16K	
						0.5 mA	3K		4K		8K		12K		16K		20K		24K		28K			
						v_g → range	2.5–6		3–8		6–16		10–24		12–30		20–40		20–40		20–50		20–60	
R_{Lw} at 0.5 mA	R_{Lw} at 1 mA	R_{Lw} at 2 mA	R_{Lw} at 4 mA	R_{Lw} at 10 mA	R_{Lw} at 20 mA	R_{Lw} at 40 mA	E_{RL}	Mag	E_{RL}	Mag	E_{RL}	Mag	E_{RL}	Mag	E_{RL}	Mag	E_{RL}	Mag	E_{RL}	Mag	E_{RL}	Mag	E_{RL}	Mag
2K	1K	500	250	100	50	25	*	*	*	*	*	*	*	*	*	*	*	*	*	*	*	*	*	*
4K	2K	1K	500	200	100	50	6	3.5	*	*	*	*	*	*	*	*	*	*	*	*	*	*	*	*
6K	3K	1.5K	750	300	150	75	3	4	6	5.3	*	*	*	*	*	*	*	*	*	*	*	*	*	*
8K	4K	2K	1K	400	200	100	2.4	5.5	4	5.6	*	*	*	*	*	*	*	*	*	*	*	*	*	*
10K	5K	2.5K	1250	500	250	125	2.1	6	3.3	7.2	20	8	*	*	*	*	*	*	*	*	*	*	*	*
12K	6K	3K	1.5K	600	300	150	2.0	6	3	8.0	12	10	*	*	*	*	*	*	*	*	*	*	*	*
14K	7K	3.5K	1750	700	350	175	1.9	6	2.8	8	9.3	10	42	14	*	*	*	*	*	*	*	*	*	*
16K	8K	4K	2K	800	400	200	1.8	6	2.7	8	8	10	24	14	*	*	*	*	*	*	*	*	*	*
18K	9K	4.5K	2250	900	450	225	1.8	6	2.6	8	7.2	12	18	16	72	15	*	*	*	*	*	*	*	*
20K	10K	5K	2.5K	1K	500	250	1.7	6	2.5	8	6.7	14	15	16	40	15	*	*	*	*	*	*	*	*
24K	12K	6K	3K	1200	600	300	1.7	6	2.4	8	6.0	16	12	16	24	20	60	20	*	*	*	*	*	*
30K	15K	7.5K	3750	1.5K	700	350	1.6	6	2.3	8	5.5	16	10	20	17	20	30	25	60	20	210	30	*	*
40K	20K	10K	5K	2K	1K	500	1.6	6	2.2	8	5.0	16	8.5	24	13	25	20	25	30	30	47	30	80	30
50K	25K	12.5K	6250	2.5K	1250	625	1.6	6	2.2	8	4.7	16	7.9	24	11	30	17	35	23	30	32	30	44	
100K	50K	25K	12.5K	5K	2.5K	1250	1.5	6	2	8	4.3	16	6.8	24	9.5	30	12.5	40	16	40	19	50	24	60
200K	100K	50K	25K	10K	5K	2.5K	1.5	6	2	8	4	16	6.4	24	8.7	30	11	40	13.5	40	17	50	19	60

* An asterisk signifies an impossible condition.

† Practical design limits.

TABLE 7-3. TABLE OF POSSIBILITIES FOR THE COMMON-COLLECTOR CIRCUIT

Minimum values of QI_c, mA

	E_o p-p Millivolts					E_o p-p Volts									
R_{Lac}, Ω	**50**	**100**	**200**	**400**	**800**	**1.6**	**2**	**4**	**8**	**12**	**16**	**20**	**24**	**28**	**32**
0.25	100	200	400	800	1.6 A	3.2 A	4 A	8 A	16 A	48 A	*	*	*	*	*
0.3	80	166	333	666	1.4 A	2.7 A	3 A	7 A	13 A	20 A	27 A	34 A	*	*	*
0.6	40	80	160	320	640	1.4 A	2 A	3 A	7 A	10 A	13 A	17 A	20 A	*	*
1.0	25	50	100	200	400	800	1 A	2 A	4 A	6 A	8 A	10 A	12 A	14 A	16 A
2.5	10	20	40	80	160	320	400	800	2 A	2 A	3 A	4 A	5 A	6 A	6 A
5.0	5	10	20	40	80	160	200	400	800	1 A	2 A	2 A	2 A	3 A	3 A
8.0	3	6	12	25	50	100	125	250	500	750	1 A	1 A	2 A	2 A	2 A
10	2.5	5	10	20	40	80	100	200	400	600	800	1 A	1 A	1 A	2 A
16	1.6	3	6	12.5	25	50	63	125	250	380	500	700	750	900	1 A
50	0.5	1.2	2	4	8	16	20	40	80	120	160	200	240	280	320
80	0.3	0.6	1.2	2.5	5.0	10	12.5	25	50	75	100	130	150	175	200
100	0.25	0.5	1	2	4	8	10	20	40	60	80	100	120	140	160
160	0.16	0.3	0.6	1.2	2.5	5	6.3	13	25	38	50	70	75	90	100
250	0.1	0.2	0.4	0.8	1.6	3.2	4	8	16	24	32	40	48	56	64
500	0.05	0.12	0.2	0.4	0.8	1.6	2	4	8	12	16	20	24	28	32
1,000	0.025	0.05	0.1	0.2	0.4	0.8	1	2	4	6	8	10	12	14	16
5,000	0.005	0.012	0.02	0.04	0.08	0.16	0.2	0.4	0.8	1.2	1.6	2	2.4	2.8	3.2
10,000	0.003	0.005	0.01	0.02	0.04	0.08	0.1	0.2	0.4	0.6	0.8	1	1.2	1.4	1.6

* Impossible design conditions

TABLE 9-1. THE EXPANDED TABLE OF POSSIBILITIES FOR ITERATIVE FULLY BYPASSED CIRCUITS

Working Loads at Various Values of QI_c							**Voltage Drop Across R_L for Various Voltage Gains**		
R_{Lw} at 0.5 mA	**R_{Lw} at 1.0 mA**	**R_{Lw} at 2.0 mA**	**R_{Lw} at 4.0 mA**	**R_{Lw} at 10 mA**	**R_{Lw} at 20 mA**	**R_{Lw} at 40 mA**	**E_{RL} at $V_g = 40$**	**E_{RL} at $V_g = 60$**	**E_{RL} at $V_g = 66$**
4K	2K	1K	500	200	100	50	2.0	6.0	10

TABLE 11-2. TABLE OF POSSIBILITIES FOR THE DIFFERENTIAL AMPLIFIER CIRCUIT

Working Loads at Various Values of QI_c							Voltage Drop Across the Collector Load Resistor (R_L) for Various Voltage Gains								
R_{Lw} at 0.5 mA	R_{Lw} at 1.0 mA	R_{Lw} at 2.0 mA	R_{Lw} at 4.0 mA	R_{Lw} at 10 mA	R_{Lw} at 20 mA	R_{Lw} at 40 mA	E_{RL} at $V_g =$ **16**	E_{RL} at $V_g =$ **24**	E_{RL} at $V_g =$ **32**	E_{RL} at $V_a =$ **40**	E_{RL} at $V_g =$ **56**	E_{RL} at $V_g =$ **64**	E_{RL} at $V_g =$ **96**	E_{RL} at $V_g =$ **128**	E_{RL} at $V_g =$ **160**
2K	1K	500	250	100	50	25	*	*	*	*	*	*	*	*	*
4K	2K	1K	500	200	100	50	2.0	6	*	*	*	*	*	*	*
6K	3K	1.5K	750	300	150	75	1.5	3	6	15	*	*	*	*	*
8K	4K	2K	1K	400	200	100	1.3	2.4	4	6.6	28	*	*	*	*
10K	5K	2.5K	1.25K	500	250	125	1.25	2.1	3.3	5.0	11.6	20	*	*	*
12K	6K	3K	1.5K	600	300	150	1.2	2.0	3	4.3	8.4	12	*	*	*
14K	7K	3.5K	1.75K	700	350	175	1.16	1.9	2.8	3.9	7.0	9.3	42	*	*
16K	8K	4K	2K	800	400	200	1.1	1.8	2.7	3.6	6.2	8	24	*	*
18K	9K	4.5K	2.25K	900	450	225	1.1	1.8	2.6	3.5	5.7	7.2	18	72	*
20K	10K	5K	2.5K	1K	500	250	1.1	1.7	2.5	3.3	5.4	6.7	15	40	*
24K	12K	6K	3K	1.2K	600	300	1.1	1.7	2.4	3.1	5.0	6.0	12	24	60
30K	15K	7.5K	3.75K	1.5K	700	350	1.1	1.6	2.3	3.0	4.5	5.5	10	17	30
40K	20K	10K	5K	2K	1K	500	1.1	1.6	2.2	2.8	4.2	5.0	8.5	13	20
50K	25K	12.5K	6.25K	2.5K	1.2K	625	1.1	1.6	2.2	2.7	4.0	4.7	8.5	11	17
100K	50K	25K	12.5K	5K	2.5K	1.25K	1.0	1.6	2	2.6	3.7	4.3	6.8	9.5	12.5
200K	100K	50K	25K	10K	5K	2.5K	1.0	1.6	2	2.5	3.6	4	6.4	8.7	11

* An asterisk signifies an impossible condition.

PART 2. MISCELLANEOUS TABLES

TABLE:

TABLE 4-4. PERCENTAGE CHANGE IN QI_c WITH A 100 PERCENT CHANGE IN β FOR VARIOUS STABILITY FACTORS

	% ΔQI_c		Circuit Current Gain		
S	$\beta = 50 \rightarrow \beta = 100$	$\beta = 100 \rightarrow \beta = 200$	$\beta = 50$	$\beta = 100$	$\beta = 200$
50	25	16.6	25	33	40
20	14.3	8.3	14	16.6	18.2
10	8.3	4.5	8.3	9.0	9.5
9	7.6	4.1	7.6	8.2	8.6
8	6.9	3.7	6.9	7.4	7.6
7	6.1	3.3	6.1	6.5	6.8
6	5.4	2.8	5.4	5.6	5.8
5	4.5	2.4	4.5	4.7	4.9
4	3.7	1.9	3.7	3.8	3.9
3	2.8	1.4	2.83	2.91	2.95
2	1.9	0.98	1.92	1.96	1.98
1	0.98	0.49	0.98	0.99	1

$\mathbf{S} = \frac{R_b}{R_e}$

TABLE 4-5. CENTIGRADE TO FAHRENHEIT CONVERSION TABLE

Centigrade	Fahrenheit	Centigrade	Fahrenheit
10°	50°	55°	131°
15°	59°	60°	140°
20°	68°	65°	149°
25°	77°	70°	158°
30°	86°	75°	167°
35°	95°	80°	176°
40°	104°	85°	185°
45°	113°	90°	194°
50°	122°	95°	203°
		100°	212°

TABLE 5-1. LIST OF SYMBOLS USED

Resistances

1. R_L Collector load resistor.
   Resistor-load.
2. R_{Lw} Working load resistance (or impedance). The load being driven "worked" by the stage.
   Resistor-load-working.
3. R_{Lac} The total signal load impedance. This is composed of R_L and R_{Lw} in parallel.
   Resistor-load-ac.
4. R_{eac} The unbypassed portion of the emitter resistor.
   Resistor-emitter-ac.
5. R_{edc} The bypassed portion of the emitter resistor.
   Resistor-emitter-dc.
6. R_{es} The entire emitter resistor ($R_{eac} + R_{edc}$).
   Resistor-emitter-stabilizing.
7. R_e' The transistor's intrinsic emitter resistance.
8. R_b The resistance from base of the transistor to ground (external).
9. R_{b1} The upper base bias resistor. Generally returned to ground through a forward-biasing voltage.
10. R_{b2} The lower base bias resistor. Generally returned from base directly to ground.

Voltages

1. E_{RL} Voltage drop across the collector load (R_L).
2. E_{Res} Voltage drop across R_{es}, the emitter stability resistance. (R_{edc} and R_{eac}).
3. E_Q Voltage drop across the transistor, between collector and emitter.
4. E_o Maximum peak-to-peak ac signal output voltage.
5. E_s Dc power supply voltage.
6. E_{Rb1} The voltage drop across the upper base bias resistor R_{b1}.
7. E_{Rb2} The voltage drop across the lower base bias resistor (base stability resistor) R_{b2}.

Capacitors

1. C_{bp} Emitter bypass capacitor.
2. C_{cin} Input coupling capacitor.
3. C_{co} Output coupling capacitor.

Miscellaneous

1. V_{BE} The base-emitter junction voltage drop (see also V_j).
2. ΔV_{BE} Variations in the base-emitter voltage drop (usually due to temperature).
3. β Beta, the common-emitter current gain of the transistor.

   $$\beta = \frac{I_c}{I_b}$$

   where I_c is the collector current and I_b is the base-emitter current
4. f_o The lowest frequency of interest. The low-frequency cutoff value, where the voltage gain begins to roll off.
5. t Temperature in centigrade degrees.
6. Δt The change in temperature $t_1 - t_0$, where t_0 is generally 20°C and t_1 is the highest expected temperature.
7. V_j The junction voltage 0.6 V in silicon and 0.2 V in germanium. Also called V_{BE}.

Current

1. QI_c Quiescent collector current.
2. I_b Base current.

Impedances

1. Z_{in} The total ac signal input impedance.
   $Z_{in} = R_{b1} \| R_{b2} \| \beta \times R_{eac}$
   where $\|$ means in parallel with
2. R_{int} The dc input resistance looking into the base, but not including R_{b1}.
   $R_{int} = R_{b2} \| \beta \times R_{es}$

TABLE 5-3. TABLE OF VARIATIONS IN V_{BE} AT VARIOUS CURRENTS

QI_c	ΔV_{BE}
0–0.1 mA	−3.0 mV/C°
0.1–1.0 mA	−2.75 mV/C°
1.0–10.0 mA	−2.5 mV/C°
10.0–100 mA	−2.0 mV/C°

TABLE 5-4. STANDARD VALUES FOR RESISTORS

Tolerance		
20%	**10%**	**5%**
10	10	10
		11
	12	12
		13
15	15	15
		16
	18	18
		20
22	22	22
		24
	27	27
		30
33	33	33
		36
	39	39
		43
47	47	47
		51
	56	56
		62
68	68	68
		75
	82	82
		91
100	100	100

NOTE: All numbers are to be multiplied by the desired power of 10. For design work use resistors with highest tolerances when possible.

TABLE 6-3. TABLE OF SELECTED CAPACITIVE REACTANCE VALUES

f_o = 5 Hz – Range: 16 Ω to 649 Ω

X_C, Ω	16–31.9	32–64.9	65–124.9	125–209	210–319	320–649
C, μF	2,000	1,000	500	250	150	100

f_o = 5 Hz – Range: 650 Ω to 31.9 kΩ

X_C, Ω	650–1.24K	1.25K–3.19K	3.2K–6.49K	6.5–15.9K	16–31.9K
C, μF	50	25	10	2	1

f_o = 10 Hz – Range: 7 Ω to 649.9 Ω

X_C, Ω	7–16	16–25	25–50	50–100	100–160	160–320	320–650
C, μF	2,000	1,000	500	250	150	100	50

f_o = 10 Hz – Range: 650 Ω to 31.9 kΩ

X_C, Ω	650–1.59K	1.6K–3.19K	3.2K–7.99K	8–15.99K	16–31.9K
C, μF	25	10	5	2	1

f_o = 20 Hz – Range: 4 Ω to 799 Ω

X_C, Ω	4–7	8–15	16–31	32–54	55–79	80–159	160–319	320–799
C, μF	2,000	1,000	500	250	150	100	50	25

f_o = 20 Hz – Range: 800 Ω to 15.99 kΩ

X_C, Ω	800–1.59K	1.6K–3.99K	4K–7.99K	8K–15.99K
C, μF	10	5	2	1

f_o = 50 Hz – Range: 1.6 Ω to 64.9 Ω

X_C, Ω	1.6–2.9	3.06–4.9	6.5–11.9	12–20.9	21–31.9	32–64.9
C, μF	2,000	1,000	500	250	150	100

f_o = 50 Hz – Range: 65 Ω to 6 kΩ

X_C, Ω	65–119	120–299	300–649	650–1.59K	1.6K–3.19K	3.2K–6K
C, μF	50	25	10	5	2	1

f_o = 60 HZ – Range: 1.5 Ω to 109.9 Ω

X_C, Ω	1.5–2.5	2.6–4.9	5–10.9	11–12.9	13–25.9	26–54.9	55–109.9
C, μF	2,000	1,000	500	250	150	100	50

f_o = 60 Hz – Range: 110 Ω to 5 kΩ

X_C, Ω	110–259	260–499	500–1.29K	1.3K–2.59K	2.6K–5K
C, μF	25	10	5	2	1

TABLE 6-3. (*Cont.*)

f_o = 100 Hz – Range: 0.8 Ω to 64.9 Ω

X_C, Ω	0.8–1.59	1.6–3.19	3.2–6.49	6.5–9.9	10–15.9	16–31.9	32–64.9
C, μF	2,000	1,000	500	250	150	100	50

f_o = 100 Hz – Range: 65 Ω to 3 kΩ

X_C, Ω	65–159	160–319	320–799	800–1.59K	1.6K–3K
C, μF	25	10	5	2	1

TABLE 7-1. VOLTAGE-MODE STABILITY TABLE

E_Q = 1.6 V				E_Q = 2 V				E_Q = 2.6 V			
E_{RL}	S at $\beta = 50$	S at $\beta = 100$	S at $\beta = 200$	E_{RL}	S at $\beta = 50$	S at $\beta = 100$	S at $\beta = 200$	E_{RL}	S at $\beta = 50$	S at $\beta = 100$	S at $\beta = 200$
2	25	50	100	4	20	40	80	2	50	100	200
4	12.5	25	50	6	13.3	26.6	53.2	4	25	50	100
6	8.3	16.6	33.2	8	10	20	40	6	16.6	33.2	66.4
8	6.25	12.5	25	10	8	16	32	8	12.5	25	50
10	5	10	20	12	6.6	13.3	26.6	10	10	20	30
12	4.15	8.3	16.6	14	5.7	11.4	22.8	12	8.3	16.6	33.2
14	3.5	7.1	14.2	16	5	10	20	14	7.1	14.2	28.4
16	3.1	6.2	12.4	18	4.4	8.8	17.6	16	6.2	12.4	24.8
18	2.75	5.5	11	20	4	8	16	18	5.5	11	22
20	2.5	5.0	10	22	3.85	7.7	15.4	20	5.0	10	20
22	2.25	4.5	9	24	3.3	6.6	13.2	22	4.1	8.6	16.4
24	2.05	4.1	8.2								

TABLE 7-2. TABLE OF APPROXIMATE INPUT IMPEDANCES FOR THE PARAPHASE AMPLIFIER

	Approximate Input Impedances				
R_{Lw}	**K = 1**	**2**	**3**	**4**	**5**
200	1 kΩ	2 kΩ	3 kΩ	4 kΩ	5 kΩ
400	2.5 kΩ	5 kΩ	7.5 kΩ	10 kΩ	12.5 kΩ
500	3 kΩ	6 kΩ	9 kΩ	12 kΩ	15 kΩ
800	5 kΩ	10 kΩ	15 kΩ	20 kΩ	25 kΩ
1,000	6.5 kΩ	13 kΩ	19 kΩ	26 kΩ	32 kΩ
2,000	13 kΩ	26 kΩ	39 kΩ	52 kΩ	65 kΩ
4,000	25 kΩ	50 kΩ	75 kΩ	100 kΩ	125 kΩ
8,000	50 kΩ	100 kΩ	*	*	*
10,000	65 kΩ	130 kΩ	*	*	*

NOTE: The table is based on the following conditions:
1. $\mathbf{S} \approx 10$
2. $E_{RL} = 5$ V, $E_Q = 5$ V, $E_{Res} = 5$ V
3. $R_L = R_{es} = K \times R_{Lw}$
4. * indicates input impedances which are high enough for the collector junction impedance to become involved.

TABLE 7-4. APPROXIMATE INPUT IMPEDANCES FOR BASE-BIASED EMITTER FOLLOWERS

	Estimated Input Impedances			
β	$R_{es} =$ **10**R_{Lw}	$R_{es} =$ **20**R_{Lw}	$R_{es} =$ **25**R_{Lw}	$R_{es} =$ **50**R_{Lw}
25	$16R_{Lw}$	$20R_{Lw}$	$21R_{Lw}$	$23R_{Lw}$
50	$25R_{Lw}$	$33R_{Lw}$	$35R_{Lw}$	$40R_{Lw}$
100	$33R_{Lw}$	$50R_{Lw}$	$55R_{Lw}$	$70R_{Lw}$
150	$37R_{Lw}$	$60R_{Lw}$	$68R_{Lw}$	$90R_{Lw}$
200	$40R_{Lw}$	$66R_{Lw}$	$76R_{Lw}$	$100R_{Lw}$

$Z_{in} \approx \beta R_{Lw} \| (\mathbf{S}/2) R R_{Lw}$

$R = R_{es} | R_{Lw}$

Note: Table based on an **S** of 10. Formula and table both assume $\mathbf{E}_{Res} \approx \mathbf{E}_Q$.

TABLE 7-5. SUMMARY FOR CHAPTER 7—CURRENT-MODE CIRCUITS

Appropriate Table of Possibilities	Circuit Classification		Designed as:		Differences
a. Unbypassed, Table 5-5 b. Fully bypassed, Table 6-2	Common emitter	Base bias	Common emitter	Base bias	None
c. Split emitter, Table 6-4	Common emitter	Emitter bias	Common emitter	Emitter bias	None
Fully bypassed, Table 6-2	Common base	Base bias	Common emitter	Base bias	$Z_{in} \approx R'_e$
Fully bypassed, Table 6-2	Common base	Emitter bias	Common emitter	Emitter bias	$Z_{in} \approx R'_e$
Common collector, Table 7-3	Common collector	Base bias	Common emitter	Base bias	1. $R_{Lac} = R_{es} \| R_{Lw}$ 2. $R_{es} > 10R_{Lw}$ $\therefore R_{Lac} \approx R_{Lw}$ 3. $Z_{in} \approx R_{b1} \| R_{b2} \| \beta R_{Lw}$ 4. $R_L = 0$ 5. $E_{RL} = 0$ V 6. $V_g \approx 1$
Common collector, Table 7-3	Common collector	Emitter bias	Common emitter	Emitter bias	1. $R_{Lac} = R_{es} \| R_{Lw}$ 2. $R_{es} > 10R_{Lw}$ $\therefore R_{Lac} \approx R_{Lw}$ 3. $Z_{in} \approx \beta R_{Lw}$ or $Z_{in} \approx \beta R_{Lw} \| R_{b2}$ 4. $R_L = 0\ \Omega$ 5. $E_{RL} = 0$ V 6. $V_g \approx 1$

TABLE 9-2. RAPID-DESIGN TABLE

E_s	E_{RL}	E_Q	E_{Res}	R_{B2}	R_{b1}	R_L	Bypassed Z_{in}	Unbypassed Z_{in}	Bypassed V_g	Unbypassed V_g	R_{es}
6	2	3	1	10K	27K	2K	1.8K	6.8K	38	1.5	1K
9	5	3	1	10K	43K	5K	2K	7.5K	56	3	1K
9	4	3	2	20K	43K	4K	2K	16K	53	3.2	2K
12	8	3	1	10K	62K	8K	2K	9K	64	4	1K
12	7	4	1	10K	62K	7K	2K	9K	60	4	1K
12	7	3	2	20K	68K	7K	2K	16K	60	4	2K
18	13	4	1	10K	91K	13K	2K	9K	66	5.3	1K
18	12	4	2	20K	110K	12K	2K	16K	66	4	2K
21	16	4	1	10K	110K	16K	2K	9K	66	5.6	1K
21	15	4	2	20K	130K	15K	2K	16K	66	4	2K
24	19	4	1	10K	130K	19K	2K	9K	66	6	1K
24	18	4	2	20K	150K	18K	2K	16K	66	4	2K
28	23	4	1	10K	150K	23K	2K	9K	66	6	1K
28	22	4	2	20K	180K	22K	2K	16K	66	4	2K
32	27	4	1	10K	160K	27K	2K	9K	66	6	1K
32	26	4	2	20K	200K	26K	2K	16K	66	4	2K
48	42	4	2	20K	300K	42K	2K	16K	66	5	2K

TABLE 11-1. TABLE OF BASE-EMITTER VOLTAGE VARIATIONS WITH TEMPERATURE

Δt	R_{EF}**20°C** Temp., C°	**3 MV/C° at a QI_c of 0–0.1 mA ΔV_{BE} MV**	**2.75 MV/C° at a QI_c of 0.1–1.0 mA ΔV_{BE} MV**	**2.5 MV/C° at a QI_c of 1.0–10 mA ΔV_{BE} MV**	**2 MV/C° at a QI_c of 10–100 mA ΔV_{BE} MV**
5	25	15	13.75	12.5	10
10	30	30	27.50	25.0	20
15	35	45	41.25	37.5	30
20	40	60	55.00	50.0	40
25	45	75	68.75	62.5	50
30	50	90	82.50	75.0	60
35	55	105	96.25	87.5	70
40	60	120	110.00	100.0	80
45	65	135	123.75	112.5	90
50	70	150	137.50	125.0	100
55	75	165	151.25	137.5	110
60	80	180	165.00	150.00	120
65	85	195	178.75	162.5	130
70	90	210	192.50	175.0	140
75	95	225	206.25	187.5	150
80	100	240	220.00	200.0	160

PART 3. SAMPLE DESIGN AND ANALYSIS SHEETS

1. Universal design sheet (current mode)
2. Universal analysis sheet (current mode)
3. Voltage-mode design sheet (case 1)
4. Voltage-mode design sheet (case 2)
5. Universal analysis sheet (voltage mode)
6. Active emitter resistor design sheet
7. Active emitter resistor analysis sheet

UNIVERSAL DESIGN SHEET

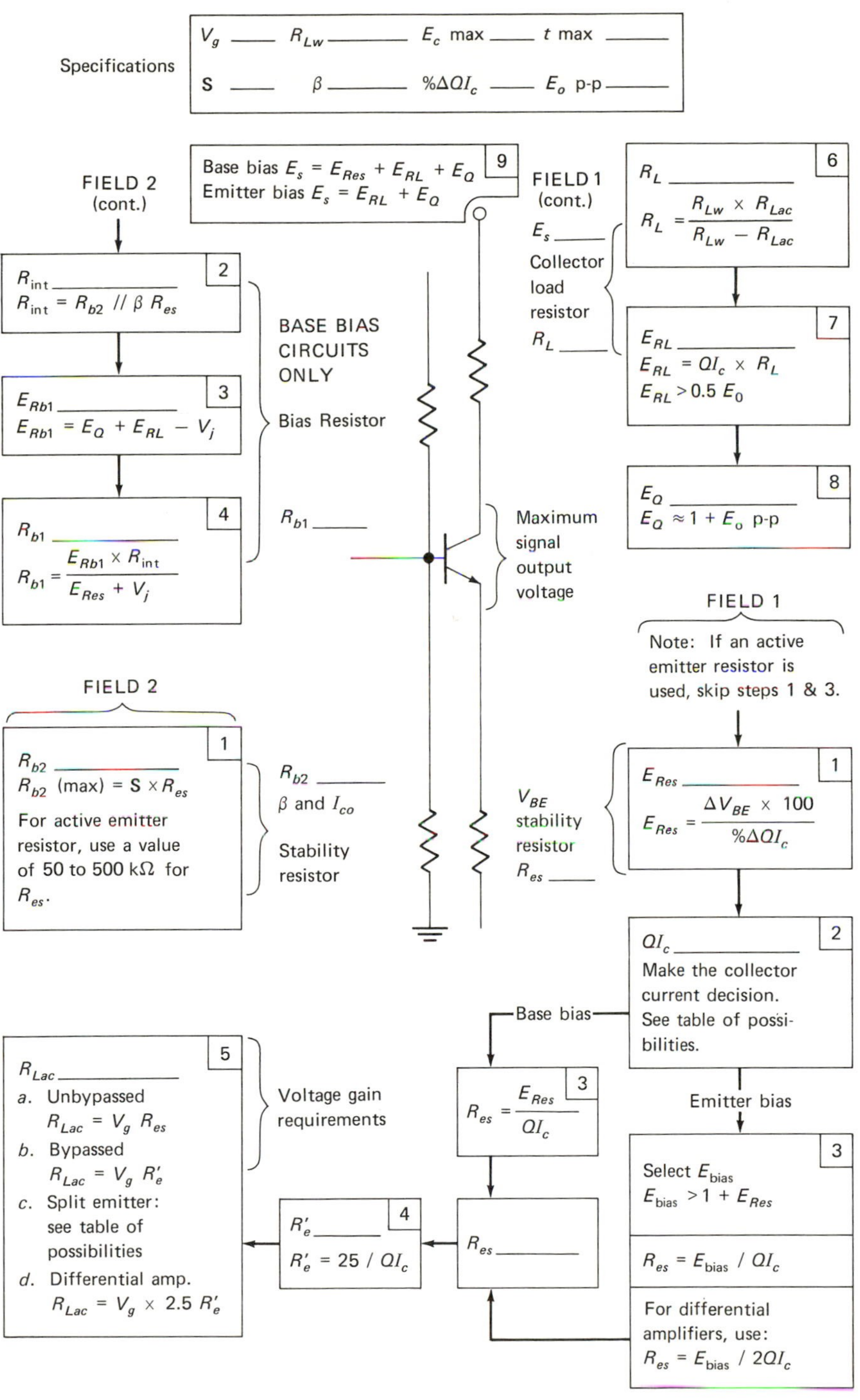

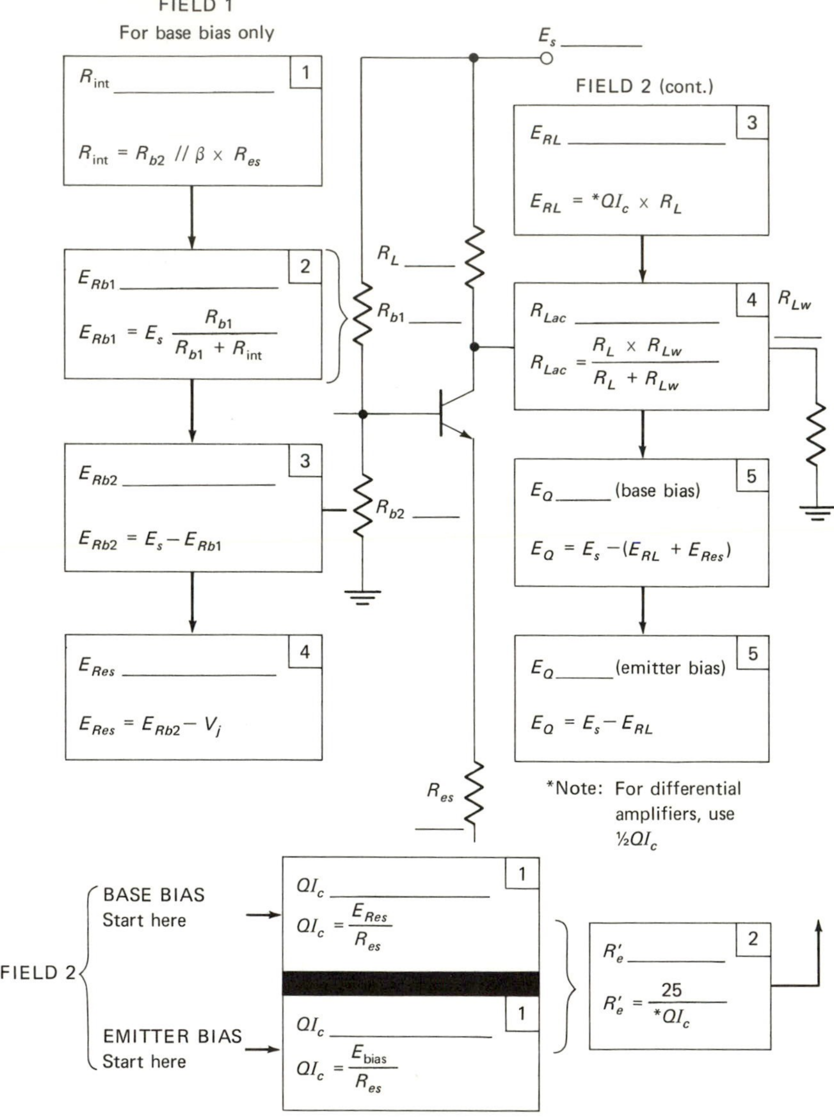

UNIVERSAL ANALYSIS SHEET
Current Mode
FIELD 1
For base bias only
1
R_{int} ______
$R_{int} = R_{b2} \,//\, \beta \times R_{es}$
2
E_{Rb1} ______
$E_{Rb1} = E_s \dfrac{R_{b1}}{R_{b1} + R_{int}}$
3
E_{Rb2} ______
$E_{Rb2} = E_s - E_{Rb1}$
4
E_{Res} ______
$E_{Res} = E_{Rb2} - V_j$
E_s ______
R_L ______
R_{b1} ______
R_{b2} ______
R_{es} ______
FIELD 2 (cont.)
3
E_{RL} ______
$E_{RL} = {}^*QI_c \times R_L$
4
R_{Lac} ______
$R_{Lac} = \dfrac{R_L \times R_{Lw}}{R_L + R_{Lw}}$
R_{Lw} ______
5
E_Q ______ (base bias)
$E_Q = E_s - (E_{RL} + E_{Res})$
5
E_Q ______ (emitter bias)
$E_Q = E_s - E_{RL}$
*Note: For differential amplifiers, use ½QI_c
FIELD 2
BASE BIAS
Start here
1
QI_c ______
$QI_c = \dfrac{E_{Res}}{R_{es}}$
EMITTER BIAS
Start here
1
QI_c ______
$QI_c = \dfrac{E_{bias}}{R_{es}}$
2
R'_e ______
$R'_e = \dfrac{25}{{}^*QI_c}$

UNIVERSAL ANALYSIS SHEET
Current Mode
Field 3

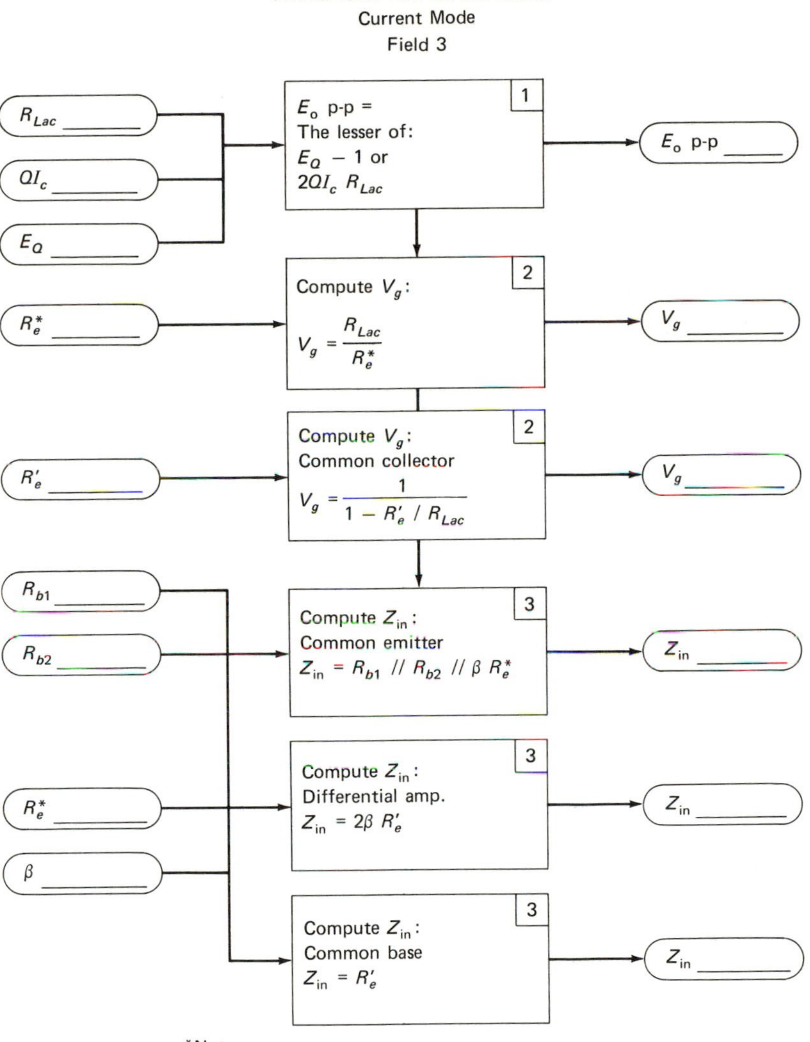

*Note:

a. For unbypassed circuits, $R_e = R_{es}$

b. For fully bypassed circuits, $R_e = R_e'$

c. For split emitter circuits, $R_e = R_{eac}$

d. For differential amplifiers, $R_e = 2.5\,R_e'$

e. For common collector circuits, $R_e = R_{es} \,||\, R_{Lw}$

VOLTAGE MODE DESIGN SHEET

Case 1

FIELD 1

Consult the fully bypassed table of possibilities for QI_c and a tentative value for E_{RL}: QI_c ______ E_{RL} ______	1
Consult the Voltage Mode Stability table, Table 7-1 for a workable value of E_Q. E_Q ______	2
Compute the intrinsic emitter resistance R'_e: $R'_e = 25/QI_c$ R'_e ______	3
Compute the required ac load resistance R_{Lac}: $R_{Lac} = V_g R'_e$ R_{Lac} ______	4

FIELD 2

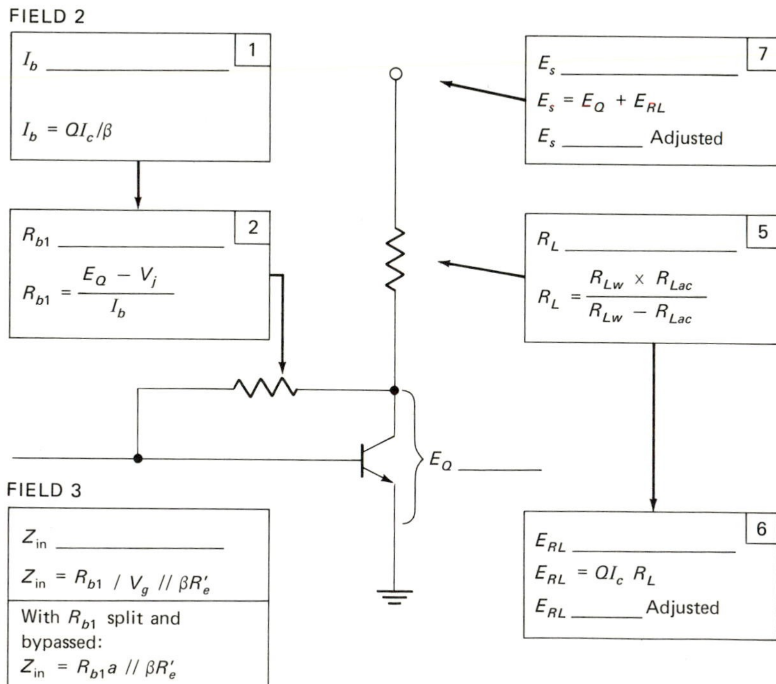

1
I_b ______
$I_b = QI_c/\beta$
2
R_{b1} ______
$R_{b1} = \frac{E_Q - V_j}{I_b}$
7
E_s ______
$E_s = E_Q + E_{RL}$
E_s ______ Adjusted
5
R_L ______
$R_L = \frac{R_{Lw} \times R_{Lac}}{R_{Lw} - R_{Lac}}$
E_Q ______
6
E_{RL} ______
$E_{RL} = QI_c R_L$
E_{RL} ______ Adjusted

FIELD 3

Z_{in} ______

$Z_{in} = R_{b1} / V_g // \beta R'_e$

With R_{b1} split and bypassed:

$Z_{in} = R_{b1}a // \beta R'_e$

VOLTAGE MODE DESIGN SHEET (case two)

SPECIFICATIONS

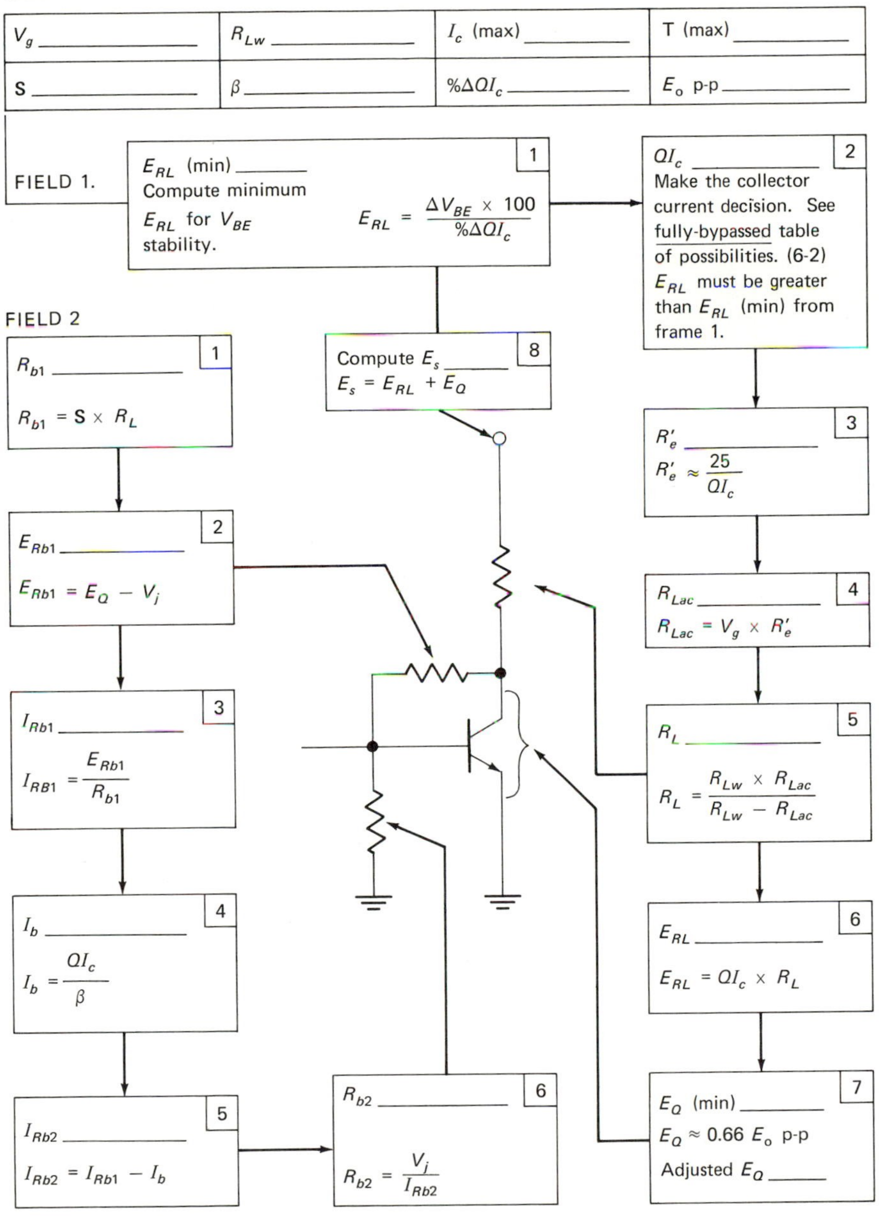

UNIVERSAL ANALYSIS SHEET
Voltage mode

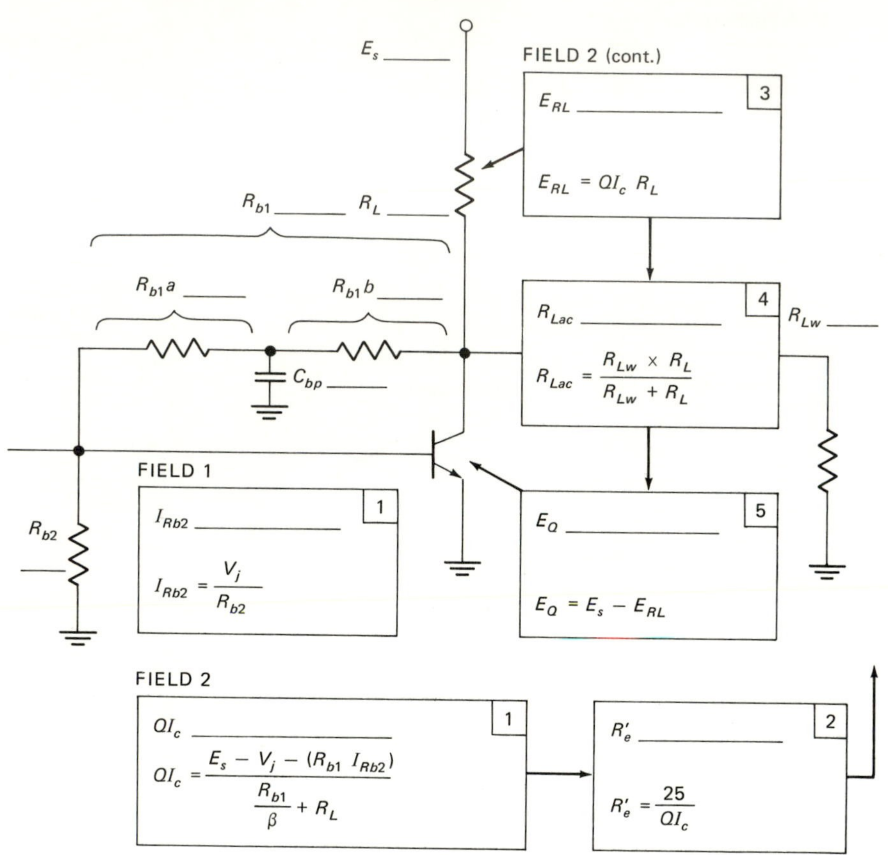

FIELD 3

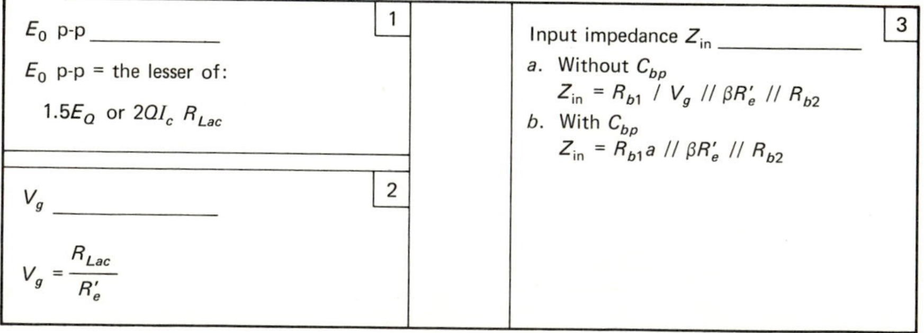

THE ACTIVE EMITTER RESISTOR DESIGN SHEET

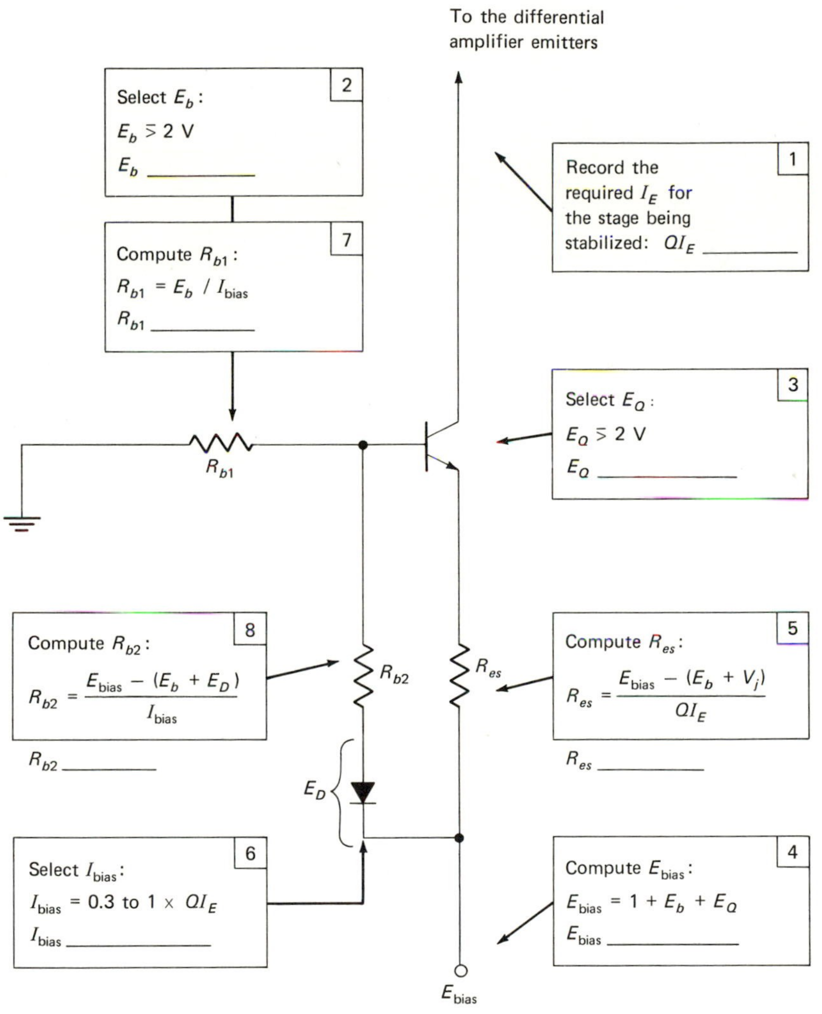

To the differential amplifier emitters
2
Select E_b:
$E_b \gtrless 2$ V
E_b ________
7
Compute R_{b1}:
$R_{b1} = E_b / I_{bias}$
R_{b1} ________
1
Record the required I_E for the stage being stabilized: QI_E ________
3
Select E_Q:
$E_Q \gtrless 2$ V
E_Q ________
R_{b1}
8
Compute R_{b2}:
$R_{b2} = \frac{E_{bias} - (E_b + E_D)}{I_{bias}}$
R_{b2} ________
R_{b2}
R_{es}
5
Compute R_{es}:
$R_{es} = \frac{E_{bias} - (E_b + V_j)}{QI_E}$
R_{es} ________
E_D
6
Select I_{bias}:
$I_{bias} = 0.3$ to $1 \times QI_E$
I_{bias} ________
4
Compute E_{bias}:
$E_{bias} = 1 + E_b + E_Q$
E_{bias} ________
E_{bias}

THE ACTIVE EMITTER RESISTOR ANALYSIS SHEET

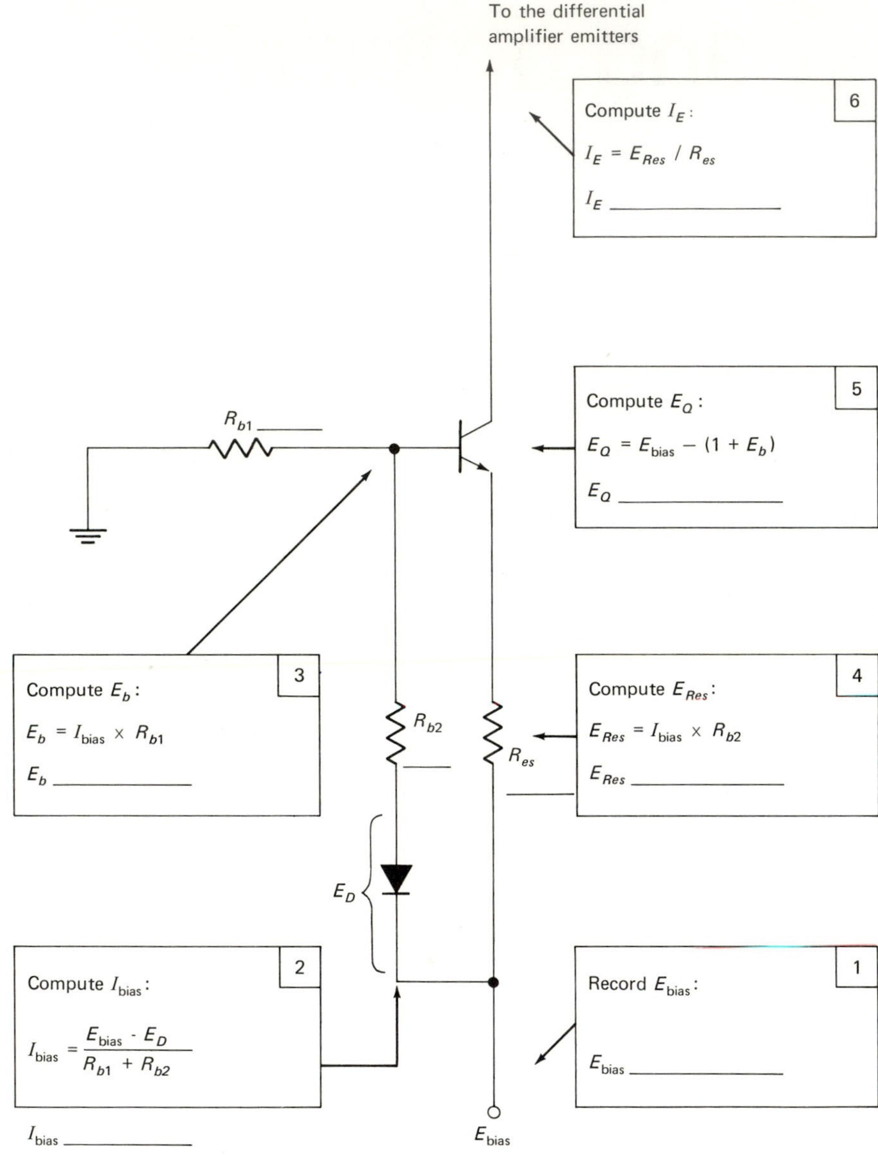

answers to selected questions and problems

CHAPTER 3

3-1 $\beta = 100$ $(\beta = I_c/I_b)$
3-2 $\alpha = 0.99$ $(\alpha = I_c/I_e)$
3-3 $R'_e \approx 25\ \Omega$ $(R'_e \approx 25/I_c)$
3-4 $V_g \approx 100$ $(V_g \approx R_L/R'_e)$
3-5 $Z_{in} \approx 1.25\ \text{k}\Omega$ $(Z_{in} \approx R_b \| (\beta R'_e)$

CHAPTER 5

5-1 $QI_c = 4$ mA, $E_{RL} = 20$ V
5-2 $QI_c = 2$ mA, $E_{RL} = 13$ V
5-3 $QI_c = 0.5$ mA, $E_{RL} = 20$ V
5-4 $QI_c = 1$ mA, $E_{RL} = 10$ V
5-5 Ans: E_Q
5-6 Ans: E_{Res}
5-7 Ans: ≈ 10

CHAPTER 6

Questions

6-1 The voltage gain is somewhat unpredictable because R'_e is not very predictable.

$$V_g = \frac{R_{Lac}}{R'_e}$$

The input impedance is somewhat unpredictable because it is a function of both β and R'_e, and β and R'_e are both somewhat unpredictable.

6-2 Yes, particularly in the unbypassed configuration, because both R'_e and β are temperature-dependent parameters. These two factors are dominant in the unbypassed circuit in terms of signal and input impedance parameters. They are involved to a lesser extent for split-emitter circuits, and have very little influence on the unbypassed circuit.

6-3 Ans: Because for signal frequencies there is no negative feedback to reduce the distortion inherent in the nonlinearity of the transfer curve. The distortion is most noticeable for large signal swings.

6-4 Ans: None

6-5 1. Unbypassed: *b*, *c*, *d*
2. Split emitter: g, *h*, *i*
3. Fully bypassed: *a*, e, *f*

CHAPTER 7

Questions

7-1 *a*. Limited E_Q (2.6 V max)
b. Very low input impedance unless R_{b1} is split and bypassed

7-2 *a*. The Miller effect
b. R_{b1} can be split and bypassed

7-3 E_o p-p (current-mode) $\approx E_Q - 1$
E_o p-p (voltage-mode) $\approx 1.5\ E_Q$

7-4 To provide two output signals 180° out of phase. The voltage gain for both outputs is unity.

7-5 $E_{RL} - E_{Res}$; E_Q is equal to or greater than E_{RL}

7-6 To elevate the impedance of R_{Lw} by a factor of from 10 to 100

7-7 Because we are no longer interested in voltage gain, but we are concerned with finding a collector current which will yield the desired E_o p-p for a given R_{Lw}.

7-8 To get the maximum input impedance by having the largest possible values for R_{b1} and R_{b2}

7-9 Greater V_{BE} stability

7-10 To increase the input impedance

7-11 Go back and look it up if you don't already know the answer

7-12 Positive

7-13 All parameters are the same except for current gain (≈ 1) and $Z_{in} \approx R_e'$

7-14 They are identical except for the signal parameters noted in Question 7

index

BOOK ONE

BOOK TWO